No. 1938
$60.00

THE ENCYCLOPEDIA OF ELECTRONIC CIRCUITS

RUDOLF F. GRAF

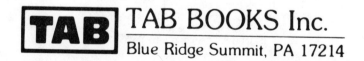

TAB BOOKS Inc.
Blue Ridge Summit, PA 17214

To Allison Nicole
A joy to behold

Disclaimer

All of the corporations represented in this book make no warranty, express or implied, that information on the products contained herein is still applicable. Specifications and availability of products herein described are subject to change without notice.

FIRST EDITION

FOURTH PRINTING

Printed in the United States of America

Library of Congress Cataloging in Publication Data

Graf, Rudolf F.
The encyclopedia of electronic circuits.

Includes index.
1. Electronic circuits. I. Title.
TK7867.G66 1985   621.3815'3   84-26772
ISBN 0-8306-0938-5
ISBN 0-8306-1938-0 (pbk.)

Contents

Acknowledgments

I gratefully acknowledge the fine cooperation of the various sources listed who permitted use of some of their material. This courtesy now makes it possible to further disseminate, in this work, some of the very best circuit ideas of our industry.

American Radio Relay League
Analog Devices, Inc.
Computer Design
Computers and Electronics
Control Engineering
CQ
Electri-Onics
Electronic Circuits
Electronics & Wireless World
Electronics Australia
Electronics Today International
Electronics Week
Fairchild Camera and Instrument Corporation
Ferranti Electric, Inc.
GE Intersil
GE Semiconductor
Ham Radio Magazine
Harris Semiconductor
Machine Design

Microwaves & RF
MITEL
Motorola Inc.
NASA
National Semiconductor
Plessey Solid State
Popular Mechanics
Precision Monolithics Incorporated
Radio-Electronics
RCA
r.f. design
73: Amateur Radio's Technical journal
SGS Technology and Service
Signetics Corporation
Siliconix Incorporated
Supertex inc.
Teledyne Semiconductor
Texas Instruments Incorporated
Western Digital Corporation
Yuasa Battery (America), Inc.

Introduction

This volume of timely and practical circuits highlights the creative work of many people. Featured here are many circuits that appeared only briefly in some of our finer periodicals or limited-circulation publications. Also included are other useful and unique circuits from more readily available sources.

The source for each circuit is given in the sources section at the back of the book. The bold figure number that appears inside the box of each circuit is the key to the source. For example, the High Stability Voltage Reference circuit shown below is Fig. 93-10. If you turn to the Sources section and look for Fig. 93-10 you will find that Precision Monolithics supplied this circuit from p. 6-142 of their Full Line Catalog.

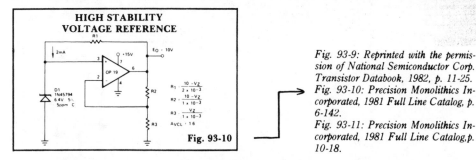

Fig. 93-9: Reprinted with the permission of National Semiconductor Corp. Transistor Databook, 1982, p. 11-25.
Fig. 93-10: Precision Monolithics Incorporated, 1981 Full Line Catalog, p. 6-142.
Fig. 93-11: Precision Monolithics Incorporated, 1981 Full Line Catalog, p. 10-18.

Many circuits are accompanied by a brief explanatory text. Those that do not have text can be readily understood from similar circuits in that chapter, or else they may be too complex to be explained briefly. The sparseness of text is deliberate so as to allow for more circuits which, after all, is what this book is all about.

The Index and Contents will be a time saver for the reader who knows exactly what he is looking for. The first page of each chapter lists the circuits in the order that they appear. The browser will surely discover many ideas and circuits that may well turn out to be most rewarding and great fun to put together.

The Common Schematic Symbols chart will help you identify circuit components.

Common Schematic Symbols

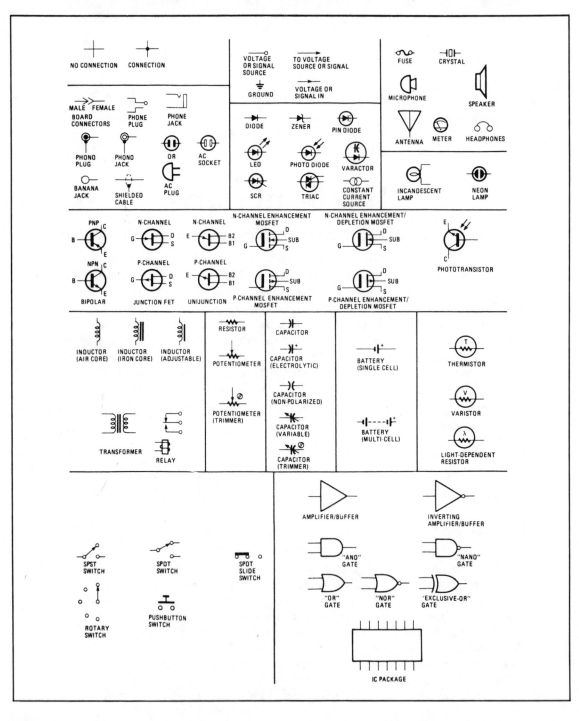

1

Alarms

The sources of the following circuits are contained in the Sources section beginning on page 730. The figure number contained in the box of each circuit correlates to the source entry in the Sources section.

Computalarm
Automotive Burglar Alarm
Security Alarm
Vehicle Security System
Home Security Monitor System
Antitheft Device
Auto Burglar Alarm
Tamper-Proof Burglar Alarm
Latching Burglar Alarm
Motion-Activated Motorcycle or Car Alarm
Boat Alarm

Blown Fuse Alarm
Auto Burglar Alarm
Continuous-Tone 2 kHz Buzzer with Bridge
    Drive, Gated on by a Logic 0
Pulsed-Tone Alarm, Gated by a High Input,
    with Direct-Drive Output
Piezoelectric Alarm
Gated 2 kHz Buzzer
Burglar Alarm
Latching Burglar Alarm
Sun -Powered Alarm

Freezer Meltdown Alarm

COMPUTALARM

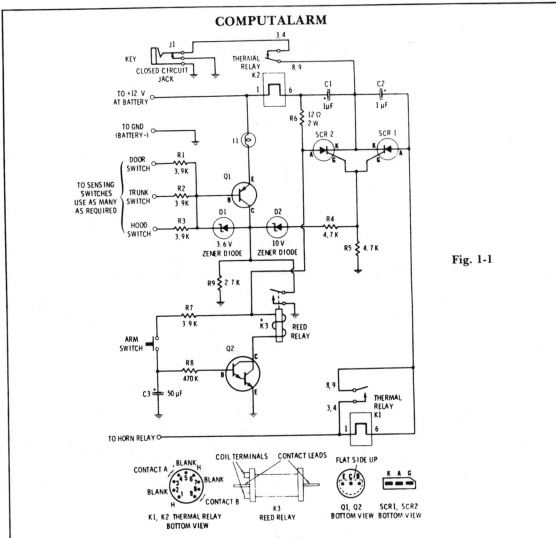

Fig. 1-1

Circuit Notes

The circuit has a built-in, self-arming feature. The driver turns off the ignition, presses the arm button on the Computalarm, and leaves the car. Within 20 seconds, the alarm arms itself—all automatically! The circuit will then detect the opening of any monitored door, the trunk lid, or the hood on the car. Once activated, the circuit remains dormant for 10 seconds. When the 10-second time delay has run out, the circuit will close the car's horn relay and sound the horn in periodic blasts (approximately 1 to 2 seconds apart) for a period of one minute. Then the Computalarm automatically shuts itself off (to save your battery) and re-arms. If a door, the trunk lid, or the hood remains ajar, the alarm circuit retriggers and another period of horn blasts occurs. The Computalarm has a "key" switch by which the driver can disarm the alarm circuit within a 10-second period after he enters the door. The key switch consists of a closed circuit jack, J1, and a mating miniature plug.

AUTOMOTIVE BURGLAR ALARM

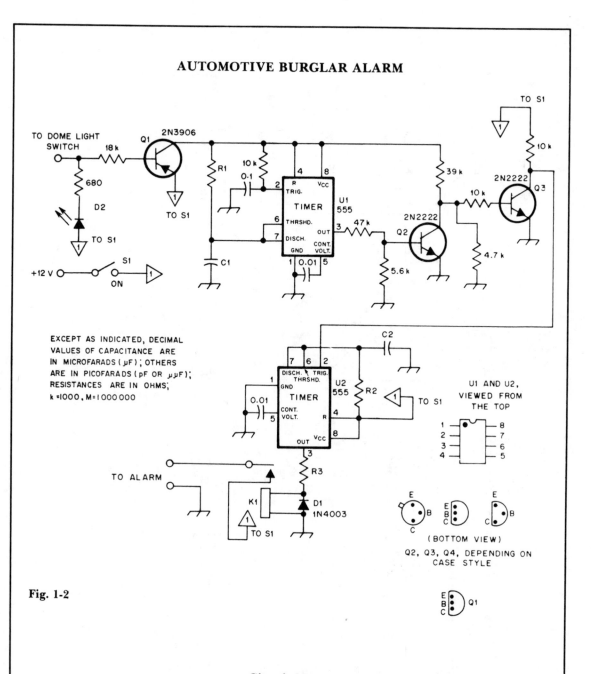

EXCEPT AS INDICATED, DECIMAL
VALUES OF CAPACITANCE ARE
IN MICROFARADS (μF); OTHERS
ARE IN PICOFARADS (pF OR μμF);
RESISTANCES ARE IN OHMS;
k =1000, M=1000000

Fig. 1-2

Circuit Notes

Alarm triggers on after a 13 second delay and stays on for 1-1½ minutes. Then it resets automatically. It can also be turned off and reset by opening and reclosing S1.

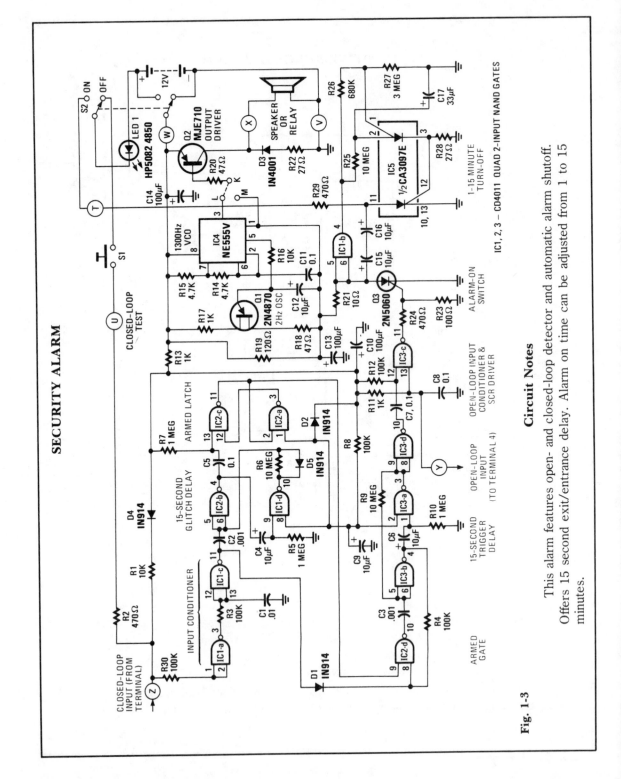

SECURITY ALARM

Fig. 1-3

Circuit Notes

This alarm features open- and closed-loop detector and automatic alarm shutoff. Offers 15 second exit/entrance delay. Alarm on time can be adjusted from 1 to 15 minutes.

VEHICLE SECURITY SYSTEM

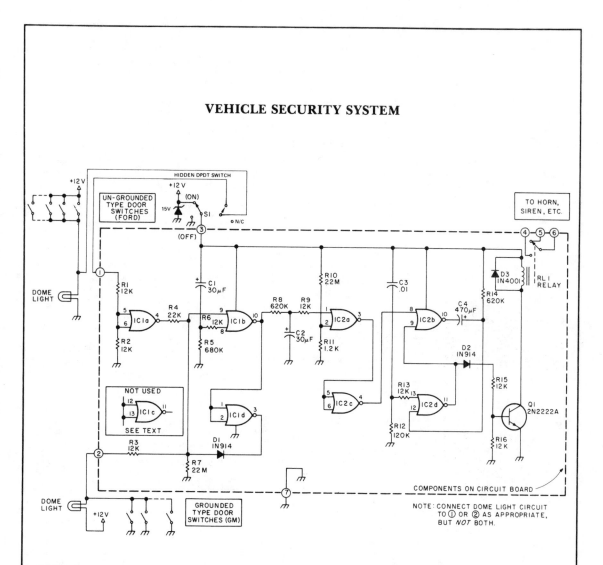

Fig. 1-4

Circuit Notes

This alarm gives a 15-20 second exit and entrance delay. After being triggered, the alarm sounds for five minutes and then shuts off. Once triggered, the sequence is automatic and is not affected by subsequent opening or closing of doors.

HOME SECURITY MONITOR SYSTEM

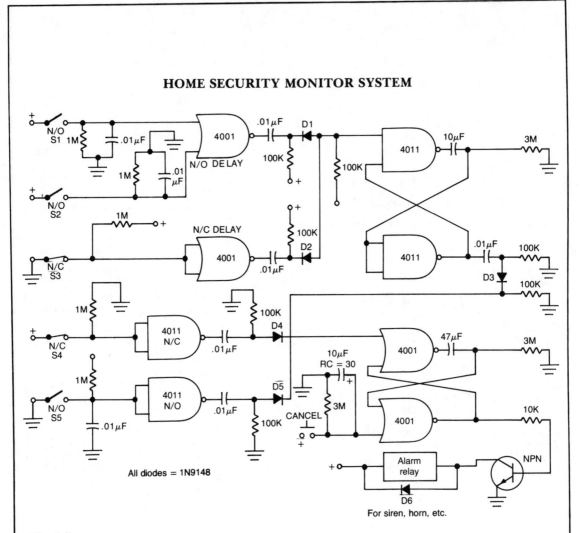

Fig. 1-5

Circuit Notes

This circuit provides normally open (NO) and normally closed (NC) contacts S1, S2, and S3 to turn on the alarm after a 30 second delay. S4 and S5 operate instantly. The CANCEL switch resets the alarm.

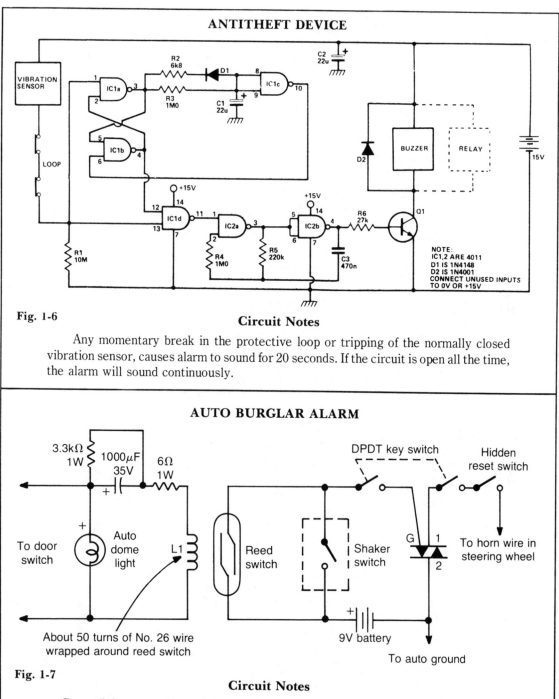

ANTITHEFT DEVICE

Fig. 1-6

Circuit Notes

Any momentary break in the protective loop or tripping of the normally closed vibration sensor, causes alarm to sound for 20 seconds. If the circuit is open all the time, the alarm will sound continuously.

AUTO BURGLAR ALARM

Fig. 1-7

Circuit Notes

Dome light current through L1 closes reed switch and sounds alarm. Shaker switch also activates alarm.

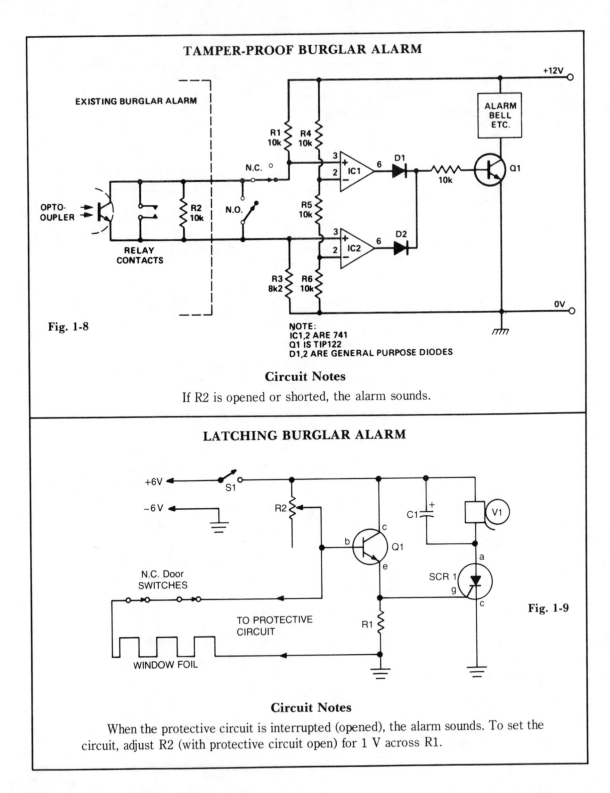

TAMPER-PROOF BURGLAR ALARM

EXISTING BURGLAR ALARM

OPTO-COUPLER

RELAY CONTACTS

R2 10k

N.C.

N.O.

R1 10k

R4 10k

R5 10k

R3 8k2

R6 10k

3
+
2
−
IC1
6

3
+
2
−
IC2
6

D1

D2

10k

ALARM BELL ETC.

Q1

+12V

0V

Fig. 1-8

NOTE:
IC1,2 ARE 741
Q1 IS TIP122
D1,2 ARE GENERAL PURPOSE DIODES

Circuit Notes

If R2 is opened or shorted, the alarm sounds.

LATCHING BURGLAR ALARM

+6V

−6V

S1

R2

N.C. Door SWITCHES

TO PROTECTIVE CIRCUIT

WINDOW FOIL

b

c

e

Q1

R1

C1

V1

SCR 1

a

g

c

Fig. 1-9

Circuit Notes

When the protective circuit is interrupted (opened), the alarm sounds. To set the circuit, adjust R2 (with protective circuit open) for 1 V across R1.

MOTION-ACTIVATED MOTORCYCLE OR CAR ALARM

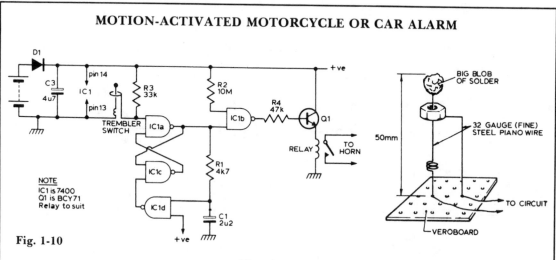

Fig. 1-10

Circuit Notes

Trembler (motion activated) switch sounds the alarm for 5 seconds. Then it goes off. Circuit is timed out for 10 seconds to allow the trembler switch to settle.

BOAT ALARM

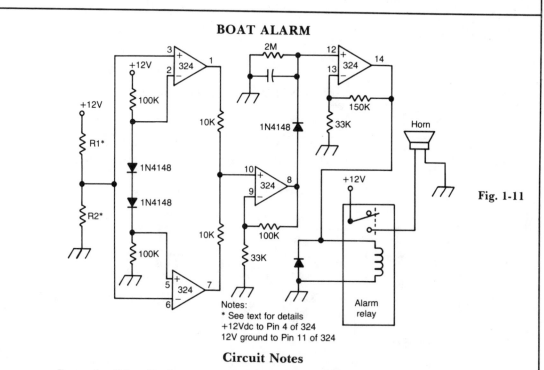

Fig. 1-11

Notes:
* See text for details
+12Vdc to Pin 4 of 324
12V ground to Pin 11 of 324

Circuit Notes

Removing R1 or R2 from the circuit (i.e., the potential thief breaks a hidden wire that connects R1 to +12 V and R2 to ground) activates the alarm for about five minutes.

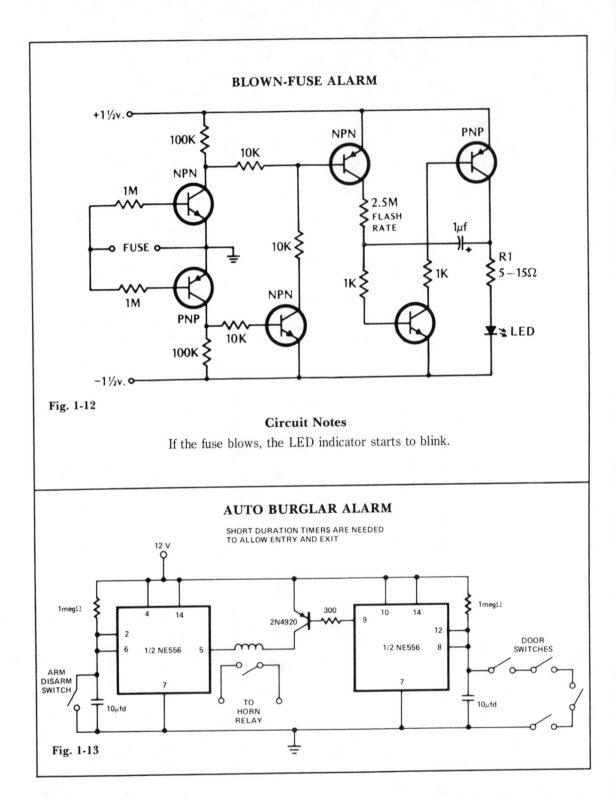

BLOWN-FUSE ALARM

Fig. 1-12

Circuit Notes

If the fuse blows, the LED indicator starts to blink.

AUTO BURGLAR ALARM

SHORT DURATION TIMERS ARE NEEDED
TO ALLOW ENTRY AND EXIT

Fig. 1-13

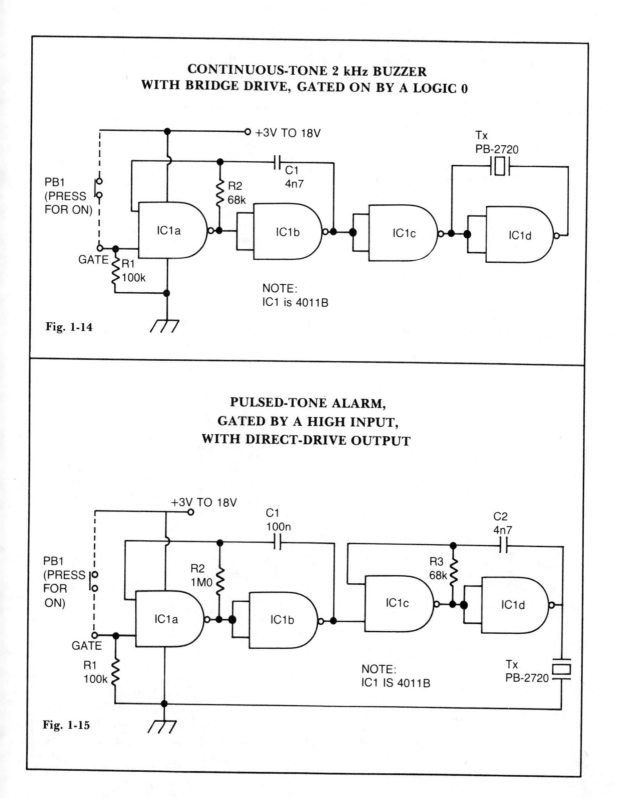

**CONTINUOUS-TONE 2 kHz BUZZER
WITH BRIDGE DRIVE, GATED ON BY A LOGIC 0**

+3V TO 18V

Tx
PB-2720

PB1
(PRESS
FOR ON)

R2
68k

C1
4n7

IC1a

IC1b

IC1c

IC1d

GATE

R1
100k

NOTE:
IC1 is 4011B

Fig. 1-14

**PULSED-TONE ALARM,
GATED BY A HIGH INPUT,
WITH DIRECT-DRIVE OUTPUT**

+3V TO 18V

C1
100n

C2
4n7

PB1
(PRESS
FOR
ON)

R2
1M0

R3
68k

IC1a

IC1b

IC1c

IC1d

GATE

R1
100k

NOTE:
IC1 IS 4011B

Tx
PB-2720

Fig. 1-15

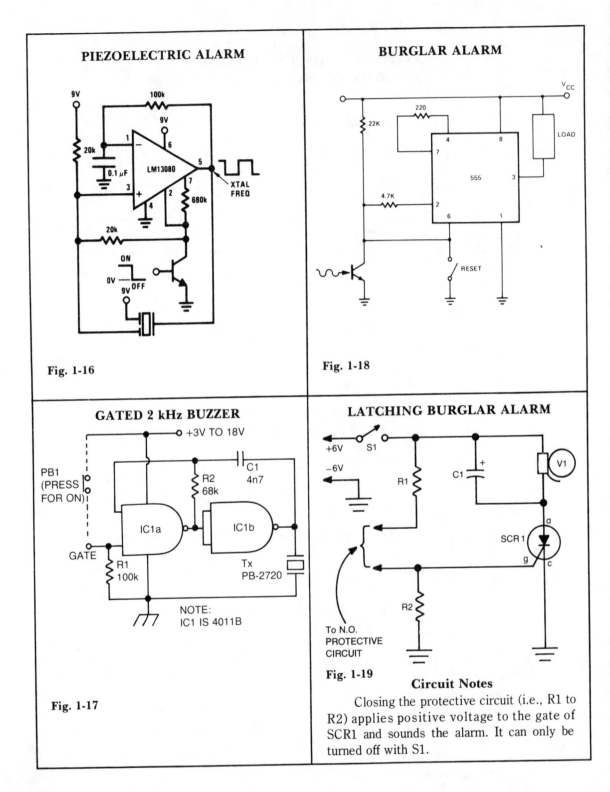

PIEZOELECTRIC ALARM

9V
100k
9V
20k
0.1 μF
LM13080
1
6
5
XTAL FREQ
3
7
2
680k
4
20k
ON
0V
OFF
9V

Fig. 1-16

BURGLAR ALARM

V_CC
22K
220
LOAD
4
7
8
555
3
4.7K
2
6
1
RESET

Fig. 1-18

GATED 2 kHz BUZZER

+3V TO 18V
PB1
(PRESS FOR ON)
C1
4n7
R2
68k
IC1a
IC1b
GATE
R1
100k
Tx
PB-2720
NOTE:
IC1 IS 4011B

Fig. 1-17

LATCHING BURGLAR ALARM

+6V
S1
−6V
R1
C1
V1
SCR1
a
g
c
R2
To N.O.
PROTECTIVE
CIRCUIT

Fig. 1-19

Circuit Notes

Closing the protective circuit (i.e., R1 to R2) applies positive voltage to the gate of SCR1 and sounds the alarm. It can only be turned off with S1.

SUN-POWERED ALARM

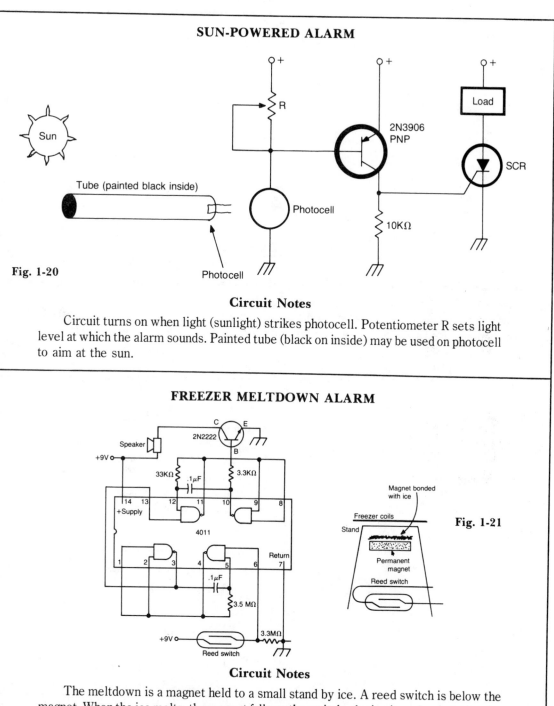

Fig. 1-20

Circuit Notes

Circuit turns on when light (sunlight) strikes photocell. Potentiometer R sets light level at which the alarm sounds. Painted tube (black on inside) may be used on photocell to aim at the sun.

FREEZER MELTDOWN ALARM

Fig. 1-21

Circuit Notes

The meltdown is a magnet held to a small stand by ice. A reed switch is below the magnet. When the ice melts, the magnet falls on the switch, closing it, and completing the alarm circuit.

2
Amateur Radio

The sources of the following circuits are contained in the Sources section beginning on page 730. The figure number contained in the box of each circuit correlates to the source entry in the Sources section.

Code Practice Oscillator Produces Automatic
   Dits and Dahs
Rf Power Meter
In-Line Wattmeter
CW Signal Processor
Two-Meter Preamplifier for Handitalkies
Repeater Beeper
Electronic Keyer
Code Practice Oscillator
Automatic Tape Recording

Self-Powered CW Monitor
Remote Rf Current Readout
Code Practice Oscillator
SWR Warning Indicator
Subaudible Tone Encoder
Audio Mixers
Rf Powered Sidetone Oscillator
Harmonic Generator
Automatic TTL Morse-Code Keyer
Remote Rf Current Readout

CODE-PRACTICE OSCILLATOR
PRODUCES AUTOMATIC DITS AND DAHS

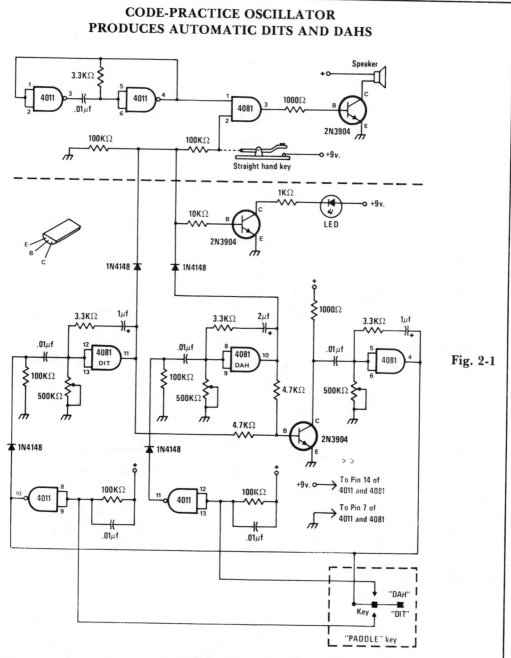

Fig. 2-1

Circuit Notes

The circuit consists of a basic oscillator (above dashed line) and an automatic keyer (below dashed line). The unit can be used with a straight hand key or a paddle key for automatic operation.

RF POWER METER

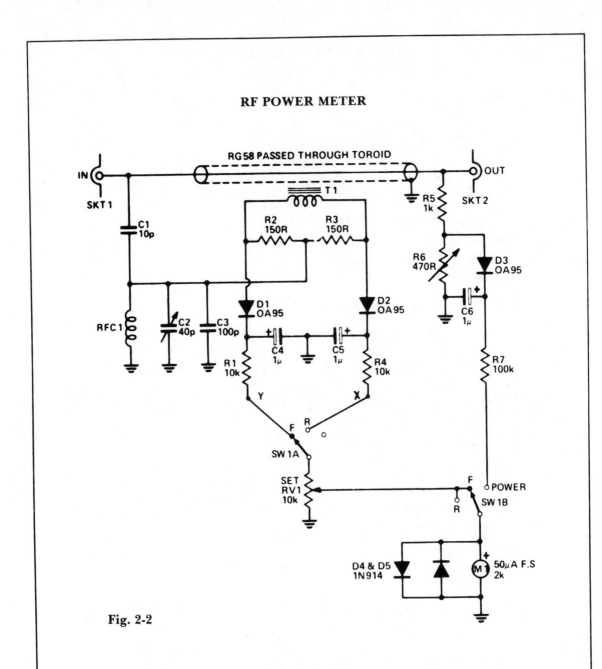

Fig. 2-2

Circuit Notes

Reflectometer (SWR Power Meter) covers three decades—from 100 kHz to 100 MHz. It can be constructed for rf powers as low as 500 mW or up to 500 watts.

IN-LINE WATTMETER

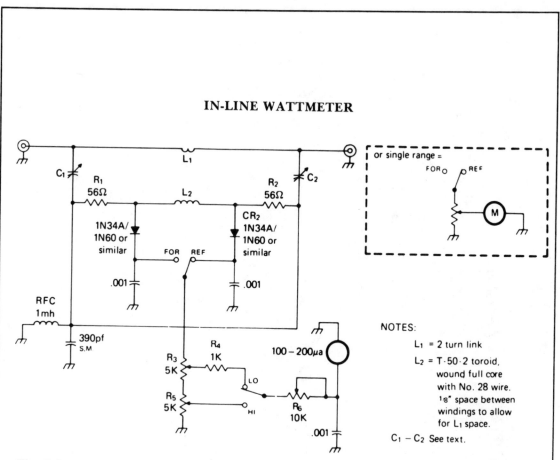

Fig. 2-3

NOTES:

L₁ = 2 turn link

L₂ = T-50-2 toroid, wound full core with No. 28 wire. 1/8" space between windings to allow for L₁ space.

C₁ – C₂ See text.

Circuit Notes

The circuit is not frequency sensitive. Its calibration will be accurate over a wide frequency spectrum, such as the entire amateur hf spectrum, if the values of L2, the voltage divider capacitors C1-2 and C3, and the resistances of R1-2 are chosen properly. R1-2 and CR1-2 should be matched for best results. Generally, R1-2 must be small compared to the reactance of L2 so as to avoid any significant effect on the L2 current which is induced by the transmission line current flowing through L1. The lower frequency limit of the bridge is set by the R1-R2/Ls ratio, and the cutoff is at the point where the value of R1-R2 becomes significant with reference to the reactance of L2 at that frequency point.

17

CW SIGNAL PROCESSOR

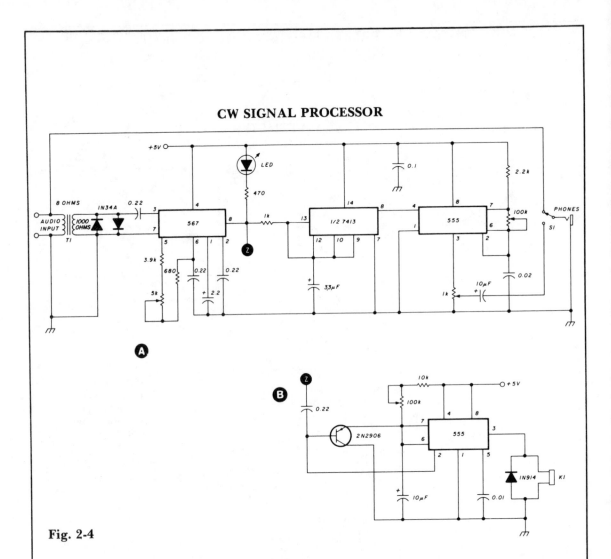

Fig. 2-4

Circuit Notes

This circuit provides interferenced rejection for the CW operator. The 567 phase-locked loop is configured to respond to tones from 500 to 1100 Hz. The Schmitt trigger reduces the weighting effect caused by the output of the PLL remaining low after removal of the audio signal. Ten to 15 millivolts of audio activate the circuit. For periods of loss of signal, circuit B will automatically switch back to live receiver audio after a suitable delay. (If a relay with a 5-volt coil is not available, the circuit can also be powered from +12 volts.) When circuit B is used, the contacts on relay K1 replace S1.

TWO-METER PREAMPLIFIER FOR HANDITALKIES

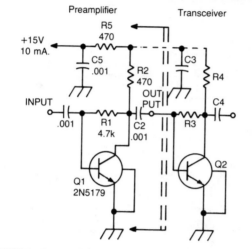

Preamplifier   **Transceiver**

Circuit Notes

This simple, inexpensive, wideband rf amplifier provides 14 dB gain on two meters without the use of tuned circuits.

Fig. 2-5

REPEATER BEEPER

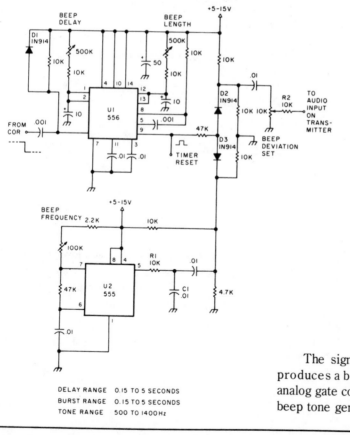

Fig. 2-6

DELAY RANGE   0.15 TO 5 SECONDS
BURST RANGE   0.15 TO 5 SECONDS
TONE RANGE   500 TO 1400 Hz

Circuit Notes

The signal from COR triggers U1 which produces a beep-gate pulse that enables the analog gate consisting of D2 and D3 to pass the beep tone generated by U2.

ELECTRONIC KEYER

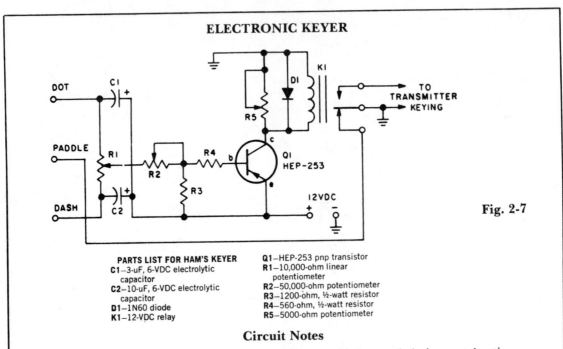

Fig. 2-7

PARTS LIST FOR HAM'S KEYER	
C1—3-uF, 6-VDC electrolytic capacitor	**Q1**—HEP-253 pnp transistor
C2—10-uF, 6-VDC electrolytic capacitor	**R1**—10,000-ohm linear potentiometer
D1—1N60 diode	**R2**—50,000-ohm potentiometer
K1—12-VDC relay	**R3**—1200-ohm, ½-watt resistor
	R4—560-ohm, ½-watt resistor
	R5—5000-ohm potentiometer

Circuit Notes

This circuit automatically produces Morse code dots and dashes set by time constants involving C1 and C2. R1 sets dot/dash ratio and R2 sets the speed. R5 sets the relay drop-out point.

CODE PRACTICE OSCILLATOR

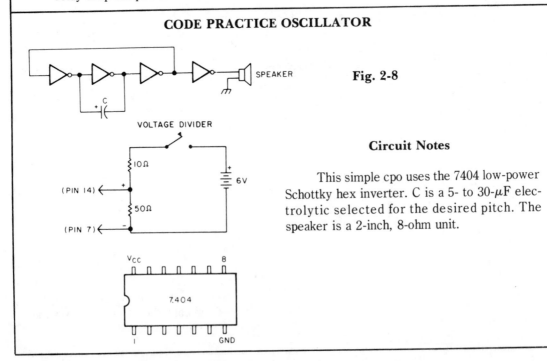

Fig. 2-8

Circuit Notes

This simple cpo uses the 7404 low-power Schottky hex inverter. C is a 5- to 30-μF electrolytic selected for the desired pitch. The speaker is a 2-inch, 8-ohm unit.

AUTOMATIC TAPE RECORDING

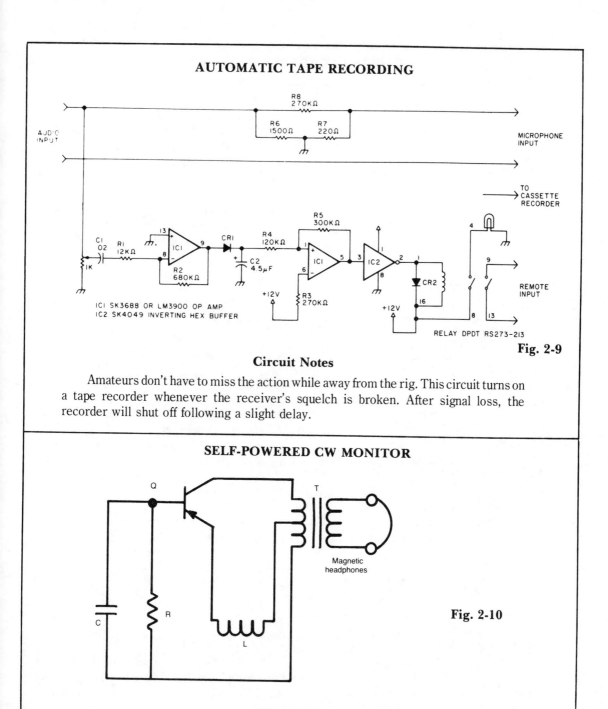

Fig. 2-9

Circuit Notes

Amateurs don't have to miss the action while away from the rig. This circuit turns on a tape recorder whenever the receiver's squelch is broken. After signal loss, the recorder will shut off following a slight delay.

SELF-POWERED CW MONITOR

Fig. 2-10

Circuit Notes

Position L near the transmitter output tank to hear the key-down tone. Then tape the coil in place. C = .047 μF, R = 8.2 K, Q = HEP 253 (or equal), T = 500: 500 ohm center tapped transformer. L = 2 to 6 turns on ½″ coil form.

REMOTE RF CURRENT READOUT

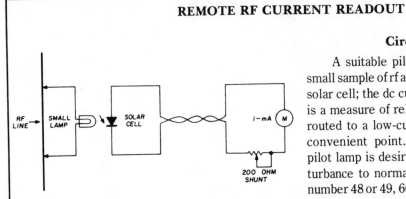

Circuit Notes

A suitable pilot lamp is illuminated by a small sample of rf and energizes an inexpensive solar cell; the dc current generated by the cell is a measure of relative rf power, and may be routed to a low-current meter located at any convenient point. A sensitive, low-current pilot lamp is desirable to cause minimum disturbance to normal rf circuit conditions. The number 48 or 49, 60 mA lamp is suitable for use with transmitters above 1-watt output.

Fig. 2-11

CODE PRACTICE OSCILLATOR

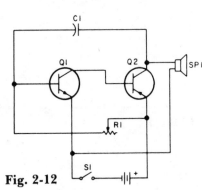

Circuit Notes

Oscillator, works with 2 to 12 Vdc (but 9 to 12 volts gives best volume and clean keying). R1 can be replaced with a 500 K pot and the circuit will sweep the entire audio frequency range.

Fig. 2-12

SWR WARNING INDICATOR

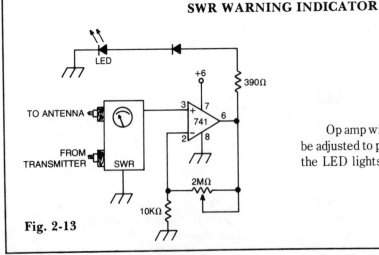

Circuit Notes

Op amp with dc input from SWR meter can be adjusted to preset the SWR reading at which the LED lights.

Fig. 2-13

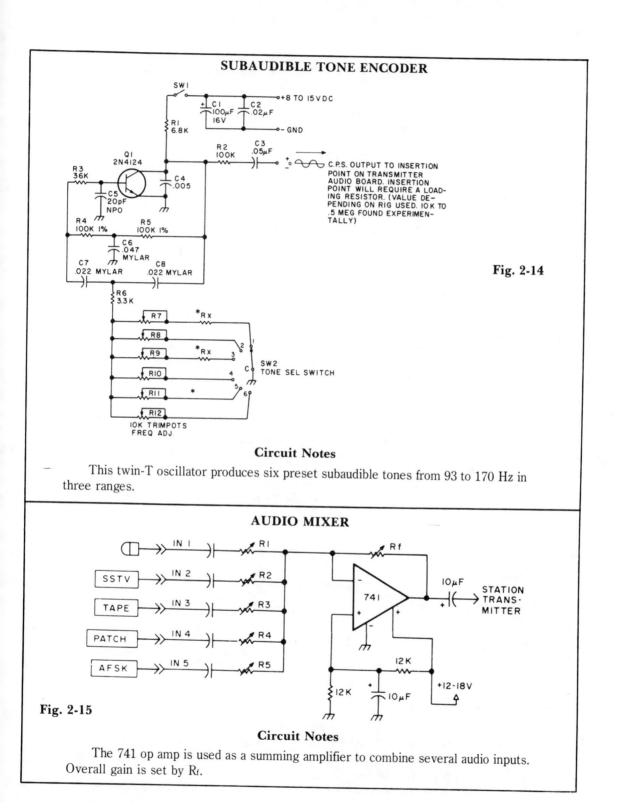

SUBAUDIBLE TONE ENCODER

SW1

+ C1
100μF
16V

C2
.02μF

+8 TO 15 VDC

— GND

R1
6.8K

Q1
2N4124

R2
100K

C3
.05μF

C.P.S. OUTPUT TO INSERTION
POINT ON TRANSMITTER
AUDIO BOARD. INSERTION
POINT WILL REQUIRE A LOAD-
ING RESISTOR. (VALUE DE-
PENDING ON RIG USED. 10 K TO
.5 MEG FOUND EXPERIMEN-
TALLY)

R3
36K

C4
.005

C5
20pF
NPO

R4
100K 1%

R5
100K 1%

C6
.047
MYLAR

C7
.022 MYLAR

C8
.022 MYLAR

Fig. 2-14

R6
3.3 K

R7

*Rx

R8

R9

*Rx

R10

R11

*

R12

SW2
TONE SEL SWITCH

10K TRIMPOTS
FREQ ADJ.

Circuit Notes

This twin-T oscillator produces six preset subaudible tones from 93 to 170 Hz in three ranges.

AUDIO MIXER

IN 1
IN 2 — SSTV
IN 3 — TAPE
IN 4 — PATCH
IN 5 — AFSK

R1
R2
R3
R4
R5

Rf

741

10μF

STATION
TRANS-
MITTER

12K

12K

10μF

+12-18V

Fig. 2-15

Circuit Notes

The 741 op amp is used as a summing amplifier to combine several audio inputs. Overall gain is set by R_f.

RF-POWERED SIDETONE OSCILLATOR

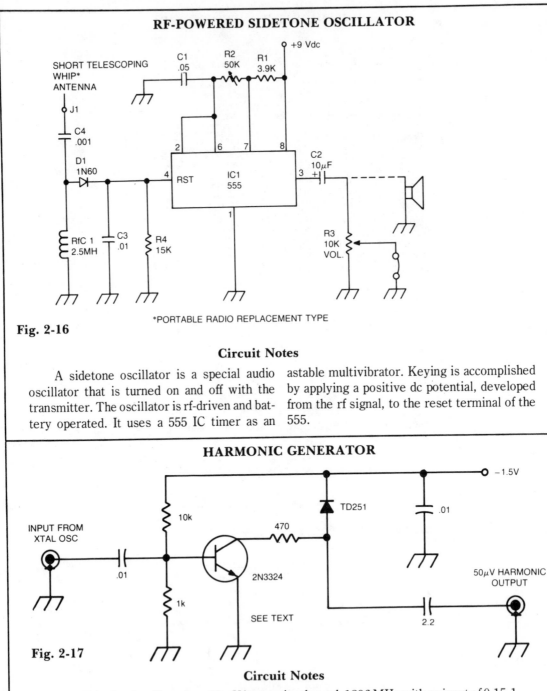

Fig. 2-16

*PORTABLE RADIO REPLACEMENT TYPE

Circuit Notes

A sidetone oscillator is a special audio oscillator that is turned on and off with the transmitter. The oscillator is rf-driven and battery operated. It uses a 555 IC timer as an astable multivibrator. Keying is accomplished by applying a positive dc potential, developed from the rf signal, to the reset terminal of the 555.

HARMONIC GENERATOR

Fig. 2-17

Circuit Notes

This circuit will produce 50 μV harmonics through 1296 MHz with an input of 0.15-1 V from a 100 or 1000 kHz crystal oscillator. With a germanium diode instead of a tunnel diode, harmonics can be heard up to about 147 MHz.

24

AUTOMATIC TTL MORSE-CODE KEYER

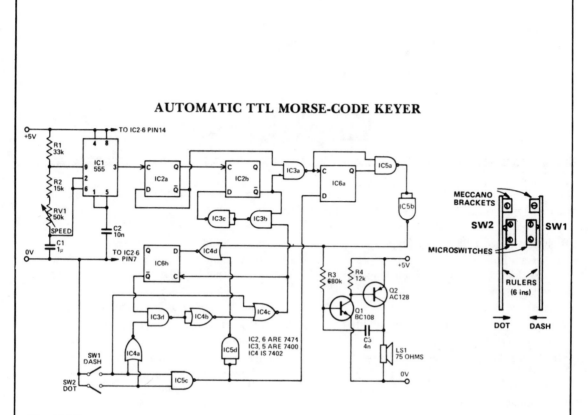

Fig. 2-18

Circuit Notes

Automatically generated dits and dahs are produced over a speed range of 11 to 39 wpm. The upper limit can be raised by decreasing R2. SW1 and SW2 can be a "home-brew" paddle operated key.

3

Amplifiers

The sources of the following circuits are contained in the Sources section beginning on page 730. The figure number contained in the box of each circuit correlates to the source entry in the Sources section.

HIGH IMPEDANCE DIFFERENTIAL AMPLIFIER

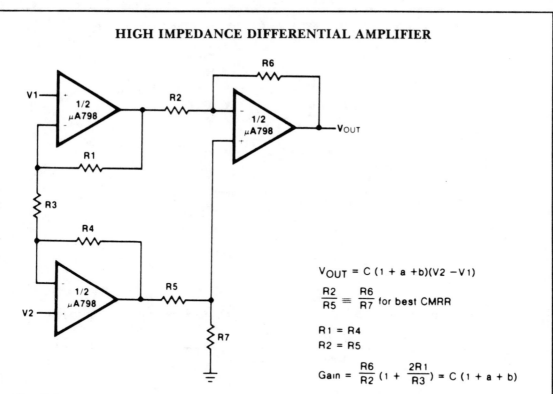

$V_{OUT} = C (1 + a + b)(V2 - V1)$

$\dfrac{R2}{R5} \equiv \dfrac{R6}{R7}$ for best CMRR

$R1 = R4$
$R2 = R5$

$Gain = \dfrac{R6}{R2} \left(1 + \dfrac{2R1}{R3}\right) = C (1 + a + b)$

Fig. 3-1

UNITY GAIN FOLLOWER

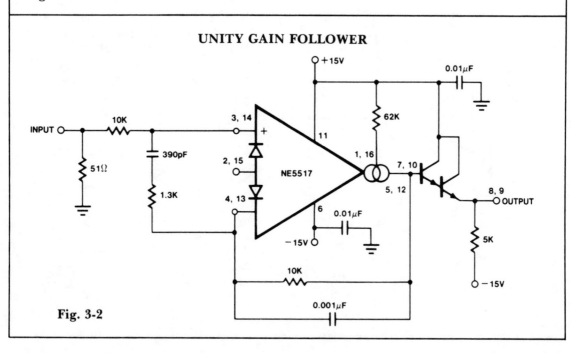

Fig. 3-2

VOLTAGE CONTROLLED VARIABLE GAIN AMPLIFIER

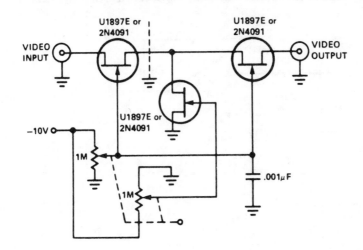

Fig. 3-3

Circuit Notes

The tee attenuator provides for optimum dynamic linear range attenuation up to 100 dB, even at f = 10.7 MHz with proper layout.

POWER BOOSTER

Circuit Notes

Power booster is capable of driving moderate loads. The circuit as shown uses a NE5535 device. Other amplifiers may be substituted only if R1 values are changed because of the I_{cc} current required by the amplifier. R1 should be calculated from the following expression:

$$R1 = \frac{600 \text{ mW}}{I_{cc}}$$

All resistor values are in ohms.

Fig. 3-4

LOGARITHMIC AMPLIFIER

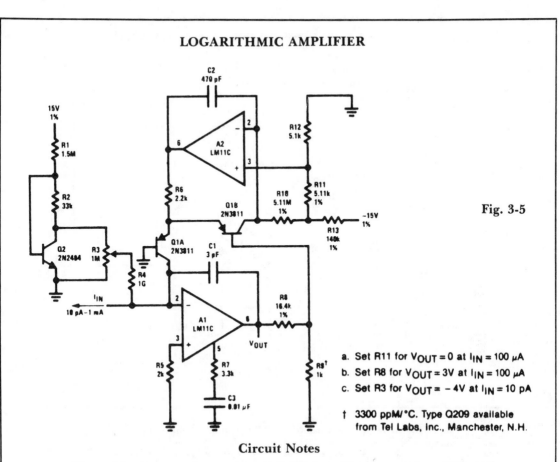

Fig. 3-5

a. Set R11 for $V_{OUT} = 0$ at $I_{IN} = 100 \ \mu A$
b. Set R8 for $V_{OUT} = 3V$ at $I_{IN} = 100 \ \mu A$
c. Set R3 for $V_{OUT} = -4V$ at $I_{IN} = 10 \ pA$

† 3300 ppM/°C. Type Q209 available
from Tel Labs, Inc., Manchester, N.H.

Circuit Notes

Unusual frequency compensation gives this logarithmic converter a 100 μs time constant from 1 mA down to 100 μA, increasing from 200 μs to 200 ms from 10 nA to 10 pA. Optional bias current compensation can give 10 pA resolution from $-55\,°C$ to $100\,°C$. Scale factor is 1 V/decade and temperature compensated.

VOLTAGE CONTROLLED VARIABLE GAIN AMPLIFIER

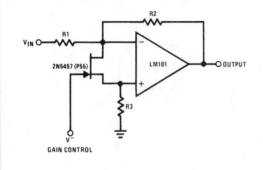

GAIN CONTROL

Circuit Notes

The 2N5457 acts as a voltage variable resistor with an $R_{ds(on)}$ of 800 ohms max. Since the differential voltage on the LM101 is in the low mV range, the 2N5457 JFET will have linear resistance over several decades of resistance providing an excellent electronic gain control.

Fig. 3-6

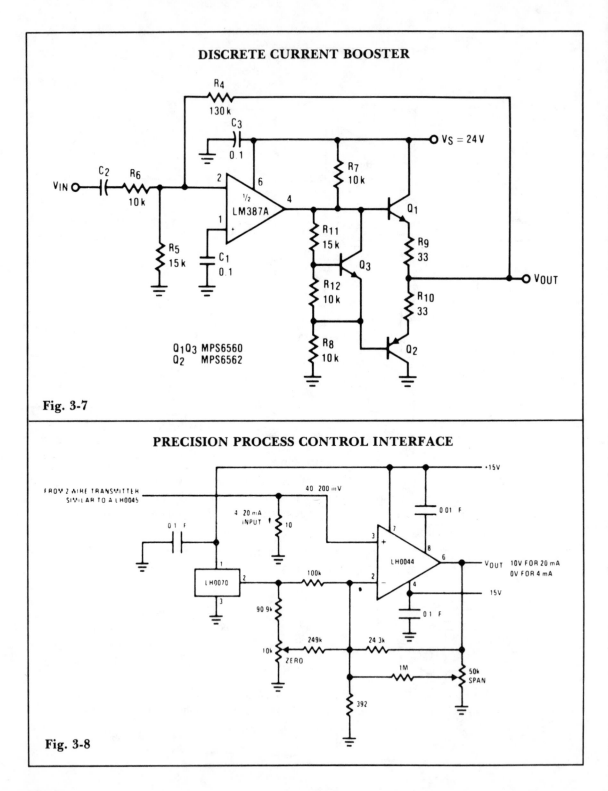

DISCRETE CURRENT BOOSTER

Fig. 3-7

PRECISION PROCESS CONTROL INTERFACE

Fig. 3-8

VOLTAGE CONTROLLED AMPLIFIER

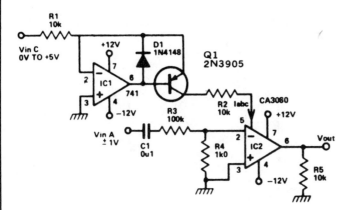

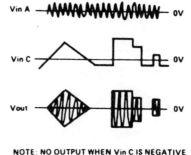

NOTE: NO OUTPUT WHEN Vin C IS NEGATIVE

Fig. 3-9

Circuit Notes

This circuit is basically an op amp with an extra input at pin 5. A current I_{ABC} is injected into this input and this controls the gain of the device linerly. Thus by inserting an audio signal (±10 mV) between pin 2 and 3 and by controlling the current on pin 5, the level of the signal output (pin 6) is controlled.

ABSOLUTE VALUE AMPLIFIER

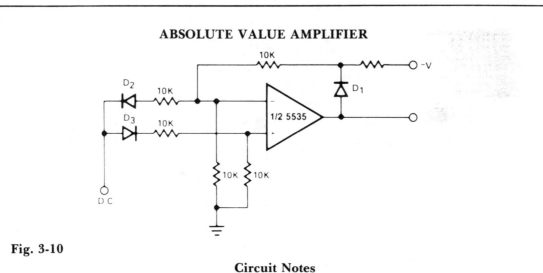

Fig. 3-10

Circuit Notes

The circuit generates a positive output voltage for either polarity of input. For positive signals, it acts as a noninverting amplifier and for negative signals, as an inverting amplifier.

The accuracy is poor for input voltages under 1 V, but for less stringent applications, it can be effective.

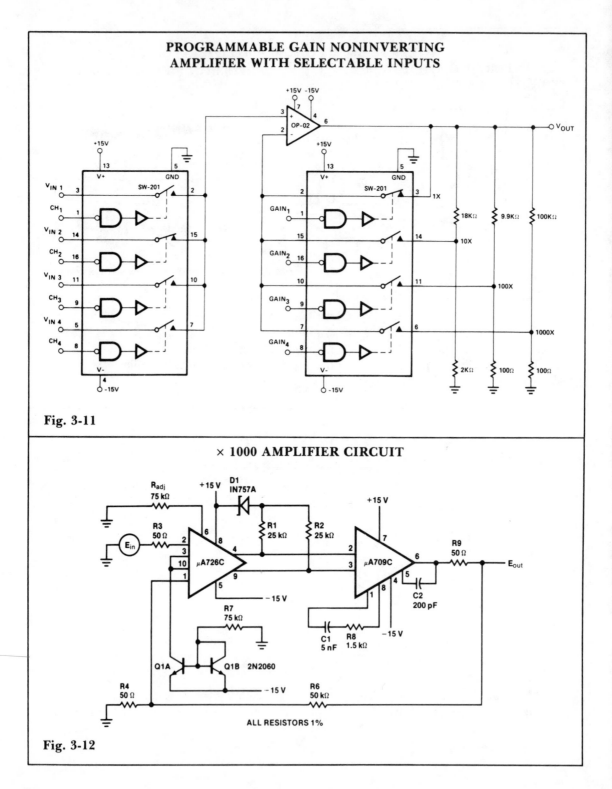

PROGRAMMABLE GAIN NONINVERTING AMPLIFIER WITH SELECTABLE INPUTS

Fig. 3-11

× 1000 AMPLIFIER CIRCUIT

ALL RESISTORS 1%

Fig. 3-12

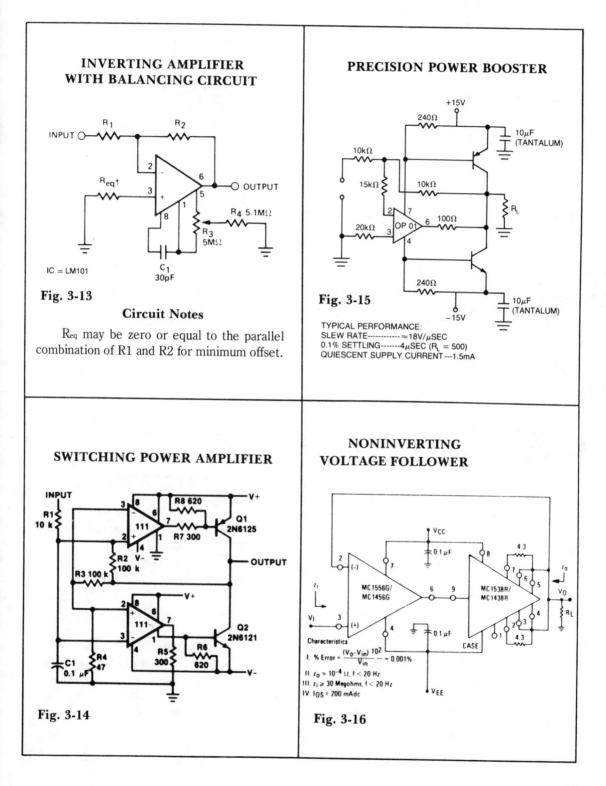

INVERTING AMPLIFIER WITH BALANCING CIRCUIT

INPUT

R_1 R_2

R_{eq}↑

R_4 5.1MΩ
R_3 5MΩ
C_1 30pF

IC = LM101

OUTPUT

Fig. 3-13

Circuit Notes

R_{eq} may be zero or equal to the parallel combination of R1 and R2 for minimum offset.

PRECISION POWER BOOSTER

+15V
240Ω
10μF (TANTALUM)
10kΩ
15kΩ   10kΩ
R_L
20kΩ   OP 01   100Ω
2 7
6
3 4
240Ω
10μF (TANTALUM)
−15V

Fig. 3-15

TYPICAL PERFORMANCE:
SLEW RATE ------------ ≈18V/μSEC
0.1% SETTLING ------- 4μSEC (R_L = 500)
QUIESCENT SUPPLY CURRENT --- 1.5mA

SWITCHING POWER AMPLIFIER

INPUT
R1 10 k
R8 620
V+
Q1 2N6125
111
R7 300
R2 V− 100 k
R3 100 k
OUTPUT
111
V+
Q2 2N6121
C1 0.1 μF
R4 47
R5 300
R6 620
V−

Fig. 3-14

NONINVERTING VOLTAGE FOLLOWER

V_{CC}
0.1 μF
4.3
z_i
V_I
MC1556G/ MC1456G
MC1538R/ MC1438R
z_0
V_O
R_L
0.1 μF
CASE
4.3
V_{EE}

Characteristics

I. % Error = $\dfrac{(V_0 - V_{in})\, 10^2}{V_{in}}$ ≈ 0.001%

II. $z_0 \approx 10^{-4}$ Ω, f < 20 Hz

III. $z_i \geq$ 30 Megohms, f < 20 Hz

IV. I_{0S} = 200 mAdc

Fig. 3-16

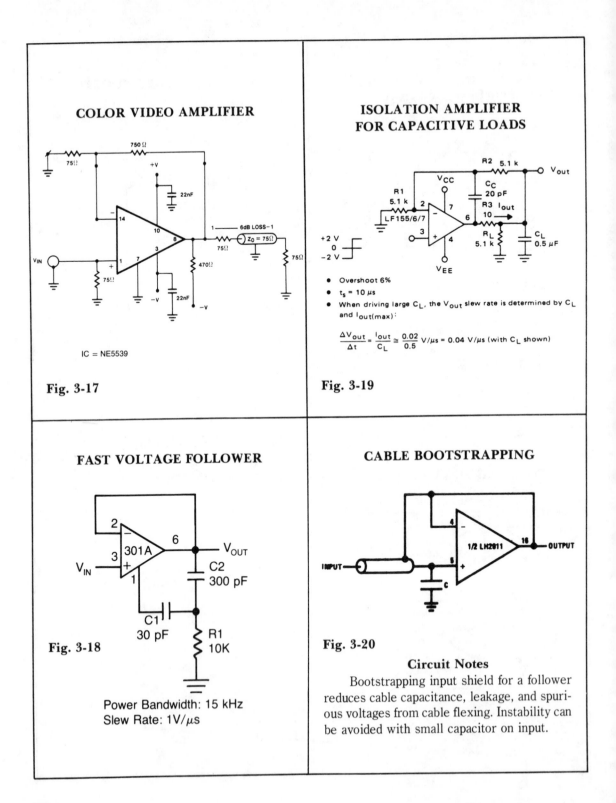

COLOR VIDEO AMPLIFIER

IC = NE5539

Fig. 3-17

ISOLATION AMPLIFIER FOR CAPACITIVE LOADS

- Overshoot 6%
- $t_s = 10 \ \mu s$
- When driving large C_L, the V_{out} slew rate is determined by C_L and $I_{out(max)}$:

$$\frac{\Delta V_{out}}{\Delta t} = \frac{I_{out}}{C_L} \cong \frac{0.02}{0.5} \ V/\mu s = 0.04 \ V/\mu s \ \text{(with } C_L \text{ shown)}$$

Fig. 3-19

FAST VOLTAGE FOLLOWER

Power Bandwidth: 15 kHz
Slew Rate: 1V/μs

Fig. 3-18

CABLE BOOTSTRAPPING

Fig. 3-20

Circuit Notes

Bootstrapping input shield for a follower reduces cable capacitance, leakage, and spurious voltages from cable flexing. Instability can be avoided with small capacitor on input.

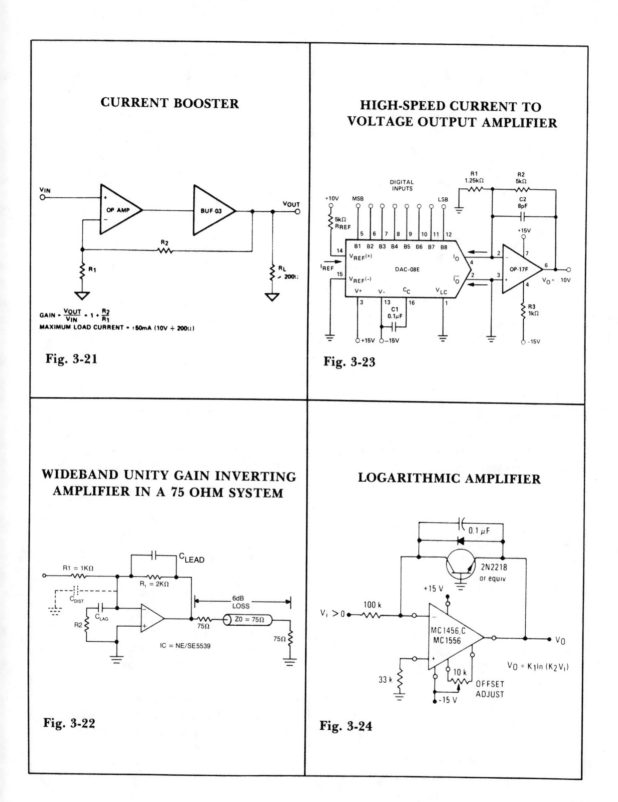

CURRENT BOOSTER

GAIN = $\frac{V_{OUT}}{V_{IN}}$ = 1 + $\frac{R_2}{R_1}$

MAXIMUM LOAD CURRENT = ±50mA (10V ÷ 200Ω)

Fig. 3-21

HIGH-SPEED CURRENT TO VOLTAGE OUTPUT AMPLIFIER

Fig. 3-23

WIDEBAND UNITY GAIN INVERTING AMPLIFIER IN A 75 OHM SYSTEM

Fig. 3-22

LOGARITHMIC AMPLIFIER

$V_O = K_1 \ln (K_2 V_i)$

Fig. 3-24

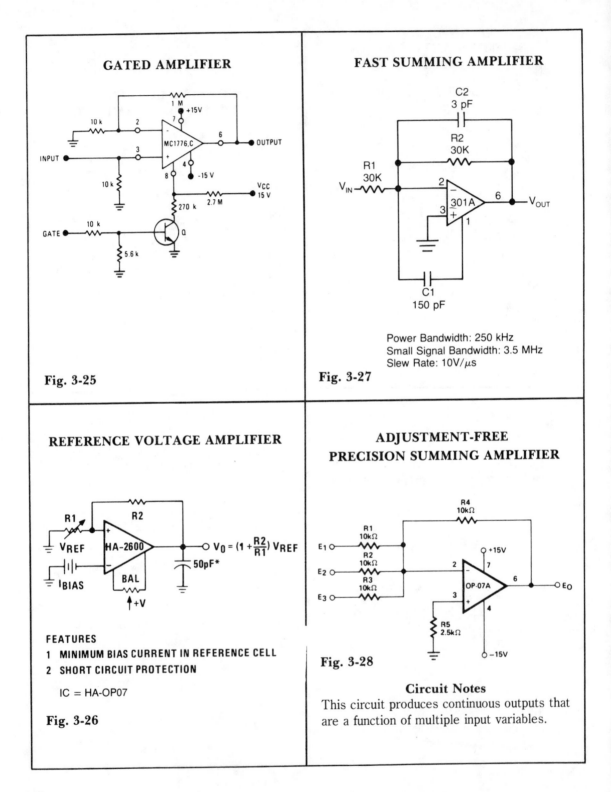

GATED AMPLIFIER

Fig. 3-25

FAST SUMMING AMPLIFIER

Power Bandwidth: 250 kHz
Small Signal Bandwidth: 3.5 MHz
Slew Rate: 10V/μs

Fig. 3-27

REFERENCE VOLTAGE AMPLIFIER

$$V_0 = (1 + \frac{R2}{R1})\, V_{REF}$$

FEATURES

1 MINIMUM BIAS CURRENT IN REFERENCE CELL
2 SHORT CIRCUIT PROTECTION

IC = HA-OP07

Fig. 3-26

ADJUSTMENT-FREE
PRECISION SUMMING AMPLIFIER

Fig. 3-28

Circuit Notes

This circuit produces continuous outputs that are a function of multiple input variables.

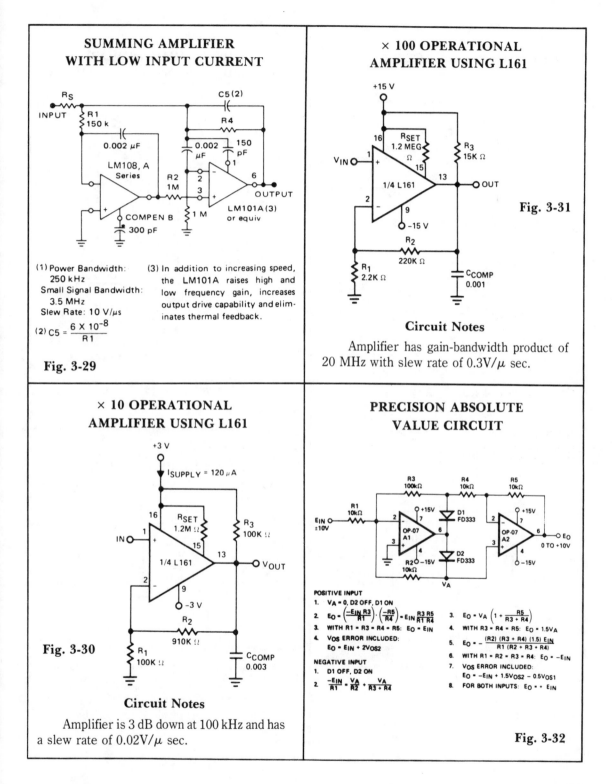

SUMMING AMPLIFIER WITH LOW INPUT CURRENT

R_S
INPUT
R1
150 k
0.002 μF
LM108, A
Series
COMPEN B
300 pF
C5 (2)
R4
150 pF
0.002 μF
R2
1M
1 M
OUTPUT
LM101A (3)
or equiv

(1) Power Bandwidth:
250 kHz
Small Signal Bandwidth:
3.5 MHz
Slew Rate: 10 V/μs

(2) $C5 = \dfrac{6 \times 10^{-8}}{R1}$

(3) In addition to increasing speed, the LM101A raises high and low frequency gain, increases output drive capability and eliminates thermal feedback.

Fig. 3-29

× 100 OPERATIONAL AMPLIFIER USING L161

+15 V
V_{IN}
R_{SET}
1.2 MEG Ω
R3
15K Ω
1/4 L161
OUT
–15 V
R2
220K Ω
R_1
2.2K Ω
C_{COMP}
0.001

Fig. 3-31

Circuit Notes

Amplifier has gain-bandwidth product of 20 MHz with slew rate of 0.3V/μ sec.

× 10 OPERATIONAL AMPLIFIER USING L161

+3 V
I_{SUPPLY} = 120 μA
IN
R_{SET}
1.2M Ω
R3
100K Ω
1/4 L161
V_{OUT}
–3 V
R2
910K Ω
R_1
100K Ω
C_{COMP}
0.003

Fig. 3-30

Circuit Notes

Amplifier is 3 dB down at 100 kHz and has a slew rate of 0.02V/μ sec.

PRECISION ABSOLUTE VALUE CIRCUIT

R3
100kΩ
R4
10kΩ
R5
10kΩ
E_{IN}
±10V
R1
10kΩ
+15V
D1
FD333
OP-07
A1
+15V
OP-07
A2
E_O
0 TO +10V
R2
10kΩ
–15V
D2
FD333
–15V
V_A

POSITIVE INPUT
1. $V_A = 0$, D2 OFF, D1 ON
2. $E_O = \left(\dfrac{-E_{IN}\ R3}{R1}\right) \cdot \left(\dfrac{-R5}{R4}\right) = E_{IN} \dfrac{R3\ R5}{R1\ R4}$
3. WITH R1 = R3 = R4 = R5: $E_O = E_{IN}$
4. V_{OS} ERROR INCLUDED:
$E_O = E_{IN} + 2V_{OS2}$

NEGATIVE INPUT
1. D1 OFF, D2 ON
2. $\dfrac{-E_{IN}}{R1} = \dfrac{V_A}{R2} + \dfrac{V_A}{R3 + R4}$

3. $E_O = V_A \left(1 + \dfrac{R5}{R3 + R4}\right)$
4. WITH R3 = R4 = R5: $E_O = 1.5V_A$
5. $E_O = -\dfrac{(R2)\ (R3 + R4)\ (1.5)\ E_{IN}}{R1\ (R2 + R3 + R4)}$
6. WITH R1 = R2 = R3 = R4: $E_O = -E_{IN}$
7. V_{OS} ERROR INCLUDED:
$E_O = -E_{IN} + 1.5V_{OS2} - 0.5V_{OS1}$
8. FOR BOTH INPUTS: $E_O = + E_{IN}$

Fig. 3-32

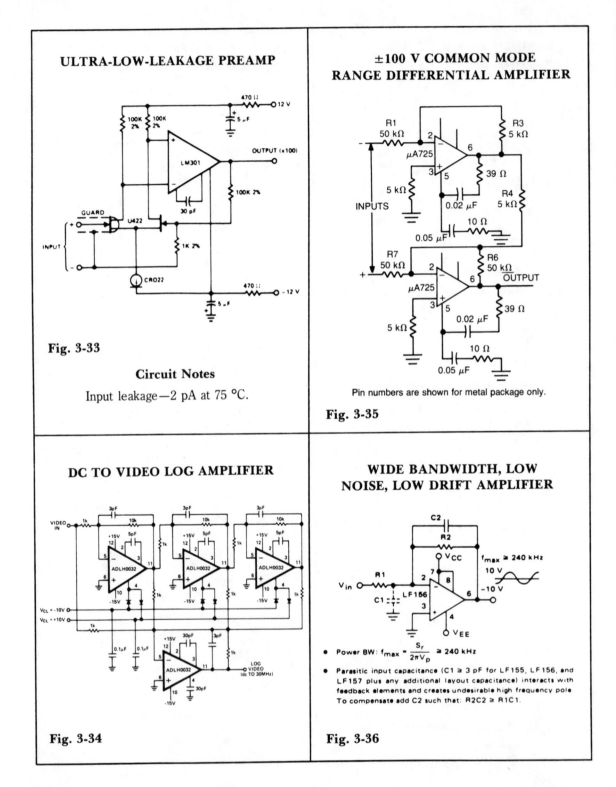

ULTRA-LOW-LEAKAGE PREAMP

470 Ω
12 V
5 μF
100K 2%
100K 2%
LM301
OUTPUT (x100)
100K 2%
30 pF
GUARD
U422
INPUT
1K 2%
CRO22
470 Ω
-12 V
5 μF

Fig. 3-33

Circuit Notes

Input leakage—2 pA at 75 °C.

±100 V COMMON MODE RANGE DIFFERENTIAL AMPLIFIER

R1
50 kΩ
2
6
μA725
3   5
R3
5 kΩ
39 Ω
R4
5 kΩ
5 kΩ
0.02 μF
10 Ω
INPUTS
0.05 μF
R7
50 kΩ
2
6
μA725
3   5
R6
50 kΩ
OUTPUT
39 Ω
5 kΩ
0.02 μF
10 Ω
0.05 μF

Pin numbers are shown for metal package only.

Fig. 3-35

DC TO VIDEO LOG AMPLIFIER

VIDEO IN
3pF   10k
1k
+15V 5pF
12   2
5   3
ADLH0032
6   11
1k
10 4
-15V
1k
3pF   10k
+15V 5pF
12   2
5   3
ADLH0032
6   11
1k
10 4
-15V
1k
3pF   10k
+15V 5pF
12   2
5   3
ADLH0032
6   11
1k
10 4
-15V
1k
V_CL = -10V
V_CL = +10V
1k
0.1μF 0.1μF
+15V 30pF
12   2
5   3
ADLH0032
6   11
10 4
-15V
30pF
3pF
1k
LOG VIDEO
(dc TO 30MHz)

Fig. 3-34

WIDE BANDWIDTH, LOW NOISE, LOW DRIFT AMPLIFIER

C2
R2
V_CC
R1
2   7   8
V_{in}
C1
LF156
3
6
4
V_EE
$f_{max} \cong 240$ kHz
10 V
-10 V

- Power BW: $f_{max} = \dfrac{S_r}{2\pi V_p} \cong 240$ kHz

- Parasitic input capacitance (C1 ≅ 3 pF for LF155, LF156, and LF157 plus any additional layout capacitance) interacts with feedback elements and creates undesirable high frequency pole. To compensate add C2 such that: R2C2 ≅ R1C1.

Fig. 3-36

38

SIGNAL DISTRIBUTION AMPLIFIER

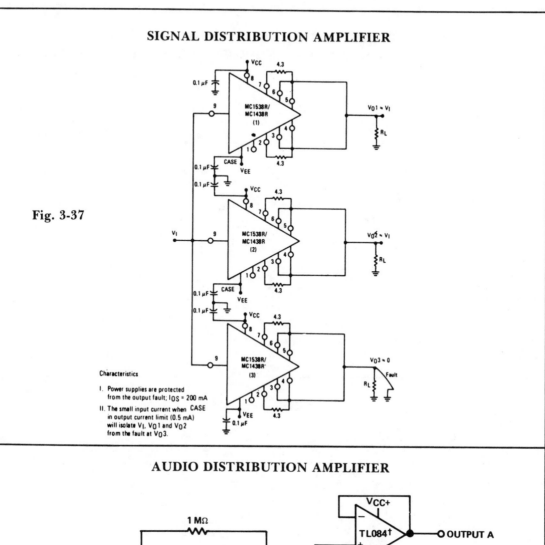

Fig. 3-37

Characteristics

I. Power supplies are protected from the output fault; I_{OS} = 200 mA

II. The small input current when CASE in output current limit (0.5 mA) will isolate V_I, $V_{O}1$ and $V_{O}2$ from the fault at $V_{O}3$.

AUDIO DISTRIBUTION AMPLIFIER

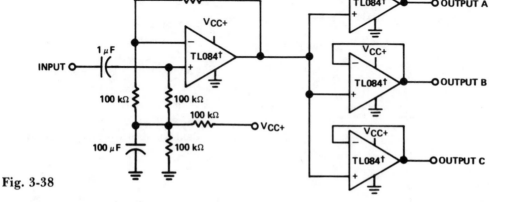

Fig. 3-38

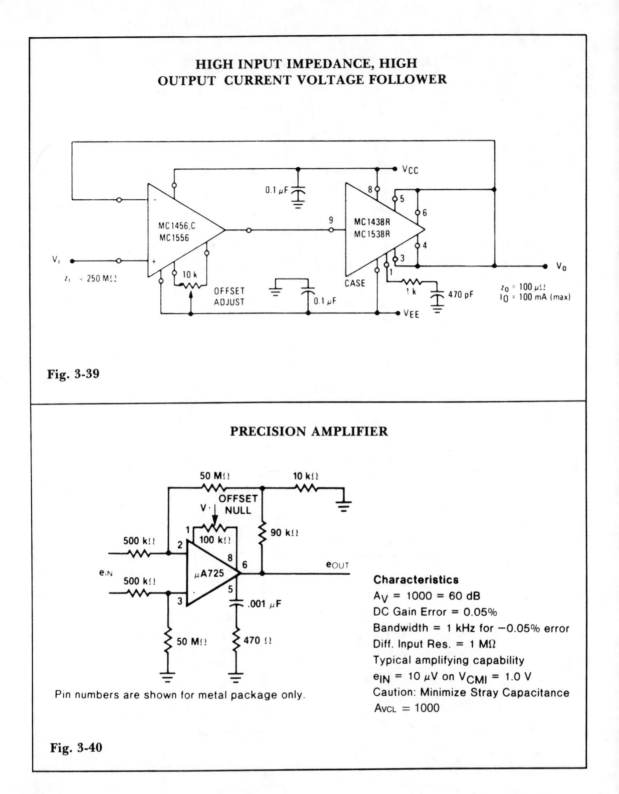

HIGH INPUT IMPEDANCE, HIGH OUTPUT CURRENT VOLTAGE FOLLOWER

MC1456,C
MC1556

MC1438R
MC1538R

$0.1 \mu F$

V_{CC}

8
5
6
9
4
3
1

CASE

$0.1 \mu F$

OFFSET ADJUST

10 k

1 k

470 pF

V_I

$z_I = 250 M\Omega$

V_O

$z_0 = 100 \mu\Omega$
$I_0 = 100 mA (max)$

V_{EE}

Fig. 3-39

PRECISION AMPLIFIER

50 MΩ

10 kΩ

OFFSET NULL

V

1
100 kΩ

90 kΩ

500 kΩ

2

8

6

e_{OUT}

$\mu A725$

e_{IN}

500 kΩ

3

5

.001 μF

50 MΩ

470 Ω

Pin numbers are shown for metal package only.

Characteristics
$A_V = 1000 = 60$ dB
DC Gain Error = 0.05%
Bandwidth = 1 kHz for −0.05% error
Diff. Input Res. = 1 MΩ
Typical amplifying capability
$e_{IN} = 10 \mu V$ on $V_{CMI} = 1.0$ V
Caution: Minimize Stray Capacitance
$A_{VCL} = 1000$

Fig. 3-40

40

PREAMPLIFIER AND HIGH-TO-LOW IMPEDANCE CONVERTER

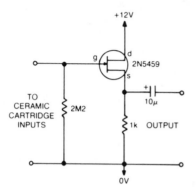

Fig. 3-41

Circuit Notes

This circuit matches the very high impedance of ceramic cartridges, unity gain, and low impedance output. By "loading" the cartridge with a 2M2 input resistance, the cartridge characteristics are such as to quite closely compensate for the RIAA recording curve. The output from this preamp may be fed to a level pot for mixing.

NONINVERTING AMPLIFIER

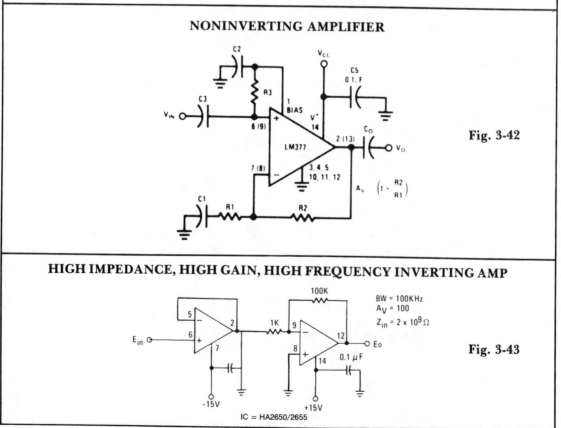

Fig. 3-42

HIGH IMPEDANCE, HIGH GAIN, HIGH FREQUENCY INVERTING AMP

Fig. 3-43

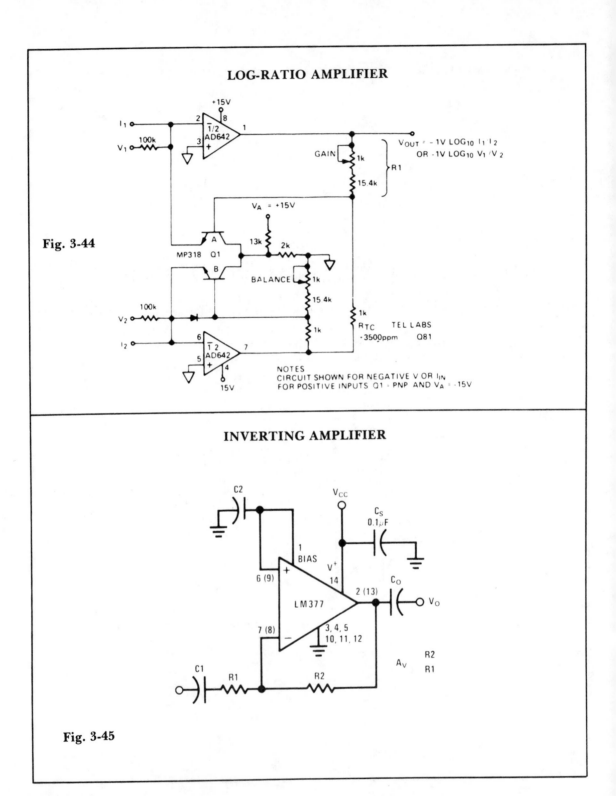

LOG-RATIO AMPLIFIER

$V_{OUT} = -1V \ LOG_{10} \ I_1 / I_2$
OR $-1V \ LOG_{10} \ V_1 / V_2$

Fig. 3-44

NOTES
CIRCUIT SHOWN FOR NEGATIVE V OR I_{IN}
FOR POSITIVE INPUTS Q1 = PNP AND $V_A = -15V$

INVERTING AMPLIFIER

Fig. 3-45

4

Analog-to-Digital Converters

The sources of the following circuits are contained in the Sources section beginning on page 730. The figure number contained in the box of each circuit correlates to the source entry in the Sources section.

8-Bit A/D Converter
Successive Approximation A/D Converter
8-Bit A/D Converter
8-Bit Tracking A/D Converter
8-Bit Successive Approximation A/D Converter
Four Channel Digitally Multiplexed Ramp

A/D Converter
Three Decade Logarithmic A/D Converter
Tracking (Servo Type) A/D Converter
3½ Digit A/D Converter with LCD Display
Fast Precision A/D Converter
High Speed 3-Bit A/D Converter
Three IC Low Cost A/D Converter

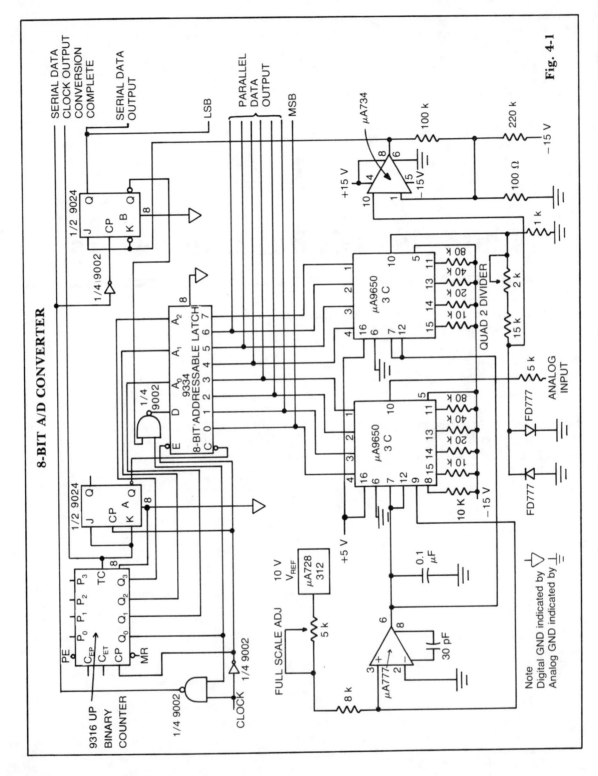

8-BIT A/D CONVERTER

Fig. 4-1

SUCCESSIVE APPROXIMATION A/D CONVERTER

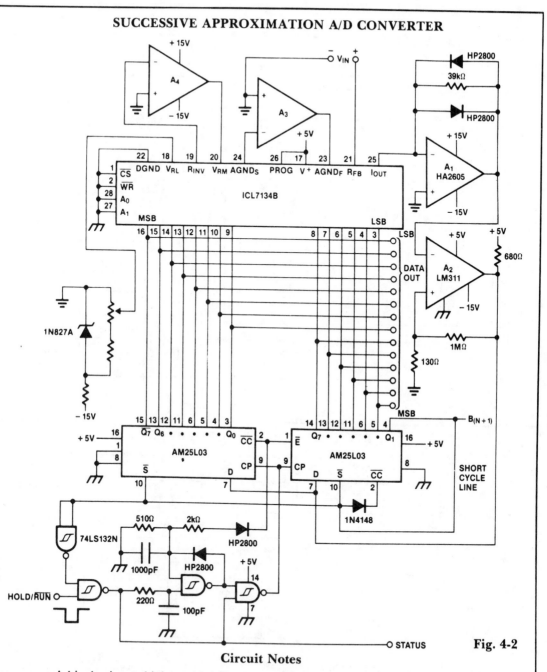

Fig. 4-2

Circuit Notes

A bipolar input, high speed A/D converter uses two AM25L03s to form a 14-bit successive approximation register. The comparator is a two-stage circuit with an HA2605 front-end amplifier used to reduce settling time problems at the summing node. Careful offset-nulling of this amplifier is needed.

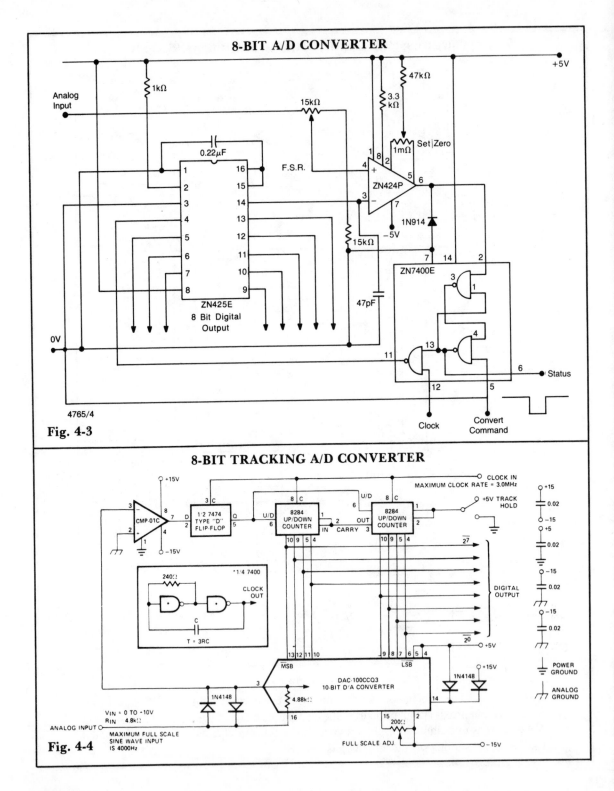

8-BIT A/D CONVERTER

Analog Input

1kΩ

15kΩ

0.22µF

F.S.R.

47kΩ

3.3 kΩ

Set|Zero

1mΩ

ZN424P

1N914

−5V

15kΩ

47pF

ZN425E
8 Bit Digital Output

ZN7400E

Status

0V

Clock

Convert Command

4765/4

Fig. 4-3

8-BIT TRACKING A/D CONVERTER

CLOCK IN
MAXIMUM CLOCK RATE = 3.0MHz

+15V

CMP-01C

−15V

1/2 7474
TYPE "D"
FLIP-FLOP

8284
UP/DOWN
COUNTER

CARRY

8284
UP/DOWN
COUNTER

U/D

OUT

+5V TRACK HOLD

240Ω

·1/4 7400

CLOCK OUT

T = 3RC

C

DAC-100CCQ3
10-BIT D/A CONVERTER

MSB

LSB

2⁷

2⁰

DIGITAL OUTPUT

+15

0.02

−15

+5

0.02

−15

0.02

−15

0.02

POWER GROUND

ANALOG GROUND

1N4148

4.88kΩ

1N4148

+5V

+15V

1N4148

VIN = 0 TO -10V
RIN 4.8kΩ

ANALOG INPUT

MAXIMUM FULL SCALE
SINE WAVE INPUT
IS 4000Hz

200Ω

FULL SCALE ADJ

−15V

Fig. 4-4

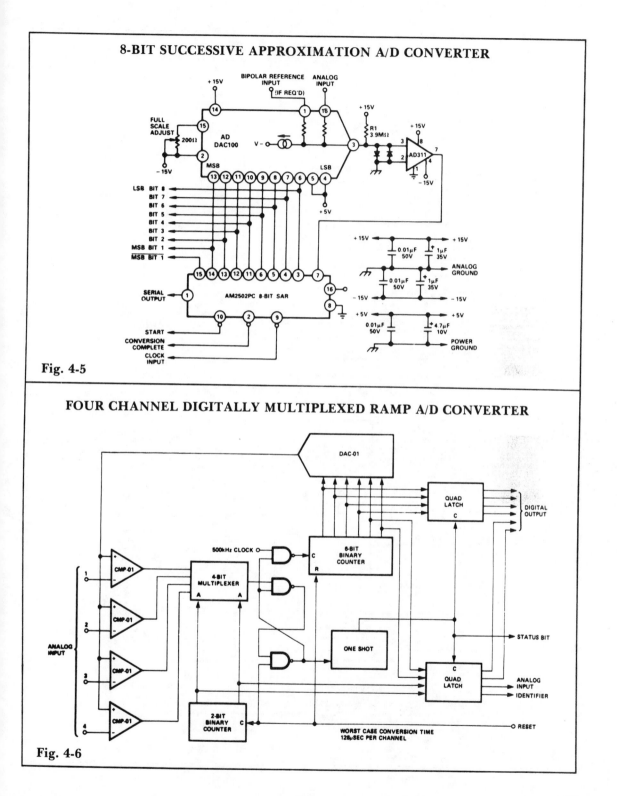

8-BIT SUCCESSIVE APPROXIMATION A/D CONVERTER

Fig. 4-5

FOUR CHANNEL DIGITALLY MULTIPLEXED RAMP A/D CONVERTER

Fig. 4-6

47

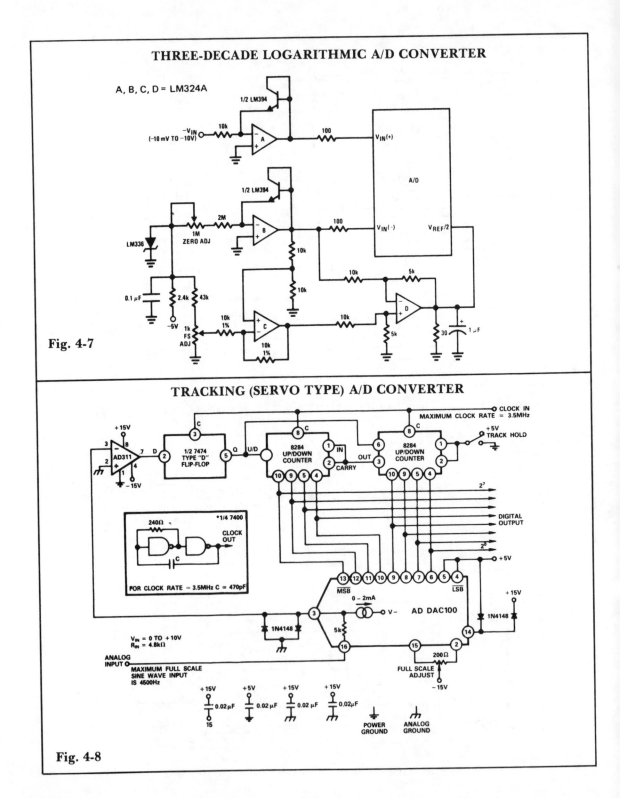

THREE-DECADE LOGARITHMIC A/D CONVERTER

Fig. 4-7

TRACKING (SERVO TYPE) A/D CONVERTER

Fig. 4-8

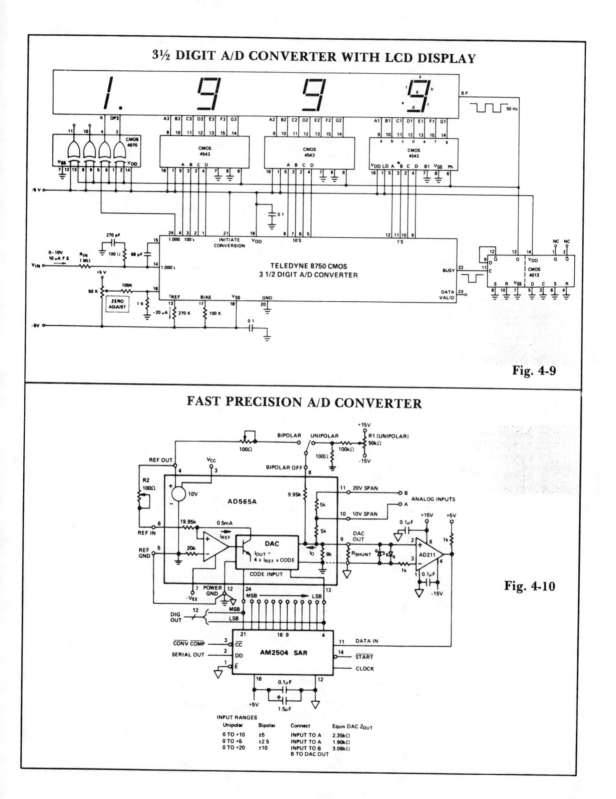

3½ DIGIT A/D CONVERTER WITH LCD DISPLAY

Fig. 4-9

FAST PRECISION A/D CONVERTER

Fig. 4-10

INPUT RANGES			
Unipolar	Bipolar	Connect	Equiv DAC Z_{OUT}
0 TO +10	±5	INPUT TO A	2.35kΩ
0 TO +5	±2.5	INPUT TO A	1.90kΩ
0 TO +20	±10	INPUT TO B B TO DAC OUT	3.08kΩ

49

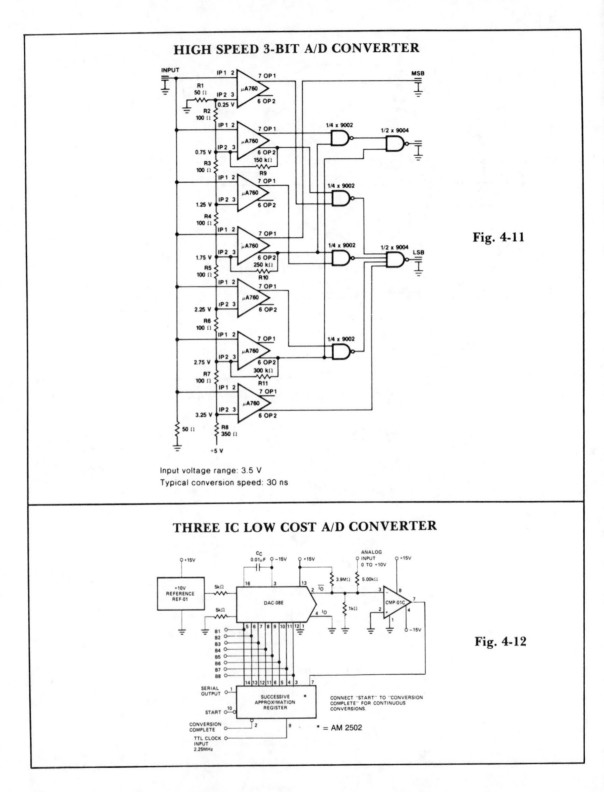

HIGH SPEED 3-BIT A/D CONVERTER

Fig. 4-11

Input voltage range: 3.5 V
Typical conversion speed: 30 ns

THREE IC LOW COST A/D CONVERTER

Fig. 4-12

5

Attenuators

The sources of the following circuits are contained in the Sources section beginning on page 730. The figure number contained in the box of each circuit correlates to the source entry in the Sources section.

Digitally Selectable Precision Attenuator
Variable Attenuator

Digitally Controlled Amplifier/Attenuator
Programmable Attenuator (1 to 0.0001)

DIGITALLY SELECTABLE PRECISION ATTENUATOR

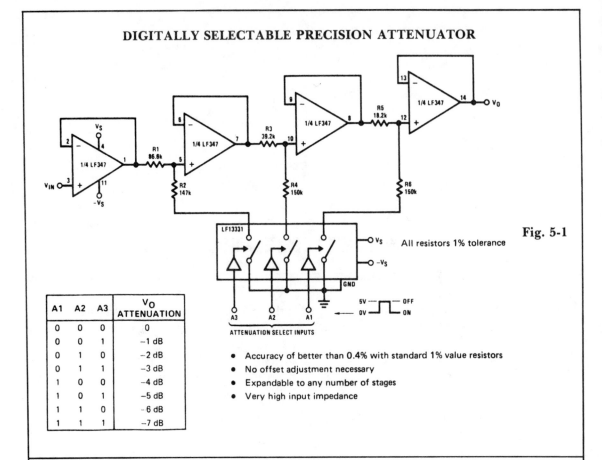

Fig. 5-1

All resistors 1% tolerance

A1	A2	A3	V_O ATTENUATION
0	0	0	0
0	0	1	−1 dB
0	1	0	−2 dB
0	1	1	−3 dB
1	0	0	−4 dB
1	0	1	−5 dB
1	1	0	- 6 dB
1	1	1	−7 dB

- Accuracy of better than 0.4% with standard 1% value resistors
- No offset adjustment necessary
- Expandable to any number of stages
- Very high input impedance

VARIABLE ATTENUATOR

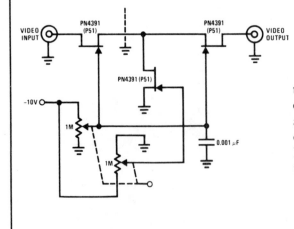

Circuit Notes

The PN4391 provides a low $R_{ds(on)}$ (less than 30 ohms). The tee attenuator provides for optimum dynamic linear range for attenuation and if complete turn-off is desired, attenuation of greater than 100 dB can be obtained at 10 MHz providing proper rf construction techniques are employed.

Fig. 5-2

DIGITALLY CONTROLLED AMPLIFIER/ATTENUATOR

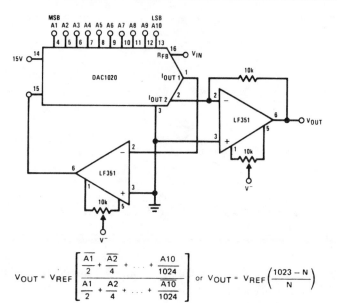

$$V_{OUT} = V_{REF} \left[\frac{\dfrac{\overline{A1}}{2} + \dfrac{\overline{A2}}{4} + \ldots + \dfrac{\overline{A10}}{1024}}{\dfrac{A1}{2} + \dfrac{A2}{4} + \ldots + \dfrac{\overline{A10}}{1024}} \right] \quad \text{or} \quad V_{OUT} = V_{REF} \left(\frac{1023 - N}{N} \right)$$

where: $0 \leq N \leq 1023$
N = 0 for A_N = all zeros
N = 1 for A10 = 1, A1–A9 = 0
.
.
.
N = 1023 for A_N = all 1's

Fig. 5-3

PROGRAMMABLE ATTENUATOR (1 TO 0.0001)

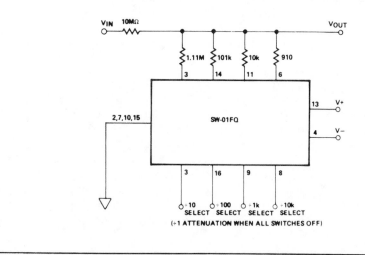

(÷1 ATTENUATION WHEN ALL SWITCHES OFF)

Fig. 5-4

6

Audio Mixers

The sources of the following circuits are contained in the Sources section beginning on page 730. The figure number contained in the box of each circuit correlates to the source entry in the Sources section.

FOUR-INPUT STEREO MIXER

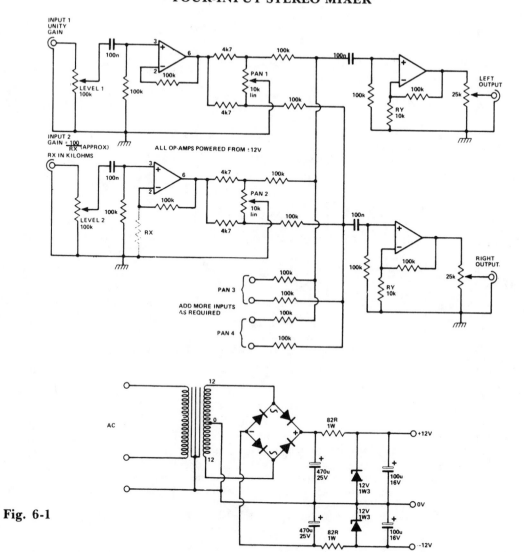

Fig. 6-1

Circuit Notes

Four (or more) inputs can be mixed and produce stereo output. Gain of each stage can be boosted by adding RX, but it should be kept below 50 (RX above 2.2 K) to avoid poor frequency, response. If more than four stages are used, decrease RX to 6.8 K for six inputs, or 4.7 K for eight inputs. The op amps are 741 or other lower noise types. The power supply circuit is also given.

HIGH-LEVEL FOUR-CHANNEL MIXER

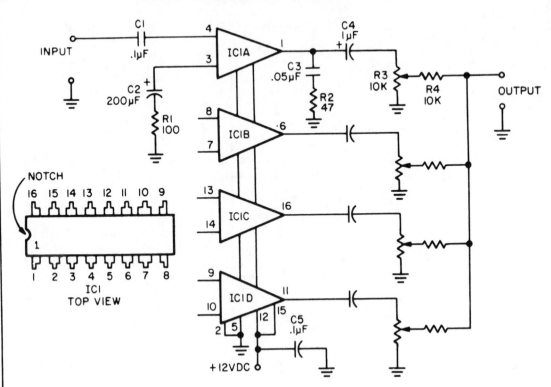

PARTS LIST FOR HI-LEVEL MIXER
C1—0.1-uF, 3 VDC capacitor
C2—200-uF, 3 VDC capacitor
C3—0.05-uF, 75 VDC disc capacitor
C4—1-uF, 15 VDC capacitor
C5—0.1-uF, 15 VDC capacitor

IC1—RCA CA 3052
R1—100-ohms, ½-watt resistor
R2—47-ohms, ½-watt resistor
R3—Potentiometer, 10,000-ohms
      audio taper
R4—10,000-ohms, ½-watt resistor

Fig. 6-2

Circuit Notes

To provide good signal-to-noise ratio, this four channel mixer amplifier controls the signal levels after the amplifiers, and then mixes them to offer a combined output. The circuit works with any 50 ohm to 50 K dynamic mi-crophone but not with crystal or ceramic mikes because the IC input impedance is low. Note that all four circuits are identical but that only one is shown complete.

TWO CHANNEL PANNING CIRCUIT

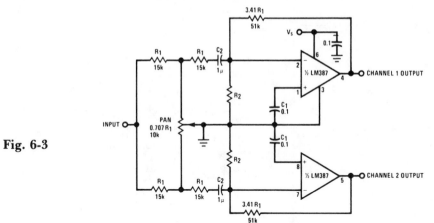

Fig. 6-3

Circuit Notes

This panning circuit (short for panoramic control circuit) provides the ability to move the apparent position of one microphone's input between two output channels. This effect is often required in recording studio mixing con-soles. Panning is how recording engineers manage to pick up your favorite pianist and "float" the sound over to the other side of the stage and back again.

CMOS MIXER

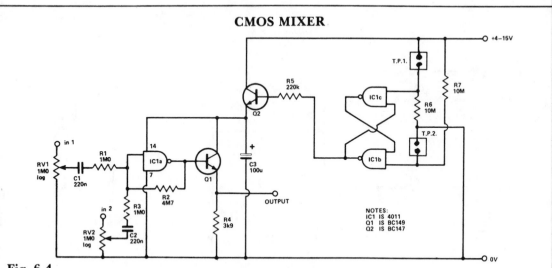

Fig. 6-4

Circuit Notes

Four inputs can be mixed by duplicating the circuit to the left of C3 and using the fourth gate of IC1. Two gates are used in a touch-operated switching circuit that controls the voltage on the base of switching transistor Q2. Touching TP1 and TP2 alternately turns the circuit on and off.

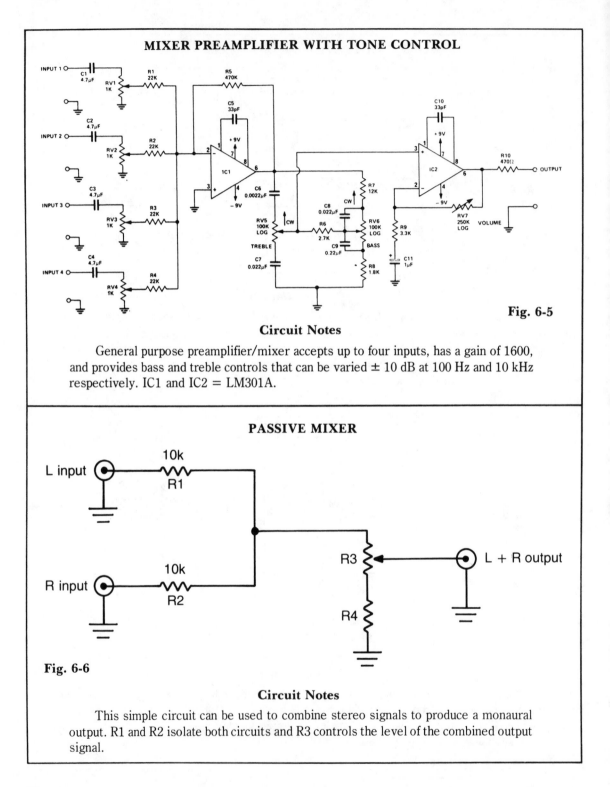

MIXER PREAMPLIFIER WITH TONE CONTROL

Fig. 6-5

Circuit Notes

General purpose preamplifier/mixer accepts up to four inputs, has a gain of 1600, and provides bass and treble controls that can be varied ± 10 dB at 100 Hz and 10 kHz respectively. IC1 and IC2 = LM301A.

PASSIVE MIXER

Fig. 6-6

Circuit Notes

This simple circuit can be used to combine stereo signals to produce a monaural output. R1 and R2 isolate both circuits and R3 controls the level of the combined output signal.

ONE TRANSISTOR AUDIO MIXER

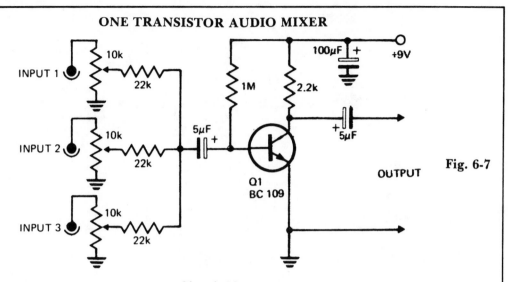

Circuit Notes

Three or more inputs with individual level controls feed into the base of Q1 that provides a voltage gain of 20.

Fig. 6-7

SILENT AUDIO SWITCHING/MIXING

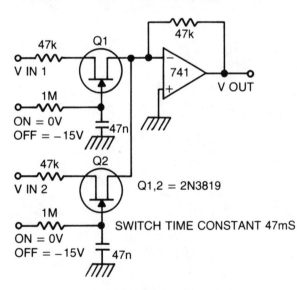

Fig. 6-8

Circuit Notes

Two or more signals can be switched and/or mixed without annoying clicks by using FETs and a low input-impedance op amp circuit.

HYBRID MIXER

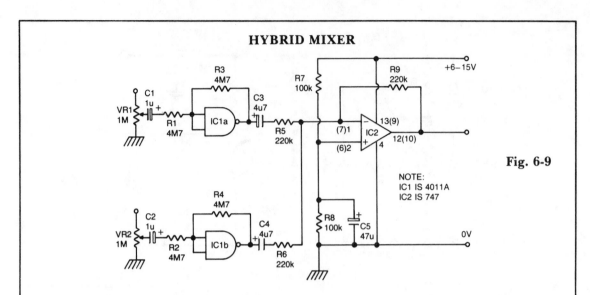

Fig. 6-9

NOTE:
IC1 IS 4011A
IC2 IS 747

Circuit Notes

IC1a and b are biased into the linear regions by R3 and R4. (IC1 must be 4011A). Outputs from gates are combined by op amp IC2, which provides low impedance output.

FOUR CHANNEL MIXER

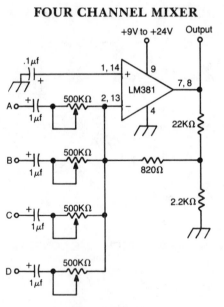

Fig. 6-10

Circuit Notes

High gain op amp combines up to four individually controlled input signals. The dc power source should be well filtered (battery is ideal), and the circuit should be well shielded to prevent hum pickup.

7

Audio Oscillators

The sources of the following circuits are contained in the Sources section beginning on page 730. The figure number contained in the box of each circuit correlates to the source entry in the Sources section.

Wien Bridge Oscillator
Wien Bridge Oscillator
Wien Bridge Oscillator
Very Low Frequency Generator
Audio Oscillator
Sine Wave Oscillator
Easily Tuned Sine/Square Wave Oscillators
Wien Bridge Sine Wave Oscillator
Phase Shift Oscillator

Tone Encoder
Feedback Oscillator
Phase Shift Oscillator
800 Hz Oscillator
Tunable Single Comparator Oscillator
Wide Range Oscillator (Frequency Range of 500 to 1)
Wien Bridge Oscillator
Wien Bridge Sine Wave Oscillator

WIEN BRIDGE OSCILLATOR

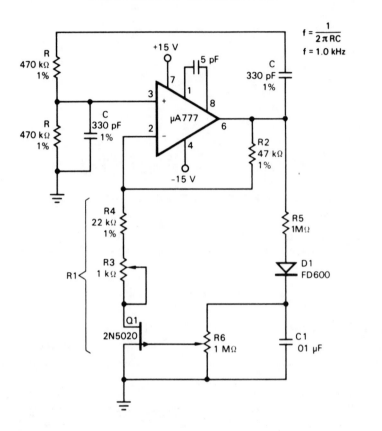

$$f = \frac{1}{2\pi RC}$$

$$f = 1.0 \text{ kHz}$$

Fig. 7-1

Circuit Notes

Field effect transistor, Q1, operates in the linear resistive region to provide automatic gain control. Because the attenuation of the RC network is one-third at the zero phase-shift oscillation frequency, the amplifier gain determined by resistor R2 and equivalent resistor R1 must be just equal to three to make up the unity gain positive feedback requirement needed for stable oscillation. Resistors R3 and R4 are set to approximately 1000 ohm less than the required R1 resistance. The FET dynamically provides the trimming resistance needed to make R1 one-half of the resistance of R2. The circuit composed of R5, D1, and C1 isolates, rectifies, and filters the output sine wave, converting it into a dc potential to control the gate of the FET. For the low drain-to-source voltages used, the FET provides a symmetrical linear resistance for a given gate-to-source voltage.

WIEN BRIDGE OSCILLATOR

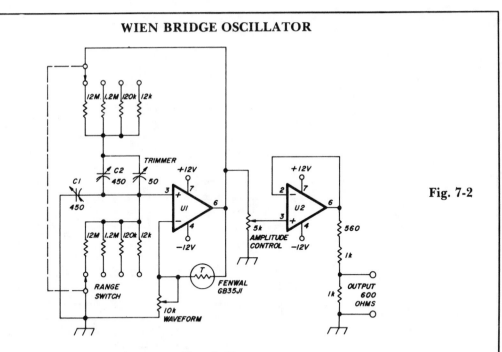

Fig. 7-2

Circuit Notes

Wien bridge sine-wave oscillator using two RCA CA3140 op amps covers 30 Hz to 100 kHz with less than 0.5 percent total harmonic distortion. The 10k pot is adjusted for the best waveform. Capacitor C1 and C2 are a two-gang, 450-pF variable with its frame isolated from ground. Maximum output into a 600-ohm load is about 1 volt rms.

WIEN BRIDGE OSCILLATOR

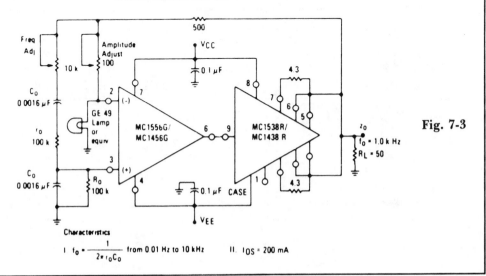

Fig. 7-3

VERY LOW FREQUENCY GENERATOR

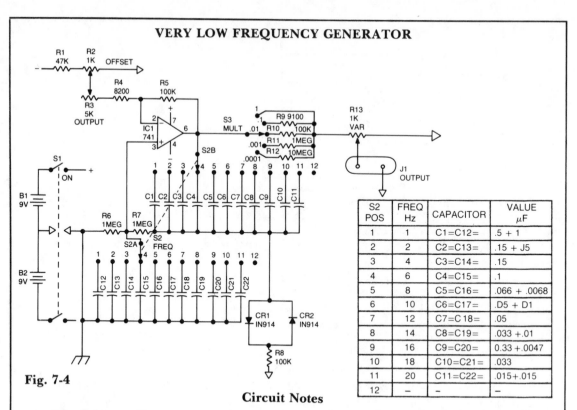

S2 POS	FREQ Hz	CAPACITOR	VALUE μF
1	1	C1=C12=	.5 + 1
2	2	C2=C13=	.15 + J5
3	4	C3=C14=	.15
4	6	C4=C15=	.1
5	8	C5=C16=	.066 + .0068
6	10	C6=C17=	.D5 + D1
7	12	C7=C18=	.05
8	14	C8=C19=	.033 +.01
9	16	C9=C20=	0.33 +.0047
10	18	C10=C21=	.033
11	20	C11=C22=	.015+.015
12	–	–	–

Fig. 7-4

Circuit Notes

Wien bridge oscillator generates frequencies of 1 Hz and 2 to 20 Hz in 2 Hz steps. Maximum output amplitude is 3 volts rms of 8.5 volts peak-to-peak. A pot-and-switch attenuator allows the output level to be set with a fair degree of precision to any value within a range of 5 decades.

AUDIO OSCILLATOR

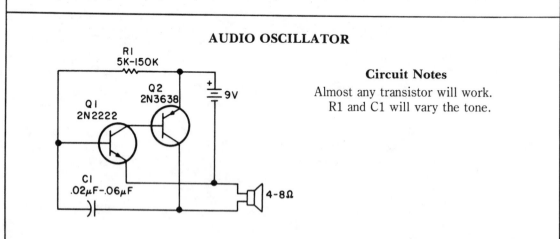

Circuit Notes

Almost any transistor will work.
R1 and C1 will vary the tone.

Fig. 7-5

SINE WAVE OSCILLATOR

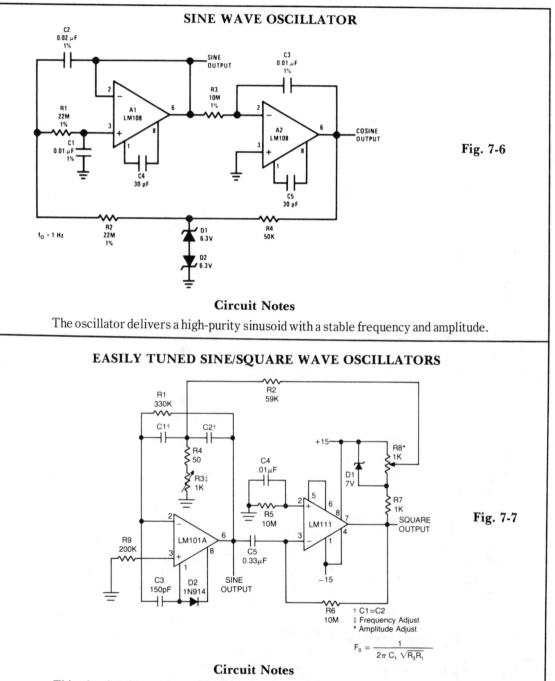

Fig. 7-6

Circuit Notes

The oscillator delivers a high-purity sinusoid with a stable frequency and amplitude.

EASILY TUNED SINE/SQUARE WAVE OSCILLATORS

Fig. 7-7

† C1=C2
‡ Frequency Adjust
* Amplitude Adjust

$$F_0 = \frac{1}{2\pi C_1 \sqrt{R_3 R_1}}$$

Circuit Notes

This circuit will provide both a sine and square wave output for frequencies from below 20 Hz to above 20 kHz. The frequency of oscillation is easily tuned by varying a single resistor.

WIEN BRIDGE SINE WAVE OSCILLATOR

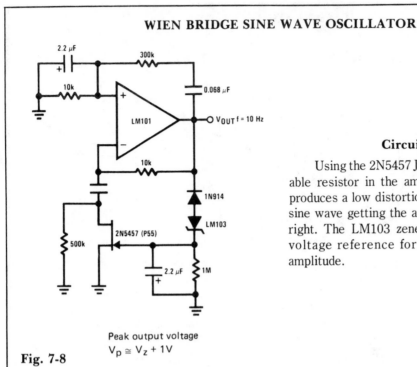

Circuit Notes

Using the 2N5457 JFET as a voltage variable resistor in the amplifier feedback loop, produces a low distortion, constant amplitude sine wave getting the amplifier loop gain just right. The LM103 zener diode provides the voltage reference for the peak sine wave amplitude.

Peak output voltage
$$V_p \cong V_z + 1V$$

Fig. 7-8

PHASE-SHIFT OSCILLATOR

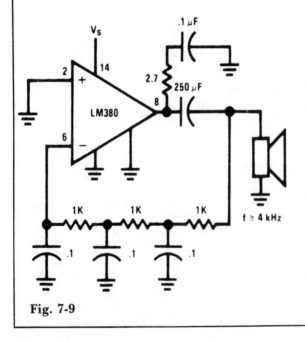

Circuit Notes

Circuit uses a simple RC network to produce an exceptionally shrill tone from a miniature speaker. With the parts values shown, the circuit oscillates at a frequency of 3.6 kHz and drives a miniature 2½" speaker with ear-piercing volume. The output waveform is a square wave with a width of 150 μs, sloping rise and fall times, and a peak-to-peak amplitude of 4.2 volts (when powered by 9 volts). Current drain of the oscillator is 90 mA at 9 volts, and total power dissipation at this voltage is 0.81 watt, which is well below the 1.25 watts the 14-pin version will absorb (at room temperature) before shutting down.

Fig. 7-9

TONE ENCODER

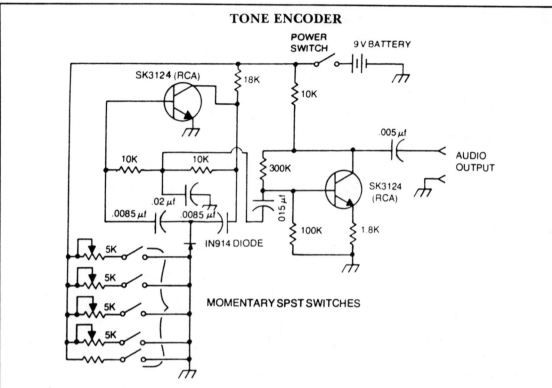

Fig. 7-10

Circuit Notes

A basic twin-T circuit uses resistors for accurately setting the frequency of the output tones, selected by pushbutton. Momentary switches produce a tone only when the button is depressed.

FEEDBACK OSCILLATOR

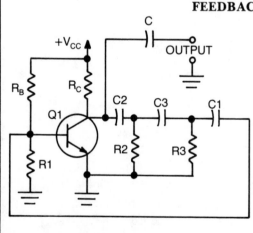

Circuit Notes

Circuit oscillates because the transistor shifts the phase of the signal 180° from the base to the collector. Each of the RC networks in the circuit is designed to shift the phase 60° at the frequency of oscillation for a total of 180°. The appropriate values of R and C for each network is found from $f = 1/2\sqrt{3}\pi RC$); that equation allows for the 60° phase shift required by the design.

Fig. 7-11

PHASE SHIFT OSCILLATOR

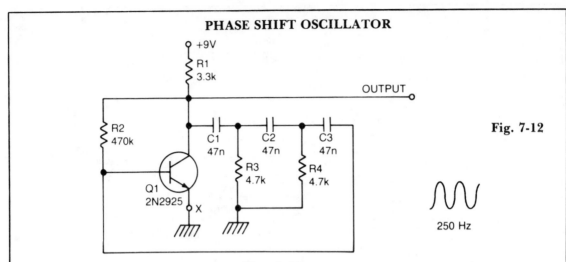

Fig. 7-12

250 Hz

Circuit Notes

A single transistor makes a simple phase shift oscillator. The output is a sine wave with distortion of about 10%. The sine wave purity can be increased by putting a variable resistor (25 ohms) in the emitter lead of Q1 (x). The resistor is adjusted so the circuit is only just oscillating, then the sine wave is relatively pure. Operating frequency may be varied by putting a 10 K variable resistor in series with R3, or by changing C1, C2, and C3. Making C1, 2, 3 equal to 100 nF will halve the operating frequency. Operating frequency can also be voltage controlled by a FET in series with R3, or optically controlled by an LDR in series with R3.

800 Hz OSCILLATOR

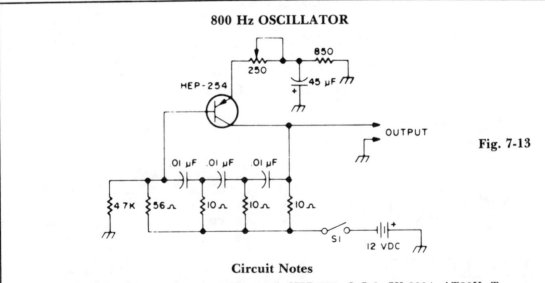

Fig. 7-13

Circuit Notes

The following transistors may be used: HEP-254, O.C-2, SK-3004, AT30H. To increase the frequency, decrease the value of the capacitors in the ladder network.

TUNABLE SINGLE COMPARATOR OSCILLATOR

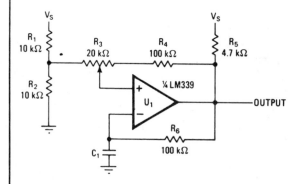

Circuit Notes

Varying the amount of this comparator circuit's hysteresis makes it possible to vary output frequencies in the 740-Hz to 2.7-kHz range smoothly. The amount of hysteresis together with time constant R6C1 determines how much time it takes for C1 to charge or discharge to the new threshold after the output voltage switches.

Fig. 7-14

WIDE RANGE OSCILLATOR (FREQUENCY RANGE OF 5000 TO 1)

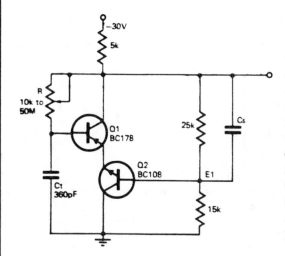

Circuit Notes

Timing resistor R may be adjusted to any value between 10 K and 50 M to obtain a frequency range from 400 kHz to 100 Hz. Returning the timing resistor to the collector of Q1 ensures that Q1 draws its base current only from the timing capacitor Ct. The timing capacitor recharges when the transistors are off, to a voltage equal to the base emitter voltage of Q2 plus the base emitter drops of Q1 and Q2. The transistors then start into conduction. Capacitor Cs is used to speed up the transition. A suitable value would be in the region of 100 pF.

Fig. 7-15

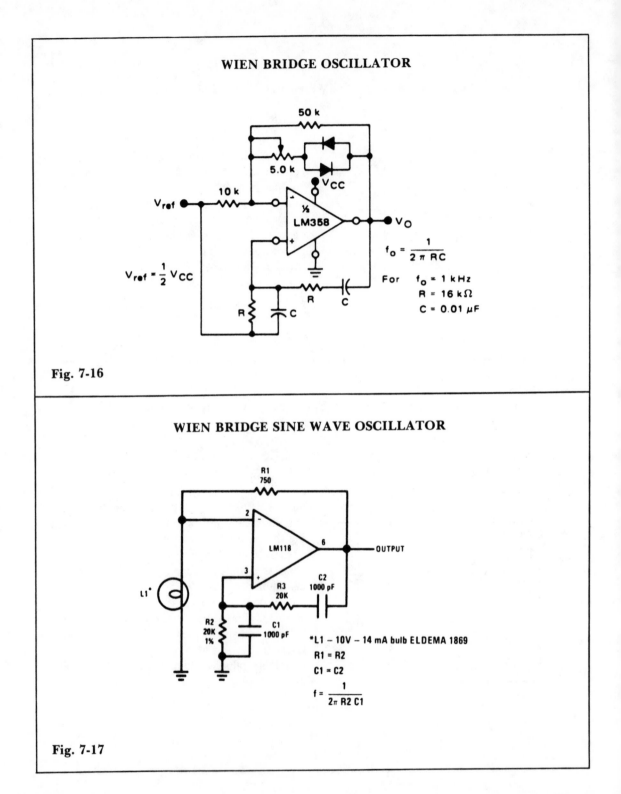

WIEN BRIDGE OSCILLATOR

V_{ref}

50 k

5.0 k

V_{CC}

10 k

V_{ref}

⅓ LM358

V_O

$f_0 = \dfrac{1}{2\pi RC}$

$V_{ref} = \dfrac{1}{2} V_{CC}$

R C

R C

For $f_0 = 1\,kHz$
     $R = 16\,k\Omega$
     $C = 0.01\,\mu F$

Fig. 7-16

WIEN BRIDGE SINE WAVE OSCILLATOR

R1
750

2

LM118

6

OUTPUT

3

R3
20K

C2
1000 pF

L1*

R2
20K
1%

C1
1000 pF

*L1 – 10V – 14 mA bulb ELDEMA 1869

R1 = R2

C1 = C2

$f = \dfrac{1}{2\pi\,R2\,C1}$

Fig. 7-17

8

Audio Power Amplifiers

The sources of the following circuits are contained in the Sources section beginning on page 730. The figure number contained in the box of each circuit correlates to the source entry in the Sources section.

Low Cost 20 W Audio Amplifier
75 Watt Audio Amplifier with Load Line
   Protection
Bridge Amplifier
Noninverting Amplifier Using Single Supply
Noninverting Amplifier Using Split Supply
6 W, 8 Ohm Output Transformerless Amplifier
12 W Low-Distortion Power Amplifier
10 W Power Amplifier
Stereo Amplifier with Av = 200
AM Radio Power Amplifier
470 mW Complementary-Symmetry
   Audio Amplifier

Novel Loudspeaker Coupling Circuit
Noninverting Ac Power Amplifier
Inverting Power Amplifier
Noninverting Power Amplifier
4 W Bridge Amplifier
Phono Amplifier with a "Common Mode"
   Volume and Tone with Control
Phono Amplifier
Phonograph Amplifier (Ceramic Cartridge)
Inverting Unity Gain Amplifier
Bridge Audio Power Amplifier
Phono Amplifier
High Slew Rate Power Op Amp/Audio Amp

16 W Bridge Amplifier

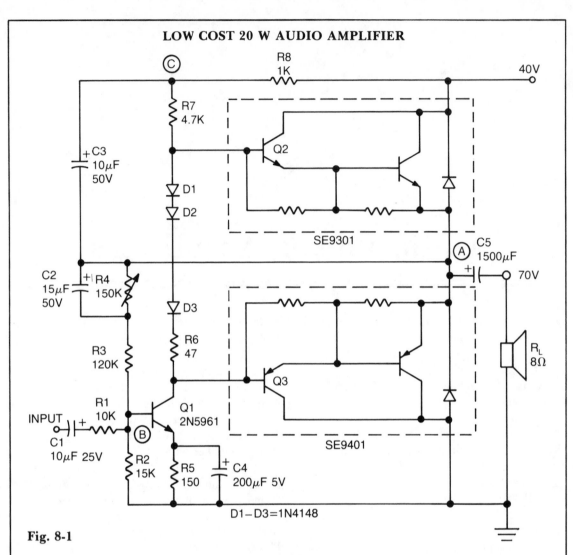

Fig. 8-1

Circuit Notes

This simple inexpensive audio amplifier can be constructed using a couple of TO-220 monolithic Darlington transistors for the push-pull output stage. Frequency response is flat within 1 dB from 30 Hz to 200 kHz with typical harmonic distortion below 0.2%. The amplifier requires only 1.2 V_{rms} for a full 20 W output into an 8 ohm load. Only one other transistor is needed, the TO-92 low-noise high-gain 2N5961 (Q1), to provide voltage gain for driving the output Darlingtons. Its base (point B) is the tie point for ac and dc feedback as well as for the signal input. Input resistance is 10 K. The center voltage at point A is set by adjusting resistor R4. A bootstrap circuit boosts the collector supply voltage of Q1 (point C) to ensure sufficient drive voltage for Q2. This also provides constant voltage across R7, which therefore acts as a current source and, together with diodes D1-D3, reduces low-signal crossover distortion.

75 WATT AUDIO AMPLIFIER WITH LOAD LINE PROTECTION

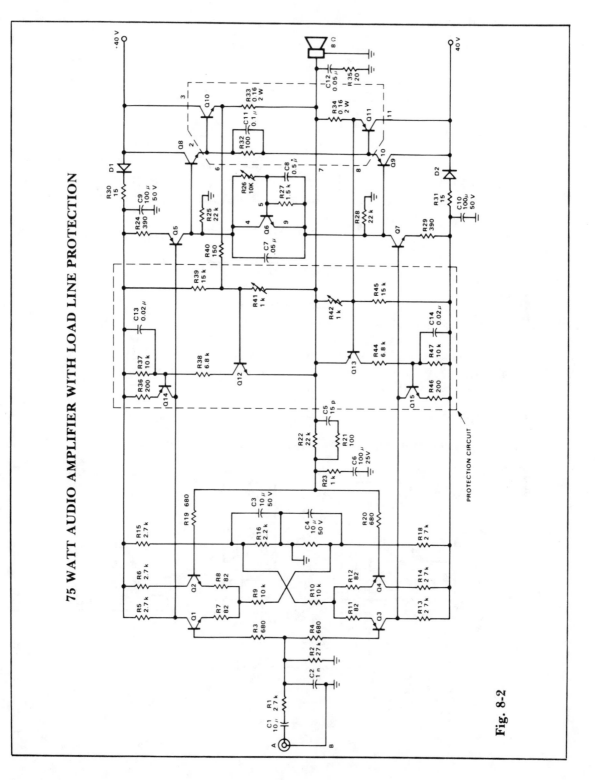

Fig. 8-2

73

BRIDGE AMPLIFIER

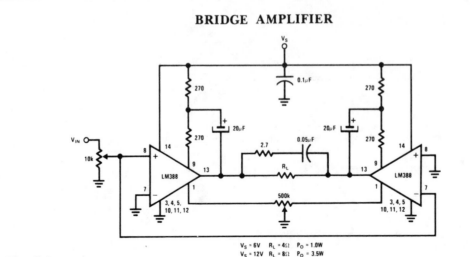

Fig. 8-3

$V_S = 6V$ $R_L = 4\Omega$ $P_O = 1.0W$
$V_S = 12V$ $R_L = 8\Omega$ $P_O = 3.5W$

Circuit Notes

This circuit is for low voltage applications requiring high power outputs. Output power levels of 1.0 W into 4 ohm from 6 V and 3.5 V into 8 ohm from 12 V are typical. Coupling capacitors are not necessary since the output dc levels will be within a few tenths of a volt of each other. Where critical matching is required the 500 K potentiometer is added and adjusted for zero dc current flow through the load.

NONINVERTING AMPLIFIER USING SINGLE SUPPLY

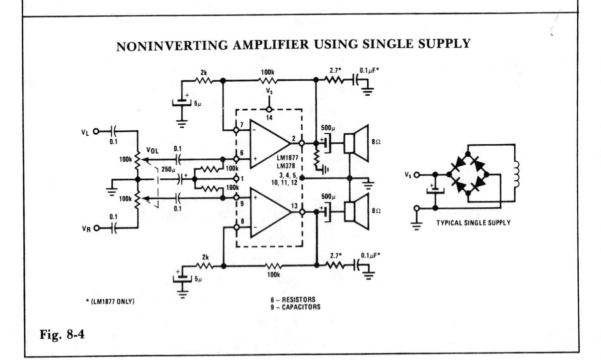

* (LM1877 ONLY)

6 – RESISTORS
9 – CAPACITORS

Fig. 8-4

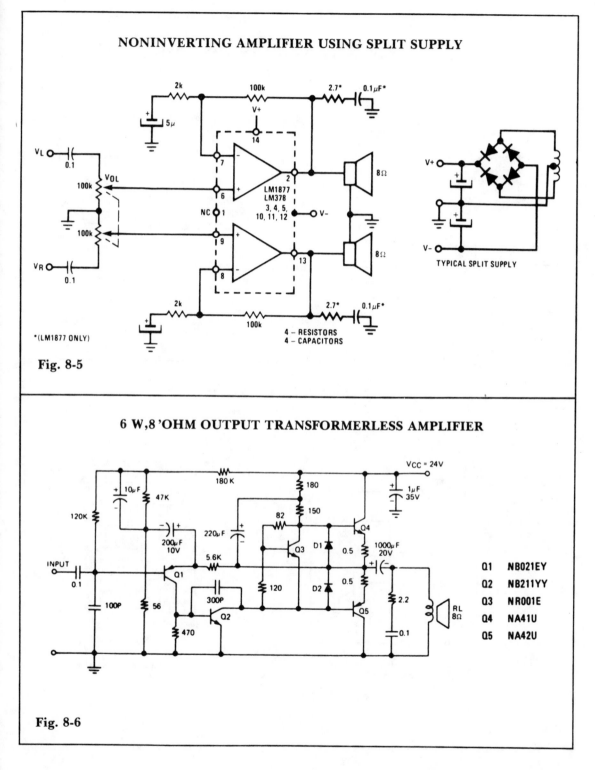

NONINVERTING AMPLIFIER USING SPLIT SUPPLY

Fig. 8-5

6 W, 8 'OHM OUTPUT TRANSFORMERLESS AMPLIFIER

Q1	NB021EY
Q2	NB211YY
Q3	NR001E
Q4	NA41U
Q5	NA42U

Fig. 8-6

12 W LOW-DISTORTION POWER AMPLIFIER

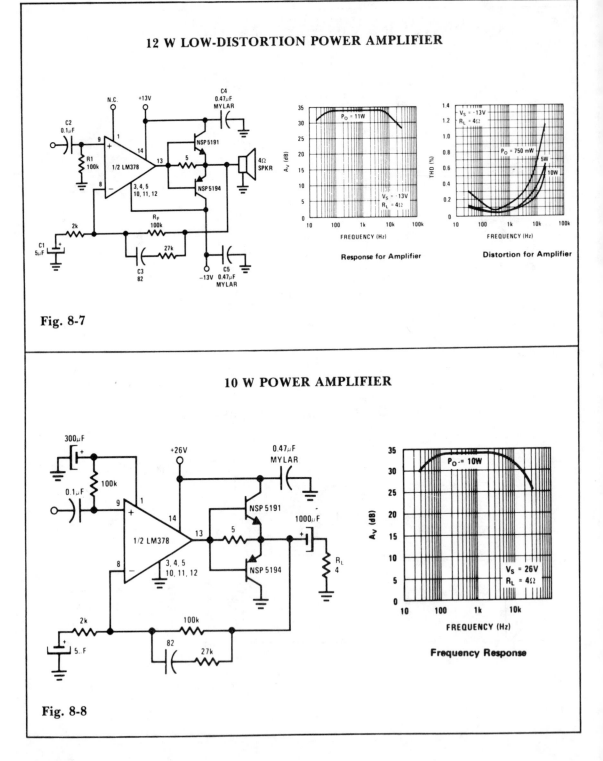

Fig. 8-7

Response for Amplifier

Distortion for Amplifier

10 W POWER AMPLIFIER

Frequency Response

Fig. 8-8

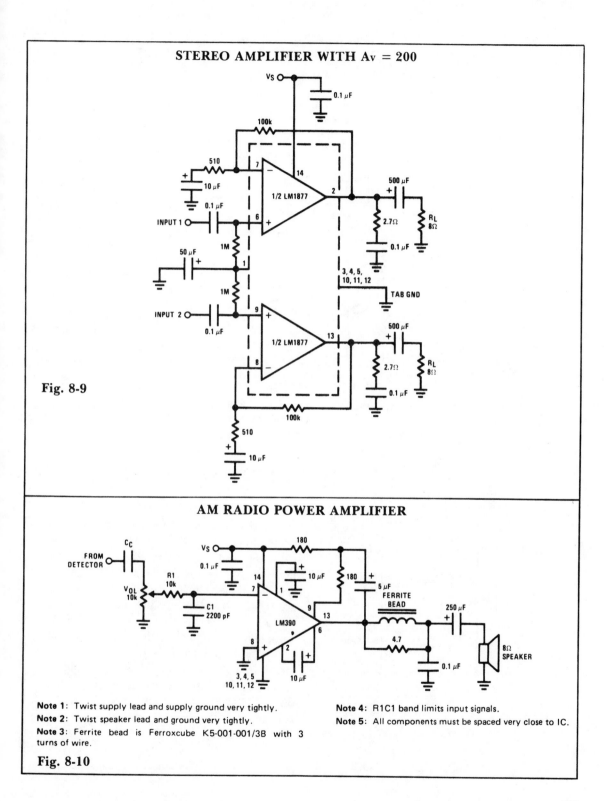

STEREO AMPLIFIER WITH Av = 200

Fig. 8-9

AM RADIO POWER AMPLIFIER

Note 1: Twist supply lead and supply ground very tightly.
Note 2: Twist speaker lead and ground very tightly.
Note 3: Ferrite bead is Ferroxcube K5-001-001/3B with 3 turns of wire.

Note 4: R1C1 band limits input signals.
Note 5: All components must be spaced very close to IC.

Fig. 8-10

77

470 mW COMPLEMENTARY-SYMMETRY AUDIO AMPLIFIER

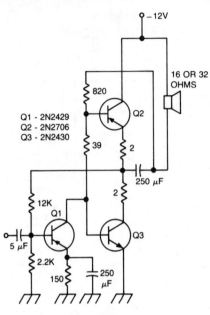

Q1 - 2N2429
Q2 - 2N2706
Q3 - 2N2430

Fig. 8-11

Circuit Notes

This circuit has less than 2% distortion and is flat within 3 dB from 15 Hz to 130 kHz.

NOVEL LOUDSPEAKER COUPLING CIRCUIT

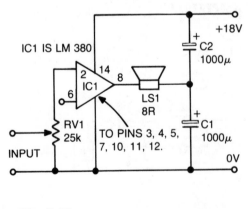

Fig. 8-12

Circuit Notes

The ground side of the speaker is connected to the junction of two equal high value capacitors (1000 μF is typical) across the supply. The amplifier output voltage will be $V_s/2$, and so will the voltage across C1 (if C1 and C2 are equal); so as the supply voltage builds up, the dc voltage across the speaker will remain zero, eliminating the switch-on surge. C1 and C2 will also provide supply smoothing. The circuit is shown with the LM380, but could be applied to any amplifier circuit, providing that the dc voltage at the output is half the supply voltage.

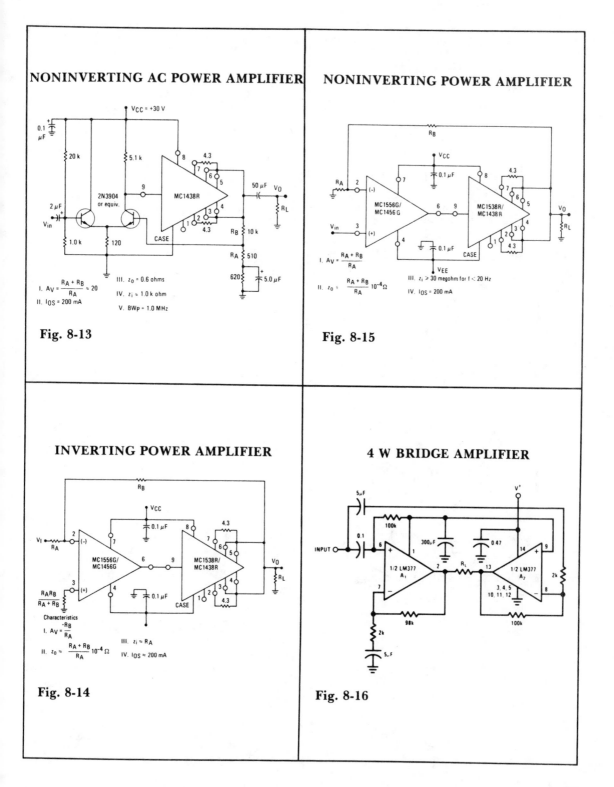

NONINVERTING AC POWER AMPLIFIER

V_{CC} = +30 V

0.1 μF

20 k

5.1 k

MC1438R

2N3904
or equiv.

2 μF

V_{in}

1.0 k

120

CASE

4.3

8

7

6

5

9

1

2

3

4

4.3

50 μF V_0

R_L

R_B 10 k

R_A 510

620

5.0 μF

I. $A_V = \dfrac{R_A + R_B}{R_A} \approx 20$

II. $I_{OS} = 200$ mA

III. $z_0 \approx 0.6$ ohms

IV. $z_i \approx 1.0$ k ohm

V. $BW_p = 1.0$ MHz

Fig. 8-13

NONINVERTING POWER AMPLIFIER

R_B

R_A 2

V_{CC}

0.1 μF

V_{in} 3

(-)

(+)

MC1556G/
MC1456 G

6   9

MC1538R/
MC1438 R

CASE

7

8

4.3

7

6

5

1

2

3

4

4.3

0.1 μF

V_{EE}

V_0

R_L

I. $A_V = \dfrac{R_A + R_B}{R_A}$

II. $z_0 \approx \dfrac{R_A + R_B}{R_A} 10^{-4} \Omega$

III. $z_i > 30$ megohm for $f < 20$ Hz

IV. $I_{OS} = 200$ mA

Fig. 8-15

INVERTING POWER AMPLIFIER

R_B

V_{CC}

0.1 μF

V_I 2

R_A

(-)

(+)

MC1556G/
MC1456G

6   9

MC1538R/
MC1438R

CASE

7

8

4.3

7

6

5

1

2

3

4

4.3

0.1 μF

V_0

R_L

$\dfrac{R_A R_B}{R_A + R_B}$

3

Characteristics

I. $A_V = \dfrac{-R_B}{R_A}$

II. $z_0 \approx \dfrac{R_A + R_B}{R_A} 10^{-4} \Omega$

III. $z_i \approx R_A$

IV. $I_{OS} \approx 200$ mA

Fig. 8-14

4 W BRIDGE AMPLIFIER

5μF

V^+

100k

0.1

INPUT

0.47

300μF

1/2 LM377
A_1

R_L

1/2 LM377
A_2

6

7

1

2

13

14

9

3, 4, 5
10, 11, 12

8

2k

2k

98k

100k

5μF

Fig. 8-16

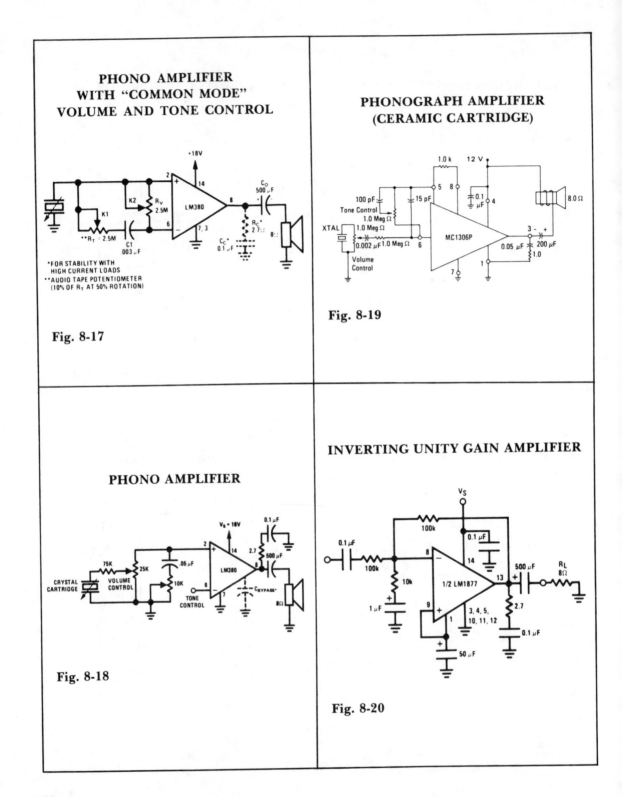

PHONO AMPLIFIER WITH "COMMON MODE" VOLUME AND TONE CONTROL

+18V

2
14
+
LM380
8
6
−
7, 3

K2
R_V 2.5M

C_O 500 μF

R_C^* 2.7Ω

8Ω

C_C^* 0.1 μF

K1

**R_T = 2.5M
C1 .003 μF

*FOR STABILITY WITH HIGH CURRENT LOADS
**AUDIO TAPE POTENTIOMETER (10% OF R_T AT 50% ROTATION)

Fig. 8-17

PHONOGRAPH AMPLIFIER (CERAMIC CARTRIDGE)

1.0 k
12 V

100 pF
Tone Control
1.0 Meg Ω

15 pF

5   8

0.1 μF   4

8.0 Ω

XTAL
1.0 Meg Ω
0.002 μF 1.0 Meg Ω
6

MC1306P

3
0.05 μF
200 μF
1.0

Volume Control

7   1

Fig. 8-19

PHONO AMPLIFIER

V_S = 18V

0.1 μF

2
14
+
LM380
8
6
−
7

2.7
500 μF

75K
VOLUME CONTROL
25K

.05 μF

10K
TONE CONTROL

C_{BYPASS}

8Ω

CRYSTAL CARTRIDGE

Fig. 8-18

INVERTING UNITY GAIN AMPLIFIER

V_S

100k

0.1 μF

0.1 μF

100k

8
−
1/2 LM1877
13
+
14

0.1 μF

10k

9

1

3, 4, 5, 10, 11, 12

500 μF

R_L 8Ω

2.7

0.1 μF

1 μF

50 μF

Fig. 8-20

BRIDGE AUDIO POWER AMPLIFIER

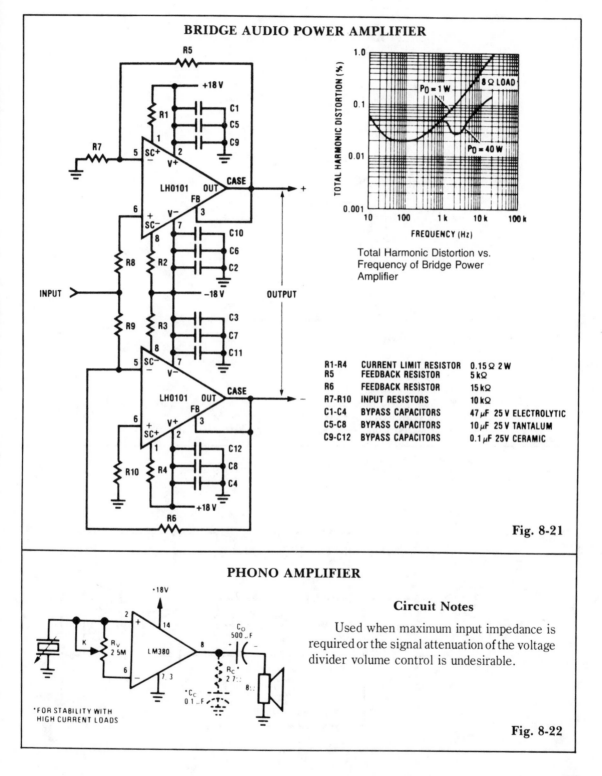

Total Harmonic Distortion vs. Frequency of Bridge Power Amplifier

R1-R4	CURRENT LIMIT RESISTOR	0.15 Ω 2 W
R5	FEEDBACK RESISTOR	5 kΩ
R6	FEEDBACK RESISTOR	15 kΩ
R7-R10	INPUT RESISTORS	10 kΩ
C1-C4	BYPASS CAPACITORS	47 μF 25 V ELECTROLYTIC
C5-C8	BYPASS CAPACITORS	10 μF 25 V TANTALUM
C9-C12	BYPASS CAPACITORS	0.1 μF 25V CERAMIC

Fig. 8-21

PHONO AMPLIFIER

Circuit Notes

Used when maximum input impedance is required or the signal attenuation of the voltage divider volume control is undesirable.

*FOR STABILITY WITH HIGH CURRENT LOADS

Fig. 8-22

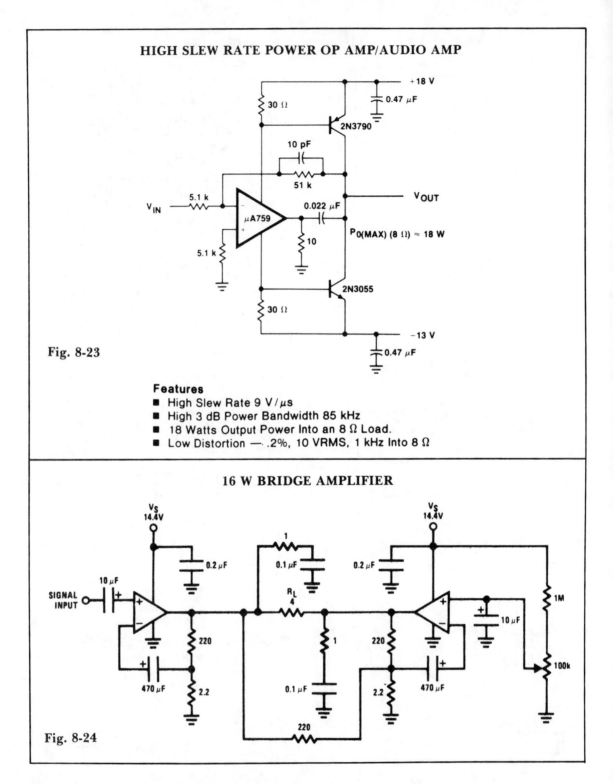

HIGH SLEW RATE POWER OP AMP/AUDIO AMP

Fig. 8-23

Features

- High Slew Rate 9 V/μs
- High 3 dB Power Bandwidth 85 kHz
- 18 Watts Output Power Into an 8 Ω Load.
- Low Distortion — .2%, 10 VRMS, 1 kHz Into 8 Ω

16 W BRIDGE AMPLIFIER

Fig. 8-24

9

Audio Signal Amplifiers

The sources of the following circuits are contained in the Sources section beginning on page 730. The figure number contained in the box of each circuit correlates to the source entry in the Sources section.

GENERAL PURPOSE PREAMPLIFIER

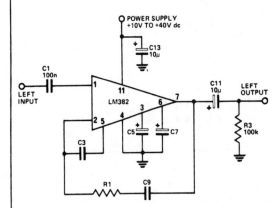
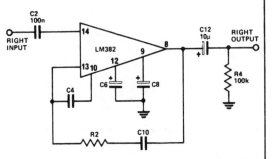

FUNCTION	C3, 4	C5, 6	C7, 8	C9, 10	R1, 2
Phono preamp (RIAA)	330n	10μF	10μF	1n5	1k
Tape preamp (NAB)	68n	10μF	10μF	—	—
Flat 40dB gain	—	—	10μF	—	—
Flat 55dB gain	—	10μF	—	—	—
Flat 80dB gain	—	10μF	10μF	—	—

Fig. 9-1

Circuit Notes

Not much can be said about how the LM382 works as most of the circuitry is contained within the IC. Most of the frequency-determining components are on the chip—only the capacitors are mounted externally. The LM382 has the convenient characteristic of rejecting ripple on the supply line by about 100 dB, thus greatly reducing the quality requirment for the power supply.

BASIC TRANSISTOR AMPLIFIER CIRCUITS

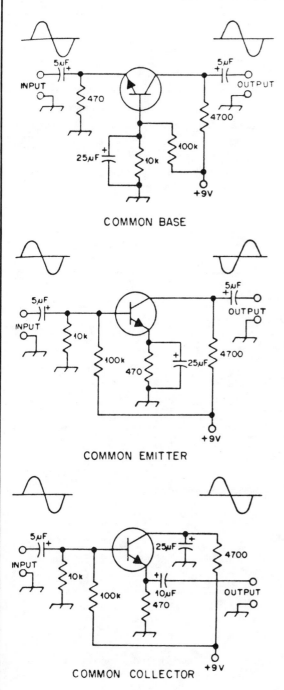

COMMON BASE

COMMON EMITTER

COMMON COLLECTOR

Circuit Notes

Typical component values are given for use at audio frequencies, where these circuits are used most often. The input and output phase relationships are shown.

Fig. 9-2

ELECTRONIC BALANCED INPUT MICROPHONE AMPLIFIER

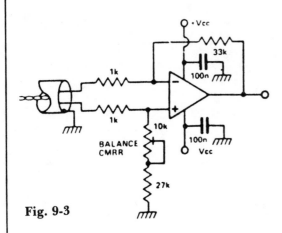

Fig. 9-3

Circuit Notes

It is possible to simulate the balanced performance of a transformer electronically with a different amplifier. By adjusting the presets, the resistor ratio can be balanced so that the best CMRR is obtained. It is possible to get a better CMRR than from a transformer. Use a RC4136 which is a quad low noise op amp.

TRANSDUCER AMPLIFIER

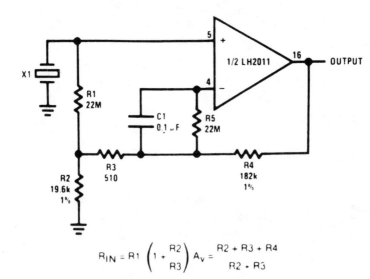

Fig. 9-4

$$R_{IN} = R1 \left(1 + \frac{R2}{R3} \right) \qquad A_v = \frac{R2 + R3 + R4}{R2 + R3}$$

Circuit Notes

This circuit is high-input-impedance ac amplifier for a piezoelectric transducer. Input resistance is 880 M, and a gain of 10 is obtained.

ULTRA-HIGH GAIN AUDIO AMPLIFIER

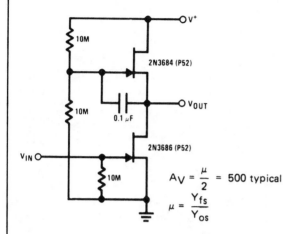

$$A_V = \frac{\mu}{2} = 500 \text{ typical}$$

$$\mu = \frac{Y_{fs}}{Y_{os}}$$

Fig. 9-5

Circuit Notes

Sometimes called the JFET μ-amp, this circuit provides a very low power, high gain amplifying function. Since μ of a JFET increases as drain current decreases, the lower drain current is, the more gain you get. Input dynamic range is sacrificed with increasing gain, however.

MICROPHONE AMPLIFIER

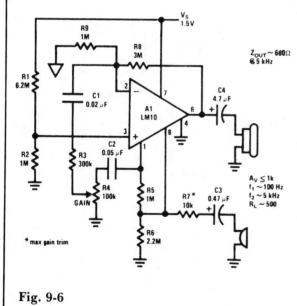

$Z_{OUT} \sim 680\Omega$
@ 5 kHz

$A_V \leq 1k$
$f_1 \sim 100$ Hz
$f_2 \sim 5$ kHz
$R_L \sim 500$

* max gain trim

Fig. 9-6

Circuit Notes

This circuit operates from a 1.5 Vdc source.

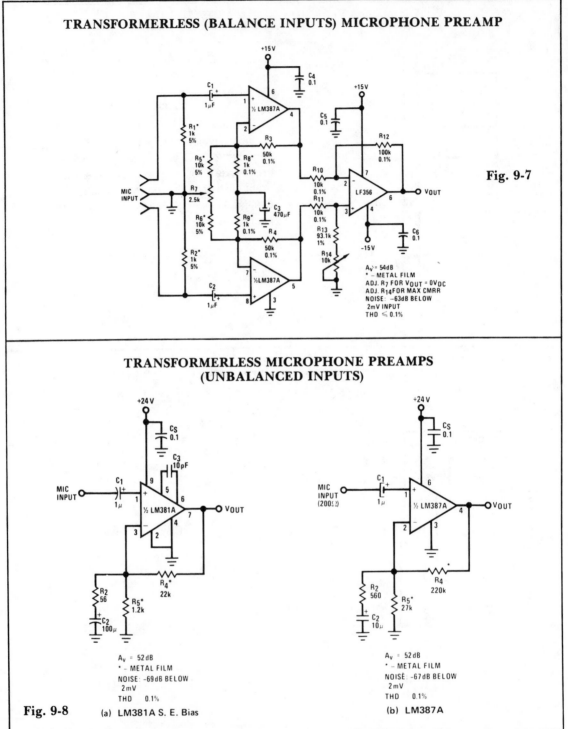

TRANSFORMERLESS (BALANCE INPUTS) MICROPHONE PREAMP

Fig. 9-7

$A_V = 54$ dB
* – METAL FILM
ADJ. R_7 FOR V_{OUT} = 0V DC
ADJ. R_{14} FOR MAX CMRR
NOISE: –63dB BELOW
2mV INPUT
THD $\leqslant$ 0.1%

**TRANSFORMERLESS MICROPHONE PREAMPS
(UNBALANCED INPUTS)**

$A_V = 52$ dB
* – METAL FILM
NOISE: –69dB BELOW
2mV
THD 0.1%

$A_V = 52$ dB
* – METAL FILM
NOISE: –67dB BELOW
2mV
THD 0.1%

Fig. 9-8   (a) LM381A S. E. Bias

(b) LM387A

MAGNETIC PICKUP PHONO PREAMPLIFIER

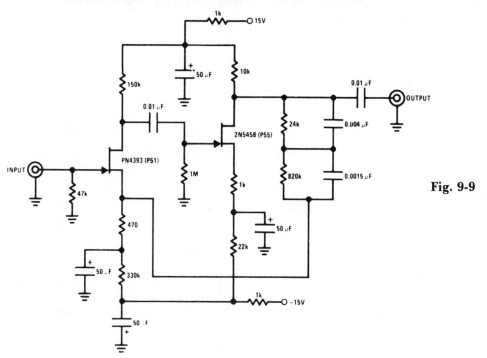

Fig. 9-9

Circuit Notes

This preamplifier provides proper loading to a reluctance phono cartridge. It provides approximately 35 dB of gain at 1 kHz (2.2 mV input for 100 mV output). It features (S + N)/N ratio of better than −70 dB (referenced to 10 mV input at 1 kHz) and has a dynamic range of 84 dB (referenced to 1 kHz). The feedback provides for RIAA equalization.

DISC/TAPE PHASE MODULATED READBACK SYSTEMS

Fig. 9-10

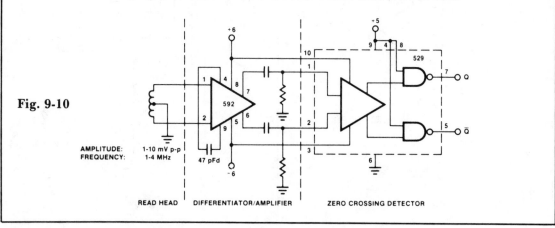

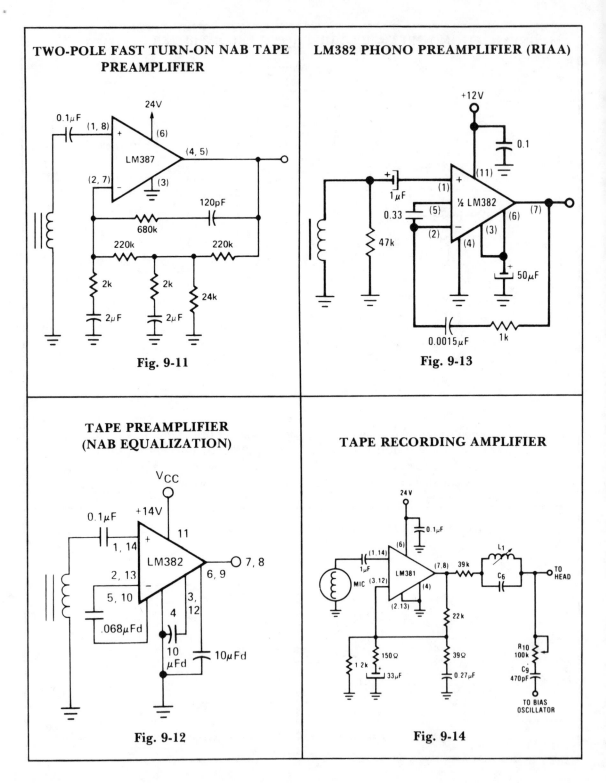

TWO-POLE FAST TURN-ON NAB TAPE PREAMPLIFIER

Fig. 9-11

LM382 PHONO PREAMPLIFIER (RIAA)

Fig. 9-13

TAPE PREAMPLIFIER (NAB EQUALIZATION)

Fig. 9-12

TAPE RECORDING AMPLIFIER

Fig. 9-14

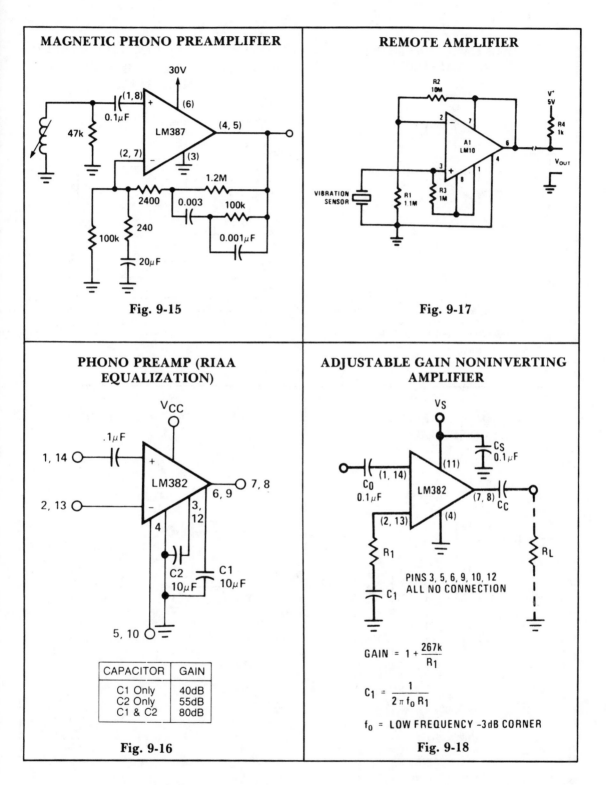

MAGNETIC PHONO PREAMPLIFIER

30V

(1, 8)
0.1μF
(6)
LM387
47k
(2, 7)
(4, 5)
(3)
1.2M
2400
0.003
100k
240
0.001μF
100k
20μF

Fig. 9-15

REMOTE AMPLIFIER

R2
10M
V⁺
5V
R4
1k
2
7
A1
LM10
6
V_OUT
3
4
8
1
VIBRATION
SENSOR
R1
1.1M
R3
1M

Fig. 9-17

PHONO PREAMP (RIAA EQUALIZATION)

V_CC

.1μF
1, 14
+
LM382
7, 8
6, 9
2, 13
−
3, 12
4
C2
10μF
C1
10μF
5, 10

CAPACITOR	GAIN
C1 Only	40dB
C2 Only	55dB
C1 & C2	80dB

Fig. 9-16

ADJUSTABLE GAIN NONINVERTING AMPLIFIER

V_S

C_S
0.1μF
(1, 14)
(11)
C_0
0.1μF
LM382
(2, 13)
(7, 8)
C_C
(4)
R_1
R_L
C_1

PINS 3, 5, 6, 9, 10, 12
ALL NO CONNECTION

$$GAIN = 1 + \frac{267k}{R_1}$$

$$C_1 = \frac{1}{2\pi f_0 R_1}$$

f_0 = LOW FREQUENCY −3dB CORNER

Fig. 9-18

91

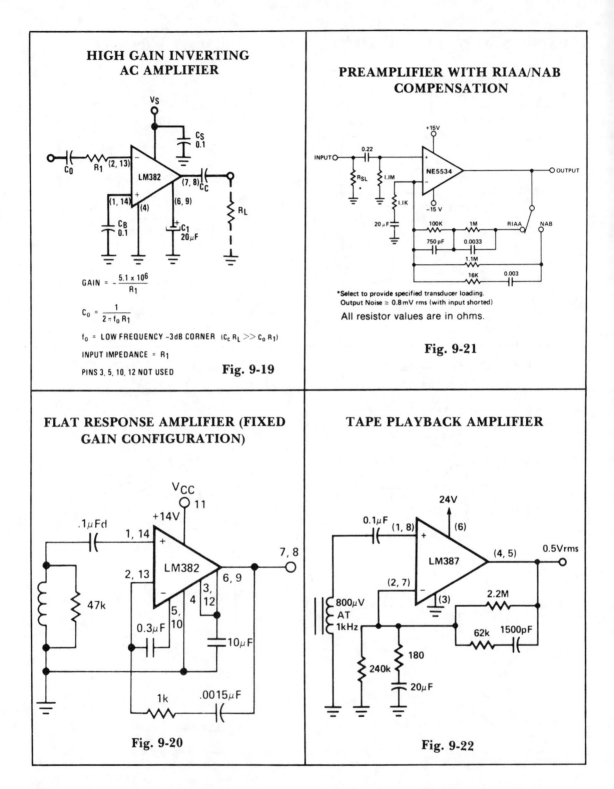

HIGH GAIN INVERTING AC AMPLIFIER

$$\text{GAIN} = -\frac{5.1 \times 10^6}{R_1}$$

$$C_0 = \frac{1}{2 \pi f_0 R_1}$$

f_0 = LOW FREQUENCY –3dB CORNER $(C_c R_L \gg C_0 R_1)$

INPUT IMPEDANCE = R_1

PINS 3, 5, 10, 12 NOT USED

Fig. 9-19

PREAMPLIFIER WITH RIAA/NAB COMPENSATION

*Select to provide specified transducer loading.
Output Noise ≥ 0.8 mV rms (with input shorted)

All resistor values are in ohms.

Fig. 9-21

FLAT RESPONSE AMPLIFIER (FIXED GAIN CONFIGURATION)

Fig. 9-20

TAPE PLAYBACK AMPLIFIER

Fig. 9-22

92

10

Automotive Circuits

The sources of the following circuits are contained in the Sources section beginning on page 730. The figure number contained in the box of each circuit correlates to the source entry in the Sources section.

Gasoline Engine Tachometer
Speed Alarm
Speed Warning Device
Universal Wiper Delay
Courtesy Light Extender
Bargraph Car Voltmeter
Tachometer
High Speed Warning Device
Breaker Point Dwell Meter
Tachometer
Capacitor Discharge Ignition System
Windshield Wiper Control

Auto Battery Current Analyzer
Speed Switch
Windshield Wiper Controller
Windshield Wiper Hesitation Control Unit
Ice Warning and Lights Reminder
Car Battery Monitor
Headlight Delay Unit
Windshield Washer Fluid Watcher
Car Battery Condition Checker
Overspeed Indicator
Sequential Flasher for Auto Turn Signals
Auto Lights-On Reminder

GASOLINE ENGINE TACHOMETER

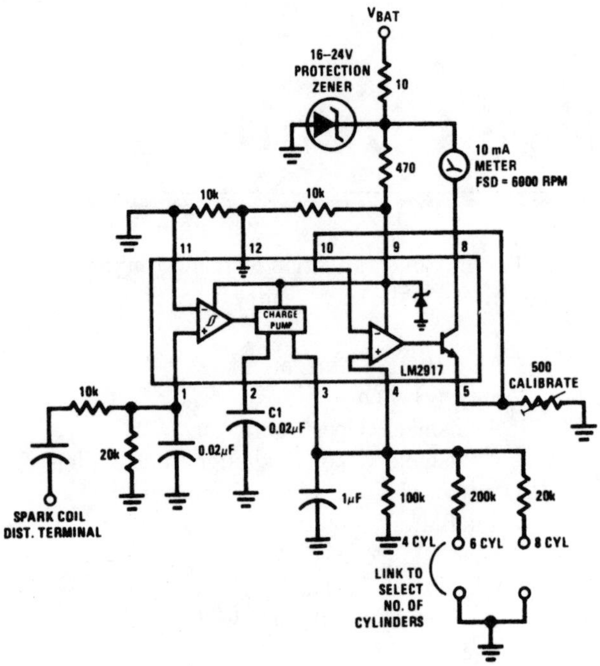

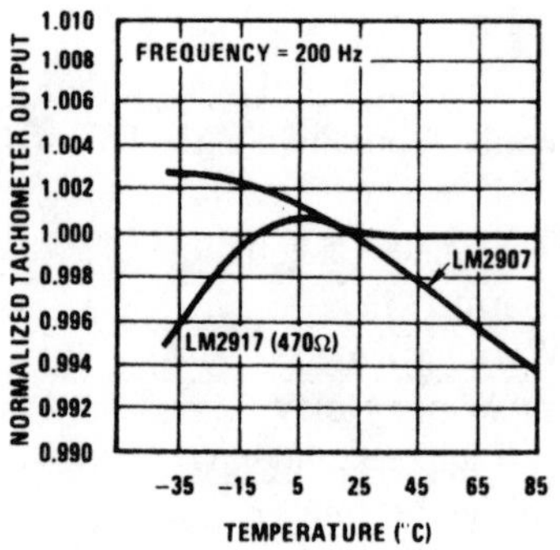

Fig. 10-1

Circuit Notes

This tachometer can be set up for any number of cylinders by linking the appropriate timing resistor as illustrated. A 500 ohm trim resistor can be used to set up final calibration.

A protection circuit composed of a 10 ohm resistor and a zener diode is also shown as a safety precaution against the transients which are to be found in automobiles.

SPEED ALARM

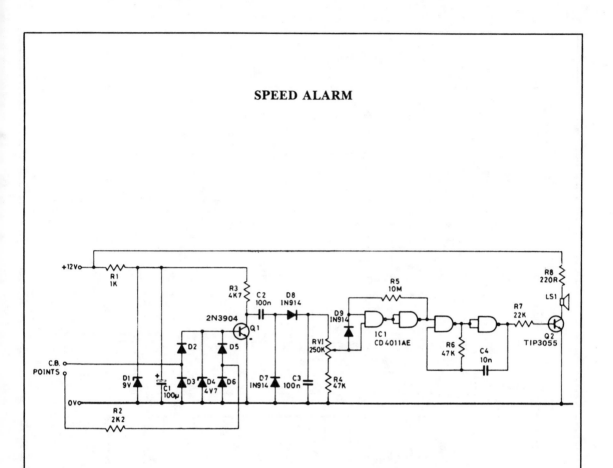

Fig. 10-2

Circuit Notes

Pulses from the distributor points are passed through a current limiting resistor, rectified, and clipped at 4.7 volts. Via Q1 and the diode pump, a dc voltage proportional to engine rpm is presented to RV1; the sharp transfer characteristic of a CMOS gate, assisted by feedback, is used to enable the oscillator formed by the remaining half of the 4011. At the pre-set speed, a nonignorable tone emits from the speaker, and disappears as soon as the speed drops by three or four mph.

SPEED WARNING DEVICE

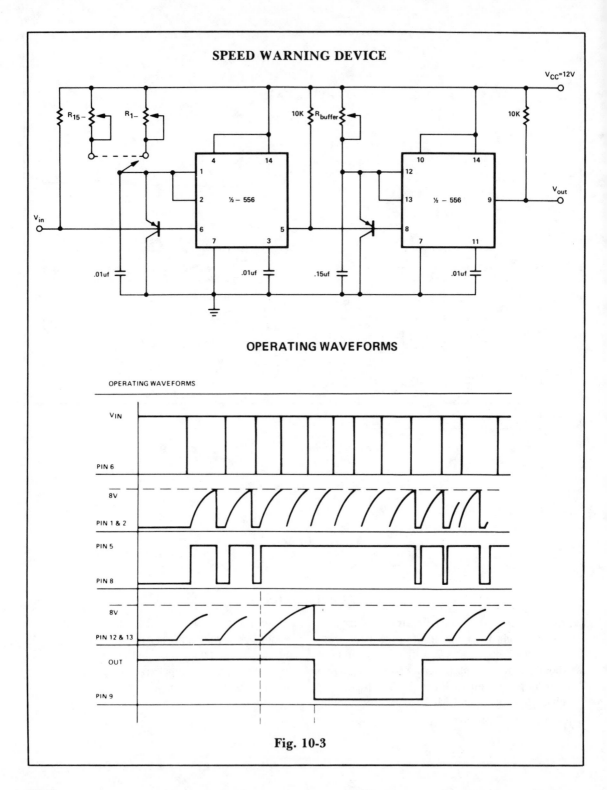

OPERATING WAVEFORMS

Fig. 10-3

UNIVERSAL WIPER DELAY

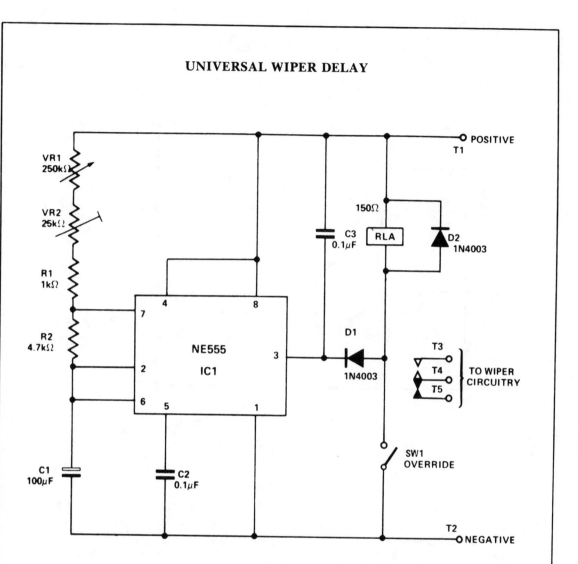

Fig. 10-4

Circuit Notes

IC1 is connected in the astable mode, driving RLA. C3, D1, and D2 prevent spikes from the relay coil and the wiper motor from triggering IC1. VR2 is adjusted to give the minimum delay time required. VR1 is the main delay control and provides a range of from about 1 second to 20 seconds. SW1 is an override switch to hold RLA permanently on (for normal wiper operation). The relay should have a resistance of at least 150 ohms and have heavy duty contacts. The suppression circuit may be needed for the protection of IC1.

COURTESY LIGHT EXTENDER

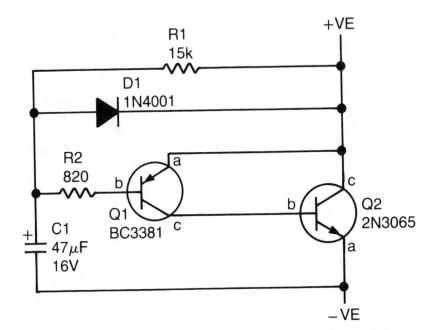

Fig. 10-5

Circuit Notes

Most car door switches are simply single-pole switches, with one side grounded. When the door is opened the switch grounds the other line thus completing the light circuit. In a car where the negative terminal of the battery is connected to the chassis, the negative wire of the unit (emitter of Q2) is connected to chassis the positive wire (case of 2N3055) is connected to the wire going to the switch. In a car having a positive ground system this connection sequence is reversed. When the switch closes (door open), C1 is discharged via D1 to zero volts, and when the switch opens, C1 charges up via R1 and R2.

Transistors Q1 and Q2 are connected as an emitter follower (Q2 just buffers Q1) therefore the voltage across Q2 increases slowly as C1 charges. Hence Q2 acts like a low resistance in parallel with the switch and keeps the lights on. The value of C1 is chosen such that a useful light level is obtained for about four seconds; therefore the light decreases until in about 10 seconds it is out completely. With different transistor gains and with variation in current drain due to a particular type of car, the timing may vary but may be simply adjusted by selecting C1.

BARGRAPH CAR VOLTMETER

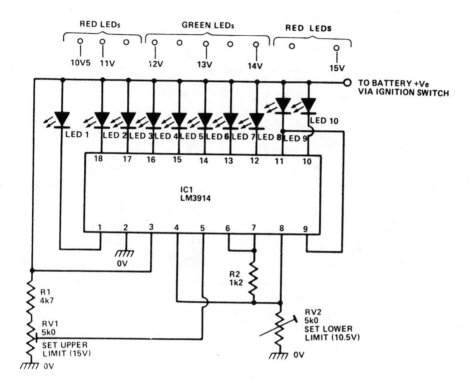

Fig. 10-6

Circuit Notes

The LM3914 acts as a LED-driving voltmeter that has its basic maximum and minimum readings determined by the values of R2 and RV2. When correctly adjusted, the unit actually covers the 2.5 volt to 3.6 volt range, but it is made to read a supply voltage span of 10-10.5 volts to 15 volts by interposing potential divider R1-RV1 between the supply line and the pin-5 input terminal of the IC. The IC is configured to give a 'dot' display, in which only one of the ten LEDs is illuminated at any given time. If the supply voltage is below 10.5 volts none of the LEDs illuminate. If the supply equals or exceeds 15 volts, LED 10 illuminates.

TACHOMETER

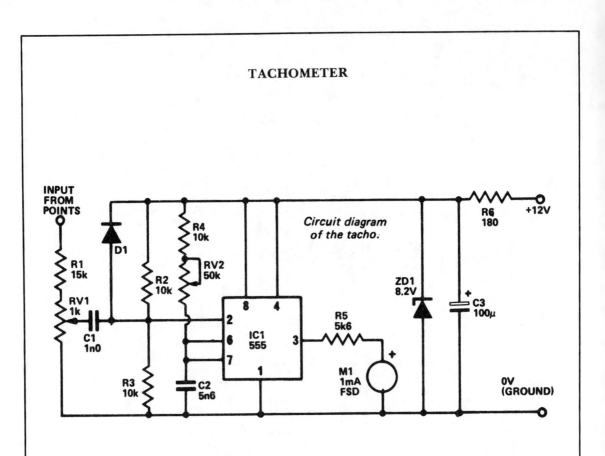

Fig. 10-7

Circuit Notes

An electrical signal taken from the low tension side of the distributor is converted into a voltage proportional to engine rpm and this voltage is displayed on a meter calibrated accordingly. The 555 timer IC is used as a monostable which, in effect, converts the signal pulse from the breaker points to a single positive pulse the width of which is determined by the value of R4 + RV2 and C2. Resistors R2 and R3 set a voltage of about 4 volts at pin 2 of IC1. The IC is triggered if this voltage is reduced to less than approximately 2.7 volts (⅓ of supply voltage), and this occurs due to the voltage swing when the breaker points open. An adjustment potentiometer RV1 enables the input level to be set to avoid false triggering. Zener diode ZD1 and the 180 ohm resistor stabilize the unit against voltage variations.

HIGH SPEED WARNING DEVICE

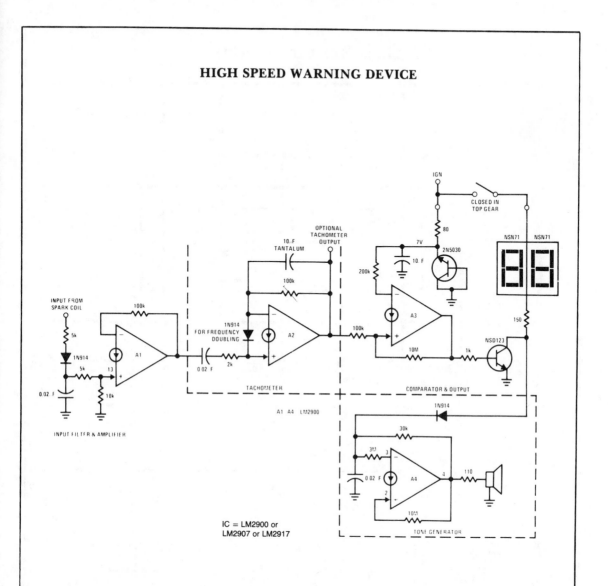

Fig. 10-8

Circuit Notes

A1 amplifies and regulates the signal from the spark coil. A2 converts frequency to voltage so that its output is a voltage proportional to engine rpm. A3 compares the tachometer voltage with the reference voltage and turns on the output transistor at the set speed. Amplifier A4 is used to generate an audible tone whenever the set speed is exceeded.

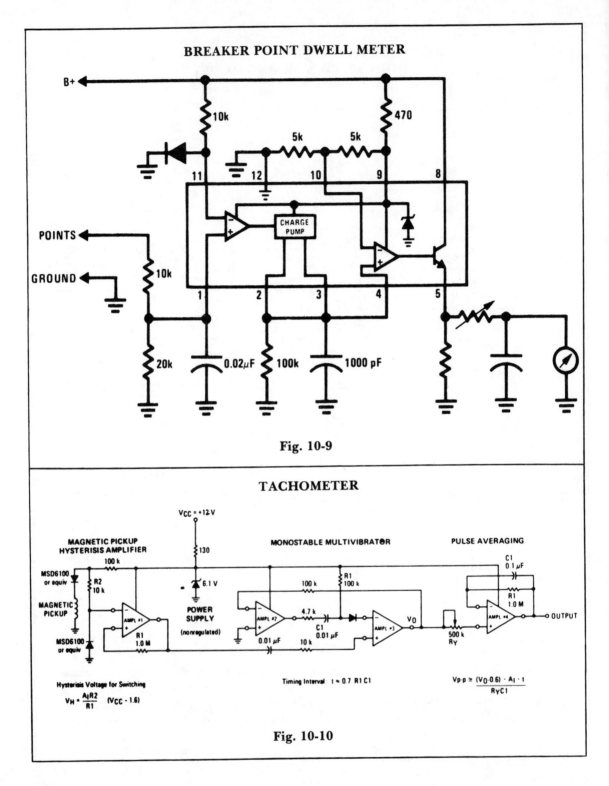

BREAKER POINT DWELL METER

B+

10k

470

5k   5k

11   12   10   9   8

CHARGE PUMP

POINTS

GROUND

1   2   3   4   5

10k

20k   0.02μF   100k   1000 pF

Fig. 10-9

TACHOMETER

V_{CC} = +12 V

MAGNETIC PICKUP HYSTERISIS AMPLIFIER

100 k

MSD6100 or equiv

R2 10 k

MAGNETIC PICKUP

MSD6100 or equiv

AMPL #1

R1 1.0 M

130

6.1 V

POWER SUPPLY
(nonregulated)

MONOSTABLE MULTIVIBRATOR

100 k

R1 100 k

4.7 k

AMPL #2

0.01 μF   10 k

C1 0.01 μF

AMPL #3

V_0

PULSE AVERAGING

C1 0.1 μF

R1 1.0 M

500 k R_Y

AMPL #4

OUTPUT

Hysterisis Voltage for Switching

$$V_H = \frac{A_1 R2}{R1} \quad (V_{CC} - 1.6)$$

Timing Interval: $t \approx 0.7\ R1\ C1$

$$V_{p\text{-}p} \cong \frac{(V_0 \text{-} 0.6) \cdot A_1 \cdot t}{R_Y C1}$$

Fig. 10-10

CAPACITOR DISCHARGE IGNITION SYSTEM

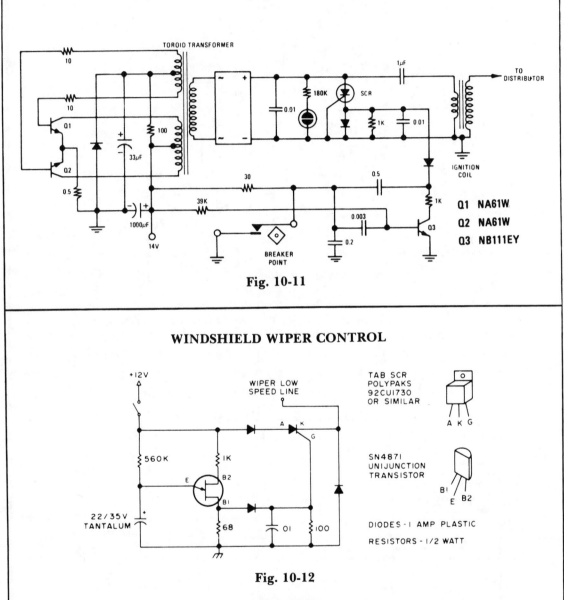

Fig. 10-11

WINDSHIELD WIPER CONTROL

Fig. 10-12

Circuit Notes

Here's a good way to set windshield wipers on an interval circuit. Only two connections to the car's wiper control, plus ground, are required. Variable control can be accomplished by substituting a 500 K pot in series with a 100 K fixed resistor in place of the 560 K.

AUTO BATTERY CURRENT ANALYZER

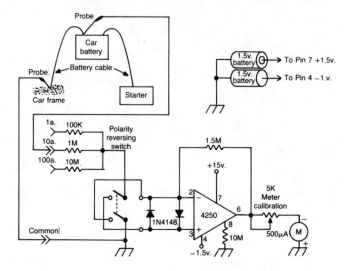

Fig. 10-13

Circuit Notes

This op-amp analyzer can measure the current drawn by any device in a car. The analyzer works by measuring the very small voltage that develops across the battery cables when current flows. To calibrate the unit, measure the current flow somewhere in the car with an accurate ammeter, then adjust the analyzer for that current reading.

SPEED SWITCH

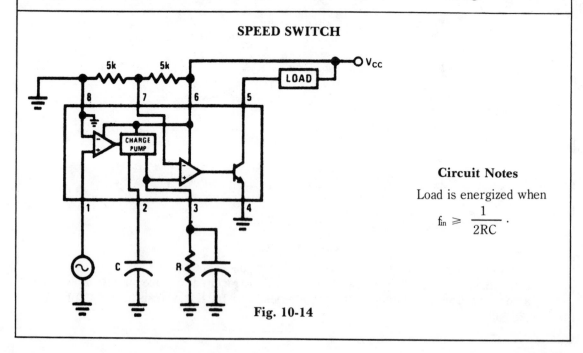

Circuit Notes

Load is energized when

$$f_{in} \geq \frac{1}{2RC}.$$

Fig. 10-14

WINDSHIELD WIPER CONTROLLER

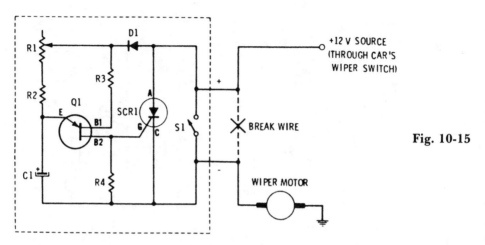

Fig. 10-15

Circuit Notes

This circuit provides complete speed control over car's windshield wipers. They can be slowed down to any rate even down to four sweeps per minute. The controller has two principal circuits: The rate-determining circuit—a unijunction transistor connected as a freerunning oscillator, and the silicon-controlled rectifier which is the actuator.

WINDSHIELD WIPER HESITATION CONTROL UNIT

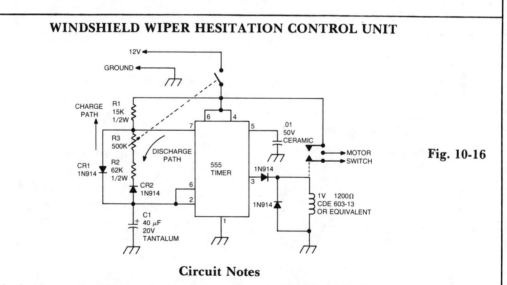

Fig. 10-16

Circuit Notes

This circuit uses the 555 timer in the astable or oscillatory mode. The length of time the timer is off is a function of the values of C1, R2, and R3. The potentiometer which controls the amount of "hesitation". (Approximately 2 to 15 seconds.) R2 provides a minimum time delay when R3 is at its zero ohms position.

105

ICE WARNING AND LIGHTS REMINDER

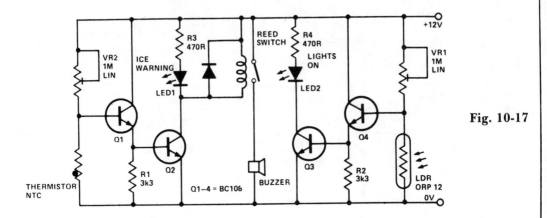

Fig. 10-17

Circuit Notes

This device will tell a driver if his lights should be on and will warn him if the outside temperature is nearing zero by lighting a LED and sounding a buzzer9 VR1 adjusts sensitivity for temperature, VR2 for light. Both thermistor and LDR should be well protected. Most high gain NPN transistors will work.

CAR BATTERY MONITOR

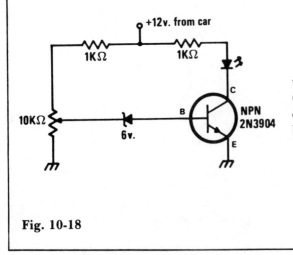

Fig. 10-18

Circuit Notes

Warning light (LED) indicates when battery voltage falls below level set by 10 K pot. Can indicate that battery is defective or needs charging if cranking drops battery voltage below preset "safe" limit.

HEADLIGHT DELAY UNIT

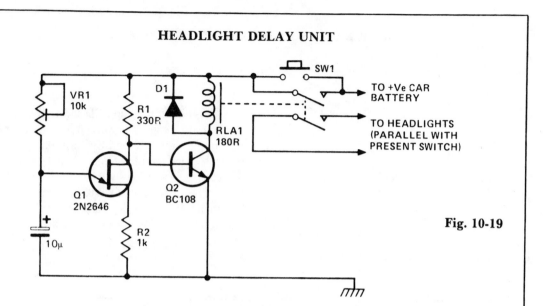

Fig. 10-19

Circuit Notes

This circuit will operate a car's headlights for a predetermined time to light up the driveway or path after the driver has left the car. SQ1 is pushed and Q2 is turned on closing the relay and turning on the car's headlights. C1 begins to charge through VR1 until Q1 turns on, turning Q2 off. The relay will then open switching off both the lights and the unit. The delay is governed by the time taken for the capacitor to charge, which is about one minute.

WINDSHIELD WASHER FLUID WATCHER

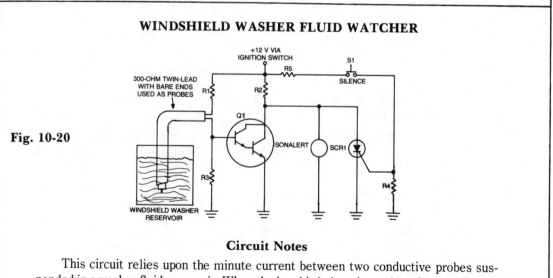

Fig. 10-20

Circuit Notes

This circuit relies upon the minute current between two conductive probes suspended in a washer fluid reservoir. When the level is below the probes, Q1 turns on and the Sonolert sounds.

CAR BATTERY CONDITION CHECKER

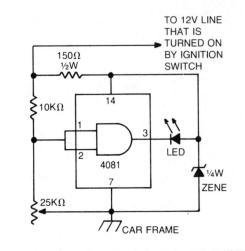

TO 12V LINE
THAT IS
TURNED ON
BY IGNITION
SWITCH

Circuit Notes

This circuit uses an LED and 4081 CMOS integrated circuit. The variable resistor sets the voltage at which the LED turns on. Set the control so that the LED lights when the voltage from the car's ignition switch drops below 13.8 volts. The LED normally will light every now and then for a short period of time. But, if it stays on for very long, your electrical system is in trouble.

Fig. 10-21

OVERSPEED INDICATOR

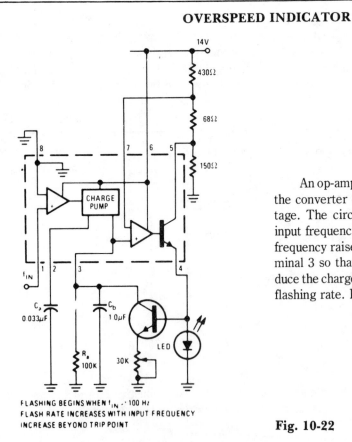

Circuit Notes

An op-amp comparator is used to compare the converter output with a dc threshold voltage. The circuit flashes the LED when the input frequency exceeds 100 Hz. Increases in frequency raise the average current out of terminal 3 so that frequencies above 100 Hz reduce the charge time of C2, increasing the LED flashing rate. IC = LM2907 or LM2917

FLASHING BEGINS WHEN f_{IN} ·' 100 Hz
FLASH RATE INCREASES WITH INPUT FREQUENCY
INCREASE BEYOND TRIP POINT

Fig. 10-22

SEQUENTIAL FLASHER FOR AUTOMOTIVE TURN SIGNALS

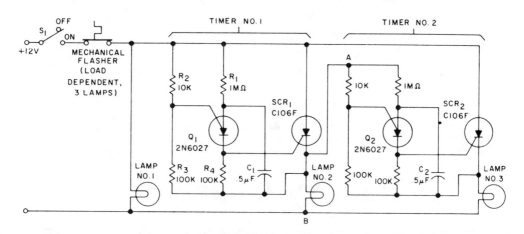

Circuit Notes

When the turn signal switch S1 is closed, lamp #1 will be activated and capacitor C1 will charge to the triggered voltage of Q1. As soon as the anode voltage on Q1 exceeds its gate voltage by 0.5 V, Q1 will switch into the low resistance mode, thereby triggering SCR1 to activate lamp #2 and the second timing circuit.

After Q2 switches into the low resistance state, SCR2 will be triggered to activate lamp #3. When the thermal flasher interrupts the current to all three lamps, SCR1 and SCR2 are commutated and the circuit is ready for another cycle.

Fig. 10-23

AUTO LIGHTS-ON REMINDER

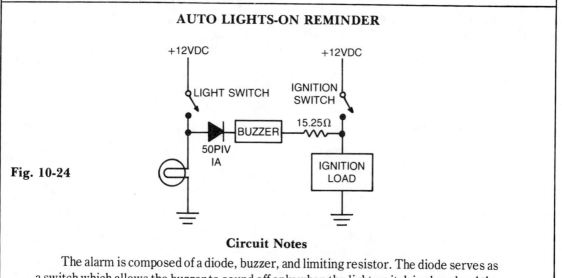

Fig. 10-24

Circuit Notes

The alarm is composed of a diode, buzzer, and limiting resistor. The diode serves as a switch which allows the buzzer to sound off only when the light switch is closed and the ignition is turned off.

11

Battery Chargers

The sources of the following circuits are contained in the Sources section beginning on page 730. The figure number contained in the box of each circuit correlates to the source entry in the Sources section.

12 V Battery Charger
Simple Ni-Cad Battery Charger
12 V Battery Charger Control (20 Amps Rms
    Max.)
Battery Charger
Automatic Shutoff Battery Charger
200 mA-Hour, 12 V Ni-Cad Battery Charger
Ni-Cad Charger with Current and Voltage
    Limiting

Automotive Charger for Ni-Cad Battery Packs
Constant Voltage, Current-Limited Charger
Ni-Cad Charger
Simple Ni-Cad Battery Zapper
Battery Charging Regulator
Low-Cost Trickle Charger for 12V Storage
    Battery
Fast Charger for Ni-Cad Batteries
Current Limited 6 V Charger

12 V BATTERY CHARGER

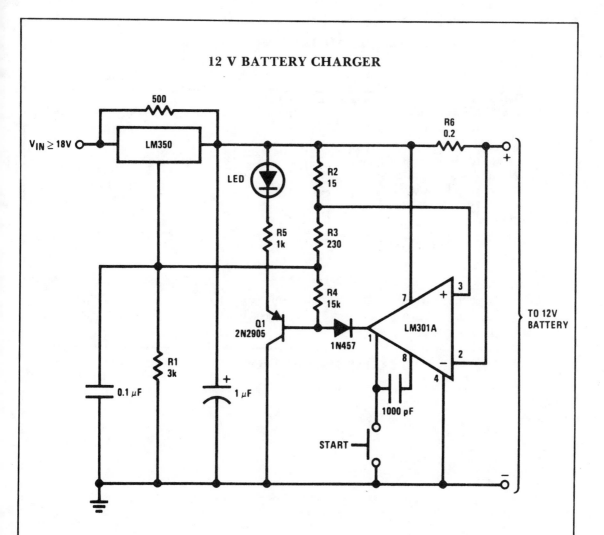

Fig. 11-1

Circuit Notes

This circuit is a high performance charger for gelled electrolyte lead-acid batteries. Charger quickly recharges battery and shuts off at full charge. Initially, charging current is limited to 2A. As the battery voltage rises, current to the battery decreases, and when the current has decreased to 150 mA, the charger switches to a lower float voltage preventing overcharge. When the start switch is pushed, the output of the charger goes to 14.5 V. As the battery approaches full charge, the charging current decreases and the output voltage is reduced from 14.5 V to about 12.5 V terminating the charging. Transistor Q1 then lights the LED as a visual indication of full charge.

111

SIMPLE NI-CAD BATTERY CHARGER

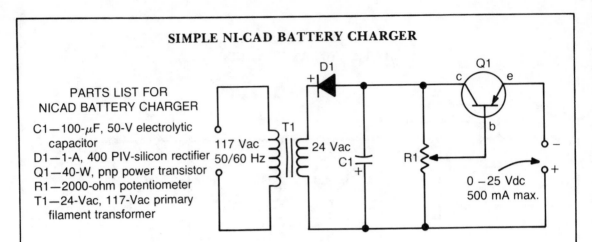

PARTS LIST FOR
NICAD BATTERY CHARGER

C1—100-μF, 50-V electrolytic
   capacitor
D1—1-A, 400 PIV-silicon rectifier
Q1—40-W, pnp power transistor
R1—2000-ohm potentiometer
T1—24-Vac, 117-Vac primary
   filament transformer

Fig. 11-2

Circuit Notes

This circuit provides an adjustable output voltage up to 35 Vdc and maximum output current of 50 mA. Transistor Q1 dissipates quite a bit of heat and must be mounted on a heatsink.

12 V BATTERY CHARGER CONTROL (20 AMPS RMS MAX.)

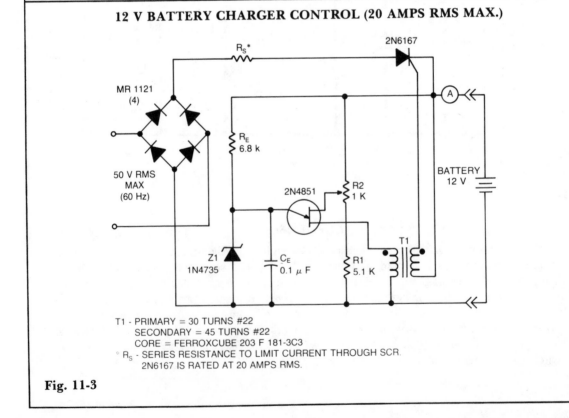

T1 - PRIMARY = 30 TURNS #22
     SECONDARY = 45 TURNS #22
     CORE = FERROXCUBE 203 F 181-3C3
* R_S - SERIES RESISTANCE TO LIMIT CURRENT THROUGH SCR.
     2N6167 IS RATED AT 20 AMPS RMS.

Fig. 11-3

BATTERY CHARGER

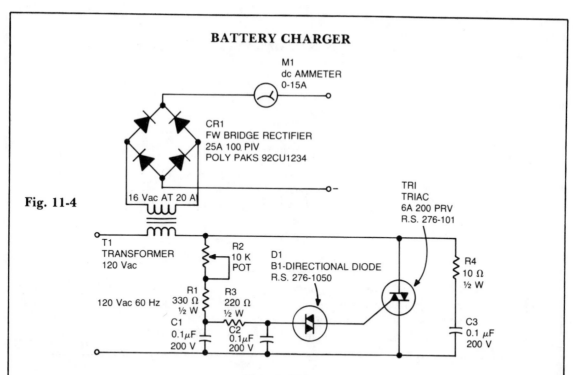

Fig. 11-4

Circuit Notes

A diac is used in the gate circuit to provide a threshold level for firing the triac. C3 and R4 provide a transient suppression network. R1, R2, R3, C1, and C2 provide a phase-shift network for the signal being applied to the gate. R1 is selected to limit the maximum charging current at full rotation of R2.

AUTOMATIC SHUTOFF BATTERY CHARGER

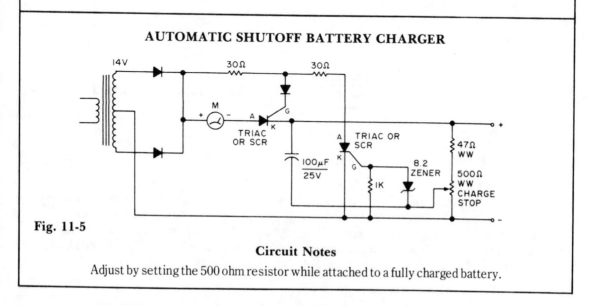

Fig. 11-5

Circuit Notes

Adjust by setting the 500 ohm resistor while attached to a fully charged battery.

200 mA-HOUR, 12 V NI-CAD BATTERY CHARGER

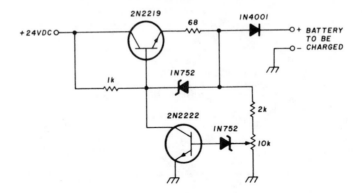

Fig. 11-6

Circuit Notes

This circuit charges the battery at 75 mA until the battery is charged, then it reduces the current to a trickle rate. It will completely recharge a dead battery in four hours and the battery can be left in the charger indefinitely. To set the shut-off point, connect a 270-ohm, 2-watt resistor across the charge terminals and adjust the pot for 15.5 volts across the resistor.

NI-CAD CHARGER WITH CURRENT AND VOLTAGE LIMITING

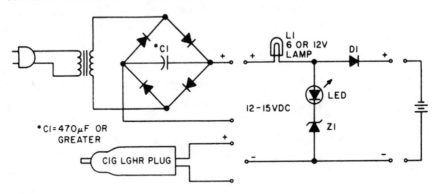

Fig. 11-7

Circuit Notes

Lamp L1 will glow brightly and the LED will be out when the battery is low and being charged, but the LED will be bright and the light bulb dim when the battery is almost ready. L1 should be a light bulb rated for the current you want (usually the battery capacity divided by 10). Diode D1 should be at least 1 A, and Z1 is a 1 W zener diode with a voltage determined by the full-charge battery voltage minus 1.5 V. After the battery is fully charged, the circuit will float it at about battery capacity divided by 100 mA.

AUTOMOTIVE CHARGER FOR NI-CAD BATTERY PACKS

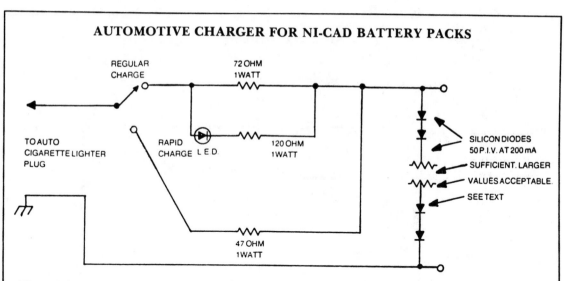

Fig. 11-8

Circuit Notes

The number of silicon diodes across the output is determined by the voltage of the battery pack. Figure each diode at 0.7 volt. For example, a 10.9- volt pack would require 10.9/0.7 = 15.57, or 16 diodes.

CONSTANT-VOLTAGE, CURRENT-LIMITED CHARGER

IC LM723C VOLTAGE REGULATOR (FOR 12V dc OUTPUT 0.42A MAX.)

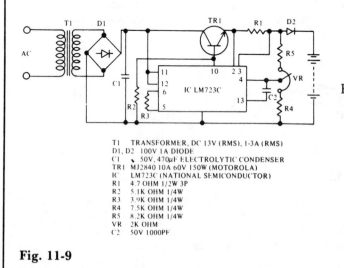

T1   TRANSFORMER, DC 13V (RMS), 1-3A (RMS)
D1, D2 100V 1A DIODE
C1   50V, 470μF ELECTROLYTIC CONDENSER
TR1 MJ2840 10A 60V 150W (MOTOROLA)
IC   LM723C (NATIONAL SEMICONDUCTOR)
R1   4.7 OHM 1/2W 3P
R2   5.1K OHM 1/4W
R3   3.9K OHM 1/4W
R4   7.5K OHM 1/4W
R5   8.2K OHM 1/4W
VR   2K OHM
C2   50V 1000PF

Circuit Notes

For 12 V sealed lead-acid batteries.

Fig. 11-9

NI-CAD CHARGER

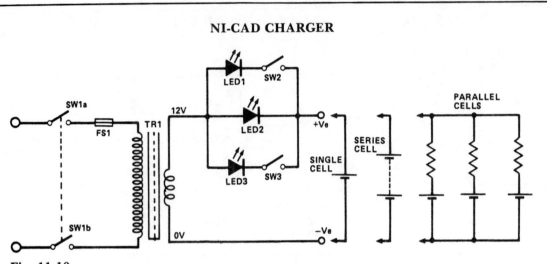

Fig. 11-10

Circuit Notes

This circuit uses constant current LEDs to adjust charging current. It makes use of LEDs that pass a constant current of about 15 mA for an applied voltage range of 2-18 V. They can be paralleled to give any multiple of 15 mA and they light up when current is flowing. The circuit will charge a single cell at 15, 30 or 45 mA or cells in series up to the rated supply voltage limit (about 14 V).

SIMPLE NI-CAD BATTERY ZAPPER

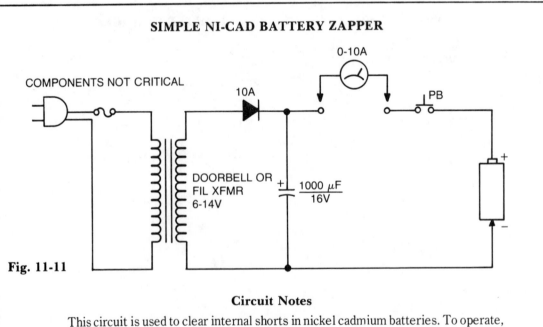

Fig. 11-11

Circuit Notes

This circuit is used to clear internal shorts in nickel cadmium batteries. To operate, connect ni-cad to output and press the pushbutton for three seconds.

BATTERY CHARGING REGULATOR

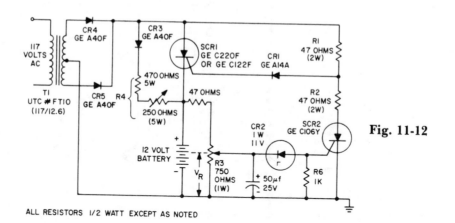

Fig. 11-12

ALL RESISTORS 1/2 WATT EXCEPT AS NOTED

Circuit Notes

The circuit is capable of charging a 12 volt battery at up to a six ampere rate. Other voltages and currents, from 6 to 600 volts and up to 300 amperes, can be accommodated by suitable component selection. When the battery voltage reaches its fully charged level, the charging SCR shuts off, and a trickle charge as determined by the value of R4 continues to flow.

LOW-COST TRICKLE CHARGER FOR 12 V STORAGE BATTERY

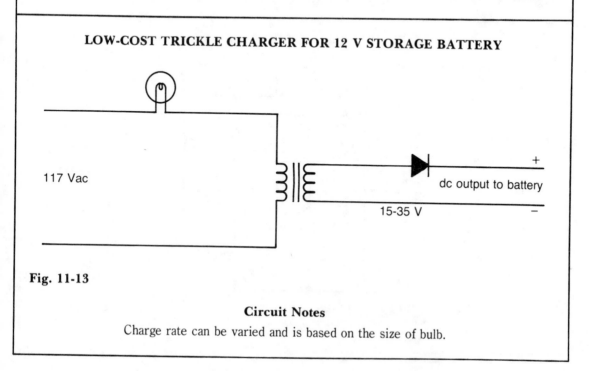

Fig. 11-13

Circuit Notes

Charge rate can be varied and is based on the size of bulb.

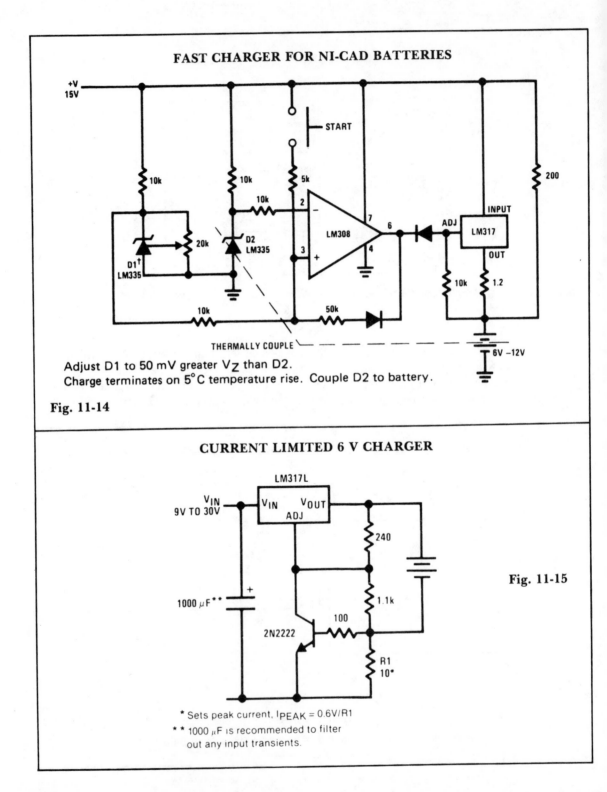

FAST CHARGER FOR NI-CAD BATTERIES

Adjust D1 to 50 mV greater V$_Z$ than D2.
Charge terminates on 5°C temperature rise. Couple D2 to battery.

Fig. 11-14

CURRENT LIMITED 6 V CHARGER

Fig. 11-15

* Sets peak current, I$_{PEAK}$ = 0.6V/R1
** 1000 μF is recommended to filter
out any input transients.

12

Battery Monitors

The sources of the following circuits are contained in the Sources section beginning on page 730. The figure number contained in the box of each circuit correlates to the source entry in the Sources section.

Solid-State Battery Voltage Indicator
Ni-Cad Discharge Limiter
Battery Condition Indicator
Equipment on Reminder
Battery Charge/Discharge Indicator
Precision Battery Voltage Monitor for HTs

Low Voltage Monitor
Undervoltage indicator for Battery Operated Equipment
Low Battery Indicator
Battery-Level Indicator
Battery-Threshold Indicator

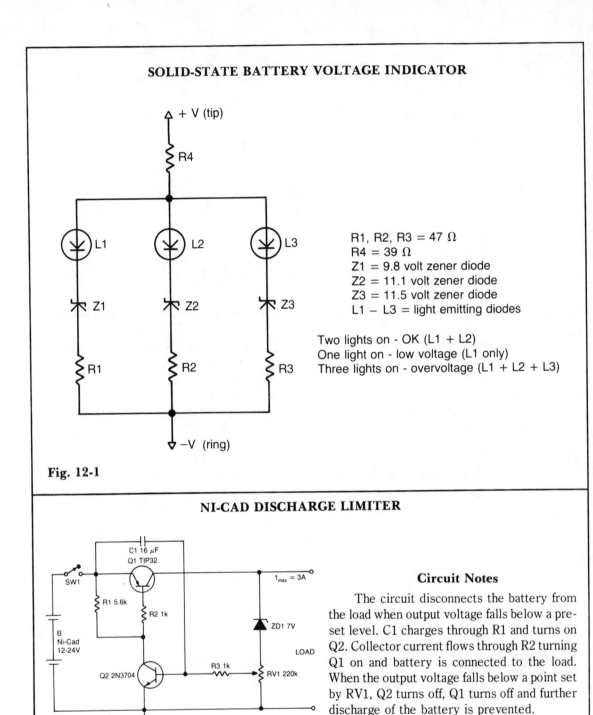

SOLID-STATE BATTERY VOLTAGE INDICATOR

+ V (tip)

R4

L1   L2   L3

Z1   Z2   Z3

R1   R2   R3

−V (ring)

R1, R2, R3 = 47 Ω
R4 = 39 Ω
Z1 = 9.8 volt zener diode
Z2 = 11.1 volt zener diode
Z3 = 11.5 volt zener diode
L1 − L3 = light emitting diodes

Two lights on - OK (L1 + L2)
One light on - low voltage (L1 only)
Three lights on - overvoltage (L1 + L2 + L3)

Fig. 12-1

NI-CAD DISCHARGE LIMITER

C1 16 μF
Q1 TIP32

SW1

R1 5.6k   R2 1k

B
Ni-Cad
12-24V

Q2 2N3704   R3 1k   RV1 220k

1_{max} = 3A

ZD1 7V

LOAD

Circuit Notes

The circuit disconnects the battery from the load when output voltage falls below a preset level. C1 charges through R1 and turns on Q2. Collector current flows through R2 turning Q1 on and battery is connected to the load. When the output voltage falls below a point set by RV1, Q2 turns off, Q1 turns off and further discharge of the battery is prevented.

Fig. 12-2

BATTERY CONDITION INDICATOR

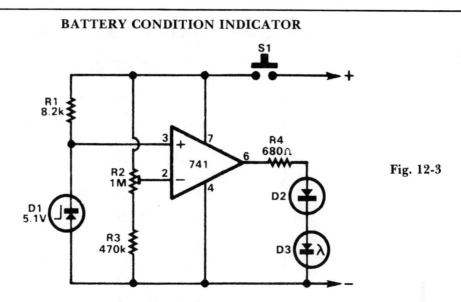

Fig. 12-3

Circuit Notes

A 741 op amp is employed as a voltage comparator. The noninverting input is connected to zener reference source. Reference voltage is 5.1V. R2 is adjusted so that the voltage at the inverting input is half the supply voltage. When supply is higher than 10.2V, the LED will not light. When the supply falls just fractionally below the 10.2V level, the IC inverting input will be slightly negative of the noninverting input, and the output will swing fully positive. The LED will light, indicating that the supply voltage has fallen to the preset threshold level. The LED can be made to light at other voltages by adjusting R2.

EQUIPMENT ON REMINDER

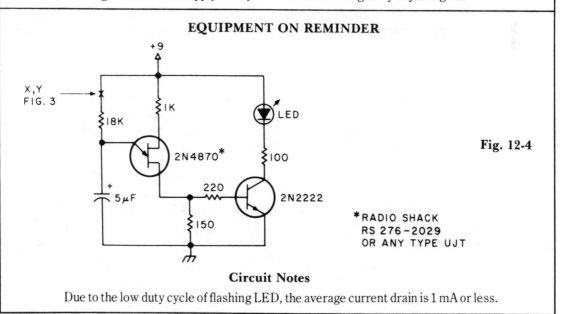

Fig. 12-4

*RADIO SHACK
RS 276-2029
OR ANY TYPE UJT

Circuit Notes

Due to the low duty cycle of flashing LED, the average current drain is 1 mA or less.

BATTERY CHARGE/DISCHARGE INDICATOR

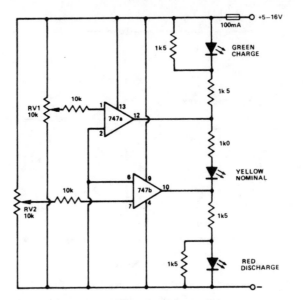

Fig. 12-5

Circuit Notes

This circuit monitors car battery voltage. It provides an indication of nominal supply voltage as well as low or high voltage. RV1 and RV2 adjust the point at which the red/yellow and yellow/green LEDs are on or off. For example the red LED comes on at 11V, and the green LED at 12V. The yellow LED is on between these values.

PRECISION BATTERY VOLTAGE MONITOR FOR HTS

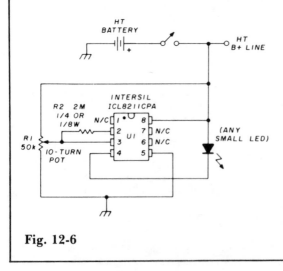

Fig. 12-6

Circuit Notes

The precision voltage-monitor chip contains a temperature-compensated voltage reference. R1 divides down the battery voltage to match the built-in reference voltage of IC1 (1.15 volts). When the voltage at pin 3 falls below 1.15 volts, pin 4 supplies a constant current of 7 mA to drive a small LED. About 0.2 volt of hysteresis is added with R2. Without hysteresis, the LED could flicker on and off when the monitored voltage varies around the set point, as might be the case on voice peaks during receive.

LOW-VOLTAGE MONITOR

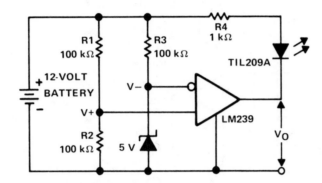

Fig. 12-7

a. SCHEMATIC OF CIRCUIT FOR LOW-VOLTAGE INDICATOR

Circuit Notes

This circuit monitors the voltage of a battery and warns the operator when the battery voltage is below a preset level by turning on an LED. The values are set for a 12V automobile battery. The preset value is 10 volts.

UNDERVOLTAGE INDICATOR FOR BATTERY OPERATED EQUIPMENT

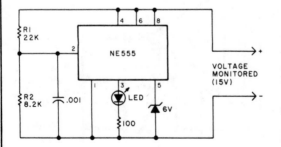

Fig. 12-8

Circuit Notes

Due to the low duty cycle of flashing LED, the average current drain is 1 mA or less. The NE555 will trigger the LED on when the monitored voltage falls to 12 volts. The ratio of R1 to R2 only needs to be changed if it is desired to change the voltage point at which the LED is triggered.

LOW BATTERY INDICATOR

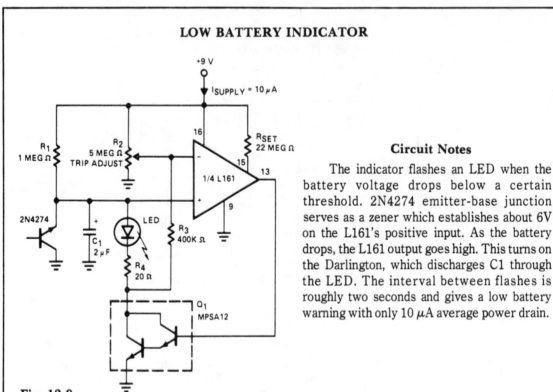

Circuit Notes

The indicator flashes an LED when the battery voltage drops below a certain threshold. 2N4274 emitter-base junction serves as a zener which establishes about 6V on the L161's positive input. As the battery drops, the L161 output goes high. This turns on the Darlington, which discharges C1 through the LED. The interval between flashes is roughly two seconds and gives a low battery warning with only 10 μA average power drain.

Fig. 12-9

BATTERY-LEVEL INDICATOR

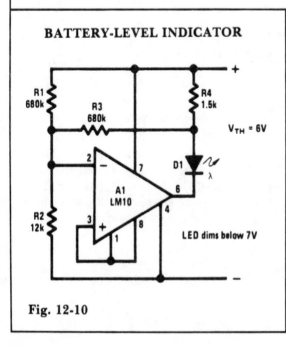

Fig. 12-10

BATTERY-THRESHOLD INDICATOR

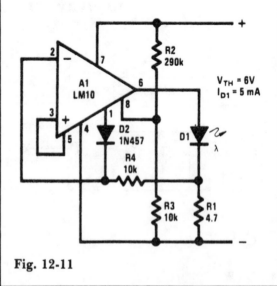

Fig. 12-11

124

13

Buffers

The sources of the following circuits are contained in the Sources section beginning on page 730. The figure number contained in the box of each circuit correlates to the source entry in the Sources section.

Sine Wave Output Buffer Amplifier
Single-Supply AC Buffer Amplifier
Single-Supply AC Buffer
High-Speed 6-Bit A/D Buffer
High Impedance, Low Capacitance

Wideband Buffer
High Resolution ADC Input Buffer
100 × Buffer Amplifier
10 × Buffer Amplifier
Stable High Impedance Buffer

High-Speed Single Supply AC Buffer

SINE WAVE OUTPUT BUFFER AMPLIFIER

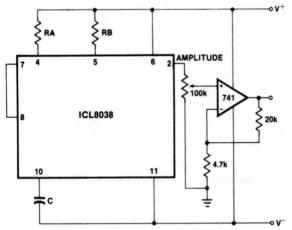

Circuit Notes

The sine wave output has a relatively high output impedance (1K typ). The circuit provides buffering, gain, and amplitude adjustment. A simple op amp follower could also be used.

Fig. 13-1

SINGLE SUPPLY AC BUFFER AMPLIFIER

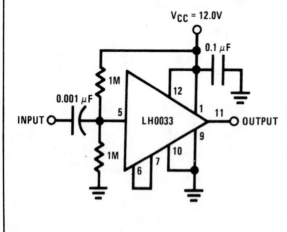

Circuit Notes

The input is dc biased to mid-operating point and is ac coupled. Its input impedance is approximately 500K at low frequencies. For dc loads referenced to ground, the quiescent current is increased by the load current set at the input dc bias voltage.

Fig. 13-2

126

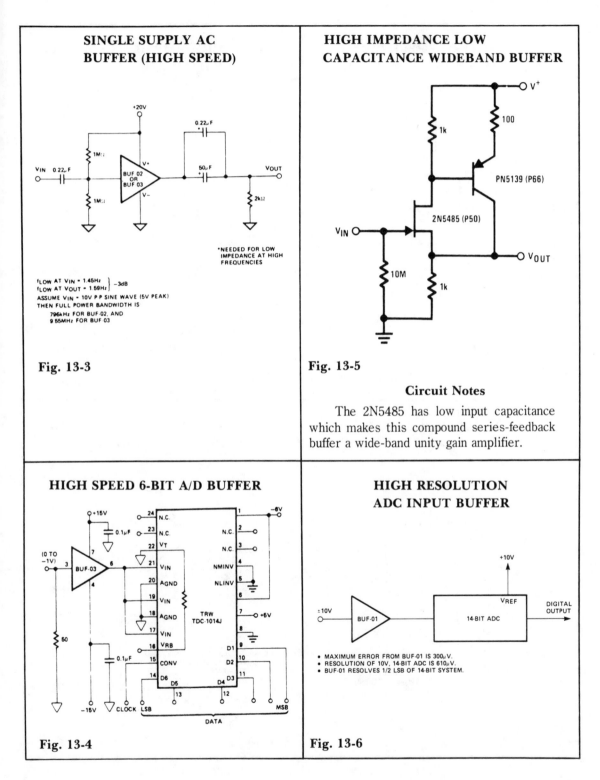

SINGLE SUPPLY AC BUFFER (HIGH SPEED)

+20V

0.22µF *

1M Ω

VIN 0.22µF

BUF 02 OR BUF 03

V+

V−

1M Ω

50µF

VOUT

2kΩ

*NEEDED FOR LOW IMPEDANCE AT HIGH FREQUENCIES

fLOW AT VIN = 1.45Hz } −3dB
fLOW AT VOUT = 1.59Hz
ASSUME VIN = 10V P-P SINE WAVE (5V PEAK)
THEN FULL POWER BANDWIDTH IS
    796kHz FOR BUF-02, AND
    9.55MHz FOR BUF-03

Fig. 13-3

HIGH IMPEDANCE LOW CAPACITANCE WIDEBAND BUFFER

V+

1k

100

PN5139 (P66)

2N5485 (P50)

VIN

VOUT

10M

1k

1k

Fig. 13-5

Circuit Notes

The 2N5485 has low input capacitance which makes this compound series-feedback buffer a wide-band unity gain amplifier.

HIGH SPEED 6-BIT A/D BUFFER

+15V

0.1µF

(0 TO −1V)

BUF-03

50

0.1µF

−15V CLOCK LSB

24 N.C.
23 N.C.
22 VT
21 VIN
20 AGND
19 VIN
18 AGND
17 VIN
16 VRB
15 CONV
14 D6 D5

TRW
TDC-1014J

−6V
1
N.C. 2
N.C. 3
NMINV 4
NLINV 5
6
7 +5V
8
D1 9
D2 10
D3 11
D4 12
13

MSB

DATA

Fig. 13-4

HIGH RESOLUTION ADC INPUT BUFFER

±10V

BUF-01

+10V

VREF

14-BIT ADC

DIGITAL OUTPUT

• MAXIMUM ERROR FROM BUF-01 IS 300µV.
• RESOLUTION OF 10V, 14-BIT ADC IS 610µV.
• BUF-01 RESOLVES 1/2 LSB OF 14-BIT SYSTEM.

Fig. 13-6

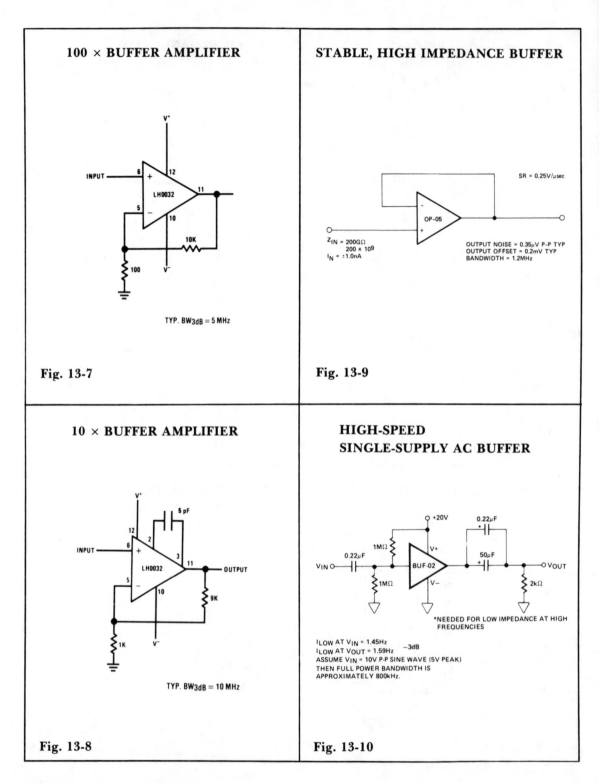

100 × BUFFER AMPLIFIER

LH0032

TYP. BW$_{3dB}$ = 5 MHz

Fig. 13-7

STABLE, HIGH IMPEDANCE BUFFER

OP-05

SR = 0.25V/μsec

Z_{IN} = 200GΩ
200 × 10^9
I_N = ±1.0nA

OUTPUT NOISE = 0.35μV P-P TYP
OUTPUT OFFSET = 0.2mV TYP
BANDWIDTH = 1.2MHz

Fig. 13-9

10 × BUFFER AMPLIFIER

LH0032

5 pF

9K

1K

TYP. BW$_{3dB}$ = 10 MHz

Fig. 13-8

HIGH-SPEED
SINGLE-SUPPLY AC BUFFER

+20V

0.22μF

1MΩ

V+

BUF-02

V–

0.22μF

V$_{IN}$

1MΩ

50μF

2kΩ

V$_{OUT}$

*NEEDED FOR LOW IMPEDANCE AT HIGH
FREQUENCIES

I_{LOW} AT V$_{IN}$ = 1.45Hz
I_{LOW} AT V$_{OUT}$ = 1.59Hz   −3dB
ASSUME V$_{IN}$ = 10V P-P SINE WAVE (5V PEAK)
THEN FULL POWER BANDWIDTH IS
APPROXIMATELY 800kHz.

Fig. 13-10

14

Capacitance (Touch) Operated Circuits

The sources of the following circuits are contained in the Sources section beginning on page 730. The figure number contained in the box of each circuit correlates to the source entry in the Sources section.

Capacitance Relay
Capacitance Operated, Battery Powered Light
Touch Sensitive Switch
Low Current Touch Switch
Capacitance Switched Light
Momentary Operation Touch Switch
Touch Triggered Bistable
Capacitance Operated Alarm to Foil Purse
    Snatchers

Self-Biased Proximity Sensor Works on De-
    tected Changing Fields
Touch Switch or Proximity Detector
Finger Touch Touch or Control Switch
Proximity Detector
Touch Circuit
CMOS Touch Switch
Latching Double-Button Touch
    Switch

CAPACITANCE RELAY

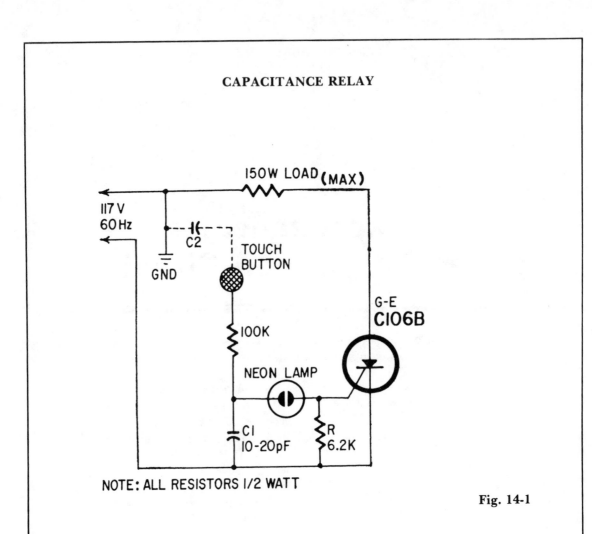

Fig. 14-1

Circuit Notes

Capacitor C1 and body capacitance (C2) of the operator form the voltage divider from the hot side of the ac line to ground. The voltage across C1 is determined by the ratio of C1 to C2. The higher voltage is developed across the smaller capacitor. When no one is close to the touch button, C2 is smaller than C1. When a hand is brought close to the button, C2 is many times larger than C1 and the major portion of the line voltage appears across C1. This voltage fires the neon lamp, C1 and C2 discharge through the SCR gate, causing it to trigger and pass current through the load. The sensitivity of the circuit depends on the area of the touch plate. When the area is large enough, the circuit responds to the proximity of an object rather than to touch. C1 may be made variable so sensitivity can be adjusted.

CAPACITANCE OPERATED, BATTERY POWERED LIGHT

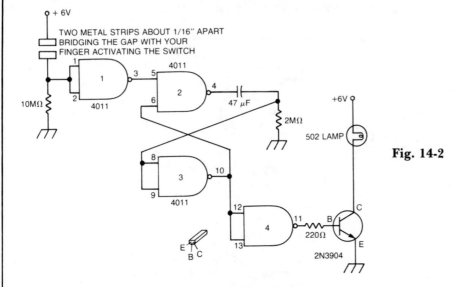

Fig. 14-2

Circuit Notes

Touch the plate and the light will go on and remain on for a time determined by the time constant of the 47 μF capacitor and the 2M resistor.

TOUCH-SENSITIVE SWITCH

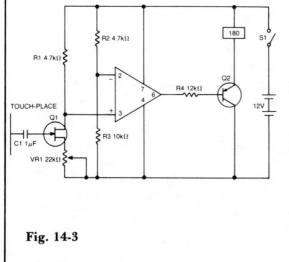

Fig. 14-3

Circuit Notes

A high impedance input is provided by Q1, a general purpose field effect transistor. 741 op amp is used as a sensitive voltage level switch which in turn operates the current Q2, a medium current PNP bipolar transistor, thereby energizing the relay which can be used to control equipment, alarms, etc.

LOW CURRENT TOUCH SWITCH

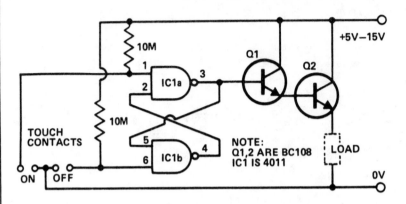

Fig. 14-4

Circuit Notes

Touching the on contacts with a finger brings pin 3 high, turning on the Darlington pair and supplying power to the load (transistor radio etc). Q1 must be a high gain transistor, and Q2 is chosen for the current required by the load circuit.

CAPACITANCE SWITCHED LIGHT

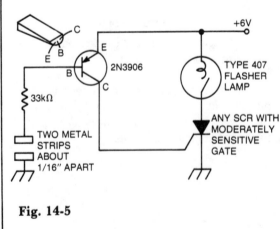

Fig. 14-5

Circuit Notes

The battery powered light turns on easily, stays on for just a few seconds, and then turns off again. The circuit is triggered when you place a finger across the gap between two strips of metal, about 1/16th inch apart. Enough current will flow through your finger to trigger the SCR after being amplified by the 2N3906. Once the SCR is fired, current will flow through the bulb until its internal bimetal switch turns it off. Once that happens, the SCR will return to its nonconducting state.

MOMENTARY OPERATION TOUCH SWITCH

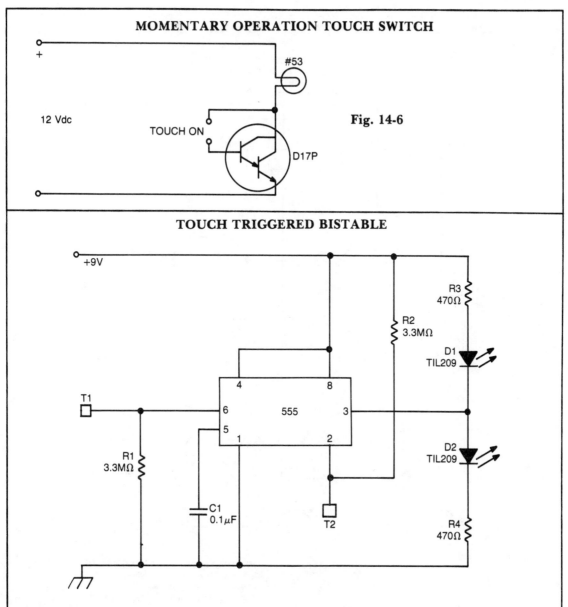

TOUCH TRIGGERED BISTABLE

Fig. 14-7

Circuit Notes

This circuit uses a 555 timer in the bistable mode. Touching T2 causes the output to go high; D2 conducts and D1 extinguishes. Touching T1 causes the output to go low; D1 conducts and D2 is cut off. The output from pin 3 can also be used to operate other circuits (e.g., a triac controlled lamp). In this case, the LEDs are useful for finding the touch terminals in the dark. C1 is not absolutely necessary but helps to prevent triggering from spurious pulses.

CAPACITANCE OPERATED ALARM TO FOIL PURSE SNATCHERS

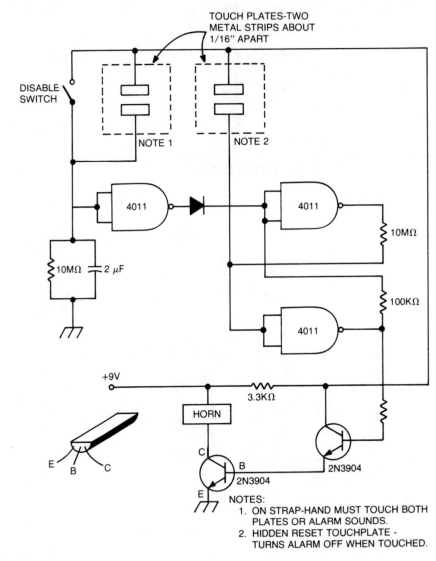

Fig. 14-8

Circuit Notes

As long as touch plates (1) are touched together, the alarm is off. If not held for about 30 seconds, the alarm goes off. The circuit can be disabled with switch or by touching the plates (2). The alarm is battery operated by a bicycle horn.

SELF-BIASED PROXIMITY SENSOR WORKS ON DETECTED CHANGING FIELD

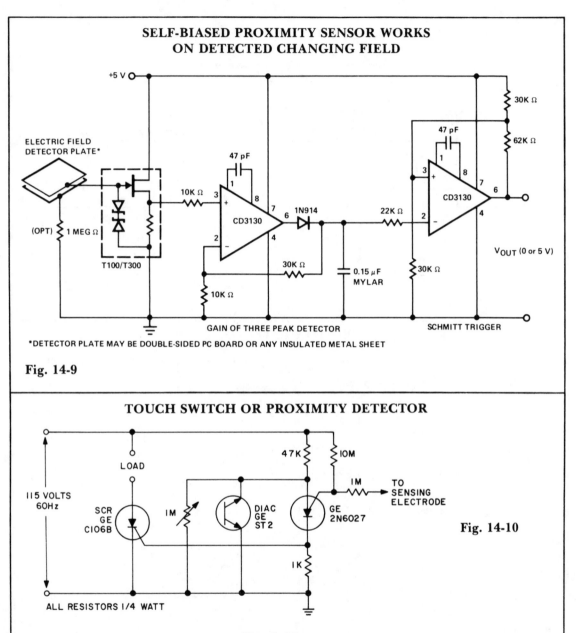

Fig. 14-9

*DETECTOR PLATE MAY BE DOUBLE-SIDED PC BOARD OR ANY INSULATED METAL SHEET

TOUCH SWITCH OR PROXIMITY DETECTOR

Fig. 14-10

Circuit Notes

This circuit is actuated by an increase in capacitance between a sensing electrode and the ground side of the line. The sensitivity can be adjusted to switch when a human body is within inches of the insulated plate used as the sensing electrode. Thus, sensitivity is adjusted with the 1 megohm potentiometer which determines the anode voltage level prior to clamping. This sensitivity will be proportional to the area of the surface opposing each other.

135

FINGER TOUCH OR CONTACT SWITCH

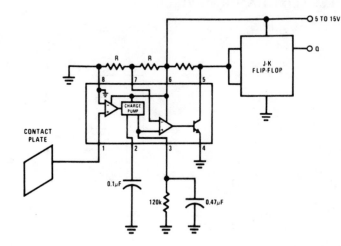

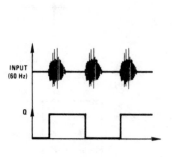

Fig. 14-11

PROXIMITY DETECTOR

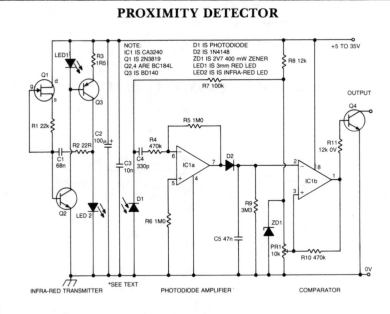

NOTE:
IC1 IS CA3240
Q1 IS 2N3819
Q2,4 ARE BC184L
Q3 IS BD140

D1 IS PHOTODIODE
D2 IS 1N4148
ZD1 IS 2V7 400 mW ZENER
LED1 IS 3mm RED LED
LED2 IS IS INFRA-RED LED

Fig. 14-12

Circuit Notes

The proximity sensor works on the principle of transmitting a beam of modulated infra-red light from the emitter diode LED2, and receiving reflections from objects passing in front of the beam with a photodiode detector D1. The circuit can be split into three distinct stages; the infra-red transmitter, the photodiode amplifier, and a variable threshold comparator.

TOUCH CIRCUIT

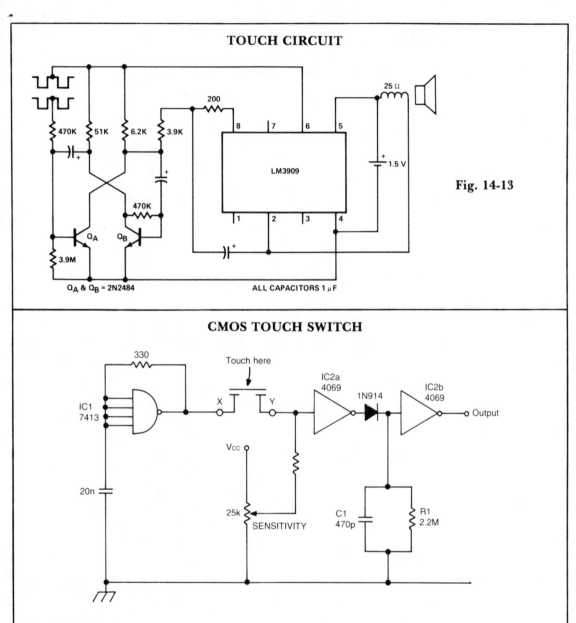

Fig. 14-13

Q_A & Q_B = 2N2484 ALL CAPACITORS 1 μF

CMOS TOUCH SWITCH

Fig. 14-14

Circuit Notes

This touch switch does not rely on mains hum for switching. It can be used with battery powered circuits. Schmitt trigger IC1 forms a 100 kHz oscillator and IC2a which is biased into the linear region, amplifies the output and charges C1 via the diode. IC2b acts as a level detector. When the sensor is touched, the oscillator signal is severely attenuated which causes C1 to discharge and IC2b to change state.

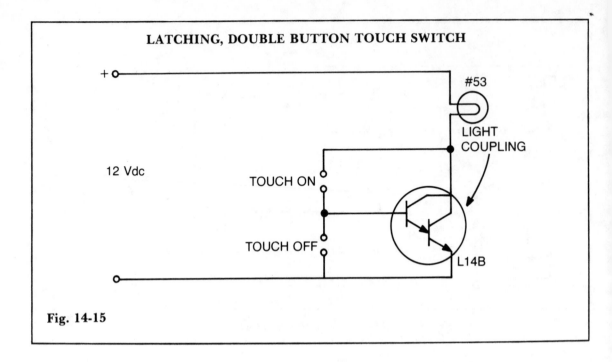

LATCHING, DOUBLE BUTTON TOUCH SWITCH

+ o———————————

#53

LIGHT
COUPLING

12 Vdc

TOUCH ON

TOUCH OFF

L14B

Fig. 14-15

15

Carrier Current Circuits

The sources of the following circuits are contained in the Sources section beginning on page 730. The figure number contained in the box of each circuit correlates to the source entry in the Sources section.

FM Carrier Current Remote Speaker
  System
200 kHz Line Carrier Transmitter with
  On/Off Carrier Modulation
Carrier Current Receiver
Carrier Current Transmitter

Carrier Current Transmitter
Integrated Circuit Current Transmitter
Single Transistor Carrier Current Receiver
IC Carrier-Current Receiver
Carrier-Current Remote Control or
    Intercom

Carrier System Transmitter

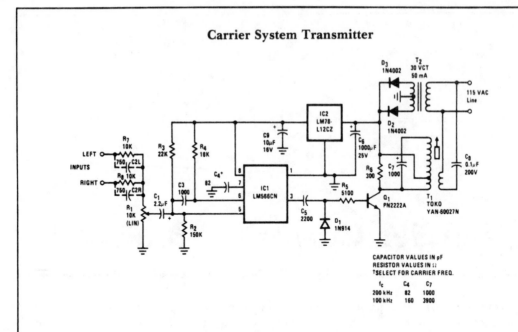

CAPACITOR VALUES IN pF
RESISTOR VALUES IN Ω
†SELECT FOR CARRIER FREQ.

f_c	C_4	C_7
200 kHz	82	1000
100 kHz	160	3900

Carrier System Receiver

FM CARRIER CURRENT REMOTE SPEAKER SYSTEM

Circuit Notes

High quality, noise free, wireless FM transmitter/receiver operates over standard power lines. Complete system is suitable for high-quality transmission of speech or music, and will operate from any ac outlet anywhere on a one-acre homesite. Frequency response is 20-20, 000 Hz and THD is under ½%. Trans-mission distance along a power line is at least adequate to include all outlets in and around a suburban home and yard.

Two input terminals are provided so that both left and right signals of a stereo set may be combined for mono transmission to a single remote speaker if desired.

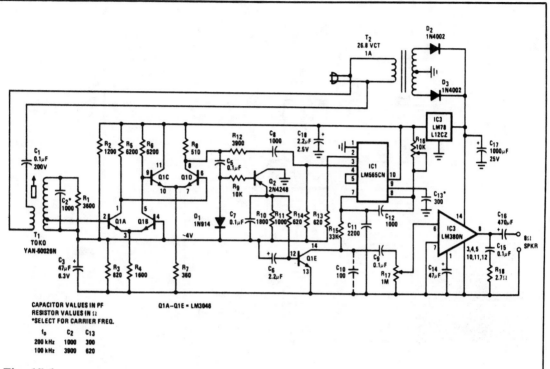

Fig. 15-1

The receiver amplifies, limits, and demodulates the received FM signal. It provides audio mute in the absence of carrier and 2.5 W output to a speaker.

200 kHz LINE CARRIER
TRANSMITTER WITH ON/OFF CARRIER MODULATOR

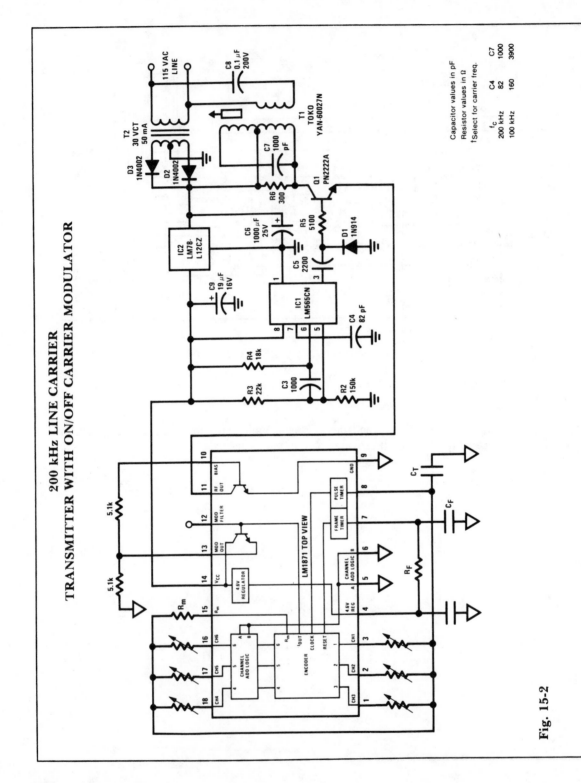

Fig. 15-2

CARRIER CURRENT RECEIVER

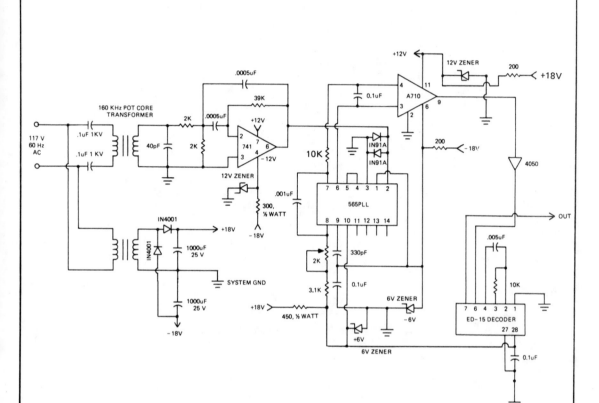

Fig. 15-3

Circuit Notes

160 kHz transformer consists of a 18 × 11mm ungapped pot core (Siemens, Ferrocube, etc.), utilizing magnetics incorporated type "F" material wound with 80½ turns of No. 35 wire for the secondary and 5½ turns for the primary. This gives a turns ratio of approximately 15 to 1.

CARRIER CURRENT TRANSMITTER

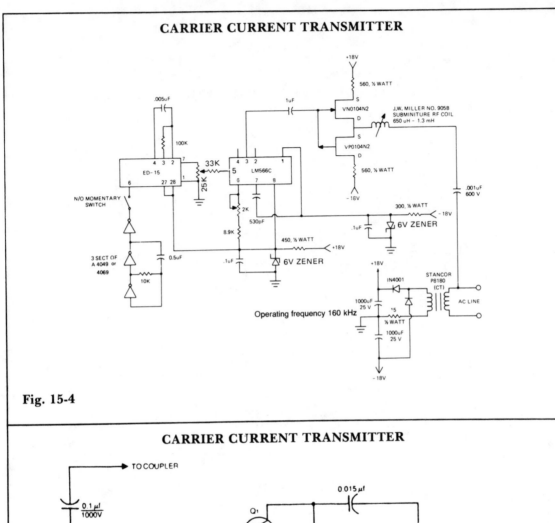

Operating frequency 160 kHz

Fig. 15-4

CARRIER CURRENT TRANSMITTER

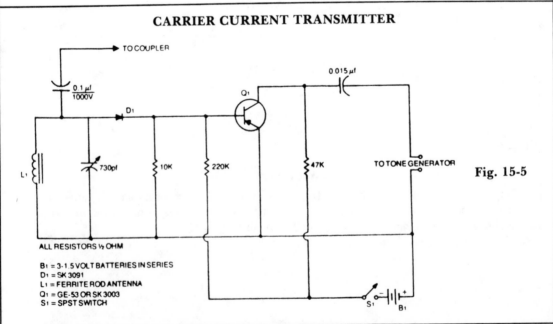

TO COUPLER

ALL RESISTORS ½ OHM

B₁ = 3-1.5 VOLT BATTERIES IN SERIES
D₁ = SK 3091
L₁ = FERRITE ROD ANTENNA
Q₁ = GE-53 OR SK 3003
S₁ = SPST SWITCH

Fig. 15-5

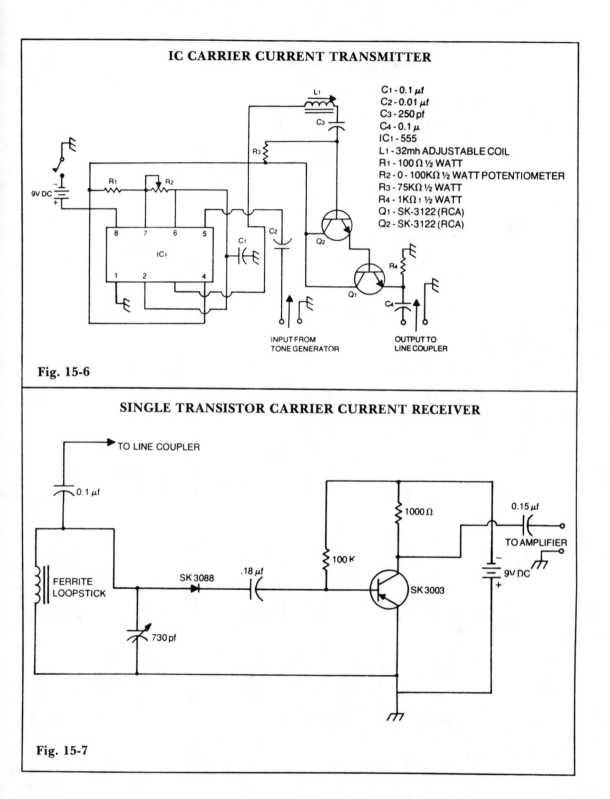

IC CARRIER CURRENT TRANSMITTER

C1 - 0.1 µf
C2 - 0.01 µf
C3 - 250 pf
C4 - 0.1 µ
IC1 - 555
L1 - 32mh ADJUSTABLE COIL
R1 - 100 Ω ½ WATT
R2 - 0 - 100KΩ ½ WATT POTENTIOMETER
R3 - 75KΩ ½ WATT
R4 - 1KΩ 1 ½ WATT
Q1 - SK-3122 (RCA)
Q2 - SK-3122 (RCA)

9V DC

INPUT FROM
TONE GENERATOR

OUTPUT TO
LINE COUPLER

Fig. 15-6

SINGLE TRANSISTOR CARRIER CURRENT RECEIVER

TO LINE COUPLER

0.1 µf

FERRITE
LOOPSTICK

730 pf

SK 3088

.18 µf

100 K

1000 Ω

SK 3003

0.15 µf

TO AMPLIFIER

9V DC

Fig. 15-7

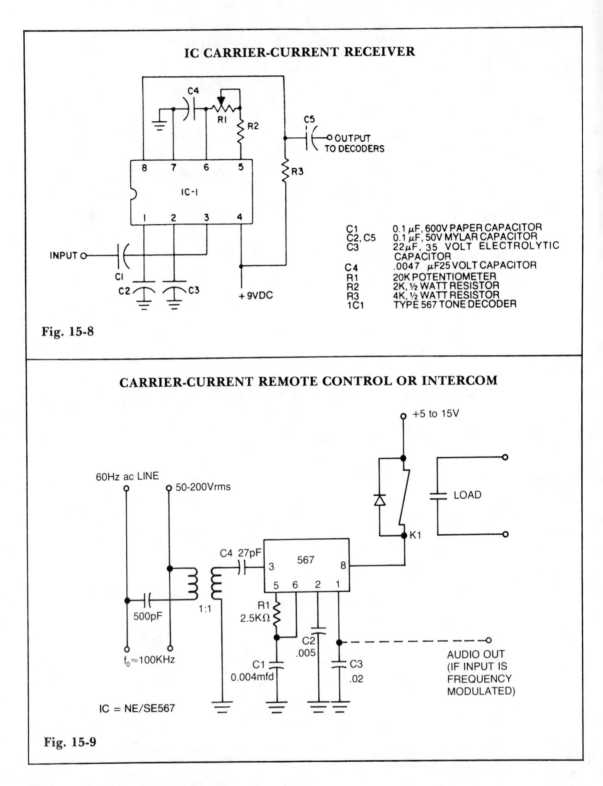

IC CARRIER-CURRENT RECEIVER

C1	0.1 μF, 600V PAPER CAPACITOR
C2, C5	0.1 μF, 50V MYLAR CAPACITOR
C3	22μF, 35 VOLT ELECTROLYTIC CAPACITOR
C4	.0047 μF 25 VOLT CAPACITOR
R1	20K POTENTIOMETER
R2	2K, ½ WATT RESISTOR
R3	4K, ½ WATT RESISTOR
1C1	TYPE 567 TONE DECODER

Fig. 15-8

CARRIER-CURRENT REMOTE CONTROL OR INTERCOM

Fig. 15-9

146

16

Comparators

The sources of the following circuits are contained in the Sources section beginning on page 730. The figure number contained in the box of each circuit correlates to the source entry in the Sources section.

Null Detector
Comparator with Variable Hysteresis
Diode Feedback Comparator
Undervoltage/Overvoltage Indicator
Dual Limit Comparator
High/Low Limit Alarm
Window Comparator
Window Comparator Driving High/Low Lamps
Comparator with Time Out
Noninverting Comparator with Hysteresis
Inverting Comparator with Hysteresis

Window Comparator
Micropower Double-Ended Limit Detector
Opposite Polarity Input Voltage Comparator
Limit Comparator
Comparator Clock Circuit
Double-Ended Limit Comparator
Limit Comparator
Precision, Dual Limit Go/No Go Tester
Comparator with Hysteresis
High Impedance Comparator
Comparator

NULL DETECTOR

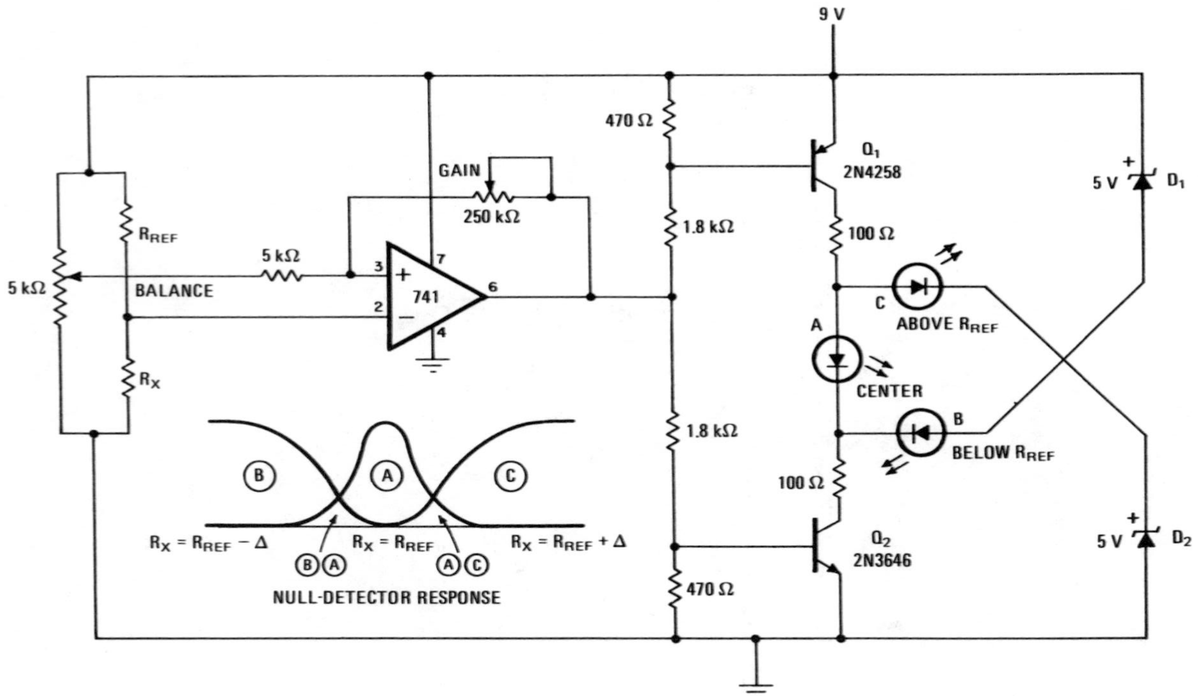

Fig. 16-1

Circuit Notes

Null detector uses simple LED readout to indicate if test resistor Rx is below, equal to, or greater than test resistance Rref. If Rx = Rref, the 741 output sits at midpoint value of 4.5 volts and LED A lights. Otherwise, the output of the 741 turns off one transistor and diverts current from the other transistor through B or C, depending on the polarity of the input voltage difference. Null-detector response is illustrated.

COMPARATOR WITH VARIABLE
HYSTERESIS (WITHOUT SHIFTING INITIAL TRIP POINT)

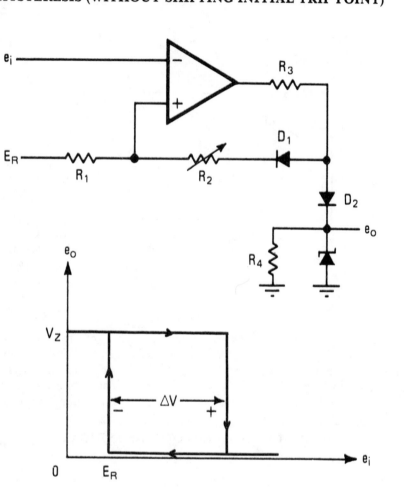

Fig. 16-2

Circuit Notes

An operational amplifier can be used as a convenient device for analog comparator applications that require two different trip points. The addition of a positive-feedback network introduces a precise variable hysteresis into the usual comparator switching action. Such feedback develops two comparator trip points centered about the initial trip point or reference point. The voltage difference, ΔV, between the trip points can be adjusted by varying resistor R2. When the output voltage is taken from the zener diode, as shown, it switches between zero and V_Z, the zener voltage.

DIODE FEEDBACK COMPARATOR

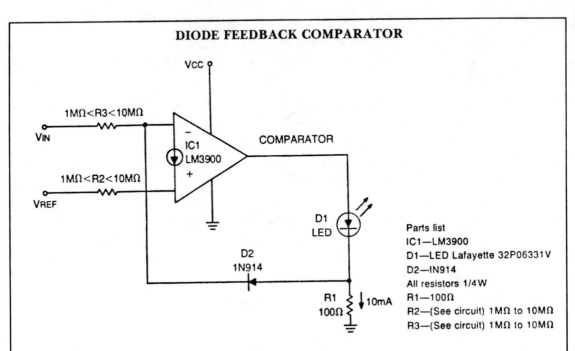

Fig. 16-3

Parts list
IC1—LM3900
D1—LED Lafayette 32P06331V
D2—IN914
All resistors 1/4W
R1—100Ω
R2—(See circuit) 1MΩ to 10MΩ
R3—(See circuit) 1MΩ to 10MΩ

Circuit Notes

This circuit can drive an LED display with constant current independently of wide power supply voltage changes. It can operate with a power supply range of at least 4V to 30V. With 10M resistances for R2 and R3 and the inverting input of the comparator grounded, the circuit becomes an LED driver with very high input impedance. The circuit can also be used in many other applications where a controllable constant current source is needed.

UNDERVOLTAGE/OVERVOLTAGE INDICATOR

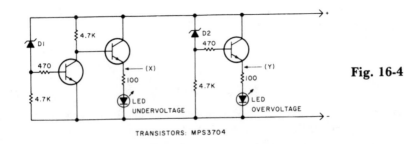

Fig. 16-4

Circuit Notes

This circuit will make the appropriate LED glow if the monitored voltage goes below or above the value determined by zener diodes D1 and D2.

DUAL LIMIT COMPARATOR

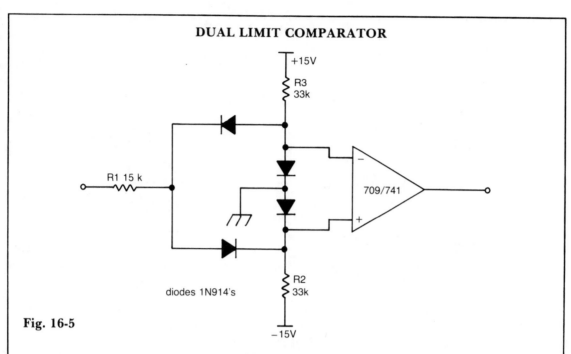

Fig. 16-5

Circuit Notes

This circuit gives a positive output when the input voltage exceeds 8.5 volts. Between these limits the output is negative. The positive limit point is determined by the ratio of R1, R2, and the negative point by R1, R3. The forward voltage drop across the diodes must be allowed for. The output may be inverted by reversing the inputs to the op amp. The 709 is used without frequency compensation.

HIGH/LOW LIMIT ALARM

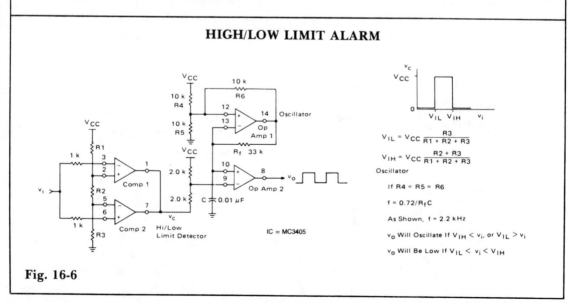

Fig. 16-6

WINDOW COMPARATOR

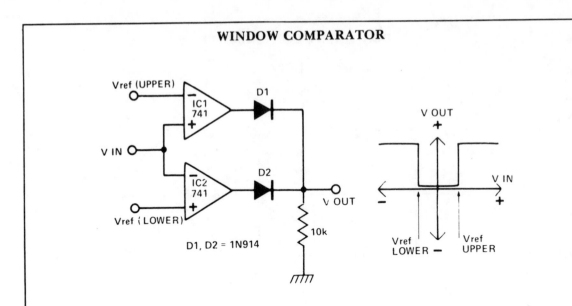

Fig. 16-7

Circuit Notes

This circuit gives an output (which in this case is 0V) when an input voltage lies in between two specified voltages. When it is outside this window, the output is positive. The two op amps are used as voltage comparators. When Vin is more positive than Vref (upper) the output of IC1 is positive and D1 is forward biased. Otherwise the output is negative, D1 reverse biased and hence Vout is 0V. Similarly, when Vin is more negative than Vref (lower), the output of IC2 is positive; D2 is forward biased and this Vout is positive. Otherwise Vout is 0V. When Vin lies within the window set by the reference voltages, Vout is 0V.

WINDOW COMPARATOR DRIVING HIGH/LOW LAMPS

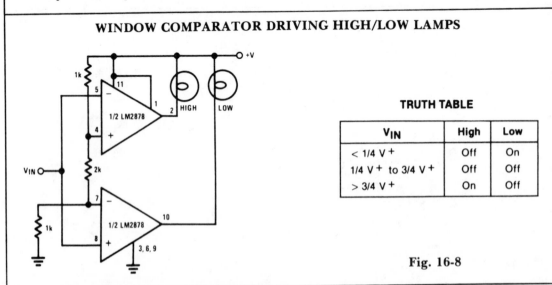

TRUTH TABLE

V_{IN}	High	Low
$< 1/4\ V^+$	Off	On
$1/4\ V^+$ to $3/4\ V^+$	Off	Off
$> 3/4\ V^+$	On	Off

Fig. 16-8

COMPARATOR WITH TIME OUT

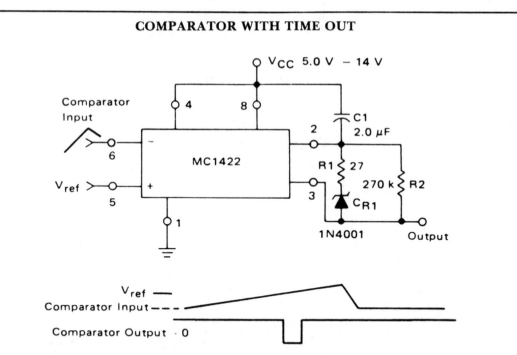

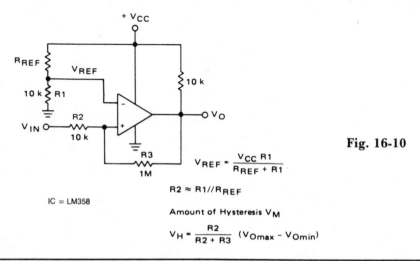

Fig. 16-9

Circuit Notes

The MC1422 is used as a comparator with the capability of a timing output pulse when the inverting input (Pin 6) is ≥ the noninverting input (Pin 5). The frequency of the pulses for the values of R2 and C1 as shown is approximately 2.0 Hz, and the pulse width 0.3 ms.

NONINVERTING COMPARATOR WITH HYSTERESIS

Fig. 16-10

$$V_{REF} = \frac{V_{CC} \, R1}{R_{REF} + R1}$$

$$R2 \approx R1 // R_{REF}$$

Amount of Hysteresis V_M

$$V_H = \frac{R2}{R2 + R3} \, (V_{Omax} - V_{Omin})$$

IC = LM358

153

INVERTING COMPARATOR WITH HYSTERESIS

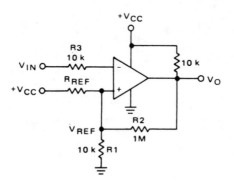

$$V_{REF} \approx \frac{V_{CC}\,R1}{R_{REF} + R1}$$

$$R3 \simeq R1 \,//\, R_{REF} \,//\, R1$$

$$V_H = \frac{R1//R_{REF}}{R1//R_{REF} + R2}\ (V_{Omax} - V_{Omin})$$

Fig. 16-11

WINDOW COMPARATOR

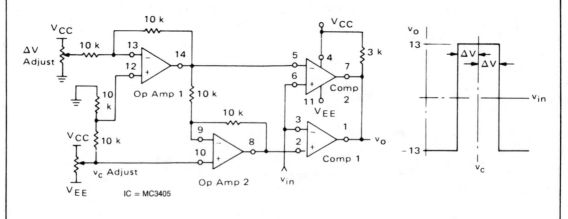

Fig. 16-12

MICROPOWER DOUBLE-ENDED LIMIT DETECTOR

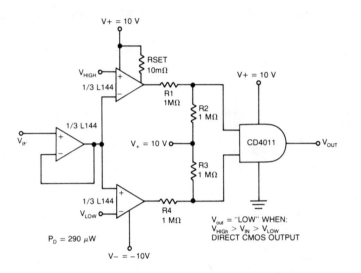

Fig. 16-13

Circuit Notes

The detector uses three sections of an L144 and a DC4011 type CMOS NAND gate to make a very low power voltage monitor. If the input voltage, V_{IN}, is above V_{HIGH} or below V_{LOW}, the output will be a logical high. If (and only if) the input is between the limits will the output be low. The 1 megohm resistors R1, R2, R3, and R4 translate the bipolar ±10V swing of the op amps to a 0 to 10V swing acceptable to the ground-referenced CMOS logic.

OPPOSITE POLARITY INPUT VOLTAGE COMPARATOR

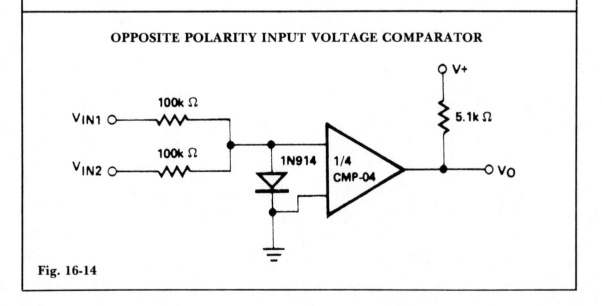

Fig. 16-14

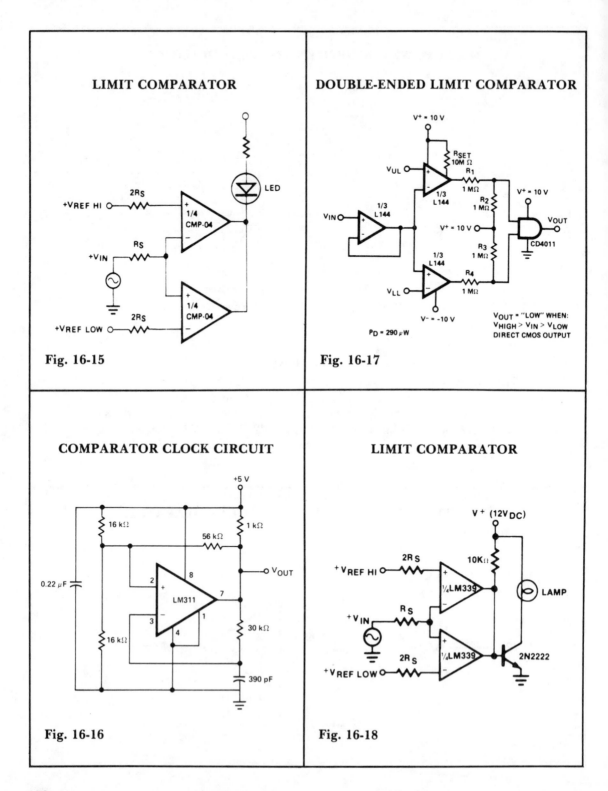

LIMIT COMPARATOR

Fig. 16-15

DOUBLE-ENDED LIMIT COMPARATOR

$V^+ = 10\ V$

R_{SET}
$10M\ \Omega$

V_{UL}

R_1
$1\ M\Omega$

$1/3$
$L144$

V_{IN}

$1/3$
$L144$

R_2
$1\ M\Omega$

$V^+ = 10\ V$

$V^+ = 10\ V$

V_{OUT}

R_3
$1\ M\Omega$

$CD4011$

$1/3$
$L144$

R_4
$1\ M\Omega$

V_{LL}

$V^- = -10\ V$

$P_D = 290\ \mu W$

V_{OUT} = "LOW" WHEN:
$V_{HIGH} > V_{IN} > V_{LOW}$
DIRECT CMOS OUTPUT

Fig. 16-17

COMPARATOR CLOCK CIRCUIT

$+5\ V$

$16\ k\Omega$

$1\ k\Omega$

$56\ k\Omega$

$0.22\ \mu F$

V_{OUT}

LM311

$16\ k\Omega$

$30\ k\Omega$

$390\ pF$

Fig. 16-16

LIMIT COMPARATOR

V^+ (12V$_{DC}$)

$2R_S$

$10K\Omega$

$+V_{REF\ HI}$

$\frac{1}{4}LM339$

LAMP

$+V_{IN}$

R_S

$\frac{1}{4}LM339$

2N2222

$2R_S$

$+V_{REF\ LOW}$

Fig. 16-18

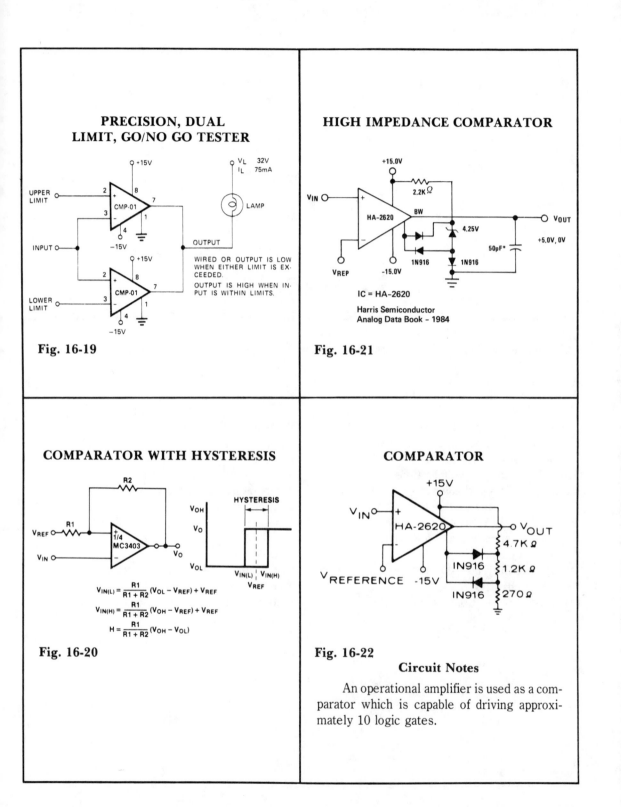

PRECISION, DUAL LIMIT, GO/NO GO TESTER

UPPER LIMIT

INPUT

LOWER LIMIT

CMP-01

CMP-01

+15V

-15V

+15V

-15V

V_L 32V
I_L 75mA

LAMP

OUTPUT

WIRED OR OUTPUT IS LOW WHEN EITHER LIMIT IS EXCEEDED.

OUTPUT IS HIGH WHEN INPUT IS WITHIN LIMITS.

Fig. 16-19

HIGH IMPEDANCE COMPARATOR

+15.0V

V_{IN}

HA-2620

2.2K Ω

BW

4.25V

50pF*

1N916   1N916

V_{REF} -15.0V

V_{OUT}

+5.0V, 0V

IC = HA-2620

Harris Semiconductor
Analog Data Book – 1984

Fig. 16-21

COMPARATOR WITH HYSTERESIS

R2

R1

V_{REF}

V_{IN}

1/4
MC3403

V_O

HYSTERESIS

V_{OH}

V_O

V_{OL}

$V_{IN(L)}$ $V_{IN(H)}$
V_{REF}

$$V_{IN(L)} = \frac{R1}{R1 + R2}(V_{OL} - V_{REF}) + V_{REF}$$

$$V_{IN(H)} = \frac{R1}{R1 + R2}(V_{OH} - V_{REF}) + V_{REF}$$

$$H = \frac{R1}{R1 + R2}(V_{OH} - V_{OL})$$

Fig. 16-20

COMPARATOR

+15V

V_{IN}

HA-2620

$V_{REFERENCE}$ -15V

V_{OUT}

4.7K Ω

1N916

1.2K Ω

1N916

270 Ω

Fig. 16-22

Circuit Notes

An operational amplifier is used as a comparator which is capable of driving approximately 10 logic gates.

17

Converters

The sources of the following circuits are contained in the Sources section beginning on page 730. The figure number contained in the box of each circuit correlates to the source entry in the Sources section.

Picoampere-to-Frequency Converter
BCD-to-Analog Converter
Resistance-to-Voltage Converter
Low Cost, μP Interfaced, Temperature-to-Digital Converter
Hi-Lo Resistance-to-Voltage Converter
Current-to-Voltage Converter
Calculator-to-Stopwatch Converter
Power Voltage-to-Current Converter
High Impedance Precision Rectifier for Ac/Dc Converter
Wide Range Current-to-Frequency Converter
Ac-to-Dc Converter
Current-to-Voltage Converter with 1% Accuracy

Polarity Converter
Voltage-to-Current Converter
Wideband, High-Crest Factor, RMS-to-Dc Converter
Light Intensity-to-Frequency Converter
Ohms-to-Volts Converter
Temperature-to-Frequency Converter
Multiplexed BCD-to-Parallel BCD Converter
Fast Logarithmic Converter
Sine Wave-to-Square Wave Converter
Self Oscillating Flyback Converter
TTL-to-MOS Logic Converter
Picoampere-to-Voltage Converter with Gain

PICOAMPERE-TO-FREQUENCY CONVERTERS

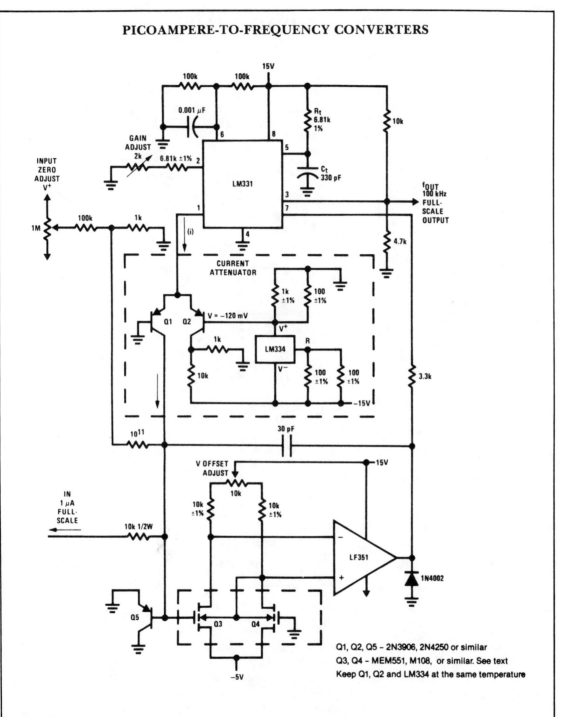

Fig. 17-1

BCD-TO-ANALOG CONVERTER

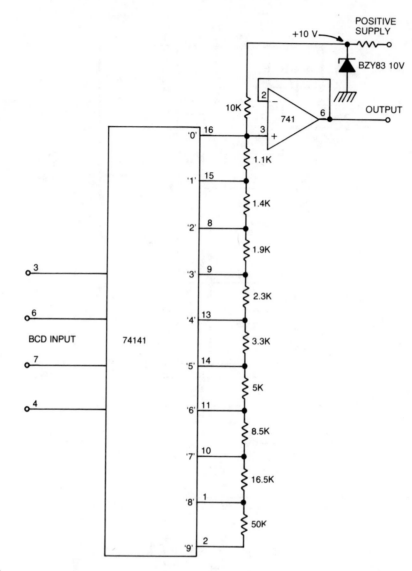

Fig. 17-2

Circuit Notes

This circuit will convert four-bit BCD into a variable voltage from 0-9 V in 1 V steps. The SN74141 is a Nixie driver, and has ten open-collector outputs. These are used to ground a selected point in the divider chain determined by the BCD code at the input, and so produce a corresponding voltage at the output. Accuracy of the circuit depends on the tolerance of the resistors and the accuracy of the reference voltage. However, presets can be used in the divider chain, with correct calibration. The 741 is used as a buffer.

RESISTANCE-TO-VOLTAGE CONVERTER

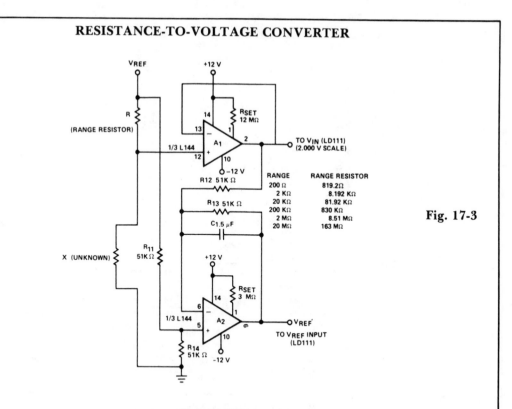

RANGE	RANGE RESISTOR
200 Ω	819.2Ω
2 KΩ	8.192 KΩ
20 KΩ	81.92 KΩ
200 KΩ	830 KΩ
2 MΩ	8.51 MΩ
20 MΩ	163 MΩ

Fig. 17-3

Circuit Notes

Circuit will measure accurately to 20 M when associated with a buffer amplifier (A1) having a low input bias current (I_{IN}) < 30 nA). The circuit uses two of the three amplifiers contained in the Siliconix L144 micropower triple op amp.

LOW-COST, μP INTERFACED, TEMPERATURE-TO-DIGITAL CONVERTER

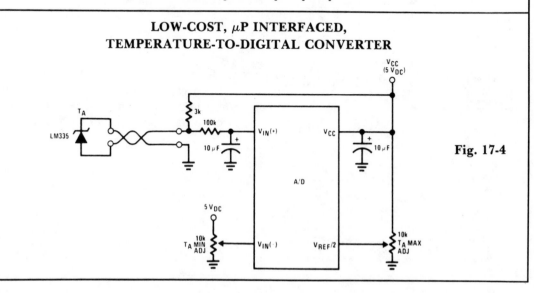

Fig. 17-4

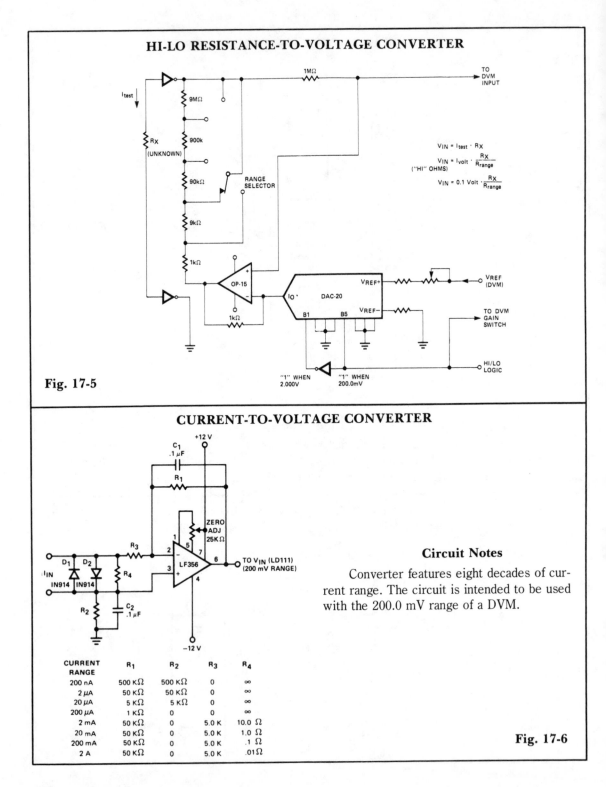

HI-LO RESISTANCE-TO-VOLTAGE CONVERTER

$V_{IN} = I_{test} \cdot R_X$

$V_{IN} = I_{volt} \cdot \dfrac{R_X}{R_{range}}$ ("HI" OHMS)

$V_{IN} = 0.1 \text{ Volt} \cdot \dfrac{R_X}{R_{range}}$

Fig. 17-5

CURRENT-TO-VOLTAGE CONVERTER

Circuit Notes

Converter features eight decades of current range. The circuit is intended to be used with the 200.0 mV range of a DVM.

CURRENT RANGE	R_1	R_2	R_3	R_4
200 nA	500 KΩ	500 KΩ	0	∞
2 µA	50 KΩ	50 KΩ	0	∞
20 µA	5 KΩ	5 KΩ	0	∞
200 µA	1 KΩ	0	0	∞
2 mA	50 KΩ	0	5.0 K	10.0 Ω
20 mA	50 KΩ	0	5.0 K	1.0 Ω
200 mA	50 KΩ	0	5.0 K	.1 Ω
2 A	50 KΩ	0	5.0 K	.01Ω

Fig. 17-6

162

CALCULATOR-TO-STOPWATCH CONVERTER

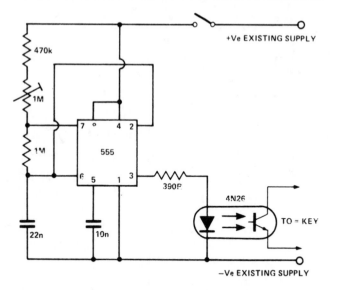

Fig. 17-7

Circuit Notes

This circuit can be fitted to any calculator with an automatic constant to enable it to be used as a stop-watch. The 555 timer is set to run at a suitable frequency and connected to the existing calculator battery via the push-on push-off switch and the existing calculator on-off switch.

POWER VOLTAGE-TO-CURRENT CONVERTER

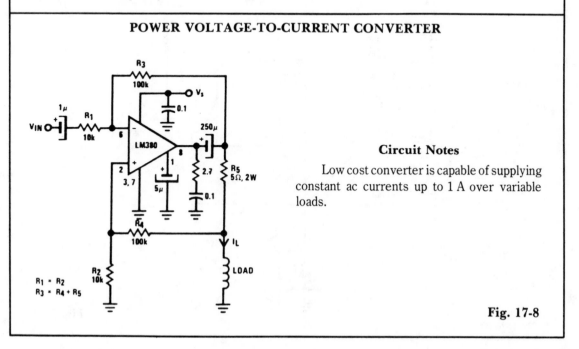

Circuit Notes

Low cost converter is capable of supplying constant ac currents up to 1 A over variable loads.

Fig. 17-8

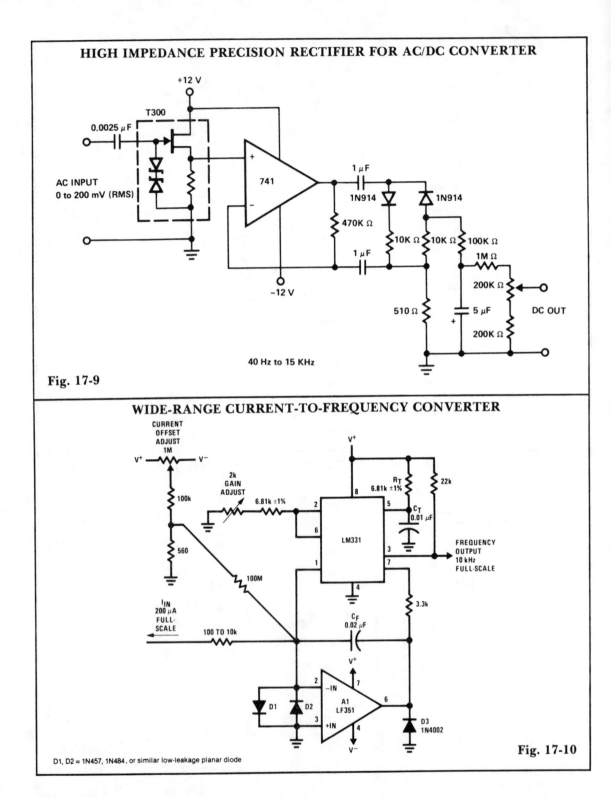

HIGH IMPEDANCE PRECISION RECTIFIER FOR AC/DC CONVERTER

+12 V

T300

0.0025 μF

AC INPUT
0 to 200 mV (RMS)

741

1 μF

1N914   1N914

470K Ω

1 μF

-12 V

10K Ω   10K Ω   100K Ω

1M Ω

200K Ω

DC OUT

510 Ω

5 μF

200K Ω

40 Hz to 15 KHz

Fig. 17-9

WIDE-RANGE CURRENT-TO-FREQUENCY CONVERTER

CURRENT
OFFSET
ADJUST
1M

V+   V−

V+

2k
GAIN
ADJUST

100k

6.81k ±1%

R_T
6.81k ±1%

22k

C_T
0.01 μF

2

8

5

560

LM331

6

3

FREQUENCY
OUTPUT
10 kHz
FULL-SCALE

1

7

100M

4

I_{IN}
200 μA
FULL-
SCALE

3.3k

100 TO 10k

C_F
0.02 μF

V+

2
−IN

7

A1
LF351

6

D1   D2

3
+IN

4

D3
1N4002

V−

D1, D2 = 1N457, 1N484, or similar low-leakage planar diode

Fig. 17-10

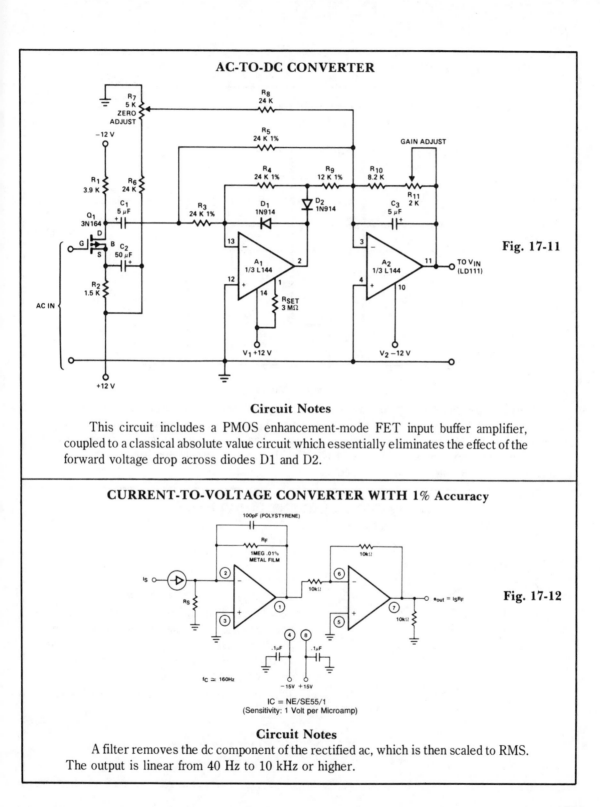

AC-TO-DC CONVERTER

Fig. 17-11

Circuit Notes

This circuit includes a PMOS enhancement-mode FET input buffer amplifier, coupled to a classical absolute value circuit which essentially eliminates the effect of the forward voltage drop across diodes D1 and D2.

CURRENT-TO-VOLTAGE CONVERTER WITH 1% Accuracy

Fig. 17-12

Circuit Notes

A filter removes the dc component of the rectified ac, which is then scaled to RMS. The output is linear from 40 Hz to 10 kHz or higher.

POLARITY CONVERTER

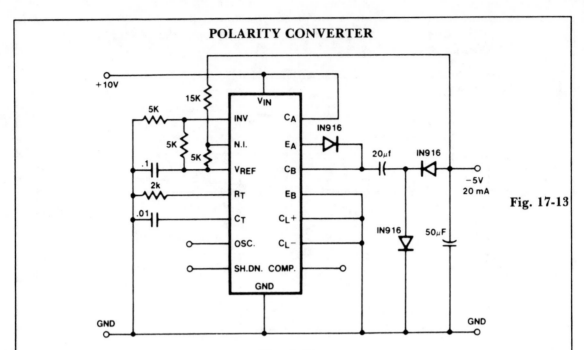

Fig. 17-13

Circuit Notes

The capacitor-diode output circuit is used here as a polarity converter to generate a −5 volt supply from +15 volts. This circuit is useful for an output current of up to 20 mA with no additional boost transistors required. Since the output transistors are current limited, no additional protection is necessary. Also, the lack of an inductor allows the circuit to be stabilized with only the output capacitor.

VOLTAGE-TO-CURRENT CONVERTER

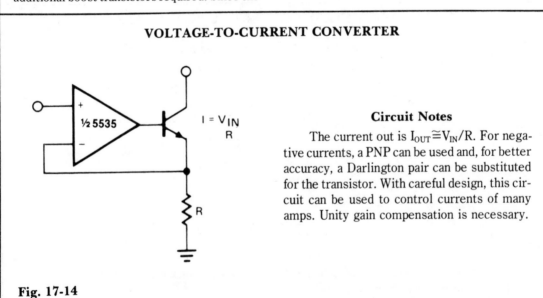

Fig. 17-14

Circuit Notes

The current out is $I_{OUT} \cong V_{IN}/R$. For negative currents, a PNP can be used and, for better accuracy, a Darlington pair can be substituted for the transistor. With careful design, this circuit can be used to control currents of many amps. Unity gain compensation is necessary.

WIDEBAND, HIGH-CREST FACTOR, RMS-TO-DC CONVERTER

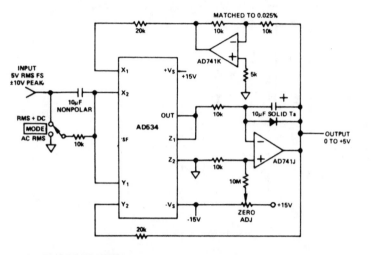

CALIBRATION PROCEDURE:

WITH 'MODE' SWITCH IN 'RMS + DC' POSITION, APPLY AN INPUT OF +1.00VDC. ADJUST ZERO UNTIL OUTPUT READS SAME AS INPUT. CHECK FOR INPUTS OF ±10V; OUTPUT SHOULD BE WITHIN ±0.05% (5mV).

ACCURACY IS MAINTAINED FROM 60Hz to 100kHz, AND IS TYPICALLY HIGH BY 0.5% AT 1MHz FOR V_{IN} = 4V RMS (SINE, SQUARE OR TRIANGULAR WAVE).

PROVIDED THAT THE PEAK INPUT IS NOT EXCEEDED, CREST-FACTORS UP TO AT LEAST TEN HAVE NO APPRECIABLE EFFECT ON ACCURACY.

INPUT IMPEDANCE IS ABOUT 10kΩ; FOR HIGH (10MΩ) IMPEDANCE, REMOVE MODE SWITCH AND INPUT COUPLING COMPONENTS.

FOR GUARANTEED SPECIFICATIONS THE AD536A AND AD636 IS OFFERED AS A SINGLE PACKAGE RMS-TO-DC CONVERTER.

Fig. 17-15

LIGHT INTENSITY-TO-FREQUENCY CONVERTER

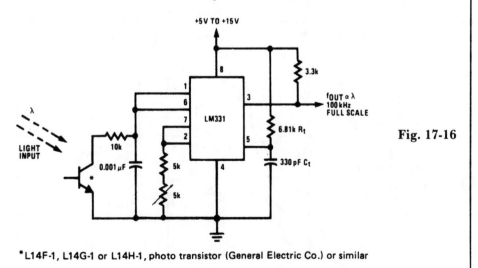

Fig. 17-16

*L14F-1, L14G-1 or L14H-1, photo transistor (General Electric Co.) or similar

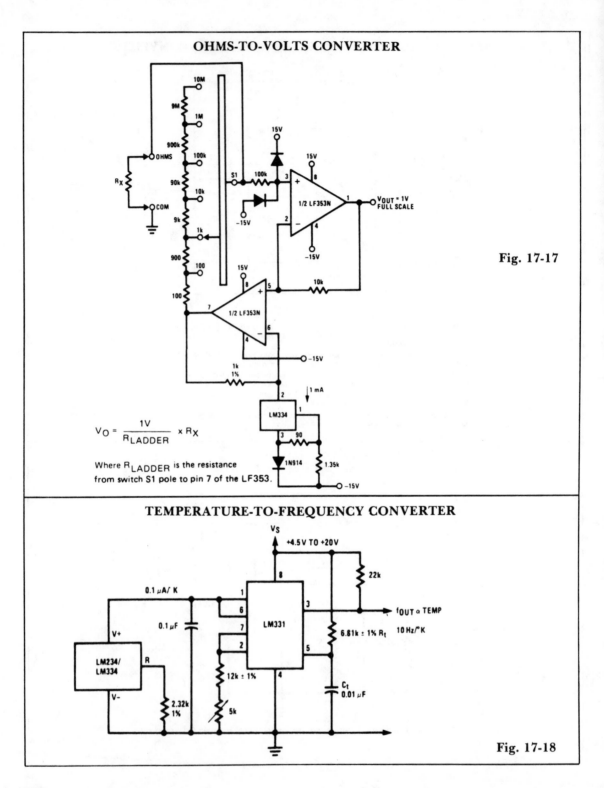

OHMS-TO-VOLTS CONVERTER

10M

9M

1M

900k

100k

90k

10k

9k

1k

900

100

100

OHMS

COM

R_X

S1

100k

15V

15V

1/2 LF353N

1

2

4

$V_{OUT} = 1V$
FULL SCALE

−15V

−15V

15V

10k

7

1/2 LF353N

5

+

6

−

4

−15V

1k
1%

2

1 mA

LM334

1

3

90

1N914

1.35k

−15V

$$V_O = \frac{1V}{R_{LADDER}} \times R_X$$

Where R_{LADDER} is the resistance
from switch S1 pole to pin 7 of the LF353.

Fig. 17-17

TEMPERATURE-TO-FREQUENCY CONVERTER

V_S

+4.5 V TO +20V

0.1 μA/ K

22k

8

1

6

7

2

LM331

3

$f_{OUT} \propto$ TEMP

10 Hz/°K

5

4

0.1 μF

V+

LM234/
LM334

R

V−

2.32k
1%

12k ± 1%

5k

6.81k ± 1% R_t

C_t
0.01 μF

Fig. 17-18

MULTIPLEXED BCD-TO-PARALLEL BCD CONVERTER

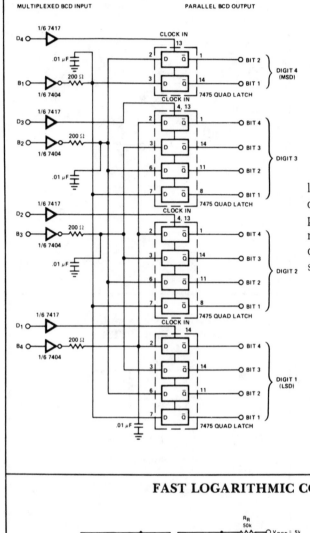

MULTIPLEXED BCD INPUT   PARALLEL BCD OUTPUT

Circuit Notes

Converter consists of four quad bistable latches activated in the proper sequence by the digit strobe output of the LD110. The complemented outputs (Q) of the quad latch set reflects the state of the bit outputs when the digit strobe goes high. It will maintain this state when the digit strobe goes low.

Fig. 17-19

FAST LOGARITHMIC CONVERTER

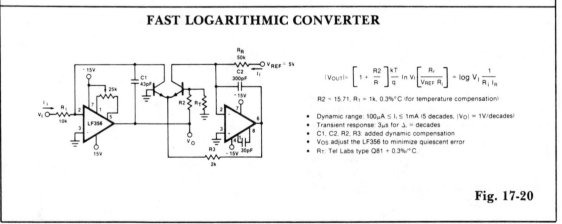

$$|V_{OUT}| = \left[1 + \frac{R2}{R}\right] \frac{kT}{q} \ln V_I \left[\frac{R_r}{V_{REF} R_1}\right] = \log V_I \frac{1}{R_i I_R}$$

R2 = 15.71, R_1 = 1k, 0.3%/°C (for temperature compensation)

- Dynamic range: $100\mu A \leq I_i \leq 1mA$ (5 decades, $|V_O|$ = 1V/decades)
- Transient response: $3\mu s$ for Δ. = decades
- C1, C2, R2, R3: added dynamic compensation
- V_{OS} adjust the LF356 to minimize quiescent error
- R_T: Tel Labs type Q81 + 0.3%/°C.

Fig. 17-20

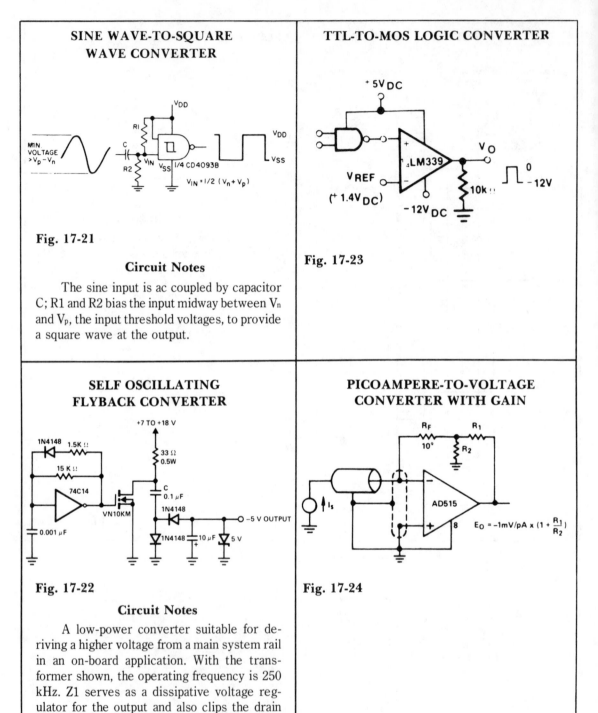

SINE WAVE-TO-SQUARE WAVE CONVERTER

Fig. 17-21

Circuit Notes

The sine input is ac coupled by capacitor C; R1 and R2 bias the input midway between V_n and V_p, the input threshold voltages, to provide a square wave at the output.

TTL-TO-MOS LOGIC CONVERTER

Fig. 17-23

SELF OSCILLATING FLYBACK CONVERTER

Fig. 17-22

Circuit Notes

A low-power converter suitable for deriving a higher voltage from a main system rail in an on-board application. With the transformer shown, the operating frequency is 250 kHz. Z1 serves as a dissipative voltage regulator for the output and also clips the drain voltage to a level below the rated VMOS breakdown voltage.

PICOAMPERE-TO-VOLTAGE CONVERTER WITH GAIN

Fig. 17-24

18

Crossover Networks

The sources of the following circuits are contained in the Sources section beginning on page 730. The figure number contained in the box of each circuit correlates to the source entry in the Sources section.

Active Crossover Network
Asymmetrical Third Order Butterworth
    Active Crossover Network

Third Order Butterworth Crossover
    Network

ACTIVE CROSSOVER NETWORK

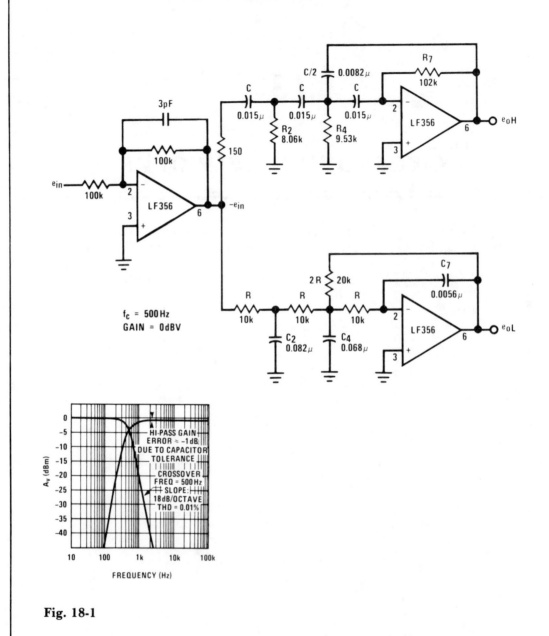

$f_c = 500\,Hz$
GAIN = 0dBV

Fig. 18-1

ASYMMETRICAL THIRD ORDER
BUTTERWORTH ACTIVE CROSSOVER NETWORK

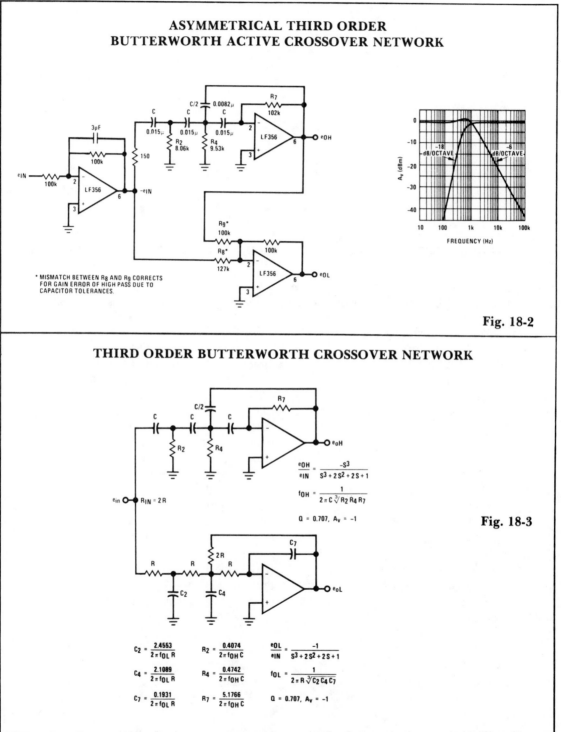

* MISMATCH BETWEEN R8 AND R9 CORRECTS
  FOR GAIN ERROR OF HIGH PASS DUE TO
  CAPACITOR TOLERANCES.

Fig. 18-2

THIRD ORDER BUTTERWORTH CROSSOVER NETWORK

$$\frac{e_{OH}}{e_{IN}} = \frac{-S^3}{S^3 + 2S^2 + 2S + 1}$$

$$f_{OH} = \frac{1}{2\pi C \sqrt[3]{R_2 R_4 R_7}}$$

$$Q = 0.707, \quad A_v = -1$$

Fig. 18-3

$$C_2 = \frac{2.4553}{2\pi f_{OL} R}$$ $$R_2 = \frac{0.4074}{2\pi f_{OH} C}$$ $$\frac{e_{OL}}{e_{IN}} = \frac{-1}{S^3 + 2S^2 + 2S + 1}$$

$$C_4 = \frac{2.1089}{2\pi f_{OL} R}$$ $$R_4 = \frac{0.4742}{2\pi f_{OH} C}$$ $$f_{OL} = \frac{1}{2\pi R \sqrt[3]{C_2 C_4 C_7}}$$

$$C_7 = \frac{0.1931}{2\pi f_{OL} R}$$ $$R_7 = \frac{5.1766}{2\pi f_{OH} C}$$ $$Q = 0.707, \quad A_v = -1$$

19

Crystal Oscillators

The sources of the following circuits are contained in the Sources section beginning on page 730. The figure number contained in the box of each circuit correlates to the source entry in the Sources section.

High Frequency Crystal Oscillator
Overtone Crystal Oscillator
Overtone Crystal Oscillator
TTL Oscillator for 1 MHz-10 MHz
Crystal Checker
96 MHz Crystal Oscillator
Simple TTL Crystal Oscillator
Crystal Oscillator
Overtone Crystal Oscillator
Schmitt Trigger Crystal Oscillator
50 MHz-150 MHz Overtone Oscillator
Fifth Overtone Oscillator
Crystal Controlled Butler Oscillator
Overtone Oscillator with Crystal Switching
Crystal Oscillator
Crystal Oscillator/Doubler
Low Frequency Crystal Oscillator
Crystal Oscillator
100 kHz Crystal Calibrator
Third Overtone Crystal Oscillator
Crystal Checker
CMOS Crystal Oscillator
Temperature-Compensated Crystal Oscillator
Crystal Controlled Transistor Oscillator

Pierce Harmonic Oscillator
Colpitts Harmonic Oscillator
International Crystal OF-1 LO Oscillator
Butler Emitter Follower Oscillator
Colpitts Harmonic Oscillator
Butler Emitter Follower Oscillator
Butler Common Base Oscillator
Pierce Harmonic Oscillator
Tube Type Crystal Oscillator
Precision Clock Generator
Miller Oscillator
Butler Emitter Follower Oscillator
Colpitts Oscillator
Crystal-Controlled Oscillator
Pierce Oscillator
Butler Aperiodic Oscillator
Parallel-mode Aperiodic Crystal Oscillator
International Crystal OF-1 HI Oscillator
Standard Crystal Oscillator for 1 MHz
TTL-Compatible Crystal Oscillator
Crystal Controlled Sine Wave Oscillator
Crystal Oscillator
Stable Low Frequency Crystal Oscillator
JFET Pierce Crystal Oscillator
CMOS Oscillator
Pierce Harmonic Oscillator

HIGH FREQUENCY CRYSTAL OSCILLATOR

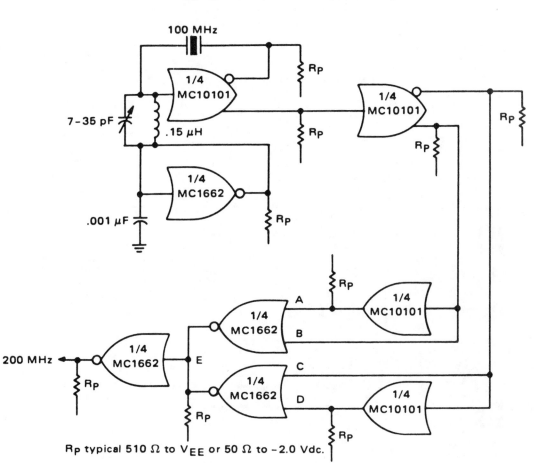

Fig. 19-1

Circuit Notes

One section of the MC10101 is connected as a 100 MHz crystal oscillator with the crystal in series with the feedback loop. The LC tank circuit tunes the 100 MHz harmonic of the crystal and may be used to calibrate the circuit to the exact frequency. A second section of the MC10101 buffers the crystal oscillator and gives complementary 100 MHz signals. The frequency doubler consists of two MC10101 gates as phase shifters and two MC1662 NOR gates. For a 50% duty cycle at the output, the delay to the true and complement 100 MHz signals should be 90°. This may be built precisely with 2.5 ns delay lines for the 200 MHz output or approximated by the two MC10101 gates as shown.

OVERTONE CRYSTAL OSCILLATOR

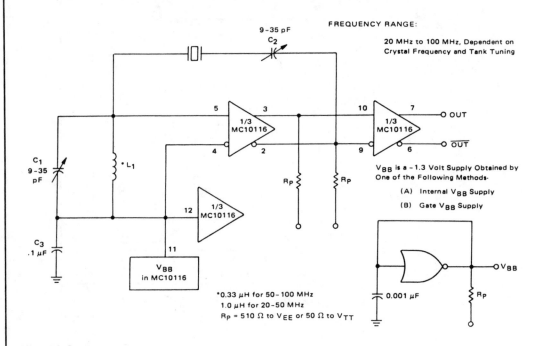

FREQUENCY RANGE:

20 MHz to 100 MHz, Dependent on
Crystal Frequency and Tank Tuning

V_{BB} is a – 1.3 Volt Supply Obtained by
One of the Following Methods.

    (A)   Internal V_{BB} Supply

    (B)   Gate V_{BB} Supply

*0.33 µH for 50–100 MHz
1.0 µH for 20–50 MHz
R_p = 510 Ω to V_{EE} or 50 Ω to V_{TT}

Fig. 19-2

Circuit Notes

This circuit employs an adjustable resonant tank circuit which insures operation at the desired crystal overtone. C1 and L1 form the resonant tank circuit, which with the values specified as a resonant frequency adjustable from approximately 50 MHz to 100 MHz. Overtone operation is accomplished by adjusting the tank circuit frequency at or near the desired frequency. The tank circuit exhibits a low impedance shunt to off-frequency oscillations and a high impedance to the desired frequency, allowing feedback from the output. Operation in this manner guarantees that the oscillator will always start at the correct overtone.

OVERTONE CRYSTAL OSCILLATOR

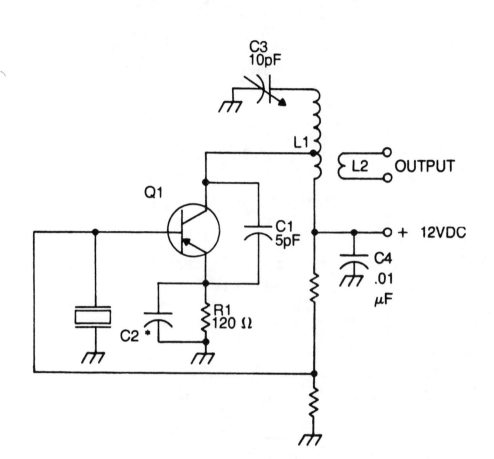

Fig. 19-3

Circuit Notes

The crystal element in this circuit is connected directly between the base and ground. Capacitor C1 is used to improve the feedback due to the internal capacitances of the transistor. This capacitor should be mounted as close as possible to the case of the transistor. The LC tank circuit in the collector of the transistor is tuned to the overtone frequency of the crystal. The emitter resistor capacitor must have a capacitive reactance of approximately 90 ohms at the frequency of operation. The tap on inductor L1 is used to match the impedance of the collector of the transistor. In most cases, the optimum placement of this tap is approximately one-third from the cold end of the coil. The placement of this tap is a trade-off between stability and maximum power output. The output signal is taken from a link coupling coil, L2, and operates by transformer action.

CRYSTAL CHECKER

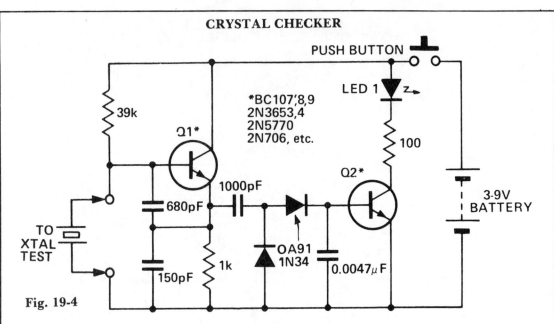

Fig. 19-4

Circuit Notes

Use this circuit for checking fundamental HF crystals on a 'Go-No-Go' basis. An untuned Colpitts oscillator drives a voltage multiplier rectifier and a current amplifier. If the crystal oscillates, Q2 conducts and the LED lights. A3 or 6V, 40mA bulb could be substituted for the LED.

TTL OSCILLATOR FOR 1 MHz-10 MHz

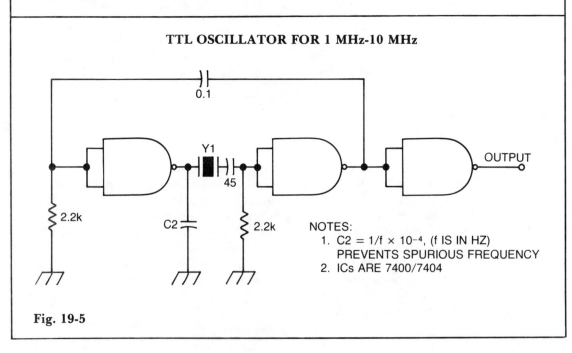

NOTES:
1. C2 = 1/f × 10⁻⁴, (f IS IN HZ) PREVENTS SPURIOUS FREQUENCY
2. ICs ARE 7400/7404

Fig. 19-5

96 MHz CRYSTAL OSCILLATOR

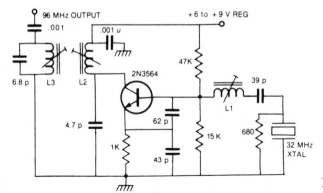

96 MHz OUTPUT

.001

.001 u

+6 to +9 V REG

47K

2N3564

6.8 p   L3   L2

4.7 p

62 p

39 p

L1

15 K   680

1K   43 p

32 MHz XTAL

L1, 4 mm former.   F29 slug (Neosid AZ assembly)
30 turns   .4 mm enamel wire.
L2,L3   7300 CAN TWO 722/1 FORMERS   F29 SLUGS
(Neosid double assembly)   12 turns   .63 mm enamel
wire.

Fig. 19-6

Circuit Notes

By using a crystal between 27.5 and 33 MHz, the 3rd harmonic will deliver between 82.5 and 99 MHz.

SIMPLE TTL CRYSTAL OSCILLATOR

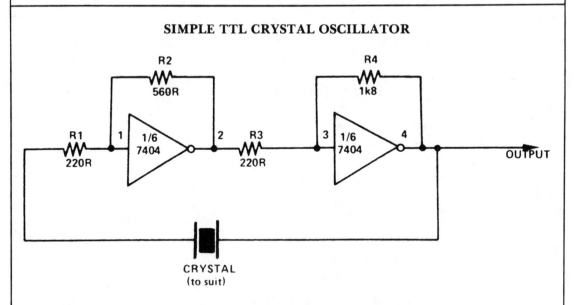

R2
560R

R4
1k8

R1
220R

1   1/6
7404   2

R3
220R

3   1/6
7404   4

OUTPUT

CRYSTAL
(to suit)

Fig. 19-7

Circuit Notes

This simple and cheap crystal oscillator comprises one third of a 7404, four resistors and a crystal. The inverters are biased into their linear regions by R1 to R4, and the crystal provides the feedback. Oscillation can only occur at the crystals fundamental frequency.

CRYSTAL OSCILLATOR

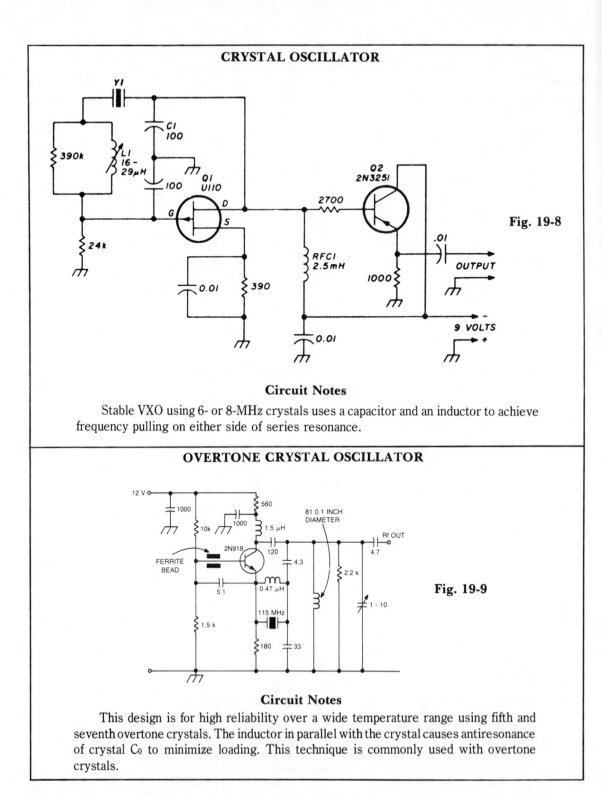

Fig. 19-8

Circuit Notes

Stable VXO using 6- or 8-MHz crystals uses a capacitor and an inductor to achieve frequency pulling on either side of series resonance.

OVERTONE CRYSTAL OSCILLATOR

Fig. 19-9

Circuit Notes

This design is for high reliability over a wide temperature range using fifth and seventh overtone crystals. The inductor in parallel with the crystal causes antiresonance of crystal C_0 to minimize loading. This technique is commonly used with overtone crystals.

SCHMITT TRIGGER CRYSTAL OSCILLATOR

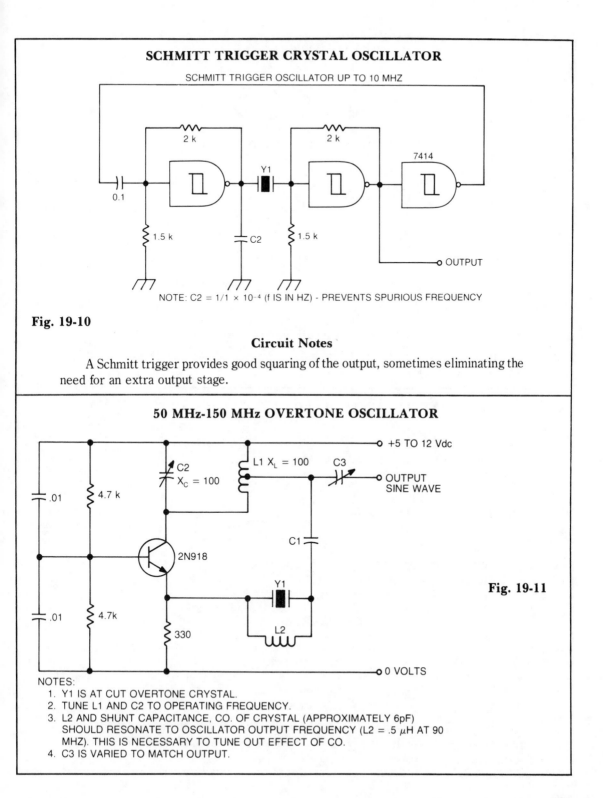

SCHMITT TRIGGER OSCILLATOR UP TO 10 MHZ

NOTE: C2 = 1/1 × 10^{-4} (f IS IN HZ) - PREVENTS SPURIOUS FREQUENCY

Fig. 19-10

Circuit Notes

A Schmitt trigger provides good squaring of the output, sometimes eliminating the need for an extra output stage.

50 MHz-150 MHz OVERTONE OSCILLATOR

Fig. 19-11

NOTES:
1. Y1 IS AT CUT OVERTONE CRYSTAL.
2. TUNE L1 AND C2 TO OPERATING FREQUENCY.
3. L2 AND SHUNT CAPACITANCE, CO. OF CRYSTAL (APPROXIMATELY 6pF) SHOULD RESONATE TO OSCILLATOR OUTPUT FREQUENCY (L2 = .5 μH AT 90 MHZ). THIS IS NECESSARY TO TUNE OUT EFFECT OF CO.
4. C3 IS VARIED TO MATCH OUTPUT.

FIFTH-OVERTONE OSCILLATOR

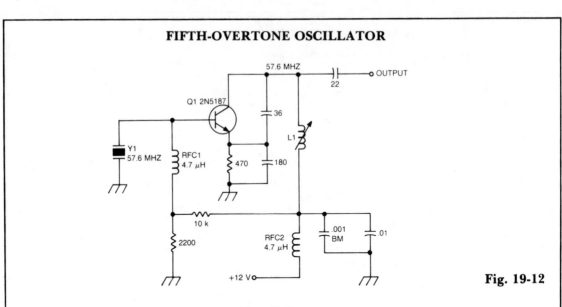

Fig. 19-12

Circuit Notes

This circuit isolates the crystal from the dc base supply with an rf choke for better starting characteristics.

CRYSTAL CONTROLLED BUTLER OSCILLATOR

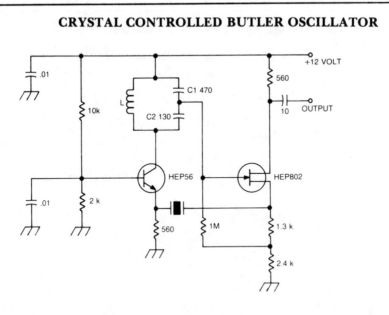

Fig. 19-13

Circuit Notes

A typical Butler oscillator (20-100 MHz) uses an FET in the second stage; the circuit is not reliable with two bipolars. Sometimes two FETs are used. Frequency is determined by LC values.

OVERTONE OSCILLATOR WITH CRYSTAL SWITCHING

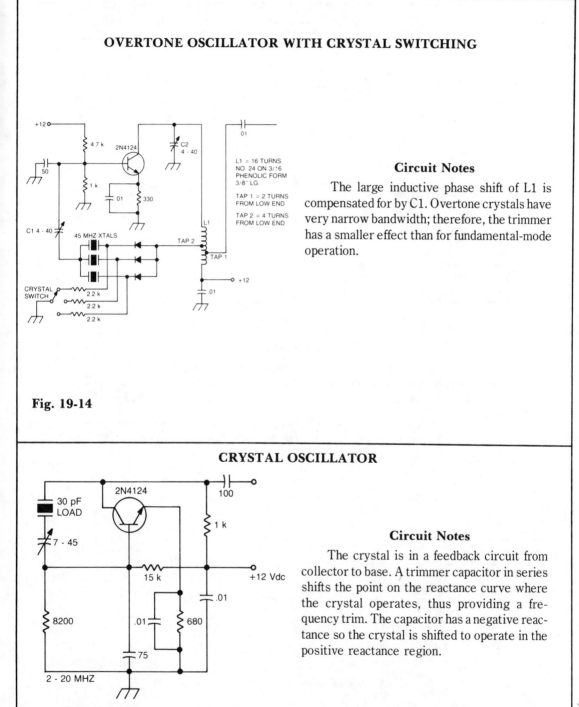

L1 = 16 TURNS
NO. 24 ON 3/16
PHENOLIC FORM
3/8" LG

TAP 1 = 2 TURNS
FROM LOW END

TAP 2 = 4 TURNS
FROM LOW END

Circuit Notes

The large inductive phase shift of L1 is compensated for by C1. Overtone crystals have very narrow bandwidth; therefore, the trimmer has a smaller effect than for fundamental-mode operation.

Fig. 19-14

CRYSTAL OSCILLATOR

Circuit Notes

The crystal is in a feedback circuit from collector to base. A trimmer capacitor in series shifts the point on the reactance curve where the crystal operates, thus providing a frequency trim. The capacitor has a negative reactance so the crystal is shifted to operate in the positive reactance region.

Fig. 19-15

CRYSTAL OSCILLATOR/DOUBLER

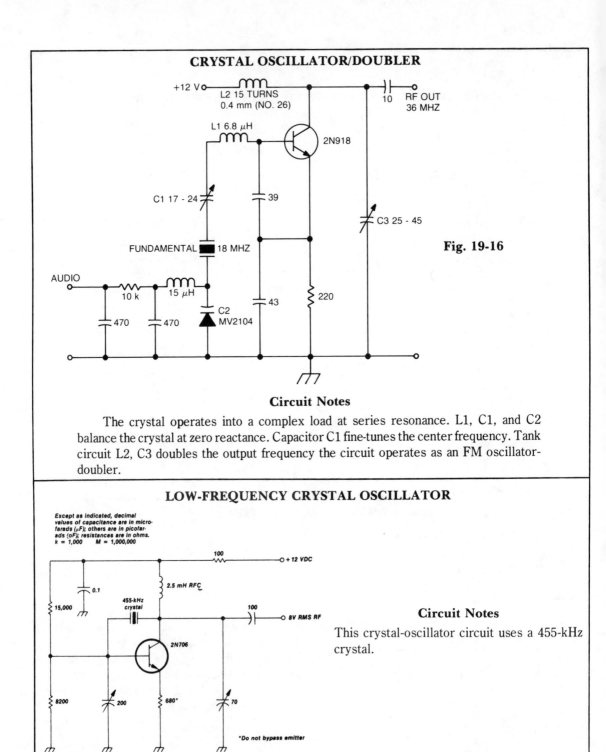

Fig. 19-16

Circuit Notes

The crystal operates into a complex load at series resonance. L1, C1, and C2 balance the crystal at zero reactance. Capacitor C1 fine-tunes the center frequency. Tank circuit L2, C3 doubles the output frequency the circuit operates as an FM oscillator-doubler.

LOW-FREQUENCY CRYSTAL OSCILLATOR

Except as indicated, decimal values of capacitance are in micro-farads (μF); others are in picofar-ads (oF); resistances are in ohms.
k = 1,000   M = 1,000,000

Circuit Notes

This crystal-oscillator circuit uses a 455-kHz crystal.

Fig. 19-17

CRYSTAL OSCILLATOR

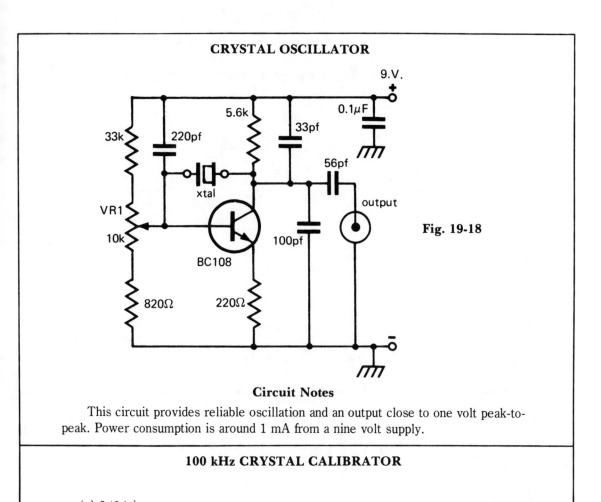

Fig. 19-18

Circuit Notes

This circuit provides reliable oscillation and an output close to one volt peak-to-peak. Power consumption is around 1 mA from a nine volt supply.

100 kHz CRYSTAL CALIBRATOR

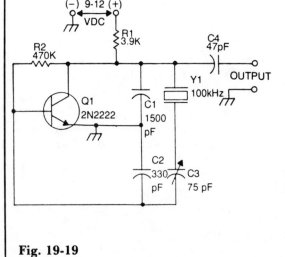

Circuit Notes

This circuit is often used by amateur radio operations, shortwave listeners, and other operators of shortwave receivers to calibrate the dial pointer. The oscillator operates at a fundamental frequency of 100 kHz, and the harmonics are used to locate points on the shortwave dial, provided that the output of the calibrator is coupled to the antenna circuit of the receiver. The crystal shunts the feedback voltage divider, and is in series with a variable capacitor (C3) that is used to set the actual operating frequency of the calibrator.

Fig. 19-19

THIRD-OVERTONE CRYSTAL OSCILLATOR

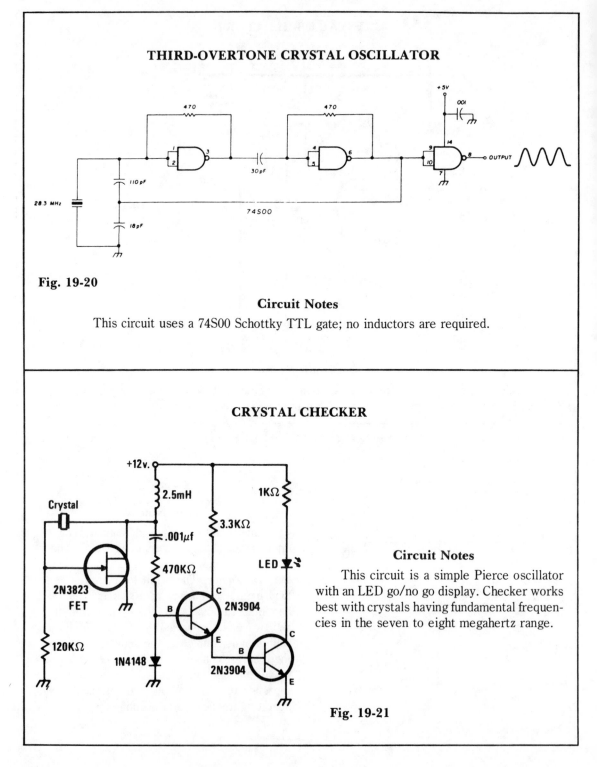

Fig. 19-20

Circuit Notes

This circuit uses a 74S00 Schottky TTL gate; no inductors are required.

CRYSTAL CHECKER

Circuit Notes

This circuit is a simple Pierce oscillator with an LED go/no go display. Checker works best with crystals having fundamental frequencies in the seven to eight megahertz range.

Fig. 19-21

186

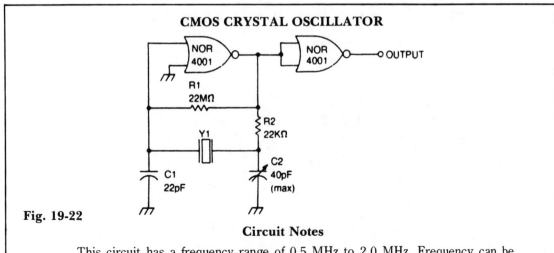

CMOS CRYSTAL OSCILLATOR

Fig. 19-22

Circuit Notes

This circuit has a frequency range of 0.5 MHz to 2.0 MHz. Frequency can be adjusted to a precise value with trimmer capacitor C2. The second NOR gate serves as an output buffer.

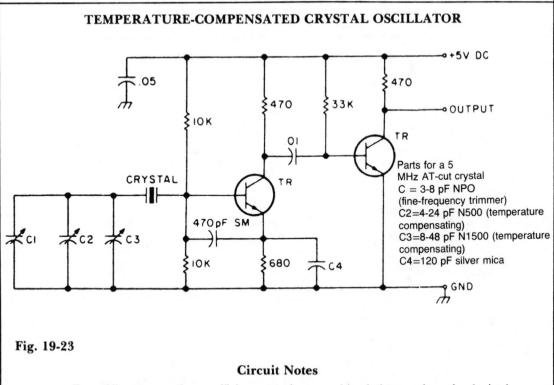

TEMPERATURE-COMPENSATED CRYSTAL OSCILLATOR

Parts for a 5 MHz AT-cut crystal
C = 3-8 pF NPO (fine-frequency trimmer)
C2 = 4-24 pF N500 (temperature compensating)
C3 = 8-48 pF N1500 (temperature compensating)
C4 = 120 pF silver mica

Fig. 19-23

Circuit Notes

Two different negative-coefficient capacitors are blended to produce the desired change in capacitance to counteract or compensate for the decrease in frequency of the "normal" AT-cut characteristics.

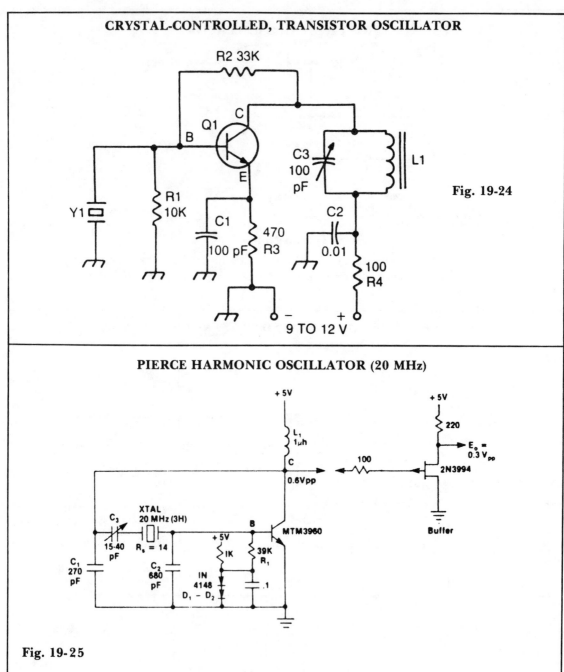

CRYSTAL-CONTROLLED, TRANSISTOR OSCILLATOR

Fig. 19-24

PIERCE HARMONIC OSCILLATOR (20 MHz)

Fig. 19-25

Circuit Notes

This circuit has excellent short term frequency stability because the external load tied across the crystal is mostly capacitive rather than resistive, giving the crystal a high in-circuit Q.

COLPITTS HARMONIC OSCILLATOR (100 MHz)

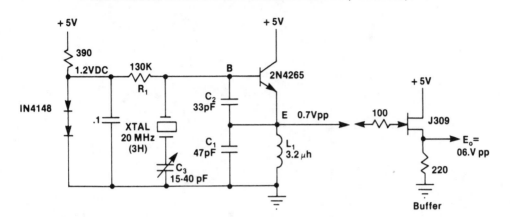

Fig. 19-26

Circuit Notes

L1C1 are selected to be resonant at a frequency below the desired crystal harmonic but above the crystal's next lower odd harmonic. C2 should have a value of 30-70 pF, independent of the oscillation frequency. There is no requirement for any specific ratio of C1/C2, but practical harmonic circuits seem to work best when C1 is approximately 1-3 times the value of C2. Diodes D1-D3 provide a simple regulated bias supply. The resistance of R1 should be as high as possible, as it affects the crystal's in-circuit Q.

INTERNATIONAL CRYSTAL OF-1 LO OSCILLATOR

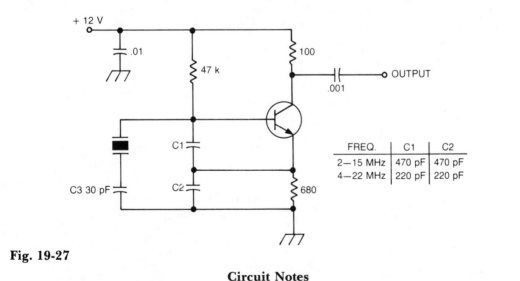

FREQ.	C1	C2
2—15 MHz	470 pF	470 pF
4—22 MHz	220 pF	220 pF

Fig. 19-27

Circuit Notes

International Crystal OF-1 LO oscillator circuit for fundamental-mode crystals.

BUTLER EMITTER FOLLOWER OSCILLATOR (100 MHz)

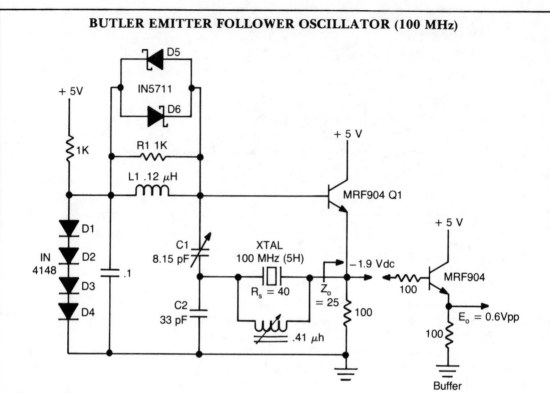

Fig. 19-28

Circuit Notes

This circuit has good performance without any parasitics because emitter follower amplifier has a gain of only one with built-in negative feedback to stabilize its gain.

COLPITTS HARMONIC OSCILLATOR (BASIC CIRCUIT)

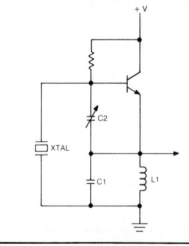

Circuit Notes

This circuit operates 30-200 ppm above series resonance. Physically simple, but analytically complex. It is inexpensive with fair frequency stability.

Fig. 19-29

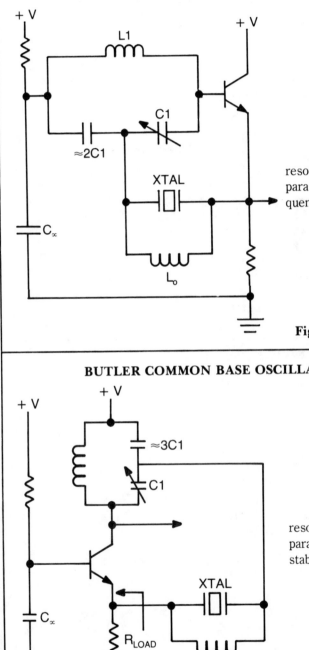

BUTLER EMITTER FOLLOWER OSCILLATOR (BASIC CIRCUIT)

Circuit Notes

This circuit operates at or near series resonance. It is a good circuit design with no parasitics. It is easy to tune with good frequency stability.

Fig. 19-30

BUTLER COMMON BASE OSCILLATOR (BASIC CIRCUIT)

Circuit Notes

This circuit operates at or near series resonance. It has fair to poor circuit design with parasitics, touch to tune, and fair frequency stability.

Fig. 19-31

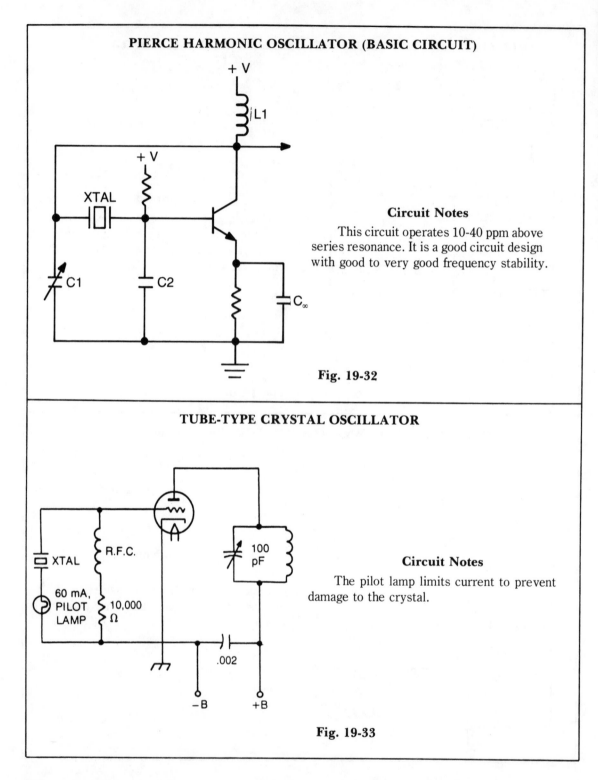

PIERCE HARMONIC OSCILLATOR (BASIC CIRCUIT)

+ V

L1

+ V

XTAL

C1

C2

C_∞

Circuit Notes

This circuit operates 10-40 ppm above series resonance. It is a good circuit design with good to very good frequency stability.

Fig. 19-32

TUBE-TYPE CRYSTAL OSCILLATOR

XTAL

R.F.C.

100 pF

60 mA, PILOT LAMP

10,000 Ω

.002

−B   +B

Circuit Notes

The pilot lamp limits current to prevent damage to the crystal.

Fig. 19-33

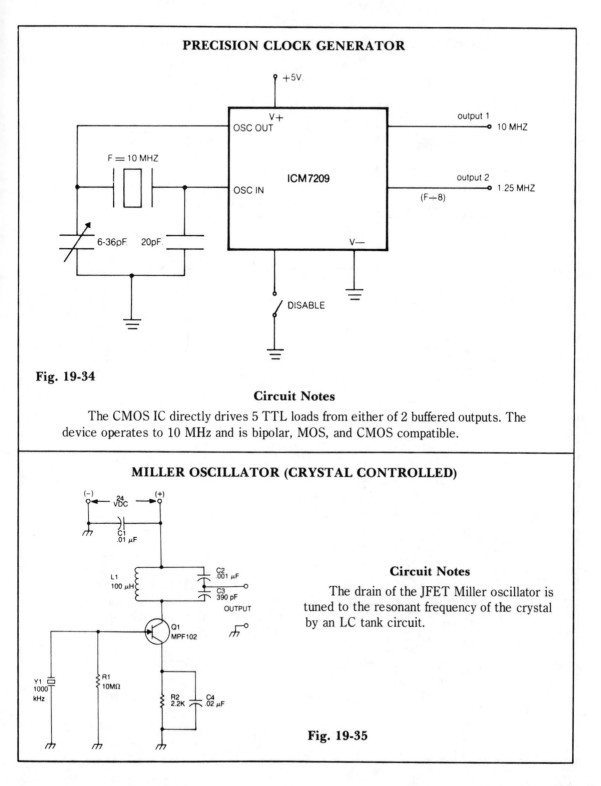

PRECISION CLOCK GENERATOR

+5V.

V+
OSC OUT

ICM 7209

F = 10 MHZ

OSC IN

6-36pF   20pF.

V—

DISABLE

output 1
10 MHZ

output 2
1.25 MHZ
(F÷8)

Fig. 19-34

Circuit Notes

The CMOS IC directly drives 5 TTL loads from either of 2 buffered outputs. The device operates to 10 MHz and is bipolar, MOS, and CMOS compatible.

MILLER OSCILLATOR (CRYSTAL CONTROLLED)

(−)   24   (+)
       VDC

C1
.01 μF

L1
100 μH

C2
.001 μF

C3
390 pF

OUTPUT

Q1
MPF102

Y1
1000
kHz

R1
10MΩ

R2
2.2K

C4
.02 μF

Circuit Notes

The drain of the JFET Miller oscillator is tuned to the resonant frequency of the crystal by an LC tank circuit.

Fig. 19-35

193

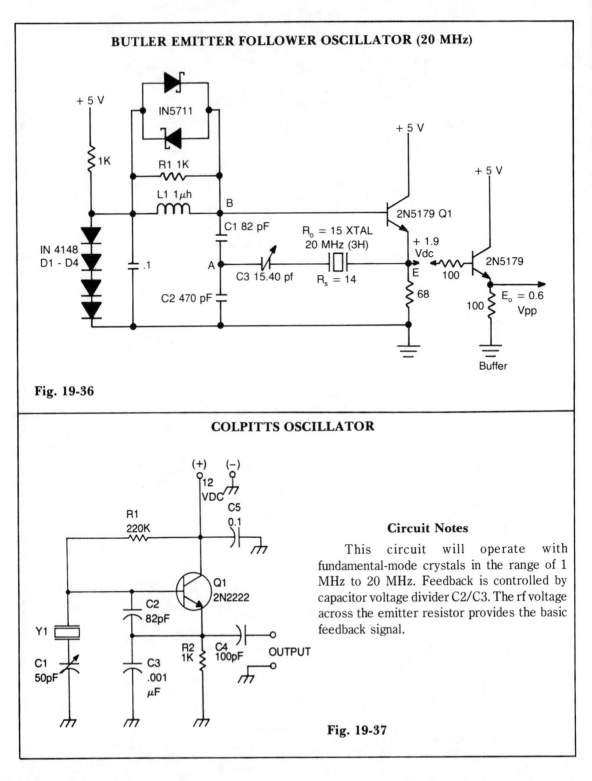

BUTLER EMITTER FOLLOWER OSCILLATOR (20 MHz)

+ 5 V

IN5711

1K

R1 1K

L1 1μh

B

IN 4148
D1 - D4

.1

A

C1 82 pF

C3 15.40 pf

C2 470 pF

R_o = 15 XTAL
20 MHz (3H)

R_s = 14

2N5179 Q1

+ 5 V

+ 1.9
Vdc

E

68

100

2N5179

100

+ 5 V

E_o = 0.6
Vpp

Buffer

Fig. 19-36

COLPITTS OSCILLATOR

(+)   (−)
12
VDC

C5
0.1

R1
220K

Q1
2N2222

C2
82pF

Y1

C1
50pF

C3
.001
μF

R2
1K

C4
100pF

OUTPUT

Circuit Notes

This circuit will operate with fundamental-mode crystals in the range of 1 MHz to 20 MHz. Feedback is controlled by capacitor voltage divider C2/C3. The rf voltage across the emitter resistor provides the basic feedback signal.

Fig. 19-37

CRYSTAL-CONTROLLED OSCILLATOR

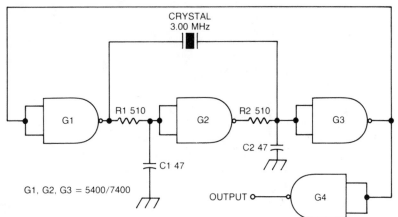

Fig. 19-38

Circuit Notes

This circuit oscillates without the crystal. With the crystal in the circuit, the frequency will be that of the crystal. The circuit has good starting characteristics even with the poorest crystals.

PIERCE OSCILLATOR

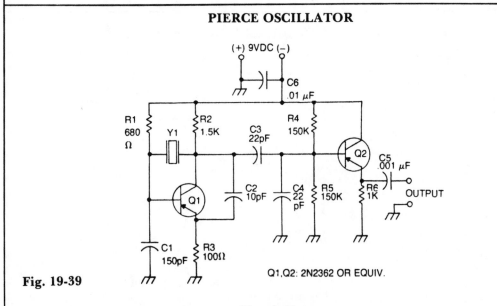

Fig. 19-39

Circuit Notes

The oscillator transistor is Q1, and the crystal is placed between the collector and base. Feedback is improved by the use of the collector-emitter capacitor C2. Transistor Q2 is used as an output buffer.

BUTLER APERIODIC OSCILLATOR

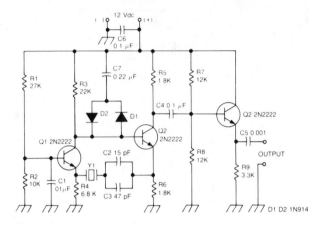

Fig. 19-40

Circuit Notes

This circuit works well in the range of 50 kHz to 500 kHz. Slight component modifications are needed for higher frequency operation. For operation over 3000 kHz, select a transistor that provides moderate gain (in the 60 to 150 range) at the frequency of operation and a gain-bandwidth product of at least 100 MHz.

PARALLEL-MODE APERIODIC CRYSTAL OSCILLATOR

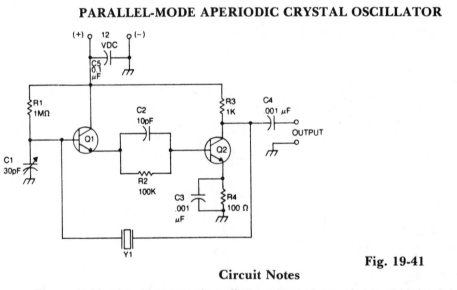

Fig. 19-41

Circuit Notes

The crystal is placed between the collector of the output stage and the base of the input stage. The frequency of oscillation can be set to a precise value with trimmer capacitor C1. The range of operation for this circuit is 500 kHz to 10 MHz. Extend the range downward (100 kHz) by increasing the value of C1 to 75 pF and increasing the value of C2 to 22pF.

INTERNATIONAL CRYSTAL OF-1 HI OSCILLATOR

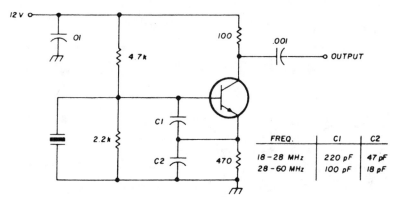

FREQ.	C1	C2
18 – 28 MHz	220 pF	47 pF
28 – 60 MHz	100 pF	18 pF

Fig. 19-42

Circuit Notes

International Crystal OF-1 HI oscillator circuit for third-overtone crystals. The circuit does not require inductors.

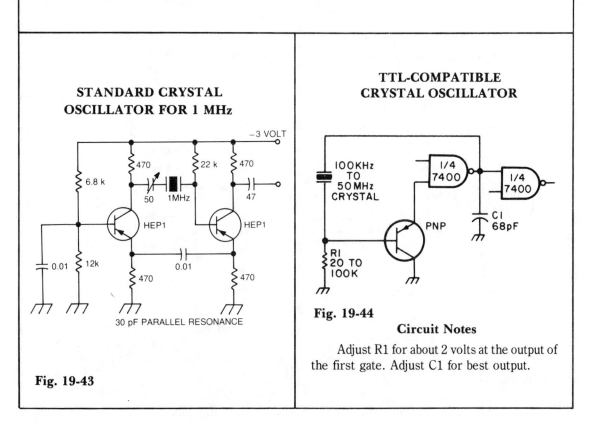

STANDARD CRYSTAL OSCILLATOR FOR 1 MHz

30 pF PARALLEL RESONANCE

Fig. 19-43

TTL-COMPATIBLE CRYSTAL OSCILLATOR

Fig. 19-44

Circuit Notes

Adjust R1 for about 2 volts at the output of the first gate. Adjust C1 for best output.

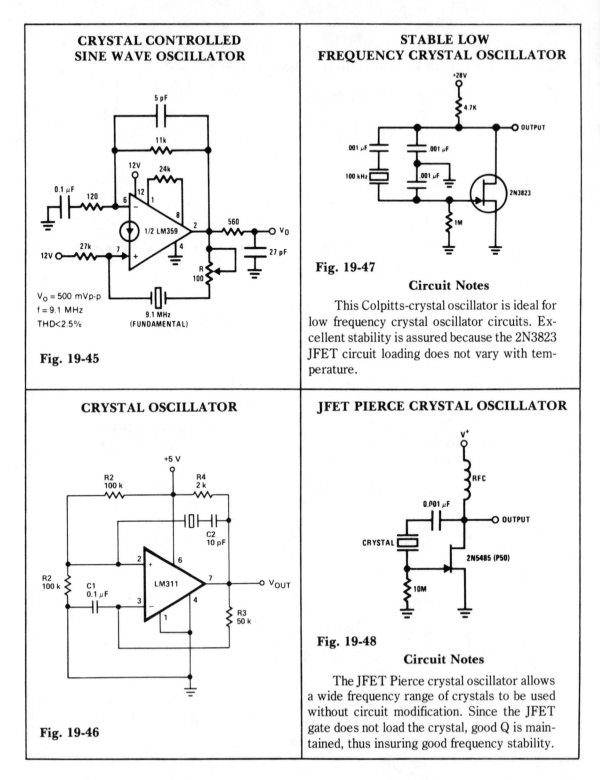

CRYSTAL CONTROLLED SINE WAVE OSCILLATOR

V_O = 500 mVp-p
f = 9.1 MHz
THD<2.5%

Fig. 19-45

STABLE LOW FREQUENCY CRYSTAL OSCILLATOR

Fig. 19-47

Circuit Notes

This Colpitts-crystal oscillator is ideal for low frequency crystal oscillator circuits. Excellent stability is assured because the 2N3823 JFET circuit loading does not vary with temperature.

CRYSTAL OSCILLATOR

Fig. 19-46

JFET PIERCE CRYSTAL OSCILLATOR

Fig. 19-48

Circuit Notes

The JFET Pierce crystal oscillator allows a wide frequency range of crystals to be used without circuit modification. Since the JFET gate does not load the crystal, good Q is maintained, thus insuring good frequency stability.

198

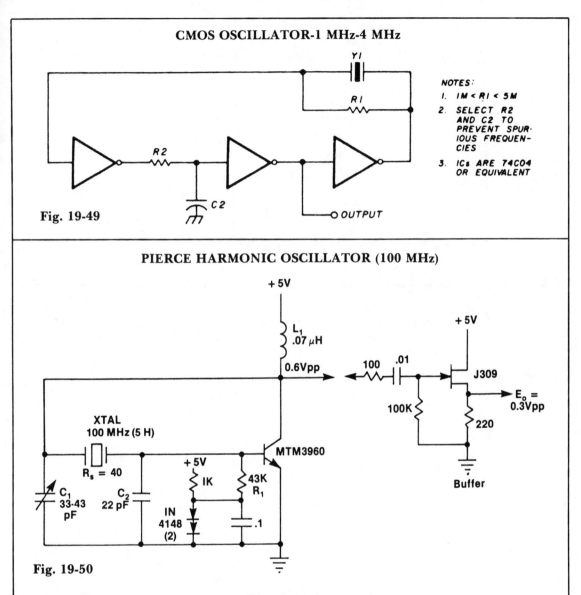

CMOS OSCILLATOR-1 MHz-4 MHz

NOTES:
1. $1M < R1 < 5M$
2. SELECT R2 AND C2 TO PREVENT SPURIOUS FREQUENCIES
3. ICs ARE 74C04 OR EQUIVALENT

Fig. 19-49

PIERCE HARMONIC OSCILLATOR (100 MHz)

Fig. 19-50

Circuit Notes

The output resistance of the transistor's collector, together with the effective value of C1, provides an RC phase lag of 30-50°. The crystal normally oscillates slightly above series resonance, where it is both resistive and inductive. Above series resonance, the crystal's internal impedance (resistive and inductive) together with C2 provides an RLC phase lag of 130-150°. The transistor inverts the signal, providing a total of 360° of phase shift around the loop. Inductor L1 is selected to resonate with C1 at a frequency between the crystal's desired harmonic and its next lower odd harmonic. Inductor L1 offsets part of the negative reactance of C1 at the oscillation frequency.

199

20

Current Measuring Circuits

The sources of the following circuits are contained in the Sources section beginning on page 730. The figure number contained in the box of each circuit correlates to the source entry in the Sources section.

Ammeter
Pico Ammeter
Nano Ammeter

Nanoampere Sensing Circuit with 100 Megohm Input Impedance
Current Monitor

AMMETER

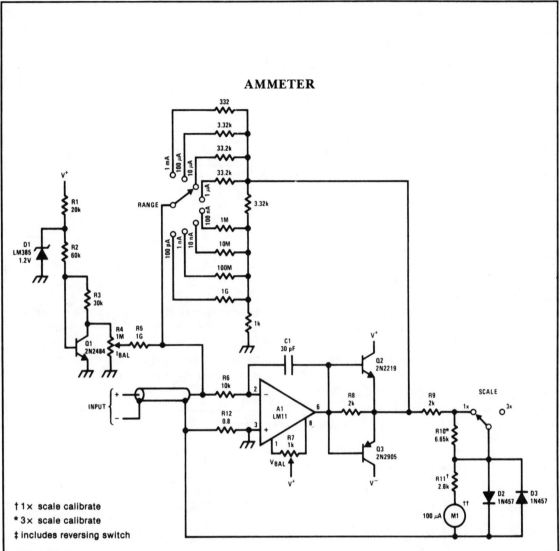

† 1× scale calibrate
* 3× scale calibrate
‡ includes reversing switch

Fig. 20-1

Circuit Notes

Current meter ranges from 100 pA to 3 mA full scale. Voltage across input is 100 μV at lower ranges rising to 3 mV at 3 mA. The buffers on the op amp are to remove ambiguity with high-current overload. The output can also drive a DVM or a DPM.

PICO AMMETER

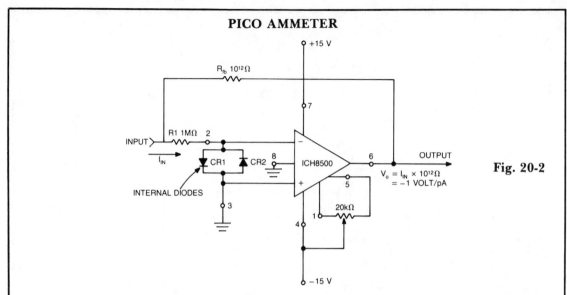

Fig. 20-2

$$V_o = I_{IN} \times 10^{12}\,\Omega$$
$$= -1 \text{ VOLT/pA}$$

Circuit Notes

A very sensitive pico ammeter ($-1\,V/pA$) employs the amplifier in the inverting or current summing mode. Care must be taken to eliminate stray currents from flowing into the current summing mode. It takes approximately 5 for the circuit to stabilize to within 1% of its final output voltage after a step function of input current has been applied. The internal diodes CR1 and CR2 together with external resistor R1 to protect the input stage of the amplifier from voltage transients.

NANO AMMETER

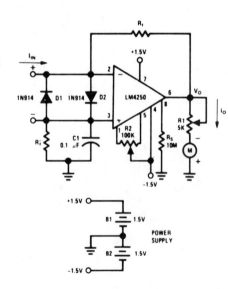

Resistance Values for DC Nano and Micro Ammeter

I FULL SCALE	$R_f\,[\Omega]$	$R_f'\,[\Omega]$
100 nA	1.5M	1.5M
500 nA	300k	300k
1 μA	300k	0
5 μA	60k	0
10 μA	30k	0
50 μA	6k	0
100 μA	3k	0

The complete meter amplifier is a differential current-to-voltage converter with input protection, zeroing and full scale adjust provisions, and input resistor balancing for minimum offset voltage.

Fig. 20-3

NANOAMPERE SENSING CIRCUIT
WITH 100 MEGOHM INPUT IMPEDANCE

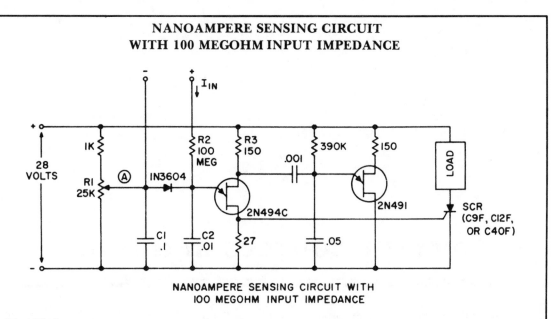

NANOAMPERE SENSING CIRCUIT WITH
IOO MEGOHM INPUT IMPEDANCE

Fig. 20-4

Circuit Notes

The circuit may be used as a sensitive current detector or as a voltage detector having high input impedance. R1 is set so that the voltage at point (A) is ½ to ¾ volts below the level that fires the 2N494C. A small input current (Iin) of only 40 nanoamperes will charge C2 and raise the voltage at the emitter to the firing level. When the 2N494C fires, both capacitors, C1 and C2, are discharged through the 27 ohm resistor, which generates a positive pulse with sufficient amplitude to trigger a controlled rectifier (SCR), or other pulse sensitive circuitry.

CURRENT MONITOR

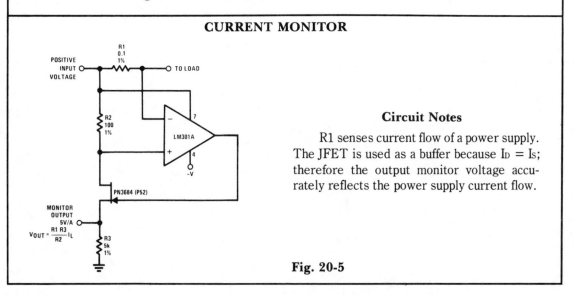

Circuit Notes

R1 senses current flow of a power supply. The JFET is used as a buffer because $I_D = I_S$; therefore the output monitor voltage accurately reflects the power supply current flow.

Fig. 20-5

21

Current Sources and Sinks

The sources of the following circuits are contained in the Sources section beginning on page 730. The figure number contained in the box of each circuit correlates to the source entry in the Sources section.

Current source

Precision Current Source

Precision 1 μA to 1 mA Current Sources

Precision Current Sink

CURRENT SOURCE

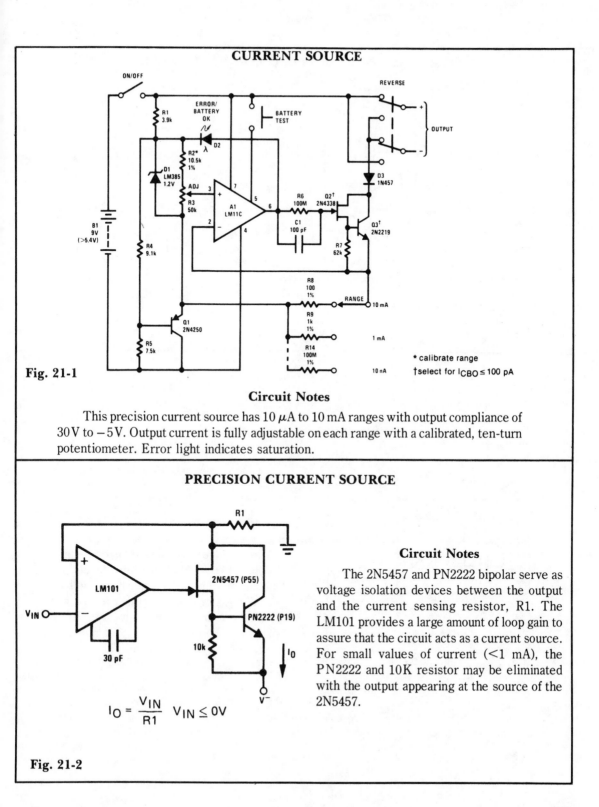

Fig. 21-1

Circuit Notes

This precision current source has 10 μA to 10 mA ranges with output compliance of 30V to −5V. Output current is fully adjustable on each range with a calibrated, ten-turn potentiometer. Error light indicates saturation.

PRECISION CURRENT SOURCE

$$I_O = \frac{V_{IN}}{R1} \quad V_{IN} \leq 0V$$

Circuit Notes

The 2N5457 and PN2222 bipolar serve as voltage isolation devices between the output and the current sensing resistor, R1. The LM101 provides a large amount of loop gain to assure that the circuit acts as a current source. For small values of current (<1 mA), the PN2222 and 10K resistor may be eliminated with the output appearing at the source of the 2N5457.

Fig. 21-2

PRECISION 1 μA to 1 mA CURRENT SOURCES

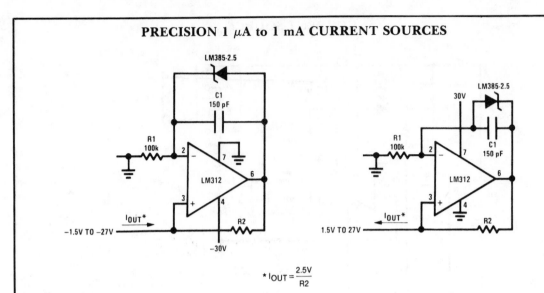

$$* I_{OUT} = \frac{2.5V}{R2}$$

Fig. 21-3

PRECISION CURRENT SINK

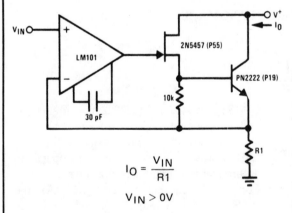

$$I_O = \frac{V_{IN}}{R1}$$

$$V_{IN} > 0V$$

Fig. 21-4

Circuit Notes

The 2N5457 JFET and PN2222 bipolar have inherently high output impedance. Using R1 as a current sensing resistor to provide feedback to the LM101 op amp provides a large amount of loop gain for negative feedback to enhance the true current sink nature of this circuit. For small current values, the 10 K resistor and PN2222 may be eliminated if the source of the JFET is connected to R1.

206

22

Dc/Dc and Dc/Ac Converters

The sources of the following circuits are contained in the Sources section beginning on page 730. The figure number contained in the box of each circuit correlates to the source entry in the Sources section.

DC-TO-DC/AC INVERTER

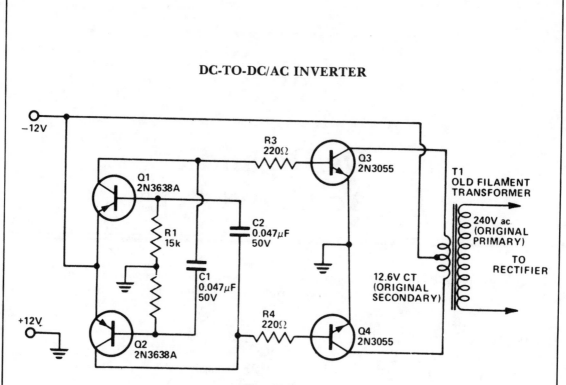

Fig. 22-1

Circuit Notes

This inverter uses no special components such as the torodial transformer used in many inverters. Cost is kept low with the use of cheap, readily available components. Essentially, it is a power amplifier driven by an astable multivibrator. The frequency is around 1200 Hz which most 50/60 Hz power transformers handle well without too much loss. Increasing the value of capacitors C1 and C2 will lower the frequency if any trouble is experienced. However, rectifier filtering capacitors required are considerably smaller at the higher operating frequency. The two 2N3055 transistor should be mounted on an adequately sized heatsink. The transformer should be rated according to the amount of output power required allowing for conversion efficiency of approximately 60%.

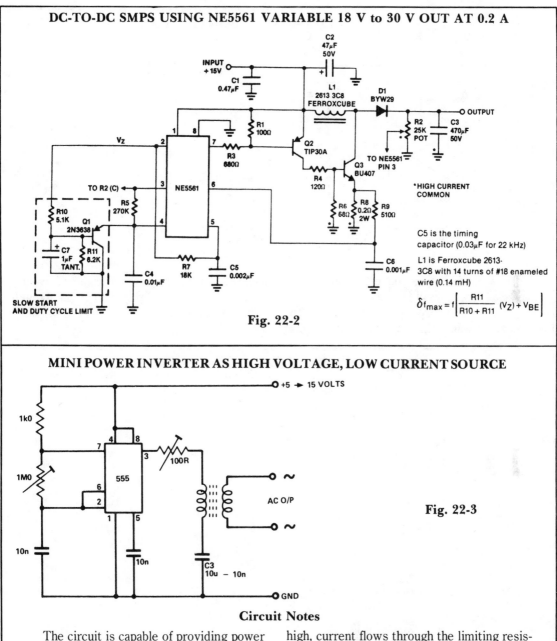

DC-TO-DC SMPS USING NE5561 VARIABLE 18 V to 30 V OUT AT 0.2 A

INPUT
+ 15V

C1
0.47μF

C2
47μF
50V

L1
2613 3C8
FERROXCUBE

D1
BYW29

O OUTPUT

R1
100Ω

Q2
TIP30A

R2
25K
POT

C3
470μF
50V

Vz

R3
680Ω

Q3
BU407

TO NE5561
PIN 3

NE5561

R4
120Ω

*HIGH CURRENT
COMMON

TO R2 (C)

R5
270K

R6
68Ω

R8
0.2Ω
2W

R9
510Ω

R10
5.1K

Q1
2N3638

C5 is the timing
capacitor (0.03μF for 22 kHz)

C7
1μF
TANT.

R11
6.2K

C4
0.01μF

R7
18K

C5
0.002μF

C6
0.001μF

L1 is Ferroxcube 2613-
3C8 with 14 turns of #18 enameled
wire (0.14 mH)

SLOW START
AND DUTY CYCLE LIMIT

$$\delta f_{max} = f\left[\frac{R11}{R10 + R11}(V_Z) + V_{BE}\right]$$

Fig. 22-2

MINI POWER INVERTER AS HIGH VOLTAGE, LOW CURRENT SOURCE

+5 → 15 VOLTS

1k0

1M0

7

4

8

3

100R

555

6

2

1

5

~

AC O/P

~

10n

10n

C3
10u – 10n

GND

Fig. 22-3

Circuit Notes

The circuit is capable of providing power for portable Geiger counters, dosimeter chargers, high resistance meters, etc. The 555 timer IC is used in its multivibrator mode, the frequency adjusted to optimize the transformer characteristics. When the output of the IC is high, current flows through the limiting resistor, the primary coil to charge C3. When the output is low, the current is reversed. With a suitable choice of frequency and C3, a good symmetric output is sustained.

209

REGULATED DC-TO-DC CONVERTER

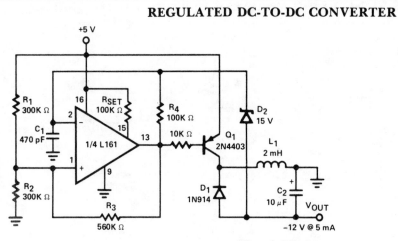

Fig. 22-4

Circuit Notes

Low power dc to dc converter obtained by adding a flyback circuit to a square wave oscillator. Operating frequency is 20 kHz to minimize the size of L1 and C2. Regulation is achieved by zener diode D2. Maximum current available before the converter drops out of regulation is 5.5 mA.

400 V, 60 W PUSH-PULL DC/DC CONVERTER

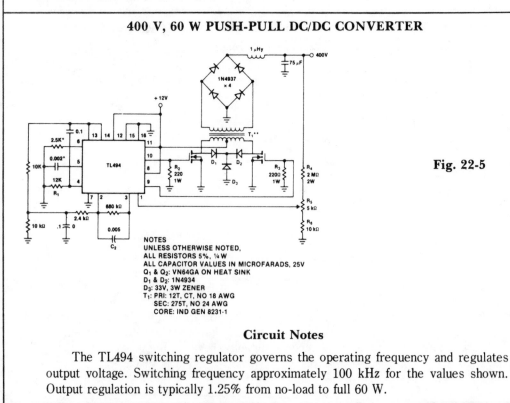

Fig. 22-5

NOTES
UNLESS OTHERWISE NOTED,
ALL RESISTORS 5%, ¼ W
ALL CAPACITOR VALUES IN MICROFARADS, 25V
Q_1 & Q_2: VN64GA ON HEAT SINK
D_1 & D_2: 1N4934
D_3: 33V, 3W ZENER
T_1: PRI: 12T, CT, NO 18 AWG
      SEC: 275T, NO 24 AWG
      CORE: IND GEN 8231-1

Circuit Notes

The TL494 switching regulator governs the operating frequency and regulates output voltage. Switching frequency approximately 100 kHz for the values shown. Output regulation is typically 1.25% from no-load to full 60 W.

DC/DC REGULATING CONVERTER

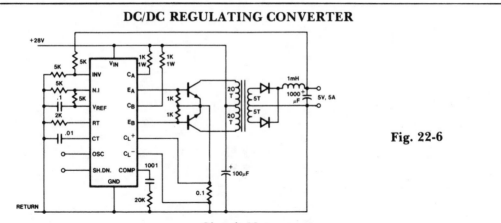

Fig. 22-6

Circuit Notes

Push-pull outputs are used in this transformer-coupled dc-dc regulating converter. Note that the oscillator must be set at twice the desired output frequency as the SG1524's internal flip-flop divides the frequency by 2 as it switches the PWM signal from one output to the other. Current limiting is done here in the primary so that the pulse width will be reduced should transformer saturation occur.

FLYBACK CONVERTER

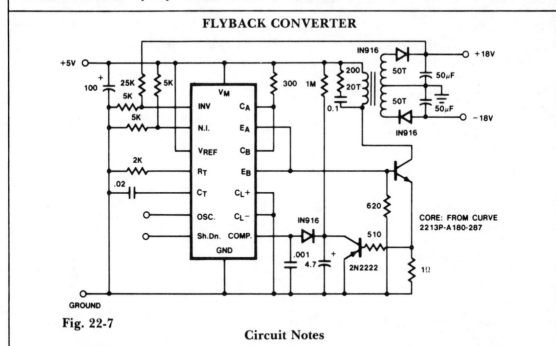

Fig. 22-7

Circuit Notes

A low-current flyback converter is used here to generate ±15 volts at 20 mA from a +5 volt regulated line. The reference generator in the SG1524 is unused with the input voltage providing the reference. Current limiting in a flyback converter is difficult and is accomplished here by sensing current in the primary line and resetting a soft-start circuit.

23

Decoders

The sources of the following circuits are contained in the Sources section beginning on page 730. The figure number contained in the box of each circuit correlates to the source entry in the Sources section.

Tone Alert Decoder
Tone Decoder with Relay Output
SCA Decoder

10.8 MHz FSK Decoder
24% Bandwidth Tone Decoder
Dual-Tone Decoder

TONE-ALERT DECODER

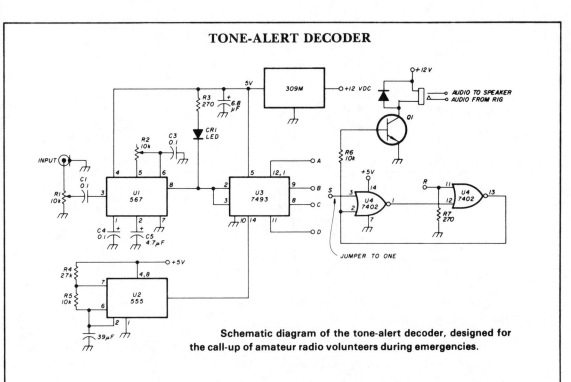

Schematic diagram of the tone-alert decoder, designed for the call-up of amateur radio volunteers during emergencies.

Fig. 23-1

Circuit Notes

PLL (U1) is set with R2 to desired tone frequency. LED lights to indicate lock-up of PLL. Reduce signal level (R1) and readjust R2 to assure lock-up. Delay is selected from counter U3 output. Circuits latches (turns on Q1 to allow audio to speaker) when proper frequency/duration signal is received. To reset latch, a positive voltage must be applied briefly to the R input of U4.

TONE DECODER WITH RELAY OUTPUT

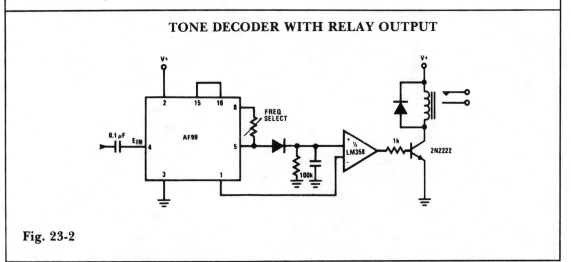

Fig. 23-2

SCA (BACKGROUND MUSIC) DECODER

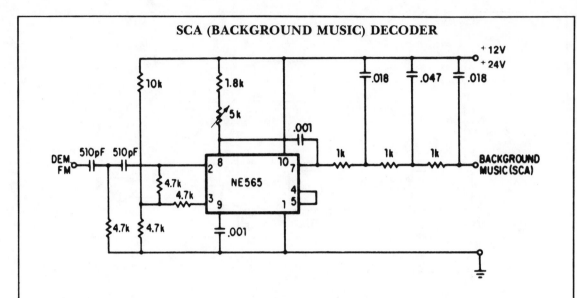

Fig. 23-3

Circuit Notes

A resistive voltage divider is used to establish a bias voltage for the input (pins 2 and 3). The demodulated (multiplex) FM signal is fed to the input through a two-stage high-pass filter, both to effect capacitive coupling and to attenuate the strong signal of the regular channel. A total signal amplitude, between 80 mV and 300 mV, is required at the input. Its source should have an impedance of less than 10,000 ohms. The Phase Locked Loop is tuned to 67 kHz with a 5000 ohm potentiometer; only approximate tuning is required, since the loop will seek the signal. The demodulated output (pin 7) passes through a three-stage low-pass filter to provide de-emphasis and attenuate the high-frequency noise which often accompanies SCA transmission. The demodulated output signal is in the rder of 50m V and the frequency response extends to 7 kHz.

10.8 MHz FSK DECODER

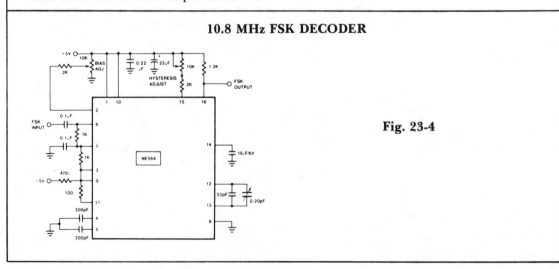

Fig. 23-4

24% BANDWIDTH TONE DECODER

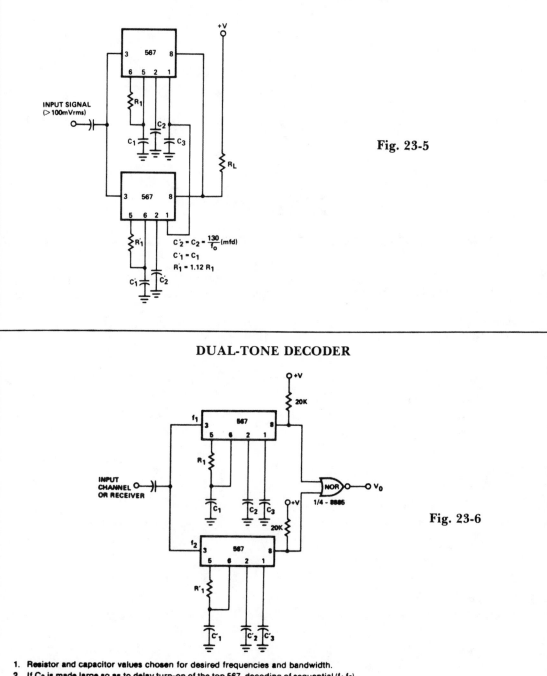

Fig. 23-5

$$C'_2 = C_2 = \frac{130}{f_0} \text{(mfd)}$$
$$C'_1 = C_1$$
$$R'_1 = 1.12 R_1$$

DUAL-TONE DECODER

Fig. 23-6

1. Resistor and capacitor values chosen for desired frequencies and bandwidth.
2. If C_3 is made large so as to delay turn-on of the top 567, decoding of sequential $(f_1 f_2)$ tones is possible.

24

Delays

The sources of the following circuits are contained in the Sources section beginning on page 730. The figure number contained in the box of each circuit correlates to the source entry in the Sources section.

Long Time Delay

Time Delay Generator

Door Chimes Delay

Time Delay Generator

Long Delay Timer Using PUT

Ultra-Precise Long Time Delay Relay

Long Duration Time Delay

Simple Time Delay Using Two SCRs

LONG TIME DELAY

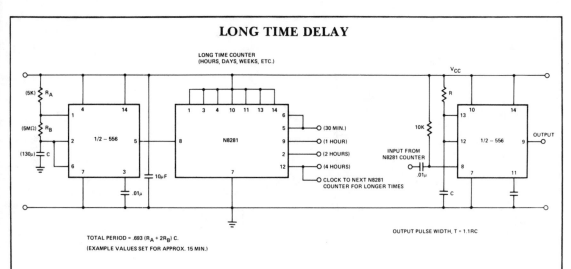

LONG TIME COUNTER
(HOURS, DAYS, WEEKS, ETC.)

TOTAL PERIOD = .693 (R_A + 2R_B) C.
(EXAMPLE VALUES SET FOR APPROX. 15 MIN.)

OUTPUT PULSE WIDTH, T = 1.1RC

Fig. 24-1

Circuit Notes

In the 556 timer, the timing is a function of the charging rate of the external capacitor. For long time delays, expensive capacitors with extremely low leakage are required. The practicality of the components involved limits the time between pulses to something in the neighborhood of 10 minutes. To achieve longer time periods, both halves of a dual timer may be connected in tandem with a "Divide-by'" network in between the first timer section operates in an oscillatory mode with a period of 1/fo. This signal is then applied to a "Divide-by-N" network to give an output with the period of N/fo. This can then be used to trigger the second half of the 556. The total time delay is now a function of N and fo.

TIME DELAY GENERATOR

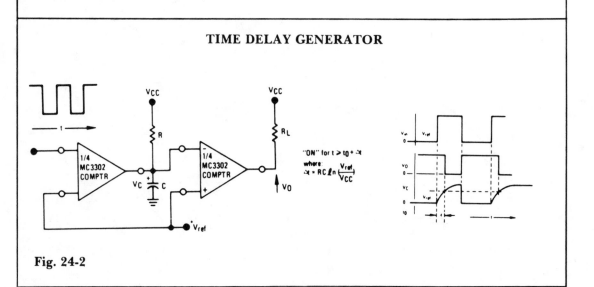

"ON" for t ≥ t_0 + △t
where:
△t = RC ℓn $\left(\frac{V_{ref}}{V_{CC}}\right)$

Fig. 24-2

DOOR CHIMES DELAY

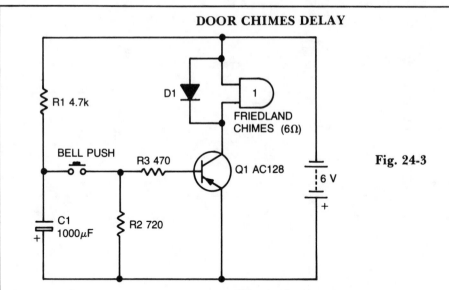

Fig. 24-3

Circuit Notes

With values shown, this simple circuit will permit one operation every 10 seconds or so. Capacitor C1 charges through R1 when the button is released. Making R1 larger will increase the delay.

TIME DELAY GENERATOR

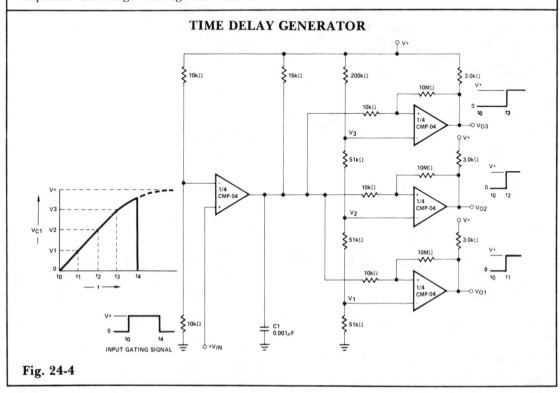

Fig. 24-4

LONG DELAY TIMER USING PUT

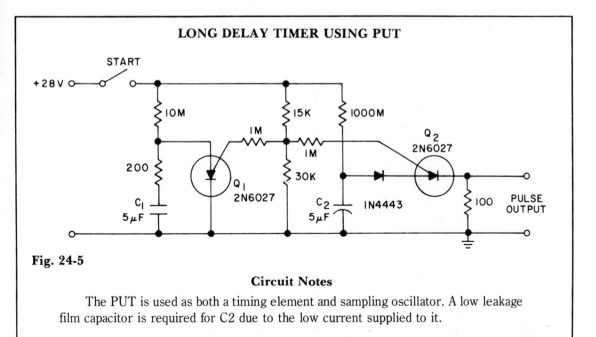

Fig. 24-5

Circuit Notes

The PUT is used as both a timing element and sampling oscillator. A low leakage film capacitor is required for C2 due to the low current supplied to it.

ULTRA-PRECISE LONG TIME DELAY RELAY

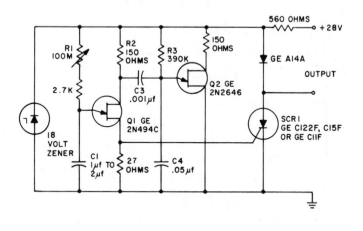

Fig. 24-6

Circuit Notes

Predictable time delays from as low as 0.3 milliseconds to over 3 minutes are obtainable without resorting to a large value electrolytic-type timing capacitor. Instead, a stable low leakage paper or mylar capacitor is used and the peak point current of the timing UJT (Q1) is effectively reduced, so that a large value emitter resistor (R1) may be substituted.

219

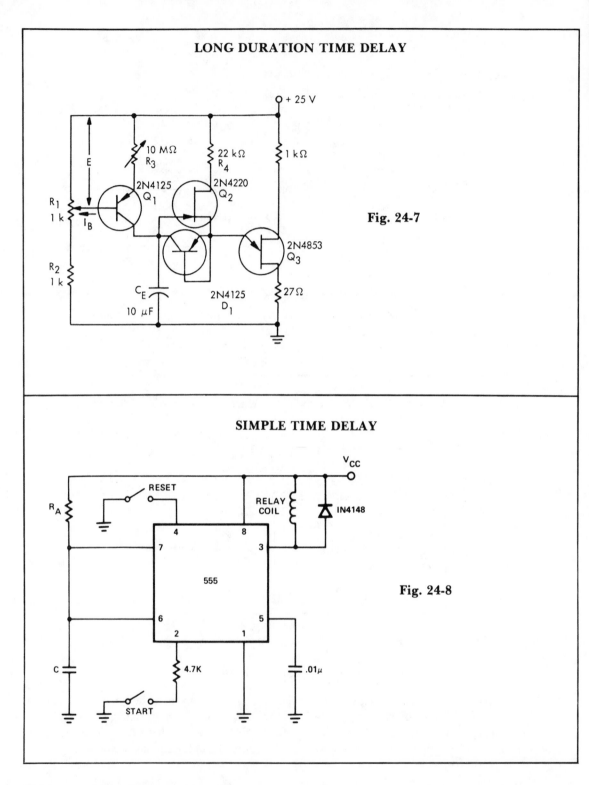

LONG DURATION TIME DELAY

+ 25 V

10 MΩ
R₃

22 kΩ
R₄

1 kΩ

2N4125
Q₁

2N4220
Q₂

R₁
1 k

I_B

2N4853
Q₃

R₂
1 k

C_E

10 μF

2N4125
D₁

27 Ω

E

Fig. 24-7

SIMPLE TIME DELAY

V_CC

R_A

RESET

RELAY
COIL

IN4148

4   8

7   3

555

6   5

2   1

C   4.7K   .01μ

START

Fig. 24-8

25

Detectors

The sources of the following circuits are contained in the Sources section beginning on page 730. The figure number contained in the box of each circuit correlates to the source entry in the Sources section.

Air-Motion Detector
Product Detector
Low Voltage Detector
Positive Peak Detector
Negative Peak Detector
Precision Peak Voltage Detector With
   Along Memory Time
Edge Detector
Ultra-Low Drift Peak Detector
Pulse Width Discriminator
True RMS Detector
Fast Half Wave Rectifier
Telemetry Demodulator
Full-Wave Rectifier and Averaging Filter
Double-Ended Limit Detector

Half-Wave Rectifier
Tone Detector
FM Tuner with a Single-Tuned Detector
   Coil
Missing Pulse Detector
High Speed Peak Detector
Detector for Magnetic Transducer
Double-Ended Limit Detector
FM Demodulator at 5 V
FM Demodulator at 12 V
Precision Full-Wave Rectifier
Negative Peak Detector
Level Detector with Hysteresis
Window Detector
Air Flow Detector

Positive Peak Detector

AIR-MOTION DETECTOR

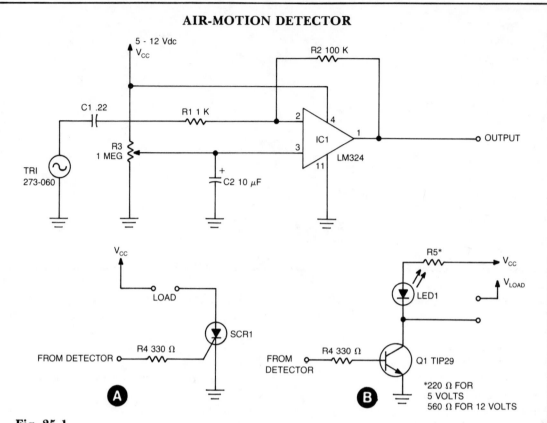

Fig. 25-1

Circuit Notes

Sensing circuit detects either steady or fluctuating air flows. The heart of the circuit is a Radio Shack piezo buzzer (P/N 273-060) and an LM324 quad op amp. (Red wire from the piezo element connects to capacitor C1, and the black wire to ground.) When a current of air hits the piezo element, a small signal is generated and is fed through C1 and R1 to the inverting input (pin 2) of one section of the LM324. That causes the output (pin 1) to go high. Resistor R3 adjusts sensitivity. The cir-cuit can be made sensitive enough to detect the wave of a hand or the sensitivity can be set so low that blowing on the element hard will produce no output. Resistor R2 is used to adjust the level of the output voltage at pin 1. The detector circuit can be used in various control applications. For example, an SCR can be used to control 117-volt AC loads as shown in A. Also, an NPN transistor, such as a TIP29, can be used to control loads as shown in B.

PRODUCT DETECTOR

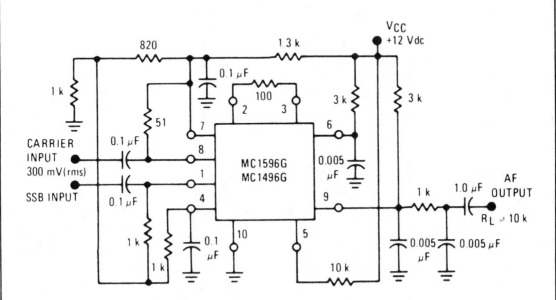

Fig. 25-2

Circuit Notes

The MC1596/MC1496 makes an excellent SSB product detector. This product detector has a sensitivity of 3.0 microvolts and a dynamic range of 90 dB when operating at an intermediate frequency of 9 MHz. The detector is broadband for the entire high frequency range. For operation at very low intermediate frequencies down to 50 kHz the 0.1 μF capacitors on pins 7 and 8 should be increased to 1.0 μF. Also, the output filter at pin 9 can be tailored to a specific intermediate frequency and audio amplifier input impedance. The emitter resistance between pins 2 and 3 may be increased or decreased to adjust circuit gain, sensitivity, and dynamic range. This circuit may also be used as an AM detector by introducing carrier signal at the carrier input and an AM signal at the SSB input. The carrier signal may be derived from the intermediate frequency signal or generated locally. The carrier signal may be introduced with or without modulation, provided its level is sufficiently high to saturate the upper quad differential amplifier. If the carrier signal is modulated, a 300 mV (rms) input level is recommended.

LOW VOLTAGE DETECTOR

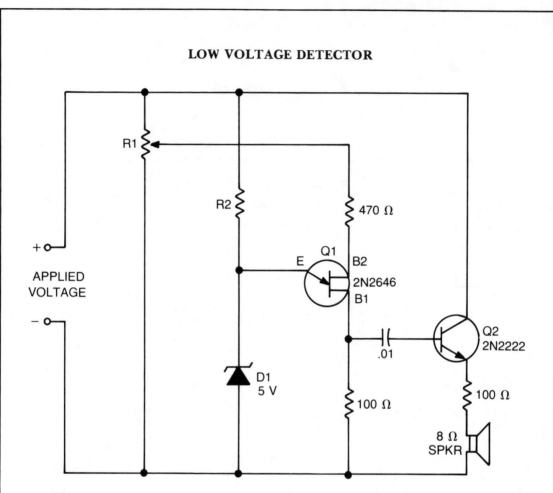

Fig. 25-3

Circuit Notes

The values of R1, R2, and D1 are selected for the voltage applied. Using a 12-volt battery, R1 = 10 K, R2 = 5.6 K and D1 is a 5-volt zener diode, or a string of forward-biased silicon rectifiers equaling about 5 volts. Transistor Q1 is a general-purpose UJT (Unijunction Transistor), and Q2 is any small-signal or switching NPN transistor. When detector is connected across the battery terminals, it draws little current and does not interfere with other devices powered by the battery. If voltage drops below the trip voltage selected with the R1 setting, the speaker beeps a warning. The frequency of the beeps is determined by the amount of undervoltage. If other voltages are being monitored, select R1 so that it draws only 1 mA or 2 mA. Zener diode D1 is about one-half of the desired trip voltage, and R2 is selected to bias it about 1 mA.

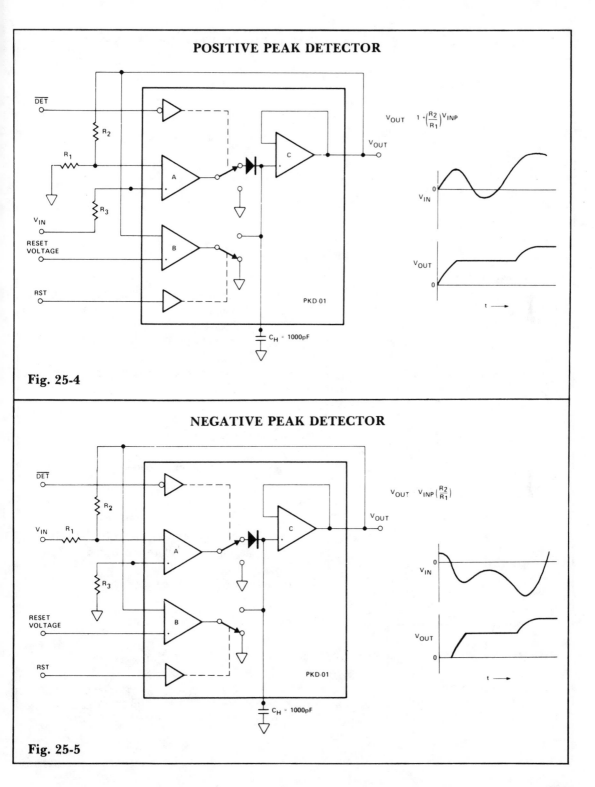

POSITIVE PEAK DETECTOR

$$V_{OUT} = \left(1 + \frac{R_2}{R_1}\right) V_{INP}$$

Fig. 25-4

NEGATIVE PEAK DETECTOR

$$V_{OUT} = -V_{INP}\left(\frac{R_2}{R_1}\right)$$

Fig. 25-5

PRECISION PEAK VOLTAGE DETECTOR WITH A LONG MEMORY TIME

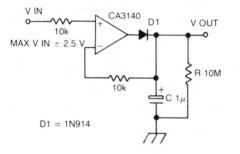

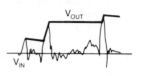

Fig. 25-6

Circuit Notes

The circuit has negative feedback only for positive signals. The inverting input can only get some feedback when diode D1 is forward biased and only occurs when the input is positive. With a positive input signal, the output of the op amp rises until the inverting input signal reaches the same potential. In so doing, the capacitor C is also charged to this potential. When the input goes negative, the diode D1 becomes reverse biased, the voltage on the capacitor remains, being slowly discharged by the op amp input bias current of 10 pico amps. Thus the discharge of the capacitor is dominantly controlled by the resistor R, giving a time constant of 10 seconds. Thus, the circuit detects the most positive peak voltage and remembers it.

EDGE DETECTOR

Circuit Notes

This circuit provides a short negative-going output pulse for every positive-going edge at the input. The input waveform is coupled to the input by capacitor C; the pulse length depends, as before, on R and C. If a negative going edge detector is required, the circuit in B should be used.

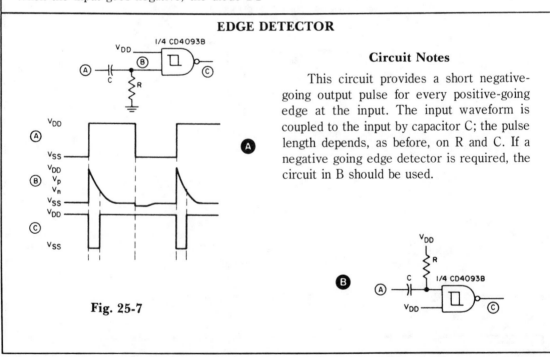

Fig. 25-7

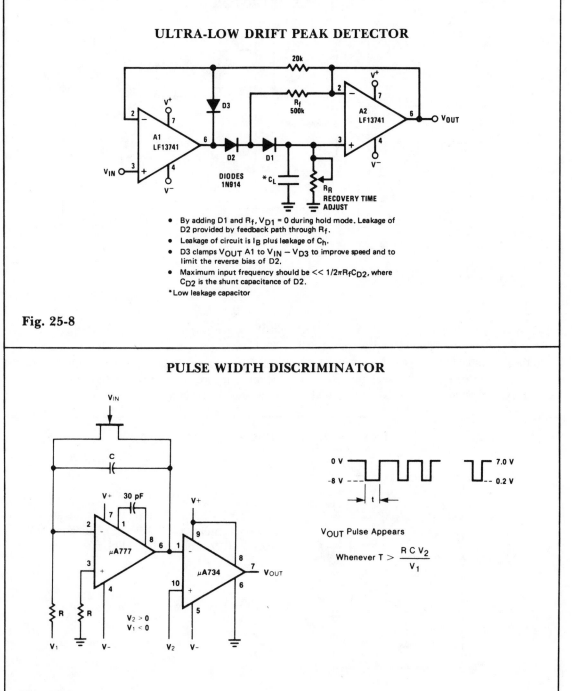

ULTRA-LOW DRIFT PEAK DETECTOR

- By adding D1 and R_f, $V_{D1} = 0$ during hold mode. Leakage of D2 provided by feedback path through R_f.
- Leakage of circuit is I_B plus leakage of C_h.
- D3 clamps V_{OUT} A1 to $V_{IN} - V_{D3}$ to improve speed and to limit the reverse bias of D2.
- Maximum input frequency should be $<< 1/2\pi R_f C_{D2}$, where C_{D2} is the shunt capacitance of D2.

*Low leakage capacitor

Fig. 25-8

PULSE WIDTH DISCRIMINATOR

V_{OUT} Pulse Appears

Whenever $T > \dfrac{R C V_2}{V_1}$

$V_2 > 0$
$V_1 < 0$

Fig. 25-9

TRUE RMS DETECTOR

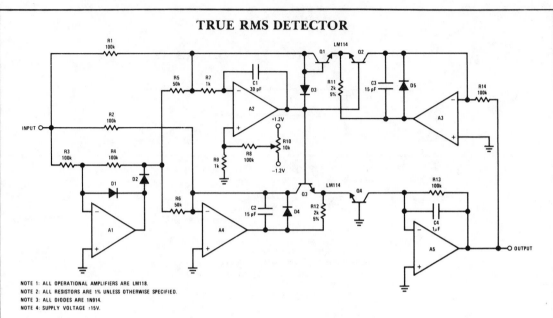

NOTE 1: ALL OPERATIONAL AMPLIFIERS ARE LM118.
NOTE 2: ALL RESISTORS ARE 1% UNLESS OTHERWISE SPECIFIED.
NOTE 3: ALL DIODES ARE 1N914.
NOTE 4: SUPPLY VOLTAGE ±15V.

Fig. 25-10

Circuit Notes

The circuit will provide a dc output equal to the rms value of the input. Accuracy is typically 2% for a 20 Vpp input signal from 50 Hz to 100 kHz, although it's usable to about 500 kHz.

The lower frequency is limited by the size of the filter capacitor. Since the input is dc coupled, it can provide the true rms equivalent of a dc and ac signal.

FAST HALF-WAVE RECTIFIER

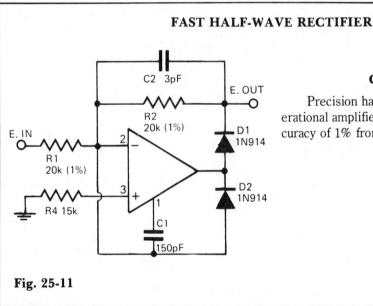

Circuit Notes

Precision half wave rectifier using an operational amplifier will have a rectification accuracy of 1% from dc to 100 kHz.

Fig. 25-11

TELEMETRY DEMODULATOR

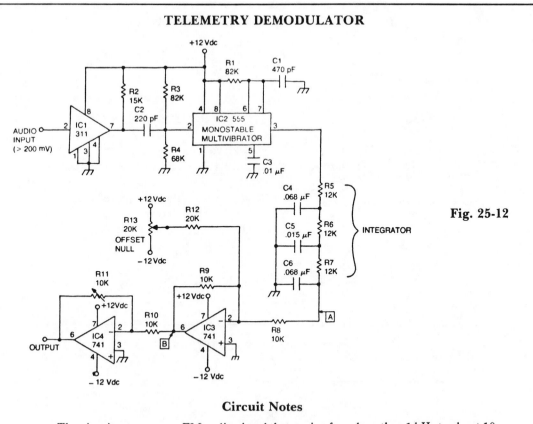

Fig. 25-12

Circuit Notes

The circuit recovers an FM audio signal that varies from less than 1 kHz to about 10 kHz.

FULL-WAVE RECTIFIER AND AVERAGING FILTER

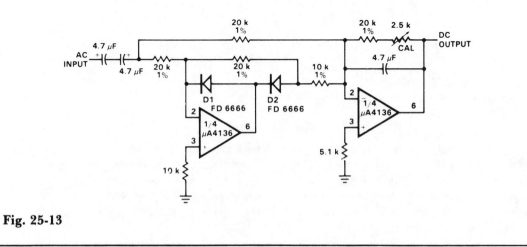

Fig. 25-13

DOUBLE-ENDED LIMIT DETECTOR

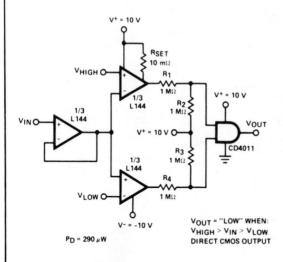

V+ = 10 V

RSET
10 mΩ

VHIGH

R1
1 MΩ

1/3
L144

VIN

1/3
L144

R2
1 MΩ

V+ = 10 V

VOUT
CD4011

V+ = 10 V

R3
1 MΩ

1/3
L144

R4
1 MΩ

VLOW

V- = -10 V

PD = 290 μW

VOUT = "LOW" WHEN:
VHIGH > VIN > VLOW
DIRECT CMOS OUTPUT

Fig. 25-14

Circuit Notes

Detector uses three sections of an L144 and a CMOS NAND gate to make a very low power voltage monitor. The 1 MΩ resistors R1, R2, R3, and R4 translate the bipolar ±10 V swing of the op amps to a 0 to 10 V swing acceptable to the ground-referenced CMOS logic. The total power dissipation is 290 μW while in limit and 330 μW while out of limit.

HALF-WAVE RECTIFIER

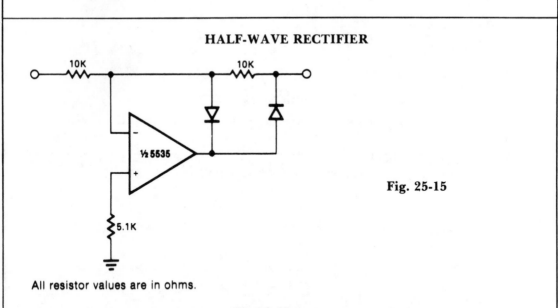

10K

10K

½ 5535

5.1K

Fig. 25-15

All resistor values are in ohms.

Circuit Notes

This circuit provides for accurate half wave rectification of the incoming signal. For positive signals, the gain is 0; for negative signals, the gain is −1. By reversing both diodes, the polarity can be inverted. This circuit provides an accurate output, but the output impedance differs for the two input polarities and buffering may be needed. The output must slew through two diode drops when the input polarity reverses. The NE5535 device will work up to 10 kHz with less ttan 5% distortion.

TONE DETECTOR

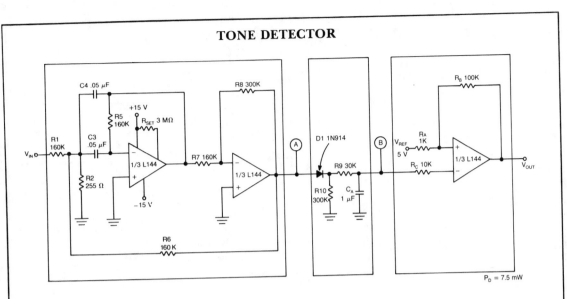

Fig. 25-16

Circuit Notes

The detector circuit is made up a two-amplifier multiple feedback bandpass filter followed by an ac-to-dc detector section and a Schmitt Trigger. The bandpass filter (with a Q of greater than 100) passes only 500 Hz inputs whch are in turn rectified by D1 and filtered by R9 and CA. This filtering action in combination with the trigger level of 5 V for the Schmitt device insures that at least 55 cycles of 500 Hz input must be present before the output will react to a tone input.

FM TUNER WITH A SINGLE-TUNED DETECTOR COIL

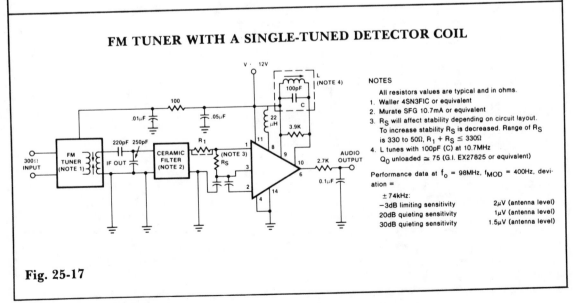

Fig. 25-17

MISSING PULSE DETECTOR

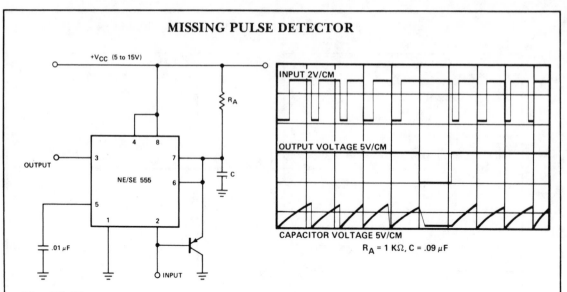

Fig. 25-18

Circuit Notes

The timing cycle is continuously reset by the input pulse train. A change in frequency, or a missing pulse, allows completion of the timing cycle which causes a change in the output level. For this application, the time delay should be set to be slightly longer than the normal time between pulses. The graph shows the actual waveforms seen in this mode of operation.

HIGH SPEED PEAK DETECTOR

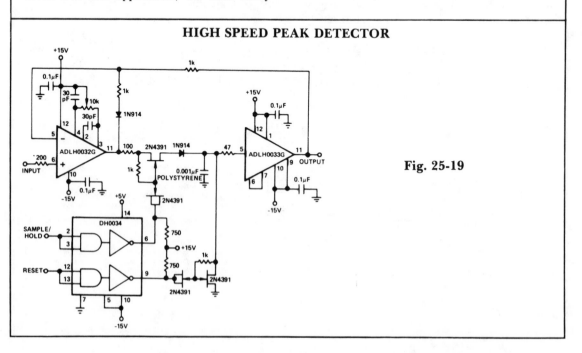

Fig. 25-19

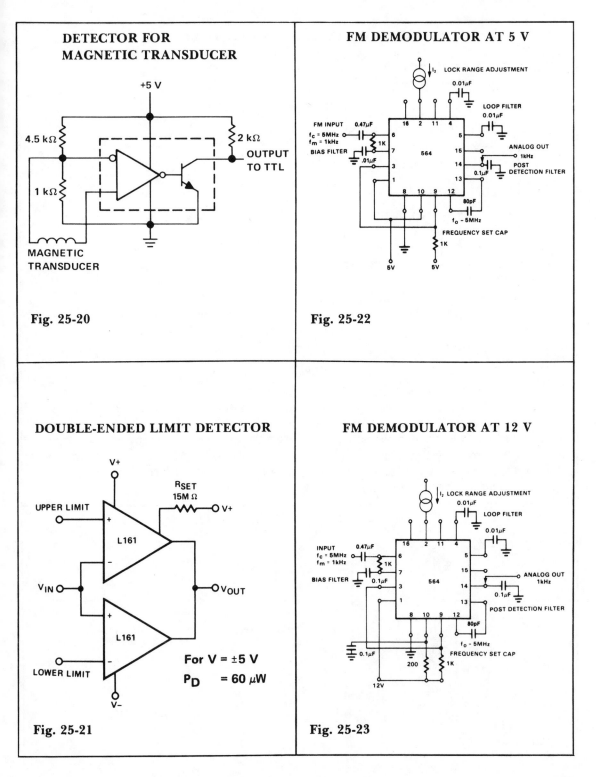

DETECTOR FOR MAGNETIC TRANSDUCER

+5 V

4.5 kΩ

2 kΩ

OUTPUT TO TTL

1 kΩ

MAGNETIC TRANSDUCER

Fig. 25-20

FM DEMODULATOR AT 5 V

I_2 LOCK RANGE ADJUSTMENT

0.01µF

LOOP FILTER
0.01µF

FM INPUT
$f_c = 5MHz$
$f_m = 1kHz$

0.47µF

1K

BIAS FILTER

.01µF

16 2 11 4

6

7

564

3

1

8 10 9 12

5

15

14

13

ANALOG OUT

1kHz

POST DETECTION FILTER

0.1µF

80pF

$f_0 - 5MHz$

FREQUENCY SET CAP

1K

5V   5V

Fig. 25-22

DOUBLE-ENDED LIMIT DETECTOR

V+

R_{SET}
15M Ω

V+

UPPER LIMIT

+

L161

−

V_{IN}

+

L161

−

V_{OUT}

LOWER LIMIT

V−

For V = ±5 V
P_D = 60 µW

Fig. 25-21

FM DEMODULATOR AT 12 V

I_2 LOCK RANGE ADJUSTMENT

0.01µF

LOOP FILTER

0.01µF

INPUT
$f_c = 5MHz$
$f_m = 1kHz$

0.47µF

1K

BIAS FILTER

0.1µF

16 2 11 4

6

7

564

3

1

8 10 9 12

5

15

14

13

ANALOG OUT
1kHz

0.1µF

POST DETECTION FILTER

80pF

$f_0 - 5MHz$

FREQUENCY SET CAP

0.1µF

200

1K

12V

Fig. 25-23

PRECISION FULL WAVE RECTIFIER

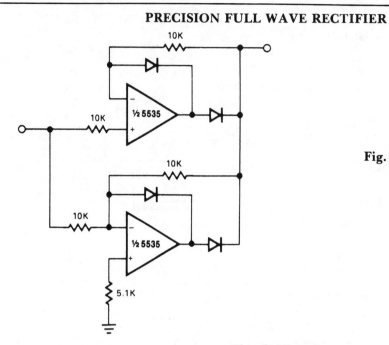

Fig. 25-24

Circuit Notes

The circuit provides accurate full wave rectification. The output impedance is low for both input polarities, and the errors are small at all signal levels. Note that the output will not sink heavy current, except a small amount through the 10 K resistors. Therefore, the load applied should be referenced to ground or a negative voltage. Reversal of all diode polarities will reverse the polarity of the output. Since the outputs of the amplifiers must slew through two diode drops when the input polarity changes, 741 type devices give 5% distortion at about 300 Hz.

NEGATIVE PEAK DETECTOR

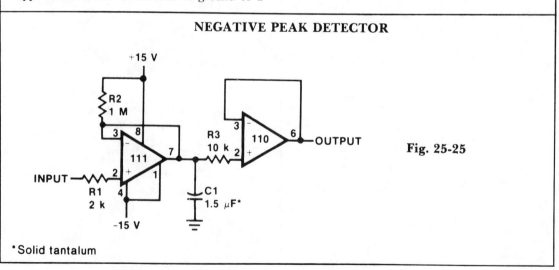

Fig. 25-25

*Solid tantalum

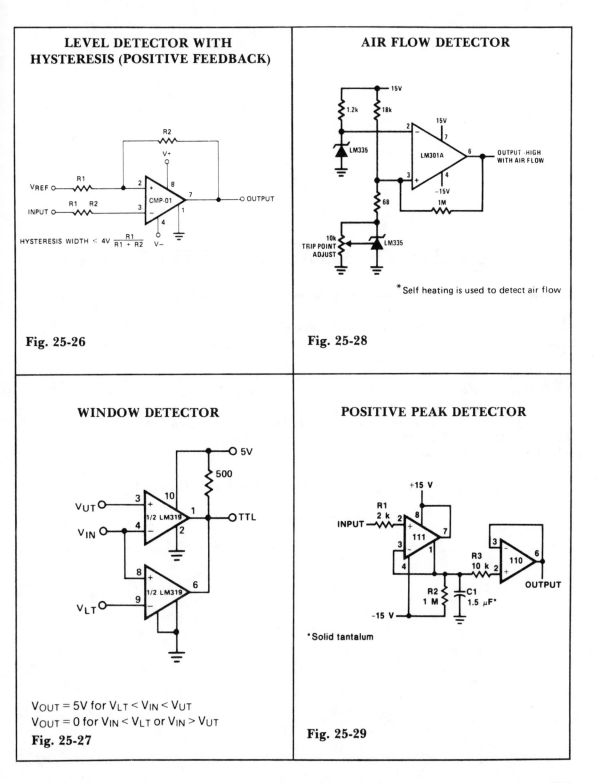

LEVEL DETECTOR WITH HYSTERESIS (POSITIVE FEEDBACK)

HYSTERESIS WIDTH $\leqslant 4V \dfrac{R1}{R1 + R2}$

Fig. 25-26

AIR FLOW DETECTOR

OUTPUT - HIGH WITH AIR FLOW

*Self heating is used to detect air flow

Fig. 25-28

WINDOW DETECTOR

$V_{OUT} = 5V$ for $V_{LT} < V_{IN} < V_{UT}$
$V_{OUT} = 0$ for $V_{IN} < V_{LT}$ or $V_{IN} > V_{UT}$

Fig. 25-27

POSITIVE PEAK DETECTOR

*Solid tantalum

Fig. 25-29

235

26

Digital-to-Analog Converters

The sources of the following circuits are contained in the Sources section beginning on page 730. The figure number contained in the box of each circuit correlates to the source entry in the Sources section.

14-BIT BINARY D/A CONVERTER (UNIPOLAR)

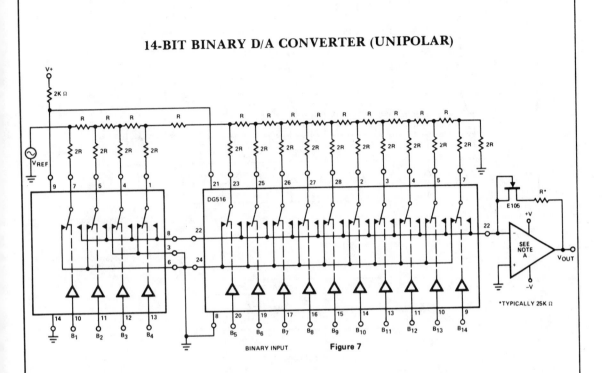

Figure 7

BINARY INPUT

NOTE:

A. Op-Amp characteristics effect D/A accuracy and settling time. The following Op-Amps, listed in order of increasing speed, are suggested:

1. LM101A   2. LF156A   3. LM118

Unipolar Binary Operation

DIGITAL INPUT	ANALOG OUTPUT
1 1 1 1 1 1 1 1 1 1 1 1 1 1	$-V_{REF}\,(1 - 2^{-14})$
1 0 0 0 0 0 0 0 0 0 0 0 0 1	$-V_{REF}\,(1/2 + 2^{-14})$
1 0 0 0 0 0 0 0 0 0 0 0 0 0	$-V_{REF}/2$
0 1 1 1 1 1 1 1 1 1 1 1 1 1	$-V_{REF}\,(1/2 - 2^{-14})$
0 0 0 0 0 0 0 0 0 0 0 0 0 1	$-V_{REF}\,(2^{-14})$
0 0 0 0 0 0 0 0 0 0 0 0 0 0	0

Fig. 26-1

237

10 BIT D/A CONVERTER

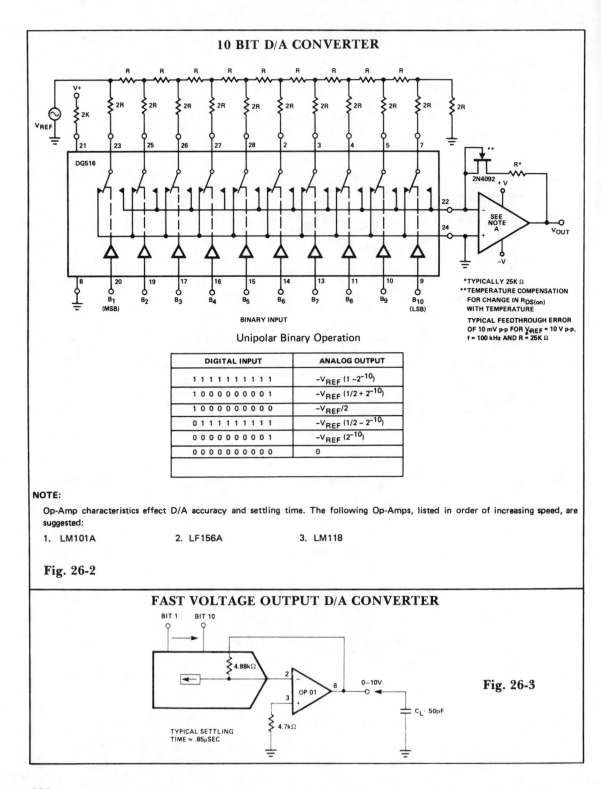

Unipolar Binary Operation

DIGITAL INPUT	ANALOG OUTPUT
1 1 1 1 1 1 1 1 1 1	$-V_{REF}\,(1 - 2^{-10})$
1 0 0 0 0 0 0 0 0 1	$-V_{REF}\,(1/2 + 2^{-10})$
1 0 0 0 0 0 0 0 0 0	$-V_{REF}/2$
0 1 1 1 1 1 1 1 1 1	$-V_{REF}\,(1/2 - 2^{-10})$
0 0 0 0 0 0 0 0 0 1	$-V_{REF}\,(2^{-10})$
0 0 0 0 0 0 0 0 0 0	0

NOTE:

Op-Amp characteristics effect D/A accuracy and settling time. The following Op-Amps, listed in order of increasing speed, are suggested:

1. LM101A   2. LF156A   3. LM118

Fig. 26-2

FAST VOLTAGE OUTPUT D/A CONVERTER

Fig. 26-3

RESISTOR TERMINATED DAC (0 TO –5 V OUTPUT)

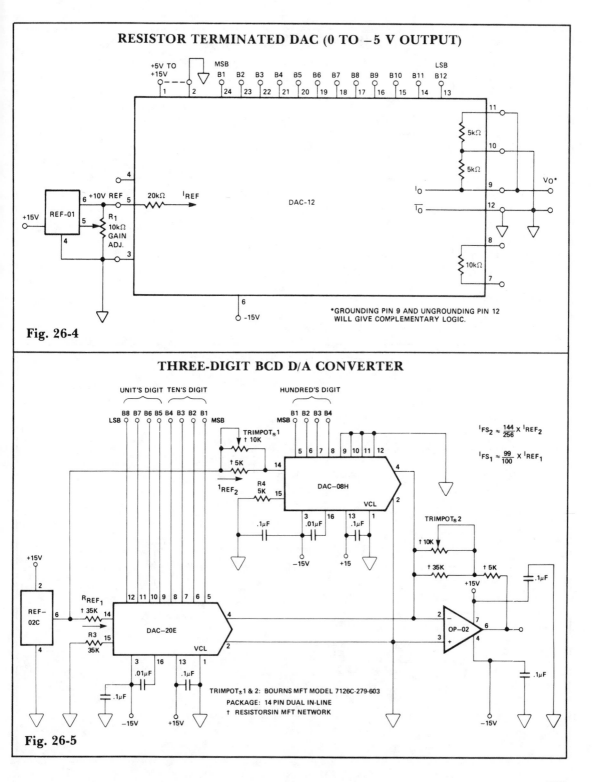

Fig. 26-4

THREE-DIGIT BCD D/A CONVERTER

$$^IFS_2 \approx \frac{144}{256} \times {}^IREF_2$$

$$^IFS_1 \approx \frac{99}{100} \times {}^IREF_1$$

TRIMPOT 1 & 2: BOURNS MFT MODEL 7126C-279-603
PACKAGE: 14 PIN DUAL IN-LINE
† RESISTORS IN MFT NETWORK

Fig. 26-5

239

8-BIT D/A CONVERTER

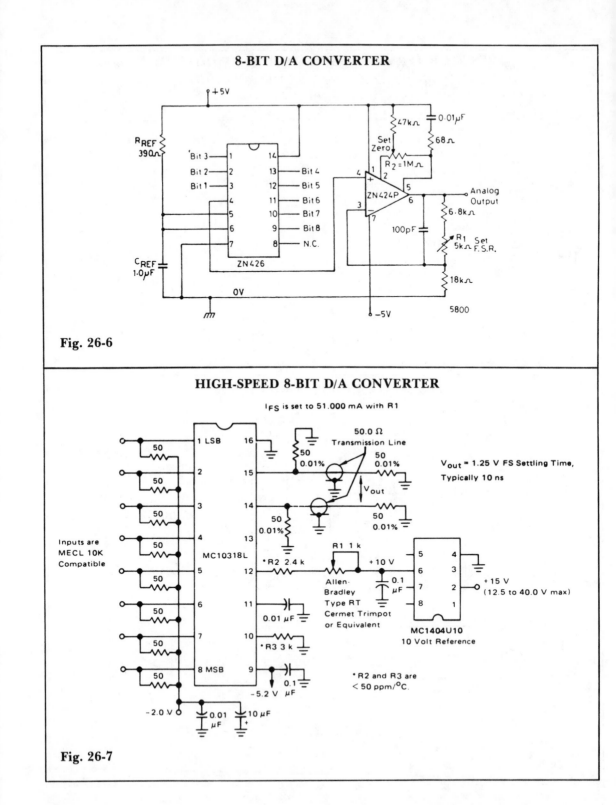

Fig. 26-6

HIGH-SPEED 8-BIT D/A CONVERTER

Fig. 26-7

10-BIT, 4 QUADRANT MULTIPLEXING
D/A CONVERTER (OFFSET BINARY CODING)

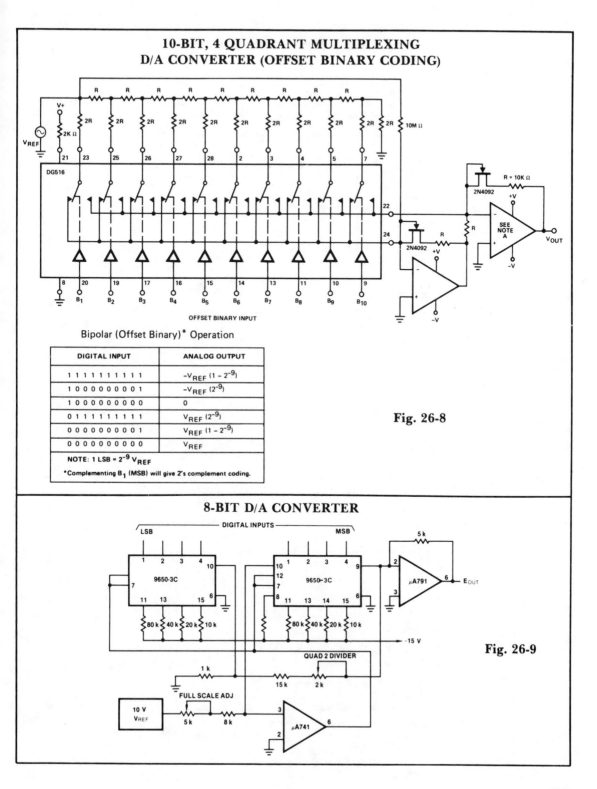

Bipolar (Offset Binary)* Operation

DIGITAL INPUT	ANALOG OUTPUT
1 1 1 1 1 1 1 1 1 1	$-V_{REF} (1 - 2^{-9})$
1 0 0 0 0 0 0 0 0 1	$-V_{REF} (2^{-9})$
1 0 0 0 0 0 0 0 0 0	0
0 1 1 1 1 1 1 1 1 1	$V_{REF} (2^{-9})$
0 0 0 0 0 0 0 0 0 1	$V_{REF} (1 - 2^{-9})$
0 0 0 0 0 0 0 0 0 0	V_{REF}

NOTE: 1 LSB = $2^{-9} V_{REF}$

*Complementing B_1 (MSB) will give 2's complement coding.

Fig. 26-8

8-BIT D/A CONVERTER

Fig. 26-9

241

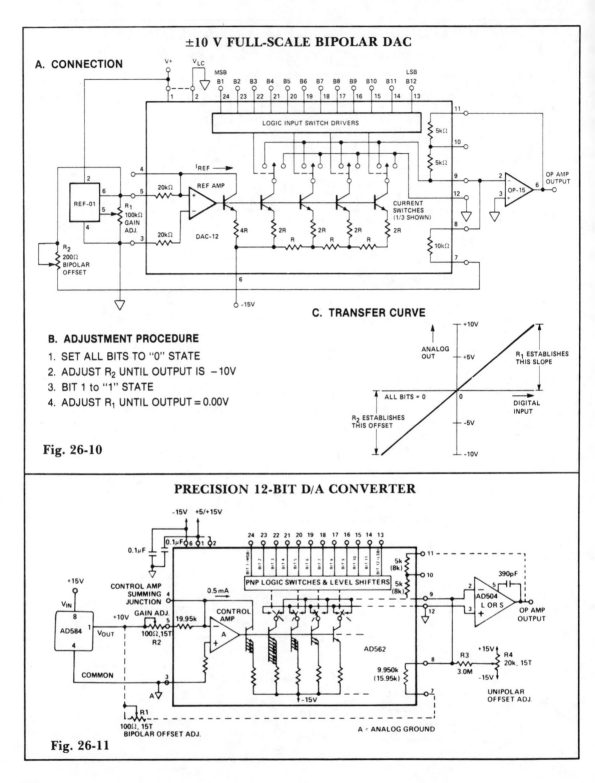

±10 V FULL-SCALE BIPOLAR DAC

A. CONNECTION

LOGIC INPUT SWITCH DRIVERS

REF-01

R_1 100kΩ GAIN ADJ.

R_2 200Ω BIPOLAR OFFSET

REF AMP

DAC-12

I_{REF}

CURRENT SWITCHES (1/3 SHOWN)

OP-15

OP AMP OUTPUT

-15V

B. ADJUSTMENT PROCEDURE

1. SET ALL BITS TO "0" STATE
2. ADJUST R_2 UNTIL OUTPUT IS −10V
3. BIT 1 to "1" STATE
4. ADJUST R_1 UNTIL OUTPUT = 0.00V

Fig. 26-10

C. TRANSFER CURVE

ANALOG OUT

R_1 ESTABLISHES THIS SLOPE

ALL BITS = 0

R_2 ESTABLISHES THIS OFFSET

DIGITAL INPUT

PRECISION 12-BIT D/A CONVERTER

CONTROL AMP SUMMING JUNCTION

PNP LOGIC SWITCHES & LEVEL SHIFTERS

GAIN ADJ.

AD584

V_{IN}

V_{OUT}

CONTROL AMP

AD562

AD504 L OR S

OP AMP OUTPUT

19.95k

100Ω,15T R2

9.950k (15.95k)

R3 3.0M

R4 20k, 15T

UNIPOLAR OFFSET ADJ.

R1 100Ω, 15T BIPOLAR OFFSET ADJ.

A = ANALOG GROUND

COMMON

Fig. 26-11

8-BIT D/A WITH OUTPUT CURRENT-TO-VOLTAGE CONVERSION

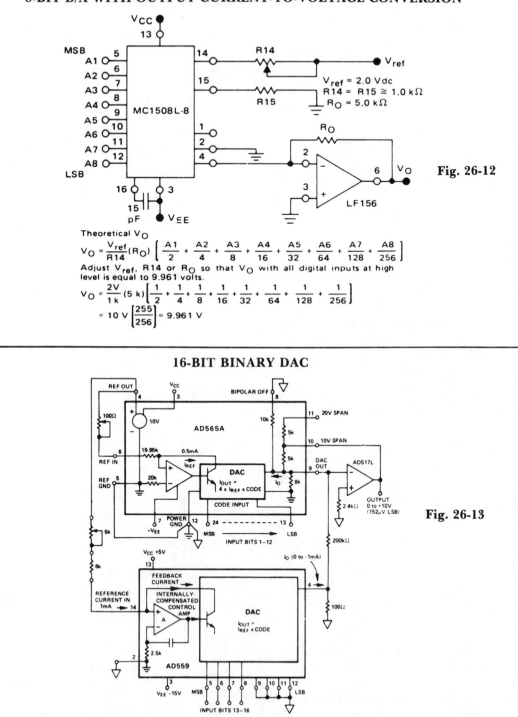

$$V_{ref} = 2.0 \text{ Vdc}$$
$$R14 = R15 \cong 1.0 \text{ k}\Omega$$
$$R_O = 5.0 \text{ k}\Omega$$

Fig. 26-12

Theoretical V_O

$$V_O = \frac{V_{ref}}{R14}(R_O)\left[\frac{A1}{2} + \frac{A2}{4} + \frac{A3}{8} + \frac{A4}{16} + \frac{A5}{32} + \frac{A6}{64} + \frac{A7}{128} + \frac{A8}{256}\right]$$

Adjust V_{ref}, R14 or R_O so that V_O with all digital inputs at high level is equal to 9.961 volts.

$$V_O = \frac{2V}{1k}(5k)\left[\frac{1}{2} + \frac{1}{4} + \frac{1}{8} + \frac{1}{16} + \frac{1}{32} + \frac{1}{64} + \frac{1}{128} + \frac{1}{256}\right]$$

$$= 10 V\left[\frac{255}{256}\right] = 9.961 \text{ V}$$

16-BIT BINARY DAC

Fig. 26-13

243

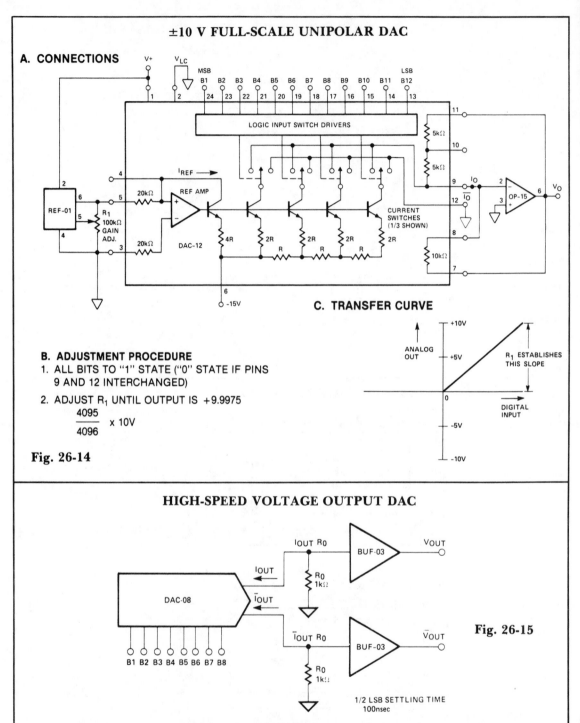

±10 V FULL-SCALE UNIPOLAR DAC

A. CONNECTIONS

LOGIC INPUT SWITCH DRIVERS

REF-01

R_1 100kΩ GAIN ADJ.

20kΩ

REF AMP

20kΩ

DAC-12

I_{REF}

4R 2R 2R 2R 2R

R R R

CURRENT SWITCHES (1/3 SHOWN)

5kΩ

5kΩ

10kΩ

OP-15

V_O

-15V

MSB
B1 B2 B3 B4 B5 B6 B7 B8 B9 B10 B11 B12 LSB

B. ADJUSTMENT PROCEDURE

1. ALL BITS TO "1" STATE ("0" STATE IF PINS 9 AND 12 INTERCHANGED)

2. ADJUST R_1 UNTIL OUTPUT IS +9.9975

$$\frac{4095}{4096} \times 10V$$

Fig. 26-14

C. TRANSFER CURVE

ANALOG OUT

+10V
+5V

R_1 ESTABLISHES THIS SLOPE

DIGITAL INPUT

-5V
-10V

HIGH-SPEED VOLTAGE OUTPUT DAC

DAC-08

I_{OUT}

$\bar{I}_{OUT}$

I_{OUT} R_0

R_0 1kΩ

BUF-03

V_{OUT}

$\bar{I}_{OUT}$ R_0

R_0 1kΩ

BUF-03

$\bar{V}_{OUT}$

Fig. 26-15

1/2 LSB SETTLING TIME 100nsec

B1 B2 B3 B4 B5 B6 B7 B8

SYSTEM WILL DRIVE CABLES OR TWISTED PAIRS.

27

Dip Meters

The sources of the following circuits are contained in the Sources section beginning on page 730. The figure number contained in the box of each circuit correlates to the source entry in the Sources section.

Dip Meter Using Dual-Gate IGFET (MOSFET)
Varicap-Tuned FET DIP Meter with 1 kHz Modulator
Dip Meter Using N-Channel IGFET (MOSFET) and Separate Diode Detector

Basic Grid-Dip Meter
Dip Meter Using Germanium PNP Bipolar Transistor with Separate Diode Detector
Gate-Dip Meter Covers 1.8 - 150 MHz

Dip Meter Using Silicon Junction FET

DIP METER USING DUAL-GATE IGFET (MOSFET)

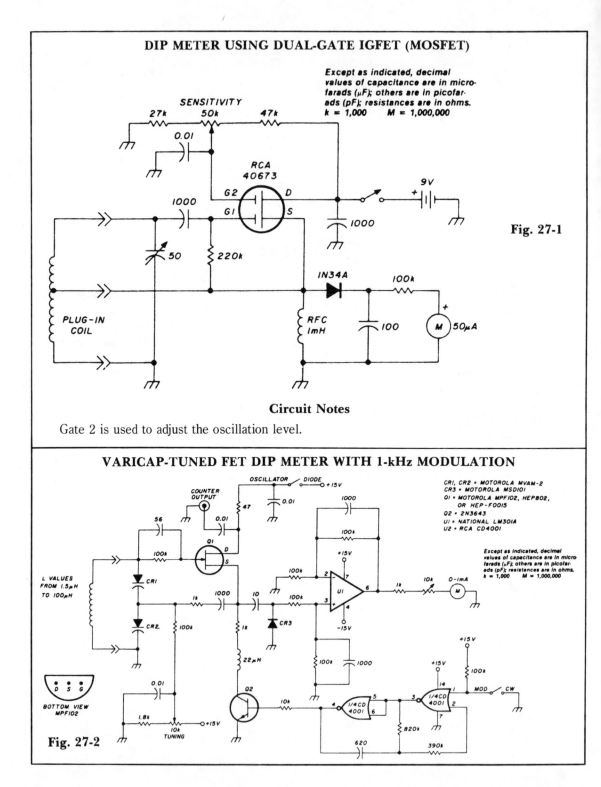

Except as indicated, decimal values of capacitance are in micro-farads (μF); others are in picofar-ads (pF); resistances are in ohms.
k = 1,000   M = 1,000,000

Fig. 27-1

Circuit Notes

Gate 2 is used to adjust the oscillation level.

VARICAP-TUNED FET DIP METER WITH 1-kHz MODULATION

CRI, CR2 = MOTOROLA MVAM-2
CR3 = MOTOROLA MSD101
QI = MOTOROLA MPF102, HEP802, OR HEP-F0015
Q2 = 2N3643
UI = NATIONAL LM301A
U2 = RCA CD4001

Except as indicated, decimal values of capacitance are in micro farads (μF); others are in picofar-ads (pF); resistances are in ohms.
k = 1,000   M = 1,000,000

Fig. 27-2

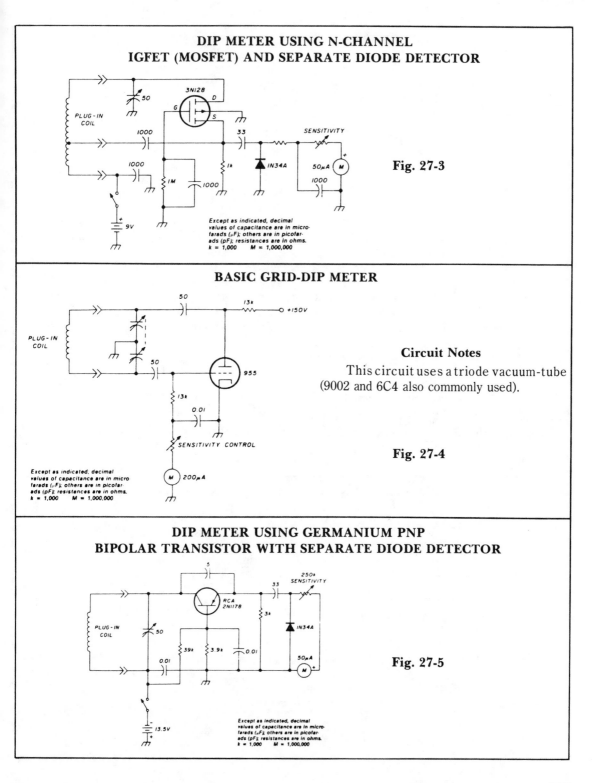

DIP METER USING N-CHANNEL
IGFET (MOSFET) AND SEPARATE DIODE DETECTOR

Fig. 27-3

Except as indicated, decimal values of capacitance are in microfarads (μF); others are in picofarads (pF); resistances are in ohms. k = 1,000   M = 1,000,000

BASIC GRID-DIP METER

Circuit Notes

This circuit uses a triode vacuum-tube (9002 and 6C4 also commonly used).

Fig. 27-4

Except as indicated, decimal values of capacitance are in micro farads (μF); others are in picofarads (pF); resistances are in ohms. k = 1,000   M = 1,000,000

DIP METER USING GERMANIUM PNP
BIPOLAR TRANSISTOR WITH SEPARATE DIODE DETECTOR

Fig. 27-5

Except as indicated, decimal values of capacitance are in microfarads (μF); others are in picofarads (pF); resistances are in ohms. k = 1,000   M = 1,000,000

GATE-DIP METER COVERS 1.8 - 150 MHz

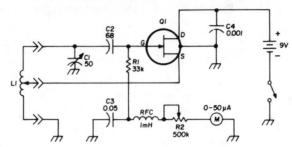

Coil data.

frequency range (MHz)	no. turns	wire size AWG	(mm)	winding length inches	(mm)	tap*	coil diameter inches	(mm)
1.8 - 3.8	82	26 enamel	(0.4)	1 9/16	(40.0)	12	1¼	(32)
3.6 - 7.3	29	26 enamel	(0.4)	9/16	(14.5)	5	1¼	(32)
7.3 - 14.4	18	22 enamel	(0.6)	3/4	(19.0)	3	1	(25)
14.4 - 32	7	22 enamel	(0.6)	1/2	(12.5)	2	1	(25)
29 - 64	3½	18 tinned	(1.0)	3/4	(19.0)	3/4	1	(25)

61 - 150 Hairpin of 16 no. AWG (1.3mm) wire, 5/8 inch (16mm) spacing, 2 3/8 inches (60mm) long including coil-form pins. Tapped at 2 inches (51mm) from ground end.

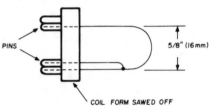

PINS

5/8" (16mm)

COIL FORM SAWED OFF

*Turns from ground-end. 1 inch (25mm) forms are Millen 45004 available from Burstein-Applebee

Fig. 27-6

DIP METER USING SILICON JUNCTION FET

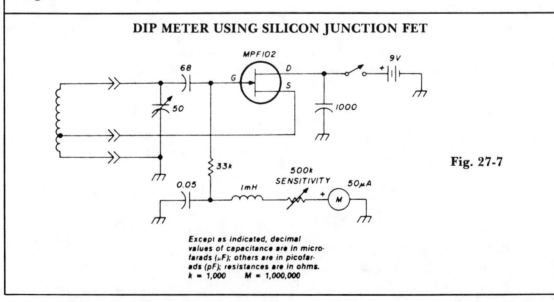

Fig. 27-7

Except as indicated, decimal values of capacitance are in microfarads (μF); others are in picofarads (pF); resistances are in ohms.
k = 1,000   M = 1,000,000

28

Displays

The sources of the following circuits are contained in the Sources section beginning on page 730. The figure number contained in the box of each circuit correlates to the source entry in the Sources section.

LED Brightness Control
LED Bar/Dot Level Meter
60 dB Dot Mode Display
Bar Display with Alarm Flasher
12-Hour Clock with Gas Discharge Displays

Precision Frequency Counter (~ 1 MHz Maximum)
Exclamation Point Display
LED Bar Peak Program Meter Display for Audio

10 MHz Universal Counter

LED BRIGHTNESS CONTROL

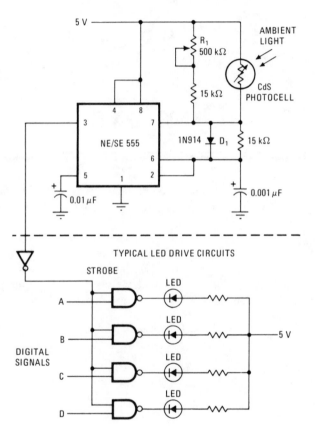

Fig. 28-1

Circuit Notes

The brightness of LED display is varied by using a photocell in place of one timing resistor in a 555 timer, and bypassing the other timing resistor to boost the timer's maximum duty cycle. The result is a brighter display in sunlight and a fainter one in the dark.

LED BAR/DOT LEVEL METER

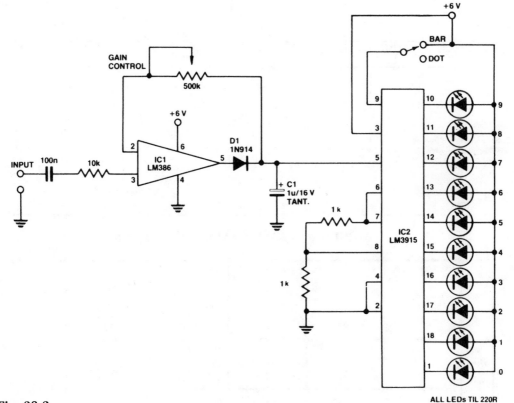

Fig. 28-2

ALL LEDs TIL 220R

Circuit Notes

A simple level of power meter can be arranged to give a bar or dot display for a hi-fi system. Use green LEDs for 0 to 7; yellow for 8 and red for 9 to indicate peak power. The gain control is provided to enable calibration on the equipment with which the unit is used. Because the unit draws some 200 mA, a power supply is advisable instead of running the unit from batteries.

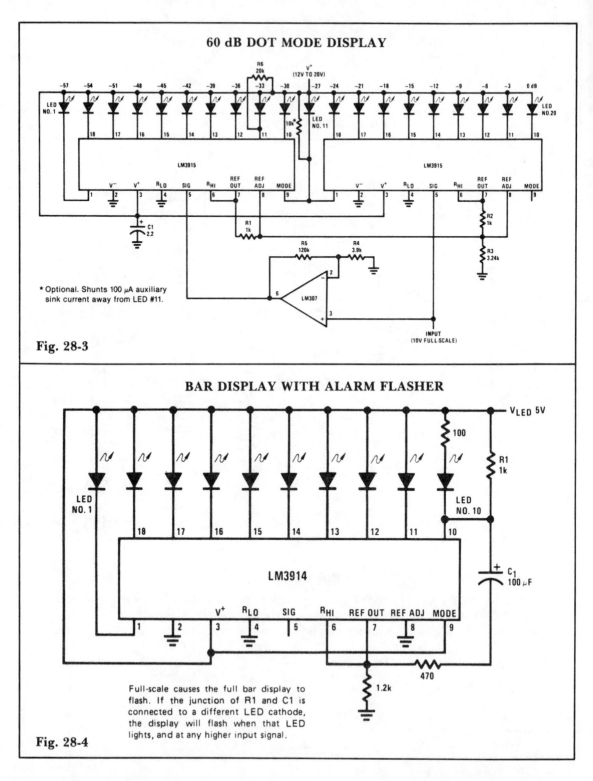

60 dB DOT MODE DISPLAY

* Optional. Shunts 100 μA auxiliary sink current away from LED #11.

Fig. 28-3

BAR DISPLAY WITH ALARM FLASHER

Full-scale causes the full bar display to flash. If the junction of R1 and C1 is connected to a different LED cathode, the display will flash when that LED lights, and at any higher input signal.

Fig. 28-4

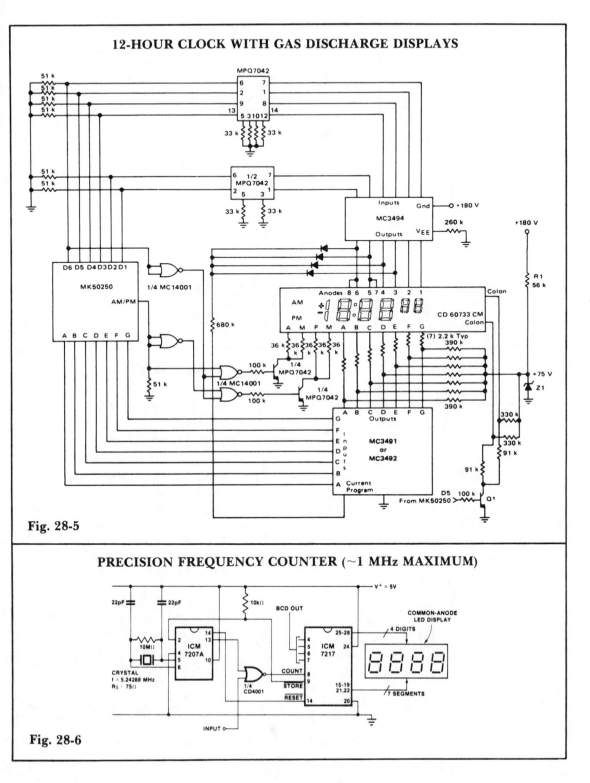

12-HOUR CLOCK WITH GAS DISCHARGE DISPLAYS

Fig. 28-5

PRECISION FREQUENCY COUNTER (~1 MHz MAXIMUM)

Fig. 28-6

EXCLAMATION POINT DISPLAY

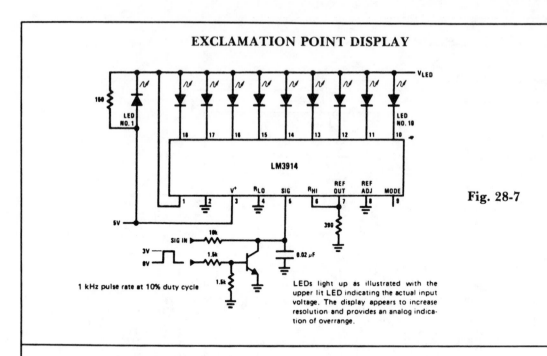

Fig. 28-7

LEDs light up as illustrated with the upper lit LED indicating the actual input voltage. The display appears to increase resolution and provides an analog indication of overrange.

1 kHz pulse rate at 10% duty cycle

LED BAR PEAK PROGRAM METER DISPLAY FOR AUDIO

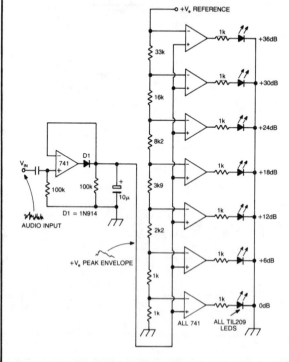

Circuit Notes

A bar column of LEDs is arranged so that as the audio signal level increases, more LEDs in the column light up. The LEDs are arranged vertically in 6 dB steps. A fast response time and a one second decay time give an accurate response to transients and a low "flicker" decay characteristic. On each of the op amps inverting inputs is a dc reference voltage, which increases in 6 dB steps. All noninverting inputs are tied together and connected to the positive peak envelope of the audio signal. Thus, as this envelope exceeds a particular voltage reference, the op amp output goes high and the LED lights up. Also, all the LEDs below this are illuminated.

Fig. 28-8

10 MHz UNIVERSAL COUNTER

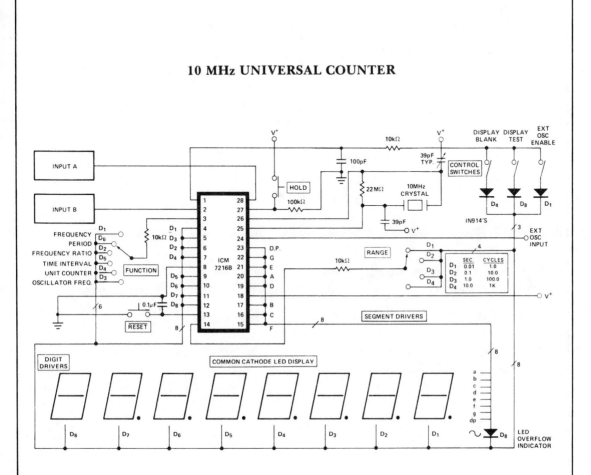

Fig. 28-9

Circuit Notes

This is a minimum component complete Universal Counter. It can use input frequencies up to 10 MHz at INPUT A and 2 MHz at INPUT B. If the signal at INPUT A has a very low duty cycle, it may be necessary to use a 74121 monostable multivibrator or similar circuit to stretch the input pulse width to be able to guarantee that it is at least 50 ns in duration.

29

Dividers

The sources of the following circuits are contained in the Sources section beginning on page 730. The figure number contained in the box of each circuit correlates to the source entry in the Sources section.

CMOS Programmable Divide-by-N Counter
Frequency Divider Chain
Frequency Divider with Transient

Free Output
Binary Divider Chain
Decade Frequency Divider

CMOS PROGRAMMABLE DIVIDE-BY-N COUNTER

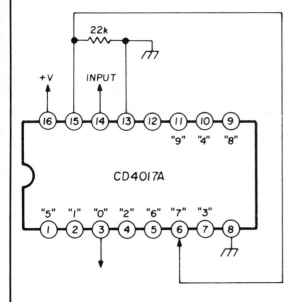

Circuit Notes

A single connection change permits division by any integer between 2 and 10. The RCA CD4017A Johnson decade counter is shown connected as a divide by 7 counter. The resistor is used to hold the reset line low. When the appropriate number is reached, that output and the reset line are driven high, resetting the counter. To divide by other integers, pin 15 should be connected to the desired output. For example, pin 1 for a divide by 5, or pin 7 for a divide by 3. The output of the divider appears on the 0 line.

Fig. 29-1

FREQUENCY DIVIDER CHAIN

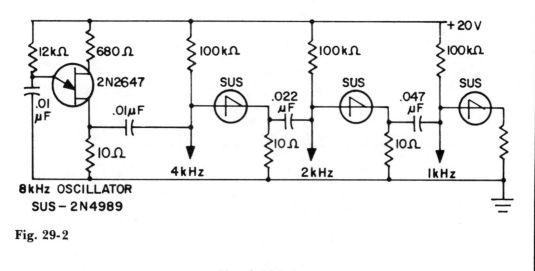

Fig. 29-2

Circuit Notes

Sawtooth output from each stage is one half frequency of preceding stage.

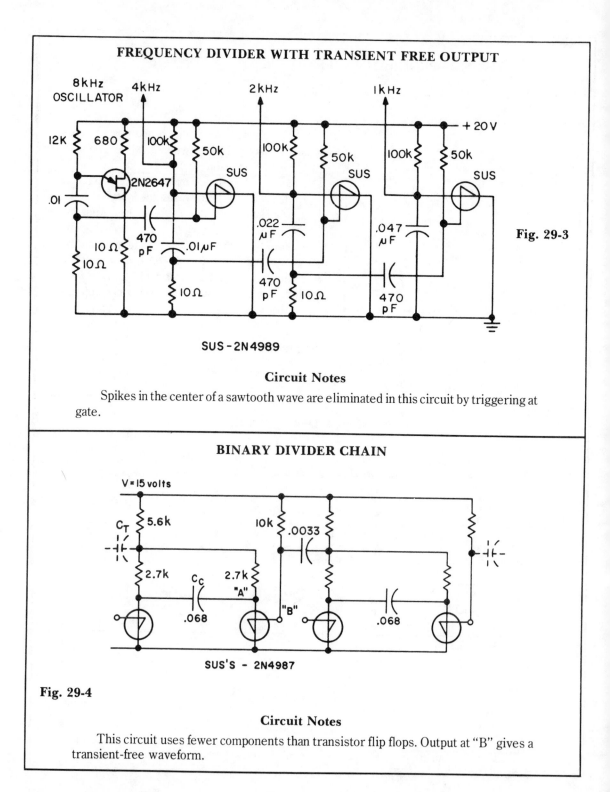

FREQUENCY DIVIDER WITH TRANSIENT FREE OUTPUT

Fig. 29-3

SUS - 2N4989

Circuit Notes

Spikes in the center of a sawtooth wave are eliminated in this circuit by triggering at gate.

BINARY DIVIDER CHAIN

SUS'S - 2N4987

Fig. 29-4

Circuit Notes

This circuit uses fewer components than transistor flip flops. Output at "B" gives a transient-free waveform.

DECADE FREQUENCY DIVIDER

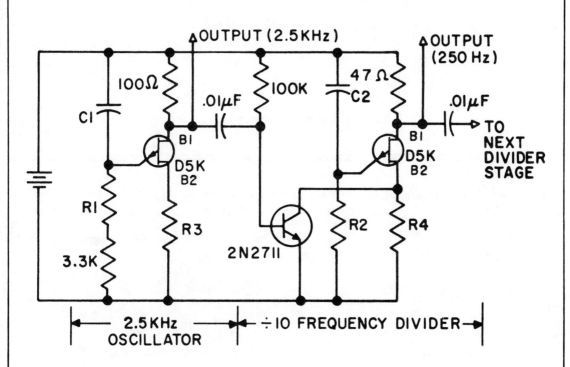

Fig. 29-5

Circuit Notes

In the next stage, the product of R2 and C2 should be 10 × that of the preceding stage (±2%). R2 should be between 27K and 10 M.

C1 & C2—.0047 μF (±1 %)
R1—100K (±1%)
R2—1M (±1%)
R3—R4—1K (may need to be adjusted for variation of R_BB of UJT)

30

Drivers

The sources of the following circuits are contained in the Sources section beginning on page 730. The figure number contained in the box of each circuit correlates to the source entry in the Sources section.

Driver Circuits
50 Ohm Driver
Line Driver
High Speed Laser Diode Driver
Capacitive Load Driver
Relay Driver
Relay Driver
BIFET Cable Driver

High Speed Line Driver for Multiplexers
High Impedance Meter Driver
CRT Deflection Yoke
CRT Yoke Driver
Solenoid Driver
Coaxial Cable Driver
High Speed Shield/Line Driver
Relay Driver with Strobe

Direct Dc Drive Interface of a Triac

DRIVER CIRCUITS

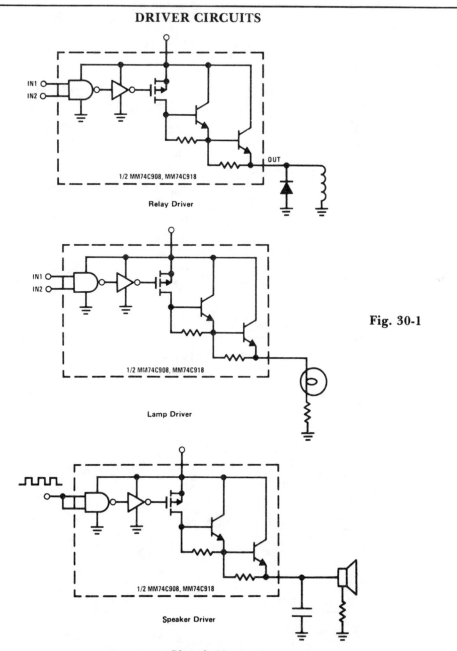

Relay Driver

Fig. 30-1

Lamp Driver

Speaker Driver

Circuit Notes

CMOS drivers for relays, lamps, speakers, etc., offers extremely low standby power. At V_{cc} = 15 V, power dissipation per package is typically 750 nW when the outputs are not drawing current. Thus, the drivers can be sitting out on line (a telephone line, for example) drawing essentially zero current until activated.

50 OHM DRIVER

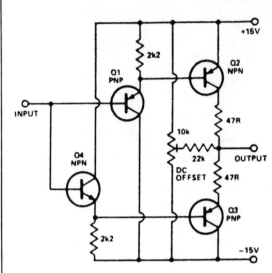

Circuit Notes

To buffer a test generator to the outside world requires an amplifier with sufficient bandwidth and power handling capability. The circuit is a very simple unity gain buffer. It has a fairly high input impedance, a 50 ohm output impedance, a wide bandwidth, and high slew rate. The circuit is simply two pairs of emitter followers. The base emitter voltages of Q1 and Q2 cancel out, and so do those of Q3 and Q4. The preset is used to zero out any small dc offsets due to mismatching in the transistors.

Fig. 30-2

LINE DRIVER

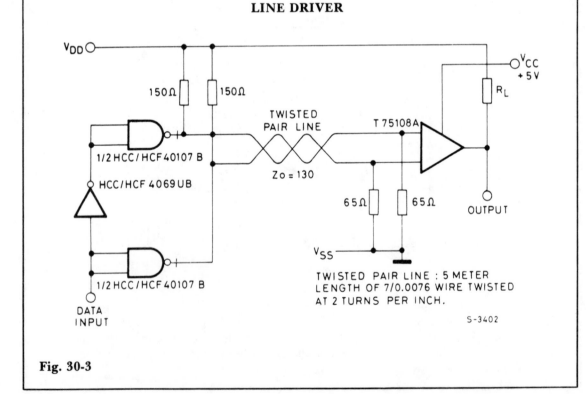

TWISTED PAIR LINE : 5 METER LENGTH OF 7/0.0076 WIRE TWISTED AT 2 TURNS PER INCH.

S-3402

Fig. 30-3

HIGH-SPEED LASER DIODE DRIVER

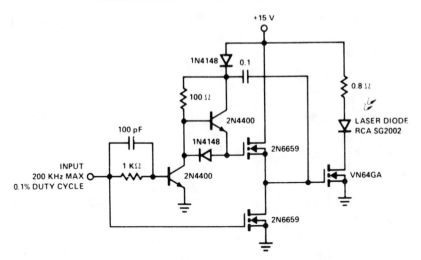

Fig. 30-4

Circuit Notes

A faster driver can supply higher peak gate current to switch the VN64GA very quickly. The circuit uses a VMOS totempole stage to drive the high power switch.

CAPACITIVE LOAD DRIVER

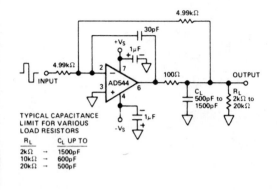

Fig. 30-5

Circuit Notes

The circuit employs a 100 ohm isolation resistor which enables the amplifier to drive capacitive loads exceeding 500 pF; the resistor effectively isolates the high frequency feedback from the load and stabilizes the circuit. Low frequency feedback is returned to the amplifier summing junction via the low pass filter formed by the 100 ohm series resistor and the load capacitance. C_L.

263

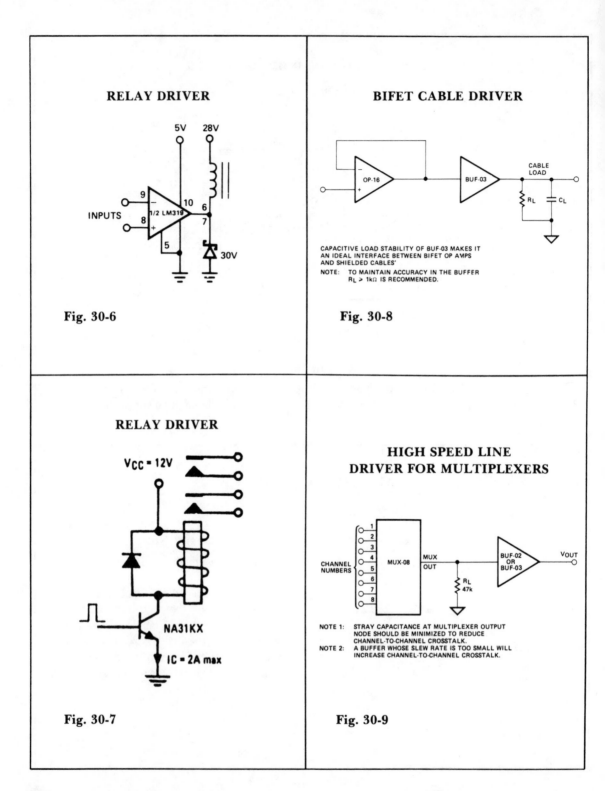

RELAY DRIVER

5V   28V

9 −
10   6
INPUTS   1/2 LM319   7
8 +
5
30V

Fig. 30-6

BIFET CABLE DRIVER

OP-16

BUF-03

CABLE
LOAD

R_L C_L

CAPACITIVE LOAD STABILITY OF BUF-03 MAKES IT
AN IDEAL INTERFACE BETWEEN BIFET OP AMPS
AND SHIELDED CABLES'
NOTE: TO MAINTAIN ACCURACY IN THE BUFFER
$R_L > 1k\Omega$ IS RECOMMENDED.

Fig. 30-8

RELAY DRIVER

$V_{CC} = 12V$

NA31KX

IC = 2A max

Fig. 30-7

HIGH SPEED LINE
DRIVER FOR MULTIPLEXERS

1
2
3
4
CHANNEL   MUX-08   MUX
NUMBERS   OUT
5
6
7
8

R_L
47k

BUF-02
OR
BUF-03

V_{OUT}

NOTE 1:   STRAY CAPACITANCE AT MULTIPLEXER OUTPUT
NODE SHOULD BE MINIMIZED TO REDUCE
CHANNEL-TO-CHANNEL CROSSTALK.
NOTE 2:   A BUFFER WHOSE SLEW RATE IS TOO SMALL WILL
INCREASE CHANNEL-TO-CHANNEL CROSSTALK.

Fig. 30-9

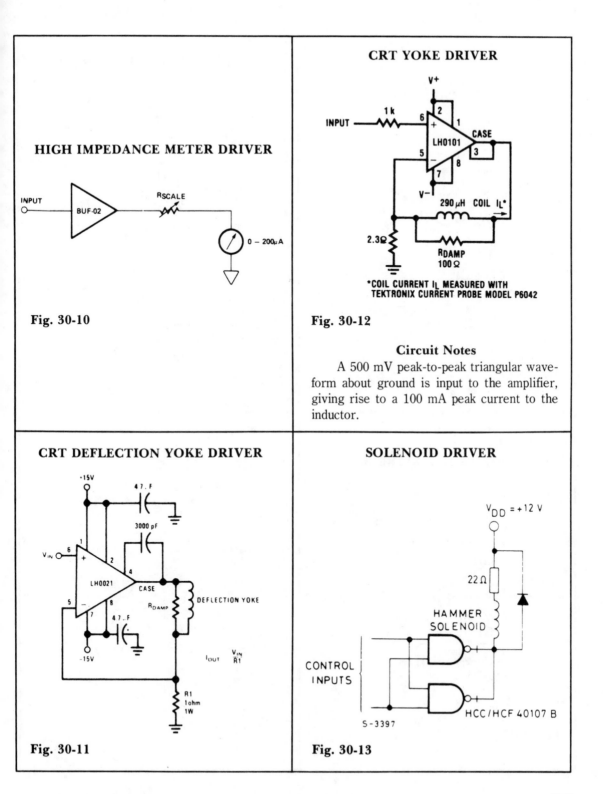

HIGH IMPEDANCE METER DRIVER

Fig. 30-10

CRT YOKE DRIVER

Fig. 30-12

Circuit Notes

A 500 mV peak-to-peak triangular wave-form about ground is input to the amplifier, giving rise to a 100 mA peak current to the inductor.

CRT DEFLECTION YOKE DRIVER

Fig. 30-11

SOLENOID DRIVER

Fig. 30-13

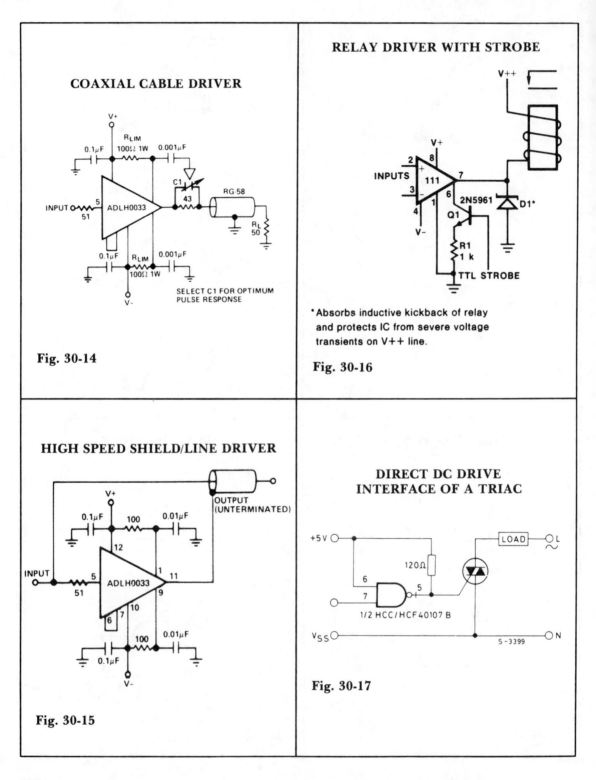

COAXIAL CABLE DRIVER

Fig. 30-14

SELECT C1 FOR OPTIMUM
PULSE RESPONSE

RELAY DRIVER WITH STROBE

*Absorbs inductive kickback of relay
and protects IC from severe voltage
transients on V++ line.

Fig. 30-16

HIGH SPEED SHIELD/LINE DRIVER

Fig. 30-15

DIRECT DC DRIVE
INTERFACE OF A TRIAC

Fig. 30-17

31

Fiber Optic Circuits

The sources of the following circuits are contained in the Sources section beginning on page 730. The figure number contained in the box of each circuit correlates to the source entry in the Sources section.

Fiber-Optics Half Duplex Information Link
Fiber-Optic Receiver, Very High Sensitivity, Low Speed, 3 nW
Fiber-Optic Link

Fiber-Optic Link Repeater
Fiber-Optic Receiver, High Sensitivity, 30 nW
Fiber-Optic Receiver, Low Sensitivity, 300 nW

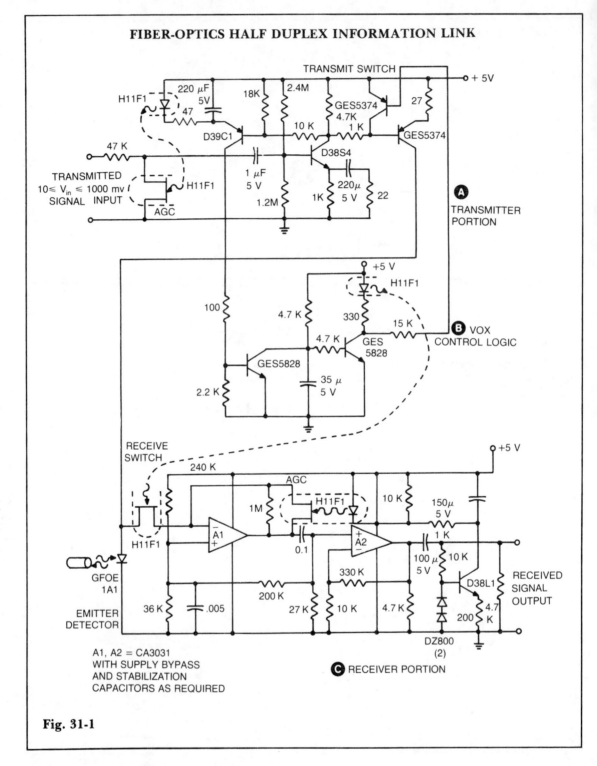

FIBER-OPTICS HALF DUPLEX INFORMATION LINK

A TRANSMITTER PORTION

B VOX CONTROL LOGIC

C RECEIVER PORTION

A1, A2 = CA3031
WITH SUPPLY BYPASS
AND STABILIZATION
CAPACITORS AS REQUIRED

Fig. 31-1

FIBER-OPTIC RECEIVER, VERY HIGH SENSITIVITY, LOW SPEED, 3nW

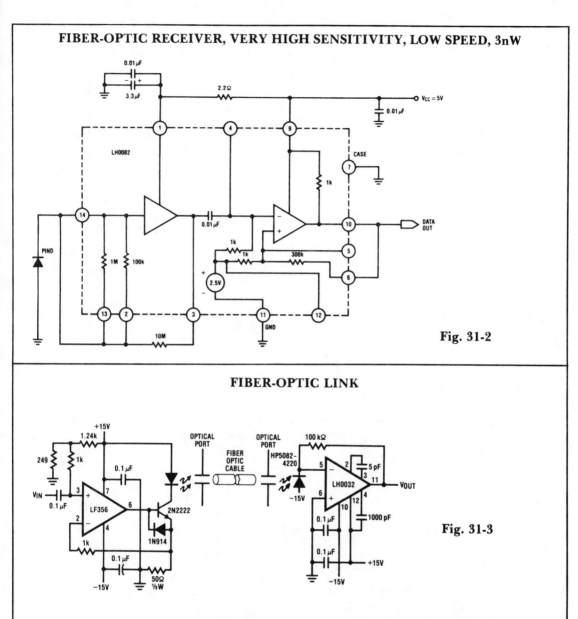

Fig. 31-2

FIBER-OPTIC LINK

Fig. 31-3

Circuit Notes

Fiber Optic applications require analog drivers and receivers operating in the megahertz region. This complete analog transmission system is suitable for optical communication applications up to 3.5 MHz. The transmitter LED is normally biased at 50 mA operating current. The input is capacitively coupled and ranges from 0 to 5 V, modulating the LED current from 0 to 100 mA. The receiver circuit is configured as a transimpedance amplifier. The photodiode with 0.5 amp per watt responsivity generates a 50 mV signal at the receiver output for 1 μW of light input.

269

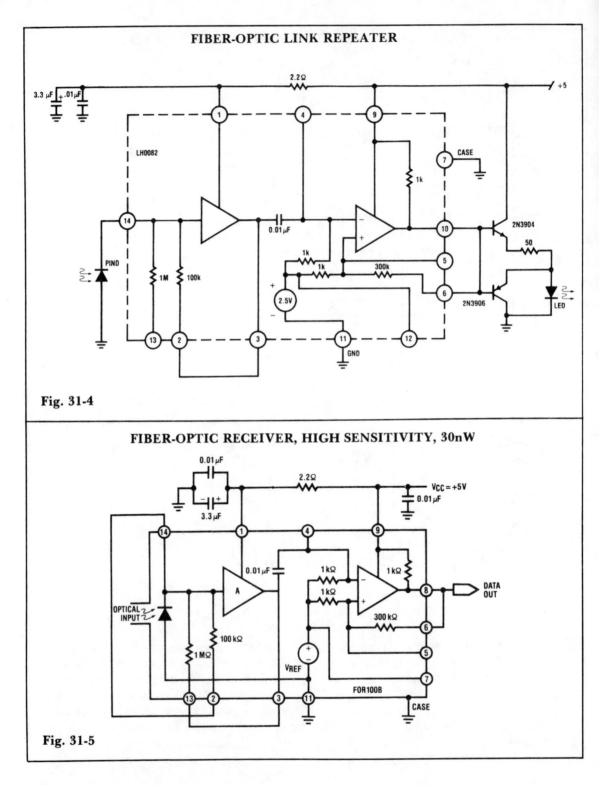

FIBER-OPTIC LINK REPEATER

Fig. 31-4

FIBER-OPTIC RECEIVER, HIGH SENSITIVITY, 30nW

Fig. 31-5

FIBER-OPTIC RECEIVER, LOW SENSITIVITY, 300nW

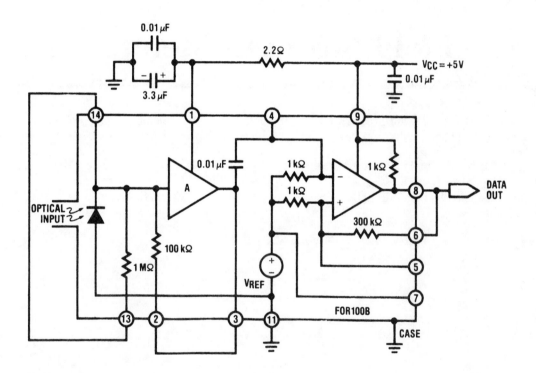

Fig. 31-6

271

32

Field Strength Meters

The sources of the following circuits are contained in the Sources section beginning on page 730. The figure number contained in the box of each circuit correlates to the source entry in the Sources section.

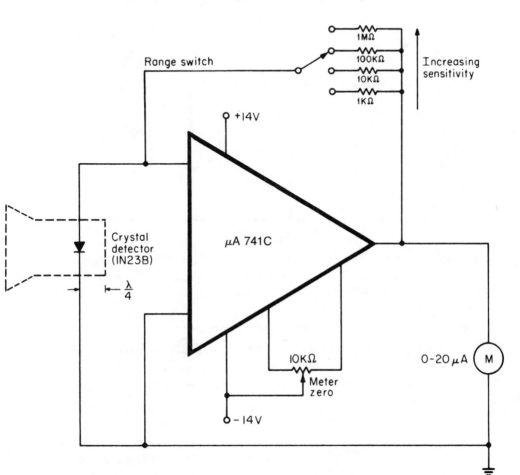

LOW COST MICROWAVE FIELD STRENGTH METER

Range switch

1MΩ
100KΩ
10KΩ
1KΩ

Increasing sensitivity

+14V

Crystal detector (IN23B)

μA 741C

$\frac{\lambda}{4}$

10KΩ
Meter zero

0-20 μA M

−14V

Fig. 32-1

Circuit Notes

When operating, a waveguide directs energy onto a crystal detector. The diode shown is for X-band operation. The waveguide is a 1½ inch piece of plastic tubing with the ends flared. The plastic is coated with an electroless copper solution to provide a conducting surface. The dimensions are not critical. For calibrated readings, the meter is placed in a known field or else compared to a calibrated meter. To operate the meter, point it away from the signal. Switch the meter to the desired range, and adjust the zero control for a 0 reading. Then point the waveguide at the signal, and read field strength directly.

SENSITIVE FIELD-STRENGTH METER

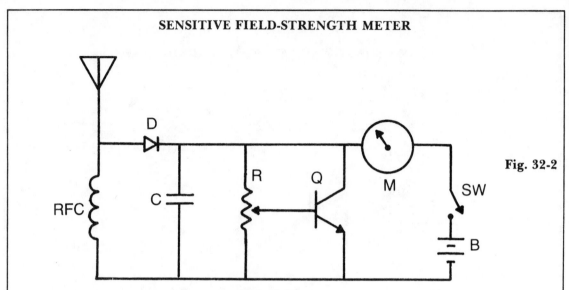

Fig. 32-2

Circuit Notes

Increased sensitivity gives field strength reading from low power transmitters. Operating range 3-30 MHz. To operate, adjust R for ⅓ to ½ scale reading. RFC = 2.5 mH choke, C = 1,000 pF, R = 50 K pot, M = 0 − 1 mA, D = 1N34 or 1N60 (Germanium), Q = NPN (RCASK3020, 2N3904 or equivalent).

ADJUSTABLE-SENSITIVITY FIELD-STRENGTH INDICATOR

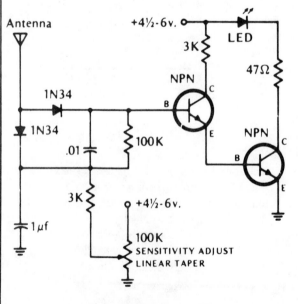

Circuit Notes

The LED lights if the rf field is higher than the pre-set field strength level. Diodes should be germanium. Transistors (NPN) = 2N2222, 2N3393, 2N3904 or equivalent.

Fig. 32-3

FIELD STRENGTH METER – 1.5 to 150 MHz

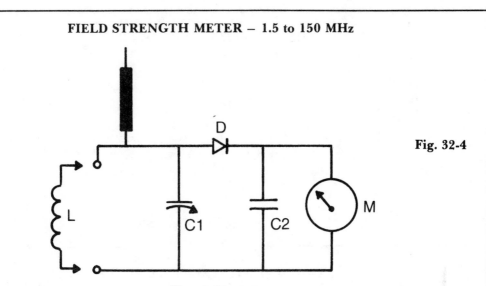

Fig. 32-4

Circuit Notes

The tuning range is determined by coil (L) dimensions and setting of C1. Coils can be plugged in for multirange use or soldered in place if only limited frequency range is of inter-est. C1 = 36 pF variable, C2 = .0047 disc, D = 1N60 (germanium) and M = 0–1 mA meter. For increased sensitivity, use 50 μA meter.

SIMPLE FIELD STRENGTH METER

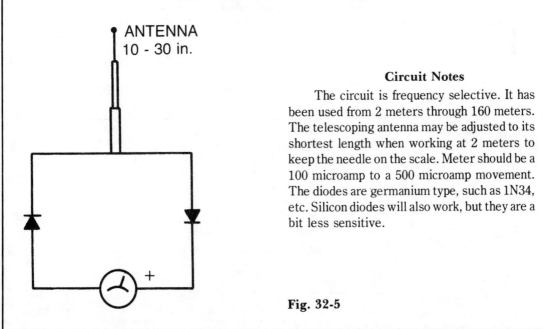

ANTENNA
10 - 30 in.

Circuit Notes

The circuit is frequency selective. It has been used from 2 meters through 160 meters. The telescoping antenna may be adjusted to its shortest length when working at 2 meters to keep the needle on the scale. Meter should be a 100 microamp to a 500 microamp movement. The diodes are germanium type, such as 1N34, etc. Silicon diodes will also work, but they are a bit less sensitive.

Fig. 32-5

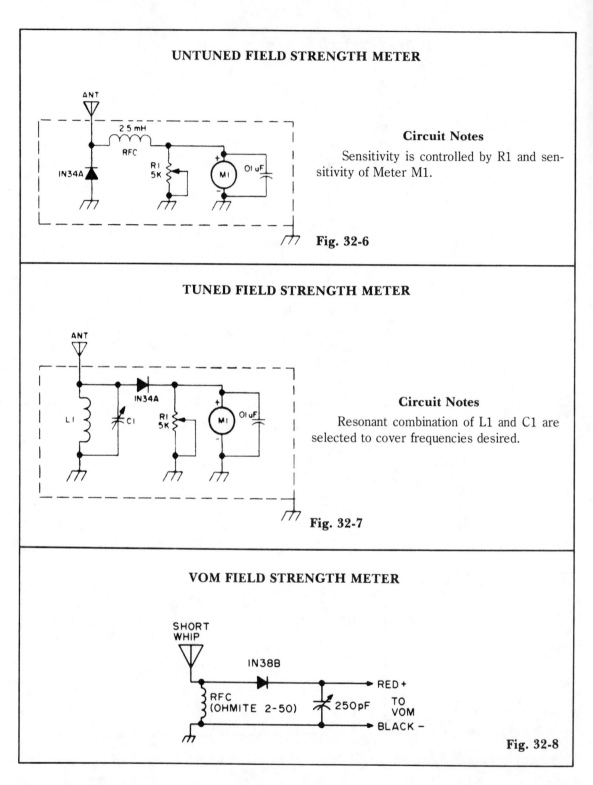

UNTUNED FIELD STRENGTH METER

ANT

2 5 mH

RFC

IN34A

R1
5K

M1

01 uF

Circuit Notes

Sensitivity is controlled by R1 and sensitivity of Meter M1.

Fig. 32-6

TUNED FIELD STRENGTH METER

ANT

L1

C1

IN34A

R1
5K

M1

01 uF

Circuit Notes

Resonant combination of L1 and C1 are selected to cover frequencies desired.

Fig. 32-7

VOM FIELD STRENGTH METER

SHORT
WHIP

IN38B

RFC
(OHMITE 2-50)

250pF

RED +

TO
VOM

BLACK −

Fig. 32-8

33

Filters

The sources of the following circuits are contained in the Sources section beginning on page 730. The figure number contained in the box of each circuit correlates to the source entry in the Sources section.

FIVE-POLE ACTIVE FILTER

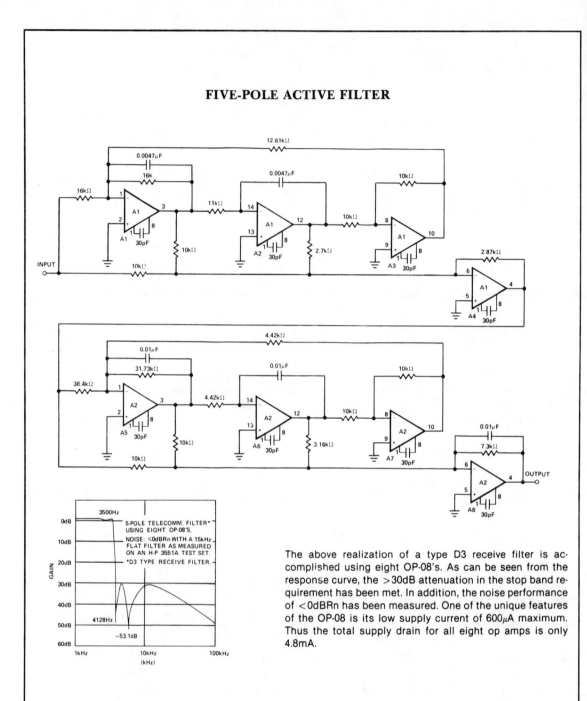

The above realization of a type D3 receive filter is accomplished using eight OP-08's. As can be seen from the response curve, the >30dB attenuation in the stop band requirement has been met. In addition, the noise performance of <0dBRn has been measured. One of the unique features of the OP-08 is its low supply current of 600μA maximum. Thus the total supply drain for all eight op amps is only 4.8mA.

Fig. 33-1

DIGITALLY TUNED LOW POWER ACTIVE FILTER

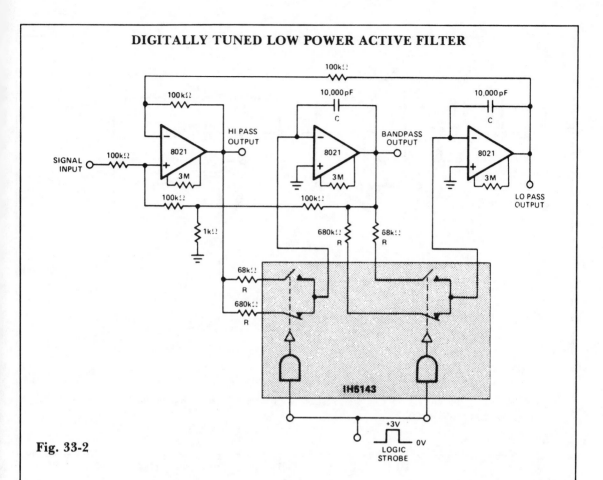

Fig. 33-2

Circuit Notes

Constant gain, constant Q, variable frequency filter which provides simultaneous low-pass, bandpass, and high-pass outputs. With the component values shown, center frequency will be 235 Hz and 23.5 Hz for high and low logic inputs respectively, Q = 100, and gain = 100.

$$f_n = \text{center frequency} = \frac{1}{2\pi RC}$$

10 kHz SALLEN-KEY LOW-PASS FILTER

Fig. 33-3

279

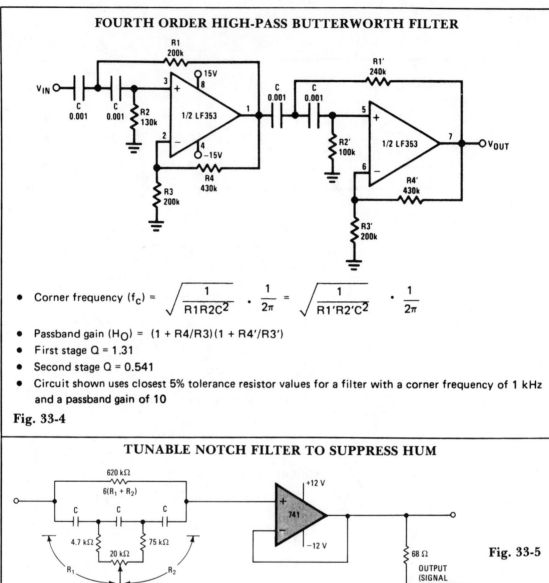

FOURTH ORDER HIGH-PASS BUTTERWORTH FILTER

- Corner frequency $(f_c) = \sqrt{\dfrac{1}{R1R2C^2}} \cdot \dfrac{1}{2\pi} = \sqrt{\dfrac{1}{R1'R2'C^2}} \cdot \dfrac{1}{2\pi}$
- Passband gain $(H_O) = (1 + R4/R3)(1 + R4'/R3')$
- First stage Q = 1.31
- Second stage Q = 0.541
- Circuit shown uses closest 5% tolerance resistor values for a filter with a corner frequency of 1 kHz and a passband gain of 10

Fig. 33-4

TUNABLE NOTCH FILTER TO SUPPRESS HUM

Fig. 33-5

C = 0.047 μF ± 10%

Circuit Notes

This narrow-stop-band filter can be tuned by the pot to place the notch at any frequency from 45 to 90 Hz. It attenuates power-line hum or other unwanted signals by at least 30 dB. Because the circuit uses wide-tolerance parts, it is inexpensive to build.

THREE-AMPLIFIER NOTCH FILTER
(OR ELLIPTIC FILTER BUILDING BLOCK)

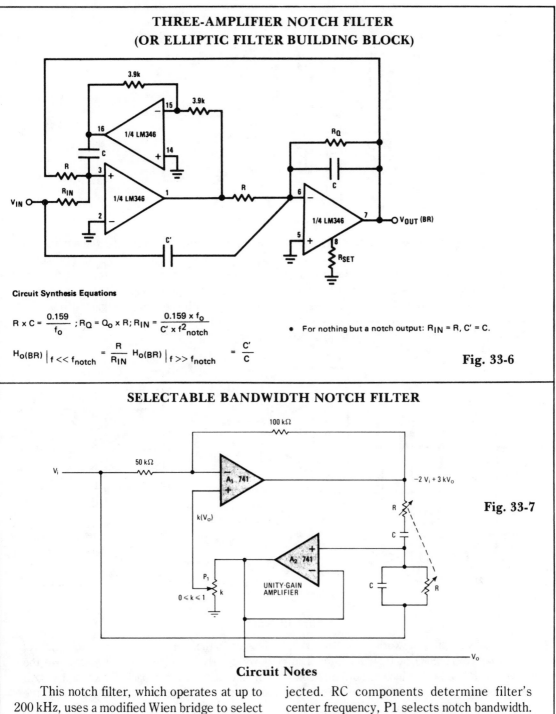

Circuit Synthesis Equations

$$R \times C = \frac{0.159}{f_o} \; ; \; R_Q = Q_o \times R; \; R_{IN} = \frac{0.159 \times f_o}{C' \times f^2_{notch}}$$

$$H_{o(BR)} \Big|_{f \ll f_{notch}} = \frac{R}{R_{IN}} \quad H_{o(BR)} \Big|_{f \gg f_{notch}} = \frac{C'}{C}$$

• For nothing but a notch output: $R_{IN} = R, C' = C$.

Fig. 33-6

SELECTABLE BANDWIDTH NOTCH FILTER

Fig. 33-7

Circuit Notes

This notch filter, which operates at up to 200 kHz, uses a modified Wien bridge to select bandwidth over which frequencies are rejected. RC components determine filter's center frequency, P1 selects notch bandwidth. Notch depth is fixed at about 60 dB.

4.5 MHz NOTCH FILTER

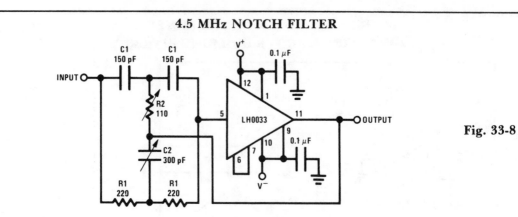

Fig. 33-8

Circuit Notes

Component value sensitivity is extremely critical, as are temperature coefficients and matching of the components. Best performance is attained when perfectly matched components are used and when the gain of the amplifier is unity. To illustrate, the quality factor Q is very high as amplifier gain approaches 1 with all components matched (in fact, theoretically it approaches ∞) but decreases to about 12.5 with the amplifier gain at 0.98.

HIGH Q NOTCH FILTER

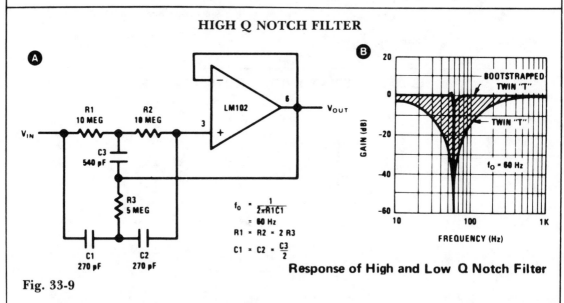

$$f_0 = \frac{1}{2\pi R1C1}$$
$$= 60 \text{ Hz}$$
$$R1 = R2 = 2 R3$$
$$C1 = C2 = \frac{C3}{2}$$

Response of High and Low Q Notch Filter

Fig. 33-9

Circuit Notes

A shows a twin-T network connected to an LM102 to form a high Q, 60 Hz notch filter. The junction of R3 and C3, which is normally connected to ground, is bootstrapped to the output of the follower. Because the output of the follower is a very low impedance, neither the depth nor the frequency of the notch change; however, the Q is raised in proportion to the amount of signal fed back to R3 and C3. B shows the response of a normal twin-T and the response with the follower added.

REJECTION FILTER

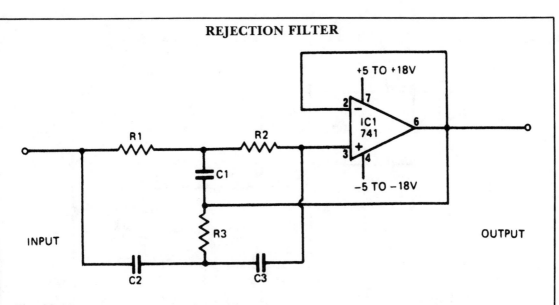

Fig. 33-10

Circuit Notes

This narrowband filter using the 741 operational amplifier can provide up to 60 dB of rejection. With resistors equal to 100 K and capacitors equal to 320 pF, the circuit will reject 50 Hz. Frequencies within the range 1 Hz to 10 kHz may be rejected by selecting components in accordance with the formula:

$$F = \frac{1}{2\pi RC}$$

To obtain rejections better than 40 dB, resistors should be matched to 0.1% and capacitors to 1%.

NOTCH FILTER USING THE μA4136 AS A GYRATOR

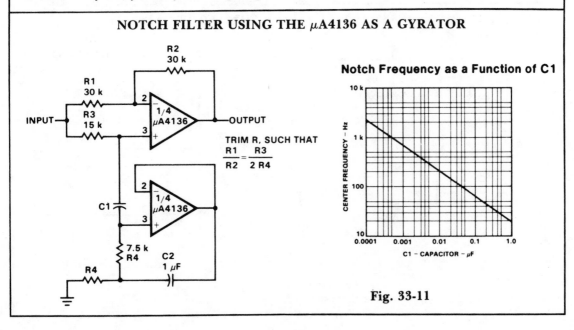

Notch Frequency as a Function of C1

Fig. 33-11

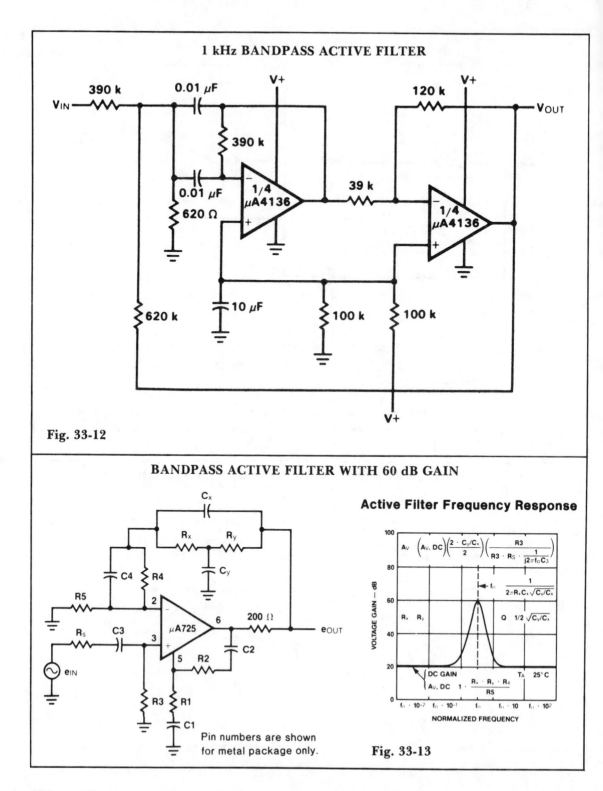

1 kHz BANDPASS ACTIVE FILTER

Fig. 33-12

BANDPASS ACTIVE FILTER WITH 60 dB GAIN

Active Filter Frequency Response

Pin numbers are shown
for metal package only.

Fig. 33-13

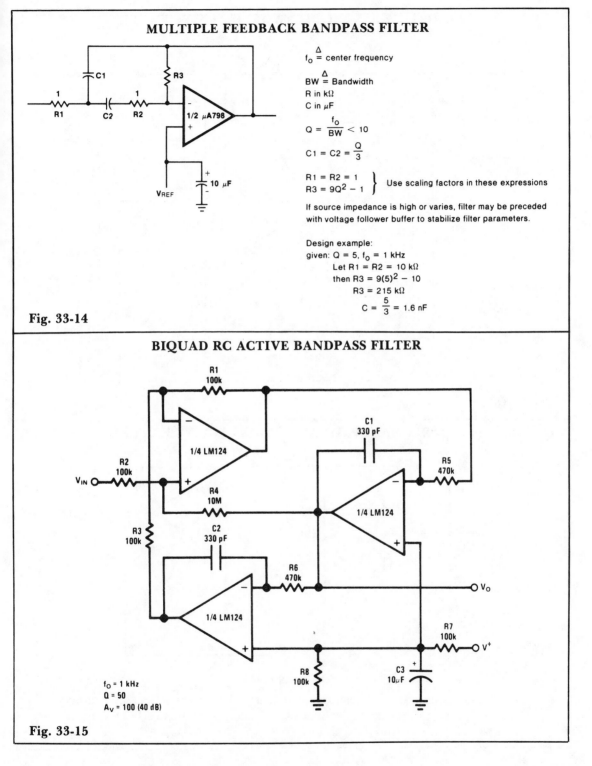

MULTIPLE FEEDBACK BANDPASS FILTER

$f_O \triangleq$ center frequency

$BW \triangleq$ Bandwidth

R in kΩ

C in μF

$$Q = \frac{f_O}{BW} < 10$$

$$C1 = C2 = \frac{Q}{3}$$

$\left.\begin{array}{l} R1 = R2 = 1 \\ R3 = 9Q^2 - 1 \end{array}\right\}$ Use scaling factors in these expressions

If source impedance is high or varies, filter may be preceded with voltage follower buffer to stabilize filter parameters.

Design example:

given: Q = 5, f_O = 1 kHz

Let R1 = R2 = 10 kΩ

then R3 = $9(5)^2 - 10$

R3 = 215 kΩ

$C = \frac{5}{3} = 1.6$ nF

Fig. 33-14

BIQUAD RC ACTIVE BANDPASS FILTER

f_O = 1 kHz

Q = 50

A_V = 100 (40 dB)

Fig. 33-15

285

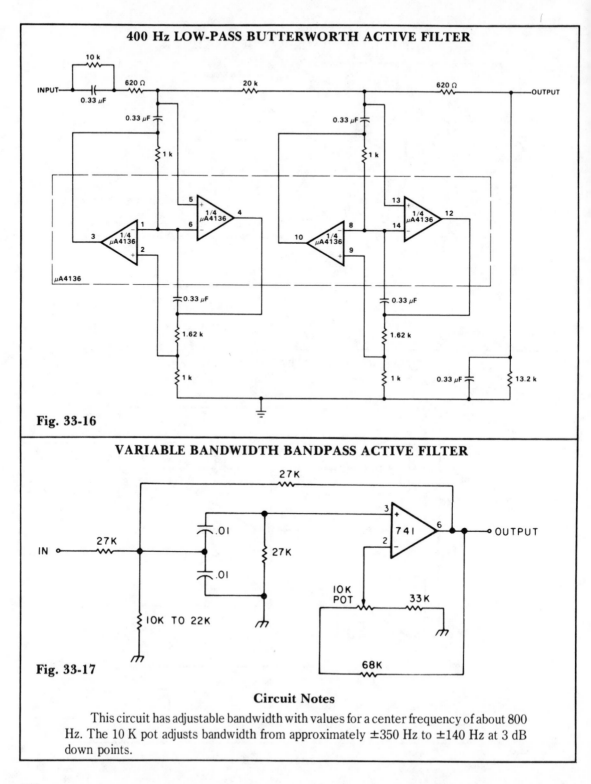

400 Hz LOW-PASS BUTTERWORTH ACTIVE FILTER

Fig. 33-16

VARIABLE BANDWIDTH BANDPASS ACTIVE FILTER

Fig. 33-17

Circuit Notes

This circuit has adjustable bandwidth with values for a center frequency of about 800 Hz. The 10 K pot adjusts bandwidth from approximately ±350 Hz to ±140 Hz at 3 dB down points.

LOW-PASS FILTER

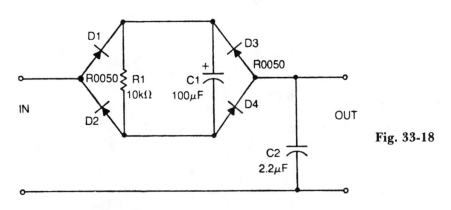

Fig. 33-18

D1, D2, D3, D4—HEP R0050   C2—2.2μF
C1—100μF, 50V electrolytic   R1—10kΩ, 1/2W

Circuit Notes

This nonlinear, passive filter circuit rejects ripple (or unwanted but fairly steady voltage) without appreciably affecting the rise time of a signal. The circuit works best when the signal level is considerably lower than the unwanted ripple, provided the ripple level is fairly constant. The circuit has characteristics similar to two peak-detecting sample-and-hold circuits in tandem with a voltage averager.

HIGH Q BANDPASS FILTER

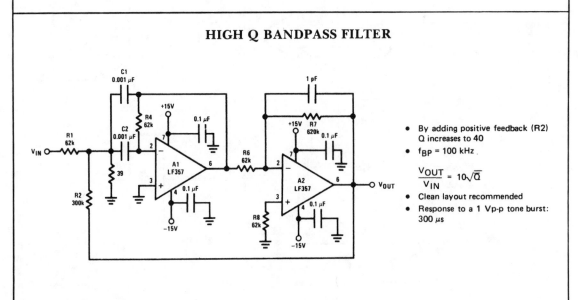

- By adding positive feedback (R2) Q increases to 40
- f_{BP} = 100 kHz

$$\frac{V_{OUT}}{V_{IN}} = 10\sqrt{Q}$$

- Clean layout recommended
- Response to a 1 Vp-p tone burst: 300 μs

Fig. 33-19

MFB BANDPASS FILTER FOR MULTICHANNEL TONE DECODER

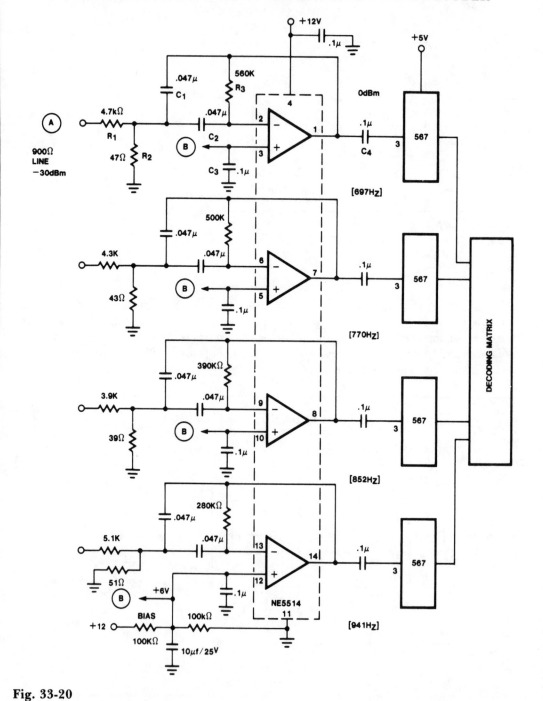

Fig. 33-20

SALLEN-KEY SECOND ORDER LOW-PASS FILTER

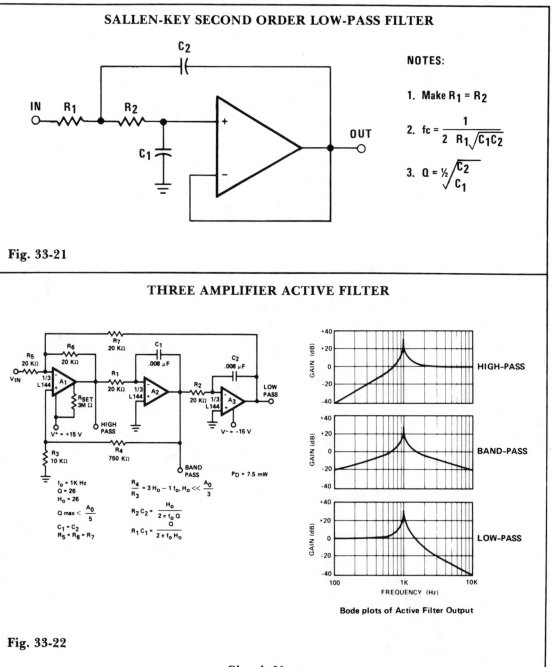

NOTES:

1. Make $R_1 = R_2$

2. $fc = \dfrac{1}{2\ R_1\sqrt{C_1 C_2}}$

3. $Q = \tfrac{1}{2}\sqrt{\dfrac{C_2}{C_1}}$

Fig. 33-21

THREE AMPLIFIER ACTIVE FILTER

$f_0 = 1K\ Hz$
$Q = 26$
$H_0 = 26$
$Q\ max < \dfrac{A_0}{5}$
$C_1 = C_2$
$R_5 = R_6 = R_7$

$\dfrac{R_4}{R_3} = 3 H_0 - 1 f_0,\ H_0 \ll \dfrac{A_0}{3}$

$R_2 C_2 = \dfrac{H_0}{2\pi f_0 Q}$

$R_1 C_1 = \dfrac{Q}{2\pi f_0 H_0}$

$P_D = 7.5\ mW$

Bode plots of Active Filter Output

Fig. 33-22

Circuit Notes

The active filter is a state variable filter with bandpass, high-pass and low-pass outputs. It is a classical analog computer method of implementing a filter using three amplifiers and only two capacitors.

BANDPASS STATE VARIABLE FILTER

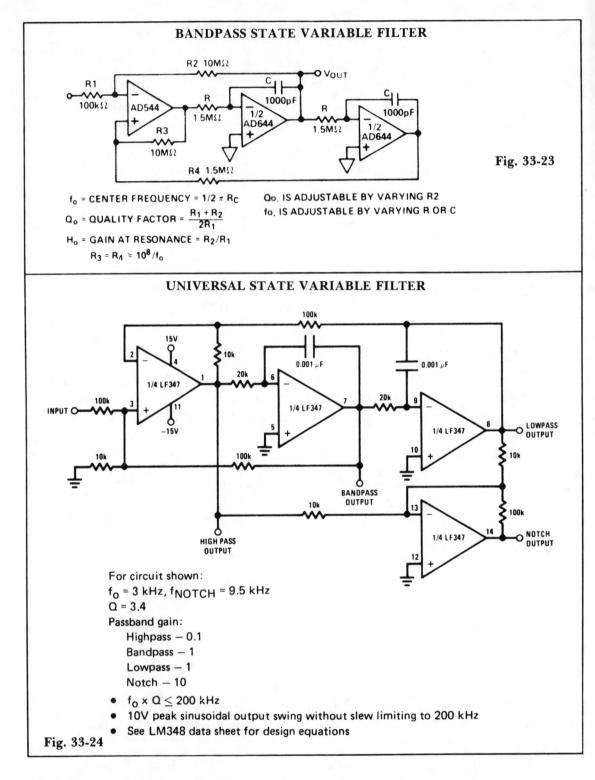

Fig. 33-23

f_o = CENTER FREQUENCY = $1/2 \pi$ R_C

Q_o = QUALITY FACTOR = $\dfrac{R_1 + R_2}{2R_1}$

H_o = GAIN AT RESONANCE = R_2/R_1

$R_3 = R_4 \approx 10^8/f_o$

Q_o, IS ADJUSTABLE BY VARYING R2

f_o, IS ADJUSTABLE BY VARYING R OR C

UNIVERSAL STATE VARIABLE FILTER

For circuit shown:

f_O = 3 kHz, f_{NOTCH} = 9.5 kHz

Q = 3.4

Passband gain:

    Highpass — 0.1

    Bandpass — 1

    Lowpass — 1

    Notch — 10

- f_O x Q $\leq$ 200 kHz
- 10V peak sinusoidal output swing without slew limiting to 200 kHz
- See LM348 data sheet for design equations

Fig. 33-24

500 Hz SALLEN-KEY BANDPASS FILTER

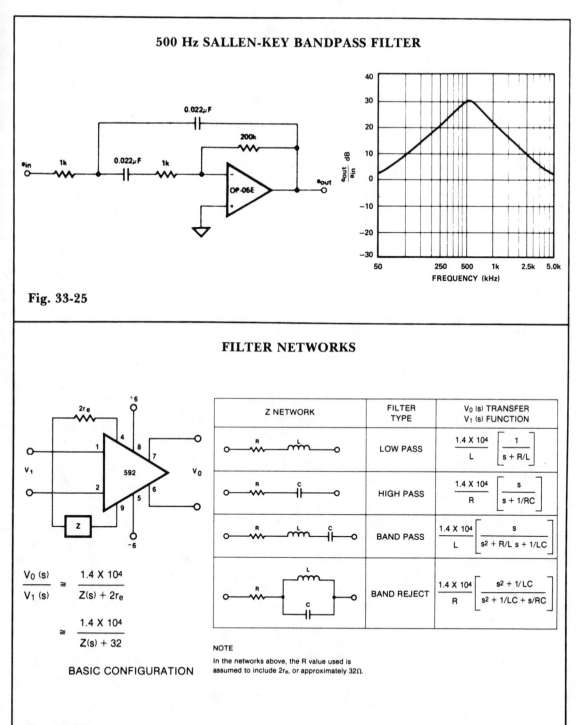

Fig. 33-25

FILTER NETWORKS

Z NETWORK	FILTER TYPE	V_0 (s) TRANSFER V_1 (s) FUNCTION
R — L	LOW PASS	$\dfrac{1.4 \times 10^4}{L}\left[\dfrac{1}{s + R/L}\right]$
R — C	HIGH PASS	$\dfrac{1.4 \times 10^4}{R}\left[\dfrac{s}{s + 1/RC}\right]$
R — L — C	BAND PASS	$\dfrac{1.4 \times 10^4}{L}\left[\dfrac{s}{s^2 + R/L\, s + 1/LC}\right]$
R — L/C	BAND REJECT	$\dfrac{1.4 \times 10^4}{R}\left[\dfrac{s^2 + 1/LC}{s^2 + 1/LC + s/RC}\right]$

$$\frac{V_0\ (s)}{V_1\ (s)} \approx \frac{1.4 \times 10^4}{Z(s) + 2r_e}$$

$$\approx \frac{1.4 \times 10^4}{Z(s) + 32}$$

BASIC CONFIGURATION

NOTE

In the networks above, the R value used is assumed to include $2r_e$, or approximately 32Ω.

Fig. 33-26

EQUAL COMPONENT SALLEN-KEY LOW-PASS FILTER

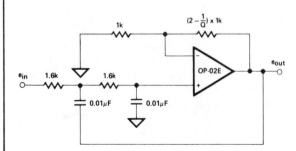

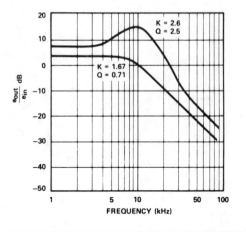

Equal R, Equal C Sallen-Key Response

Fig. 33-27

BIQUAD FILTER

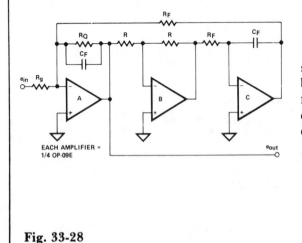

Fig. 33-28

Circuit Notes

The biquad filter, while appearing very similar to the state-variable filter, has a bandwidth that is fixed regardless of center frequency. This type of filter is useful in applications such as spectrum analyzers, which require a filter with a fixed bandwidth.

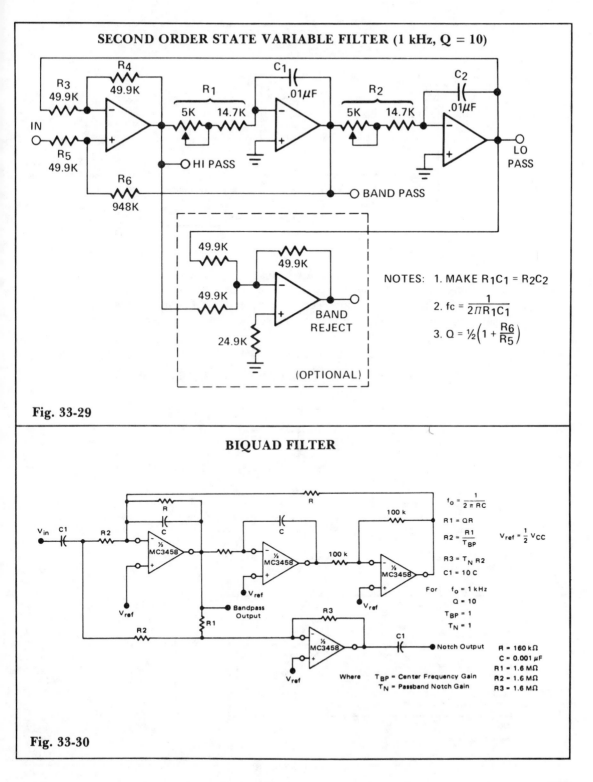

SECOND ORDER STATE VARIABLE FILTER (1 kHz, Q = 10)

R3 49.9K

R4 49.9K

R1 5K 14.7K

C1 .01μF

R2 5K 14.7K

C2 .01μF

IN

R5 49.9K

HI PASS

R6 948K

BAND PASS

LO PASS

49.9K

49.9K

49.9K

BAND REJECT

24.9K

(OPTIONAL)

NOTES: 1. MAKE $R_1C_1 = R_2C_2$

2. $f_c = \dfrac{1}{2\pi R_1 C_1}$

3. $Q = \frac{1}{2}\left(1 + \dfrac{R_6}{R_5}\right)$

Fig. 33-29

BIQUAD FILTER

V_{in} C1 R2

R

C

½ MC3458

R

100 k

½ MC3458

C

100 k

½ MC3458

Bandpass Output

R2 R1

R3

½ MC3458

C1

Notch Output

V_{ref}

$f_O = \dfrac{1}{2\pi RC}$

R1 = QR

$R2 = \dfrac{R1}{T_{BP}}$

$R3 = T_N R2$

C1 = 10 C

$V_{ref} = \frac{1}{2} V_{CC}$

For $f_O = 1$ kHz

Q = 10

$T_{BP} = 1$

$T_N = 1$

R = 160 kΩ
C = 0.001 μF
R1 = 1.6 MΩ
R2 = 1.6 MΩ
R3 = 1.6 MΩ

Where T_{BP} = Center Frequency Gain
T_N = Passband Notch Gain

Fig. 33-30

293

TUNABLE ACTIVE FILTER

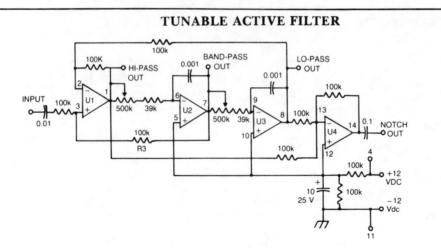

Fig. 33-31

Circuit Notes

The high-pass and low-pass outputs covering the range of 300 Hz to 3000 Hz have been summed in the fourth op amp to provide a notch output. The potentiometers must have a reverse log taper. Fixed-frequency active filter center frequency is 1 kHz, with a Q of 50.

ACTIVE RC FILTER FOR FREQUENCIES UP TO 150 kHz

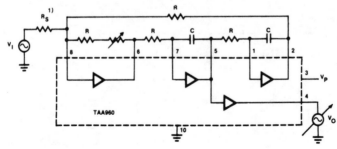

R = 10kΩ

This frequency range can be extended to 200kHz if a feed forward capacitor is connected between pin 5 and 8.

Fig. 33-32

f	Frequency	$\frac{1}{2\pi RC}$	
Vp	Supply voltage	6	V
	Filter performance		
Q	at $T_A = 25°C$	40 to 55	
Q	at $T_A = -30$ to $+65°C$	35 to 55	
Vi	Input voltage	400	mV
Vo	Output voltage	400	mV
dtot	Distortion at $V_o = 350mV$	2	%
S/N	S/N ratio at $V_o = 400mV$	50	dB
Rs	Input resistor*	470	kΩ

*NOTE

Value of input resistor to be determined for $\frac{V_o}{V_i}$ = 0.90 to 1.1.

POLE ACTIVE LOW-PASS FILTER
(BUTTERWORTH MAXIMALLY FLAT RESPONSE)

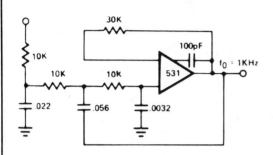

*Reference—EDN Dec. 15, 1970
Simplify 3-Pole Active Filter Design
A. Paul Brokow

RESPONSE OF 3-POLE ACTIVE BUTTERWORTH MAXIMALLY FLAT FILTER

Fig. 33-33

SPEECH FILTER (300 Hz .3 kHz BANDPASS)

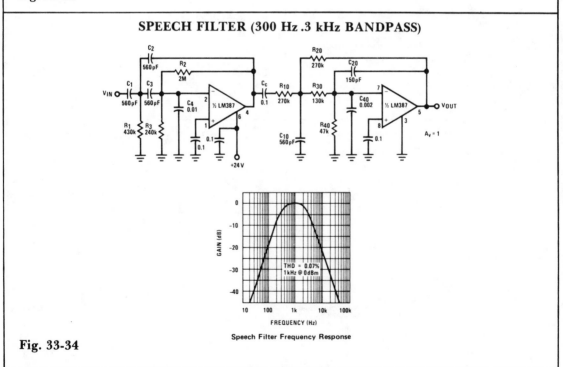

Speech Filter Frequency Response

Fig. 33-34

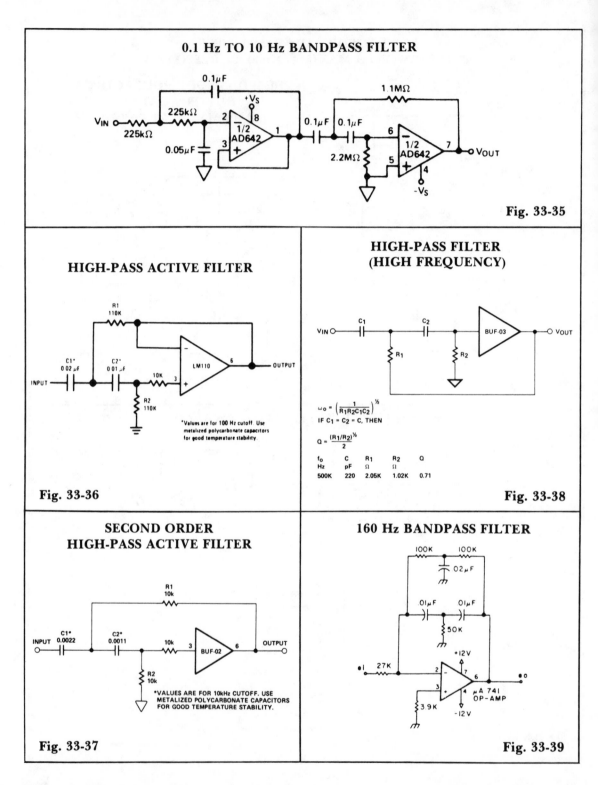

0.1 Hz TO 10 Hz BANDPASS FILTER

Fig. 33-35

HIGH-PASS ACTIVE FILTER

*Values are for 100 Hz cutoff. Use metalized polycarbonate capacitors for good temperature stability.

Fig. 33-36

HIGH-PASS FILTER (HIGH FREQUENCY)

$$\omega_o = \left(\frac{1}{R_1R_2C_1C_2}\right)^{\frac{1}{2}}$$

IF $C_1 = C_2 = C$, THEN

$$Q = \frac{(R_1/R_2)^{\frac{1}{2}}}{2}$$

f_o Hz	C pF	R_1 Ω	R_2 Ω	Q
500K	220	2.05K	1.02K	0.71

Fig. 33-38

SECOND ORDER HIGH-PASS ACTIVE FILTER

*VALUES ARE FOR 10kHz CUTOFF. USE METALIZED POLYCARBONATE CAPACITORS FOR GOOD TEMPERATURE STABILITY.

Fig. 33-37

160 Hz BANDPASS FILTER

Fig. 33-39

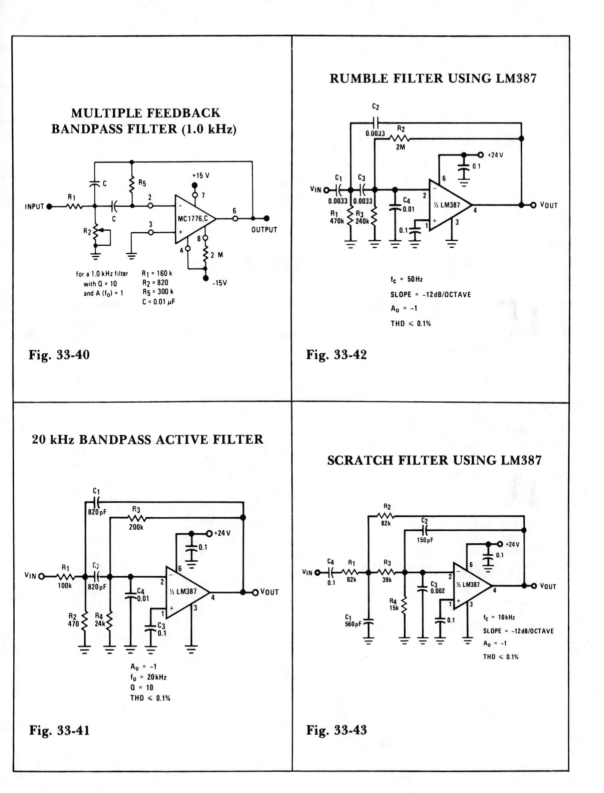

MULTIPLE FEEDBACK
BANDPASS FILTER (1.0 kHz)

INPUT

R1

C

R5

C

R2

2

3

−

+

MC1776,C

7

+15 V

6

8

4

2 M

−15V

OUTPUT

for a 1.0 kHz filter
with Q = 10
and A (f₀) = 1

$R_1 = 160 k$
$R_2 = 820$
$R_5 = 300 k$
$C = 0.01 \mu F$

Fig. 33-40

RUMBLE FILTER USING LM387

C2
0.0033

R2
2M

C1
0.0033

C3
0.0033

VIN

R1
470k

R3
240k

C4
0.01

0.1

2

½ LM387

6

+24 V

0.1

4

3

VOUT

$f_c = 50\,Hz$

SLOPE = −12dB/OCTAVE

$A_0 = -1$

THD ≤ 0.1%

Fig. 33-42

20 kHz BANDPASS ACTIVE FILTER

C1
820 pF

R3
200k

VIN

R1
100k

C2
820 pF

R2
470

R4
24k

C4
0.01

C3
0.1

2

½ LM387

1

6

+24 V

0.1

4

3

VOUT

$A_0 = -1$
$f_0 = 20\,kHz$
$Q = 10$
THD ≤ 0.1%

Fig. 33-41

SCRATCH FILTER USING LM387

R2
82k

C2
150 pF

VIN

C4
0.1

R1
82k

R3
39k

C3
0.002

R4
15k

C1
560 pF

0.1

2

½ LM387

1

6

+24 V

0.1

4

3

VOUT

$f_c = 10\,kHz$

SLOPE = −12dB/OCTAVE

$A_0 = -1$

THD ≤ 0.1%

Fig. 33-43

34

Flashers and Blinkers

The sources of the following circuits are contained in the Sources section beginning on page 730. The figure number contained in the box of each circuit correlates to the source entry in the Sources section.

Auto, Boat, or Barricade Flasher
Flip-Flop Flasher
Flashlight Finder
Low Frequency Lamp Flasher/Relay Driver
Low Cost Ring Counter
Ring Counter for Incandescent Lamps
Dual LED CMOS Flasher
Automatic Safety Flasher
Neon Blinker
Transistorized Flasher
Flasher/Light Control
Neon Tube Flasher
Dc Flasher with Adjustable On and Off Time

Low Voltage Flasher
1 A Lamp Flasher
Fast Blinker
3 V Flasher
Incandescent Bulb Flasher
Flasher for 4 Parallel LEDs
LED Booster
Safe, High Voltage Flasher
Alternating Flasher
Variable Flasher
Emergency Lantern/Flasher
High Efficiency Parallel Circuit Flasher
Minimum Power Flasher

AUTO, BOAT, OR BARRICADE FLASHER

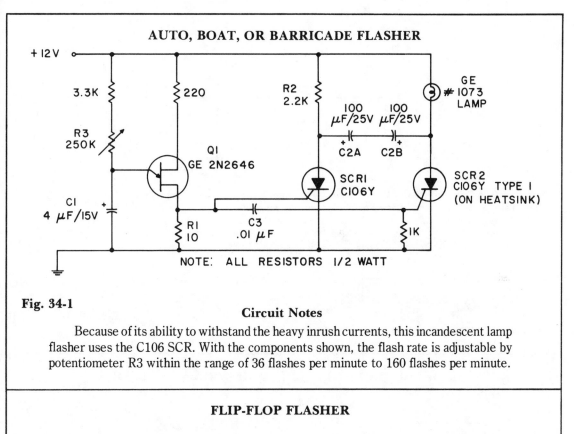

Fig. 34-1

Circuit Notes

Because of its ability to withstand the heavy inrush currents, this incandescent lamp flasher uses the C106 SCR. With the components shown, the flash rate is adjustable by potentiometer R3 within the range of 36 flashes per minute to 160 flashes per minute.

FLIP-FLOP FLASHER

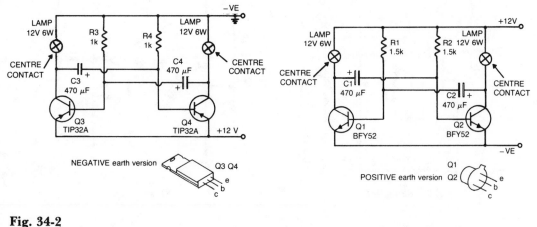

Fig. 34-2

Circuit Notes

The flashing action is provided by a simple astable multivibrator timed to give a flashing rate of about 60 flashes for each lamp per minute. Circuit for positive earth systems uses NPN transistors. The other uses PNP transistors.

FLASHLIGHT FINDER

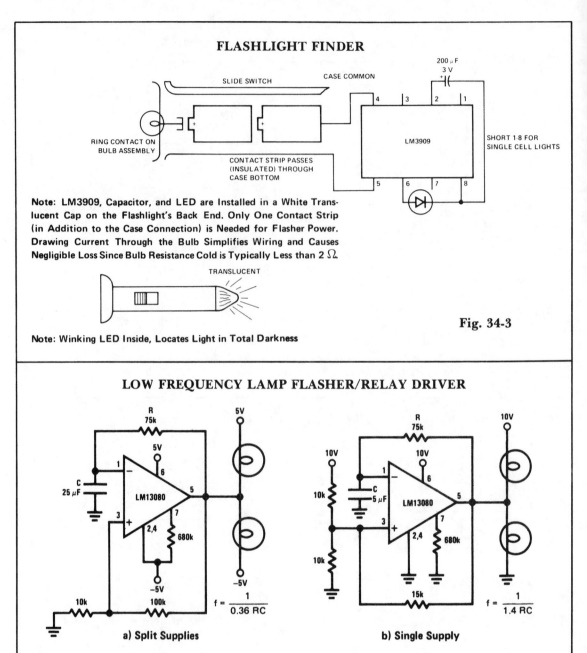

Note: LM3909, Capacitor, and LED are Installed in a White Translucent Cap on the Flashlight's Back End. Only One Contact Strip (in Addition to the Case Connection) is Needed for Flasher Power. Drawing Current Through the Bulb Simplifies Wiring and Causes Negligible Loss Since Bulb Resistance Cold is Typically Less than 2 Ω.

Note: Winking LED Inside, Locates Light in Total Darkness

Fig. 34-3

LOW FREQUENCY LAMP FLASHER/RELAY DRIVER

$$f = \frac{1}{0.36\ RC}$$

a) Split Supplies

$$f = \frac{1}{1.4\ RC}$$

b) Single Supply

Fig. 34-4

Circuit Notes

This circuit is a low frequency warning device. The output of the oscillator is a square wave that is used to drive lamps or small relays. The circuit alternately flashes two incandescent lamps.

LOW COST RING COUNTER

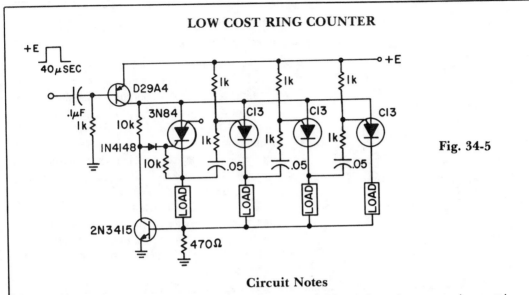

Fig. 34-5

Circuit Notes

This ring counter makes an efficient, low cost circuit featuring automatic resetting via the first stage 3N84. As many stages as desired may be cascaded.

RING COUNTER FOR INCANDESCENT LAMPS

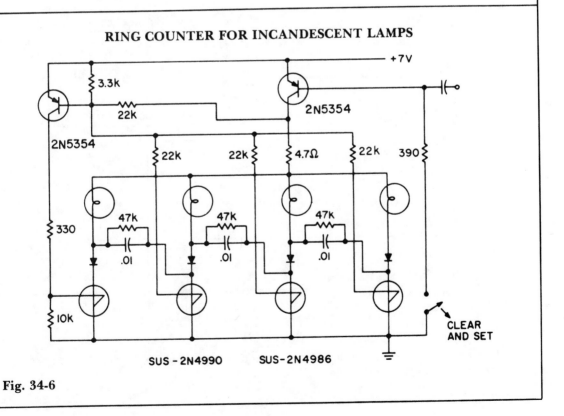

Fig. 34-6

DUAL LED CMOS FLASHER

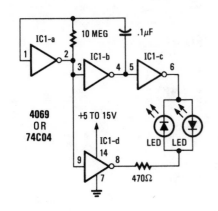

4069 OR 74C04

Fig. 34-7

Circuit Notes

Inverters IC1-a and IC1-b form a multivibrator and IC1-c is a buffer. Inverter IC1-d is connected so that its output is opposite that of IC1-c; when pin 6 is high, then pin 8 is low and vice versa. Because pins 6 and 8 are constantly changing state, first one LED and then the other is on since they are connected in reverse. The light seems to jump back and forth between the LED's. The 470-ohm resistor limits LED current. Depending upon the supply voltage used, the value of the resistor may have to be changed to obtain maximum light output. To change the switching rate, change the value of the capacitor.

AUTOMATIC SAFETY FLASHER

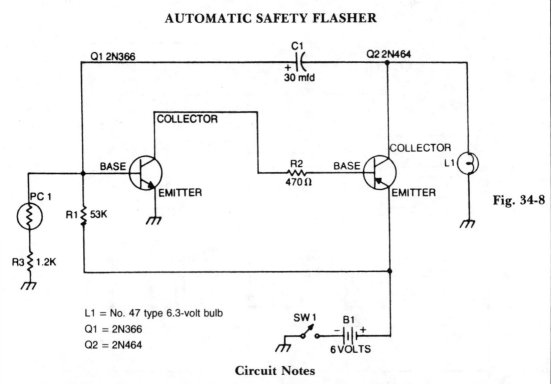

L1 = No. 47 type 6.3-volt bulb
Q1 = 2N366
Q2 = 2N464

Fig. 34-8

Circuit Notes

This flasher only comes on at night. It furnishes a bright nighttime illumination, and shuts itself off automatically as soon as the sun comes up. The photocell must be mounted on top of the unit in such a way as to detect the greatest amount of available light.

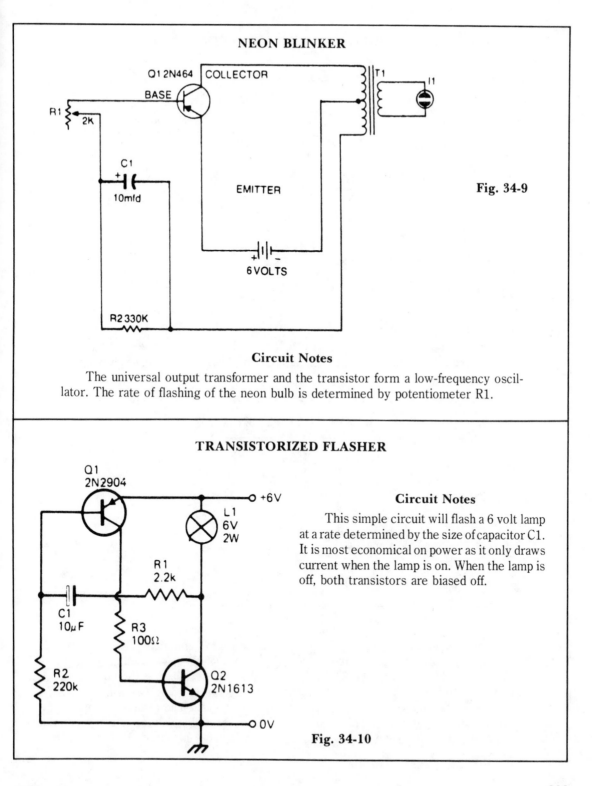

NEON BLINKER

Q1 2N464
COLLECTOR
BASE
R1 2k
C1 10mfd
EMITTER
T1
I1
6 VOLTS
R2 330K

Fig. 34-9

Circuit Notes

The universal output transformer and the transistor form a low-frequency oscillator. The rate of flashing of the neon bulb is determined by potentiometer R1.

TRANSISTORIZED FLASHER

Q1 2N2904
L1 6V 2W
+6V
R1 2.2k
C1 10μF
R3 100Ω
R2 220k
Q2 2N1613
0V

Circuit Notes

This simple circuit will flash a 6 volt lamp at a rate determined by the size of capacitor C1. It is most economical on power as it only draws current when the lamp is on. When the lamp is off, both transistors are biased off.

Fig. 34-10

303

FLASHER/LIGHT CONTROL

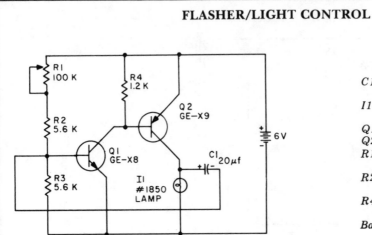

Parts List

C1 — 20-mfd, 6-volt
    electrolytic capacitor
I1 — 6-volt, GE No. 1850
    lamp and socket
Q1 — GE-X8 transistor
Q2 — GE-X9 transistor
R1 — 100K-ohm, 2-watt
    potentiometer
R2, R3 — 5.6K-ohm, 1/2-watt
    resistor
R4 — 1.2K-ohm, 1/2-watt
    resistor
Battery — 6-volt dry pack

Fig. 34-11

Circuit Notes

The circuit is a two-stage, direct-coupled transistor amplifier connected as a free-running multivibrator. Both the flash duration and flash interval can be changed by turning the potentiometer, R1.

NEON TUBE FLASHER

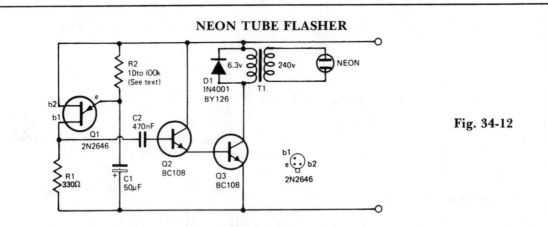

Fig. 34-12

Circuit Notes

The voltage required to ignite the neon tube is obtained by using an ordinary filament transformer (240-6.3 V) in reverse. Battery drain is quite low, around 1 to 2 milliamps for a nine volt battery. The pulses from Q1, unijunction transistor, operated as a relaxation oscillator and are applied to Q2 which in turn drives Q3 into saturation. The sharp rise in current through the 6.3 V winding of the transformer as Q3 goes into saturation induces a high voltage in the secondary winding causing the neon to flash. The diode D1 protects the transistor from high voltage spikes generated when switching currents in the transformer.

DC FLASHER WITH ADJUSTABLE ON AND OFF TIME

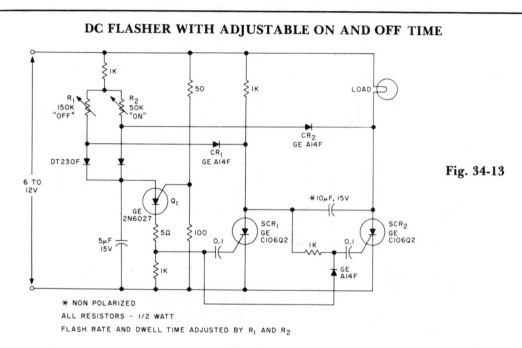

Fig. 34-13

✷ NON POLARIZED
ALL RESISTORS - 1/2 WATT
FLASH RATE AND DWELL TIME ADJUSTED BY R_1 AND R_2

Circuit Notes

This circuit utilizes a power flip-flop and programmable unijunction (PUT) to obtain adjustable on and off times.

LOW VOLTAGE FLASHER

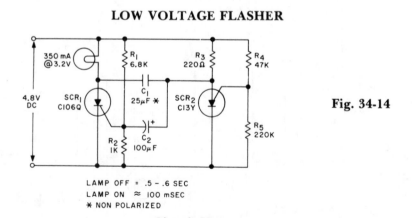

Fig. 34-14

LAMP OFF = .5 – .6 SEC
LAMP ON ≈ 100 mSEC
✷ NON POLARIZED

Circuit Notes

Applying voltage to the circuit triggers SCR1. With SCR1 on, the voltage on the anode of SCR2 rises until SCR2 triggers to commutate SCR1. The voltage on the gate of SCR1 will swing negative at this time, and only after a positive potential of ≈ 0.5 volt is once again attained, will SCR1 retrigger. The circuit could be used for higher voltage levels, but the peak negative voltage on the gate of SCR1 must be limited to less than 6 volts.

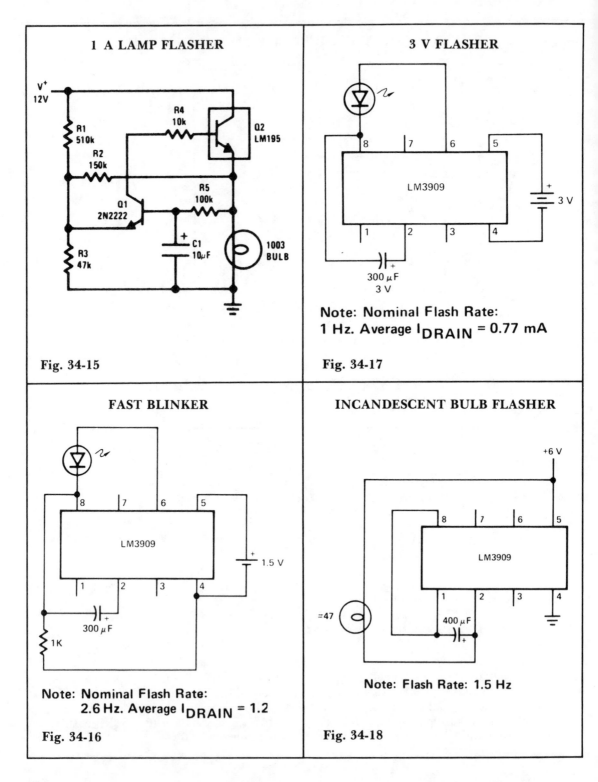

1 A LAMP FLASHER

V+
12V

R1
510k

R2
150k

R3
47k

Q1
2N2222

R4
10k

Q2
LM195

R5
100k

C1
10μF

1003
BULB

Fig. 34-15

3 V FLASHER

LM3909

8 7 6 5

1 2 3 4

3 V

300 μF
3 V

Note: Nominal Flash Rate:
1 Hz. Average I_{DRAIN} = 0.77 mA

Fig. 34-17

FAST BLINKER

LM3909

8 7 6 5

1 2 3 4

1.5 V

300 μF

1K

Note: Nominal Flash Rate:
2.6 Hz. Average I_{DRAIN} = 1.2

Fig. 34-16

INCANDESCENT BULB FLASHER

+6 V

LM3909

8 7 6 5

1 2 3 4

=47

400 μF

Note: Flash Rate: 1.5 Hz

Fig. 34-18

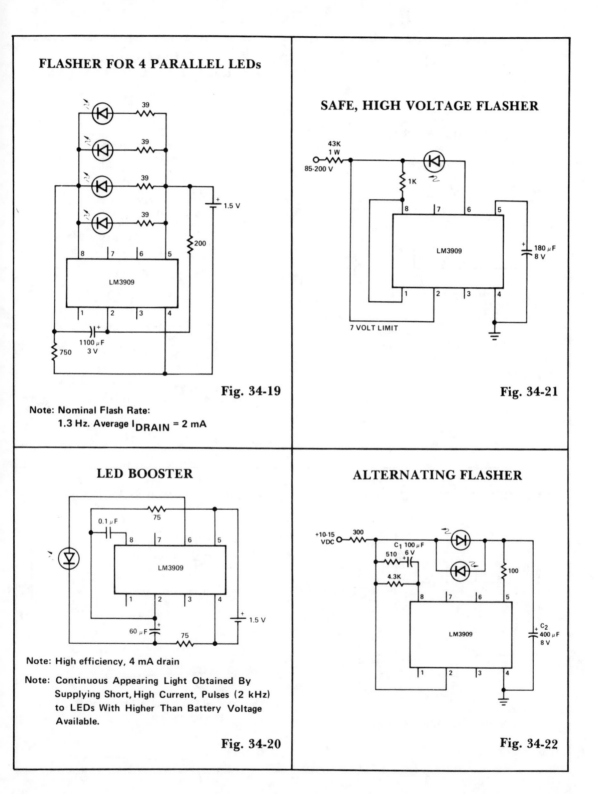

FLASHER FOR 4 PARALLEL LEDs

1.5 V

LM3909

39
39
39
39

200

1100 μF
3 V

750

Fig. 34-19

Note: Nominal Flash Rate:
1.3 Hz. Average I_{DRAIN} = 2 mA

SAFE, HIGH VOLTAGE FLASHER

43K
1 W
85-200 V

1K

LM3909

180 μF
8 V

7 VOLT LIMIT

Fig. 34-21

LED BOOSTER

0.1 μF
75

LM3909

1.5 V

60 μF
75

Note: High efficiency, 4 mA drain

Note: Continuous Appearing Light Obtained By
Supplying Short, High Current, Pulses (2 kHz)
to LEDs With Higher Than Battery Voltage
Available.

Fig. 34-20

ALTERNATING FLASHER

+10-15
VDC
300
C_1 100 μF
6 V
510

4.3K

LM3909

100

C_2
400 μF
8 V

Fig. 34-22

307

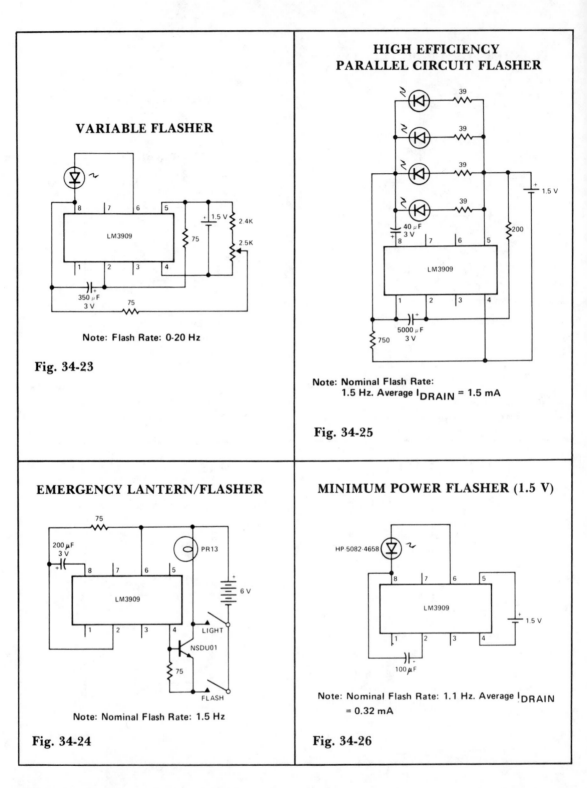

VARIABLE FLASHER

LM3909

Note: Flash Rate: 0-20 Hz

Fig. 34-23

HIGH EFFICIENCY PARALLEL CIRCUIT FLASHER

LM3909

Note: Nominal Flash Rate:
1.5 Hz. Average I_{DRAIN} = 1.5 mA

Fig. 34-25

EMERGENCY LANTERN/FLASHER

LM3909

Note: Nominal Flash Rate: 1.5 Hz

Fig. 34-24

MINIMUM POWER FLASHER (1.5 V)

LM3909

Note: Nominal Flash Rate: 1.1 Hz. Average I_{DRAIN} = 0.32 mA

Fig. 34-26

308

35

Frequency Measuring Circuits

The sources of the following circuits are contained in the Sources section beginning on page 730. The figure number contained in the box of each circuit correlates to the source entry in the Sources section.

Inexpensive Frequency Counter/
Tachometer

Audio Frequency Meter

Linear Frequency Meter
Power-Line Frequency Meter

INEXPENSIVE FREQUENCY COUNTER/TACHOMETER

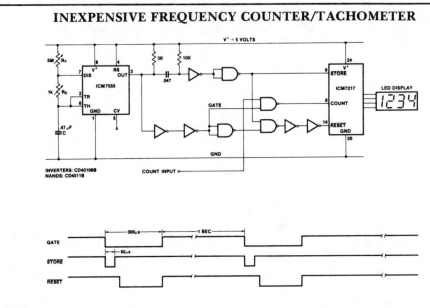

Fig. 35-1

Circuit Notes

This circuit uses the low power ICM7555 (CMOS 555) to generate the gating, $\overline{\text{STORE}}$ and $\overline{\text{RESET}}$ signals. To provide the gating signal, the timer is configured as an astable multivibrator. The system is calibrated by using a 5 M potentiometer for R_A as a coarse control and a 1 jk potentiometer for R_B as a fine control. CD40106B's are used as a monostable multivibrator and reset time delay.

LINEAR FREQUENCY METER (AUDIO SPECTRUM)

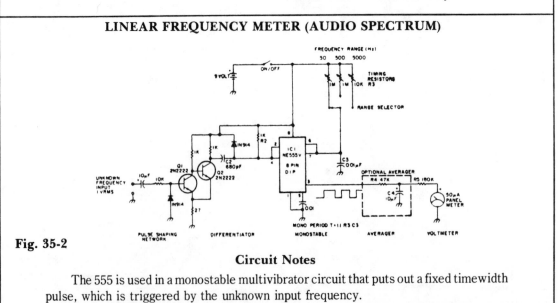

Fig. 35-2

Circuit Notes

The 555 is used in a monostable multivibrator circuit that puts out a fixed timewidth pulse, which is triggered by the unknown input frequency.

POWER-LINE FREQUENCY METER

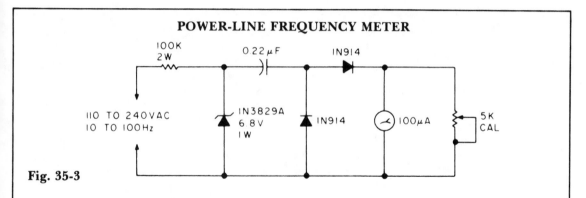

Fig. 35-3

Circuit Notes

The meter will indicate the frequency from a power generator. Incoming sine waves are converted to square waves by the 100 K resistor and the 6.8 V zener. The square wave is differentiated by the capacitor and the cur-rent is averaged by the diodes. The average current is almost exactly proportional to the frequency and can be read directly on a 100 mA meter. To calibrate, hook the circuit up to a 60 Hz power aline and adjust the 5 K pot to read 60 mA.

AUDIO FREQUENCY METER

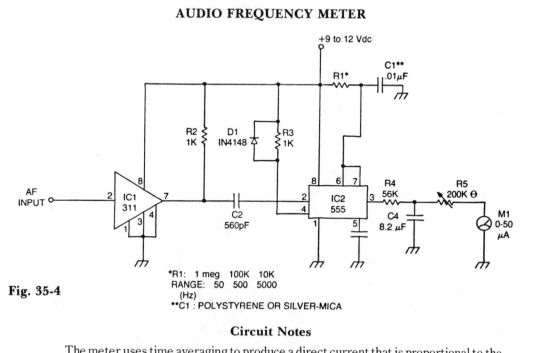

Fig. 35-4

*R1: 1 meg 100K 10K
RANGE: 50   500 5000
       (Hz)
**C1 : POLYSTYRENE OR SILVER-MICA

Circuit Notes

The meter uses time averaging to produce a direct current that is proportional to the frequency of the input signal.

36

Frequency Multipliers

The sources of the following circuits are contained in the Sources section beginning on page 730. The figure number contained in the box of each circuit correlates to the source entry in the Sources section.

Broadband Frequency Doubler
Frequency Doubler
150 to 300 MHz Doubler

Low-Frequency Doubler
Oscillator with Double Frequency
 Output

BROADBAND FREQUENCY DOUBLER

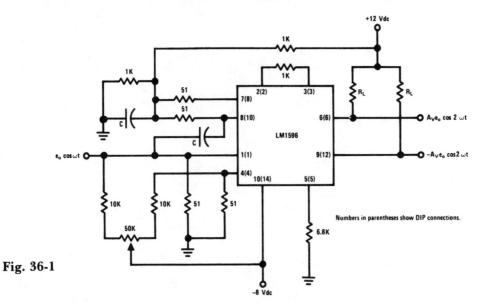

Fig. 36-1

Numbers in parentheses show DIP connections.

Circuit Notes

This circuit will double low-level signals with low distortion. The value of C should be chosen for low reactance at the operating frequency. Signal level at the carrier input must be less than 25 mV peak to maintain operation in the linear region of the switching differential amplifier. Levels to 50 mV peak may be used with some distortion of the output waveform. If a larger input signal is available, a resistive divider may be used at the carrier input with full signal applied to the signal input.

FREQUENCY DOUBLER

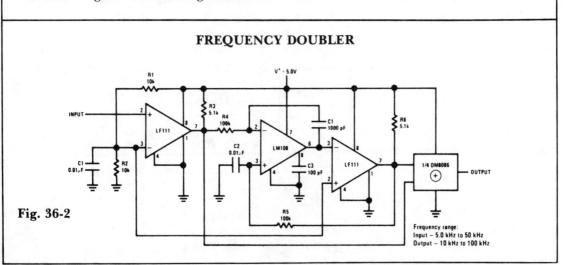

Fig. 36-2

Frequency range:
Input – 5.0 kHz to 50 kHz
Output – 10 kHz to 100 kHz

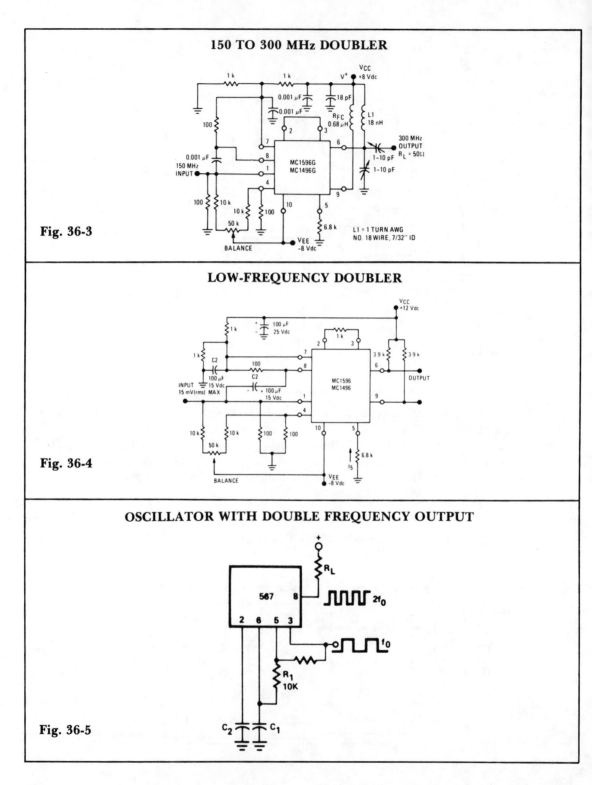

150 TO 300 MHz DOUBLER

Fig. 36-3

LOW-FREQUENCY DOUBLER

Fig. 36-4

OSCILLATOR WITH DOUBLE FREQUENCY OUTPUT

Fig. 36-5

37

Frequency-to-Voltage Converters

The sources of the following circuits are contained in the Sources section beginning on page 730. The figure number contained in the box of each circuit correlates to the source entry in the Sources section.

DC-10 kHz FREQUENCY/VOLTAGE CONVERTER

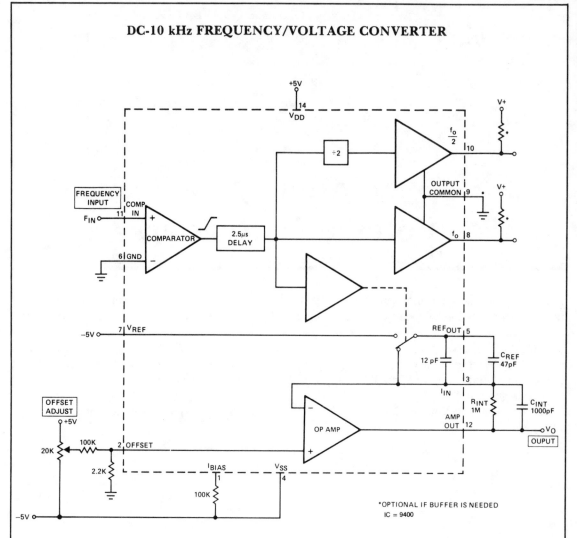

Fig. 37-1

Circuit Notes

The converter generates an output voltage which is linearly proportional to the input frequency waveform. Each zero crossing at the comparator's input causes a precise amount of change to be dispensed into the op amp's summing junction. This charge in turn flows through the feedback resistor generating voltage pulses at the output of the op amp. Capacitor (C_{INT}) across R_{INT} averages these pulses into a dc voltage which is linearly proportional to the input frequency.

FREQUENCY-TO-VOLTAGE CONVERTER
(DIGITAL FREQUENCY METER)

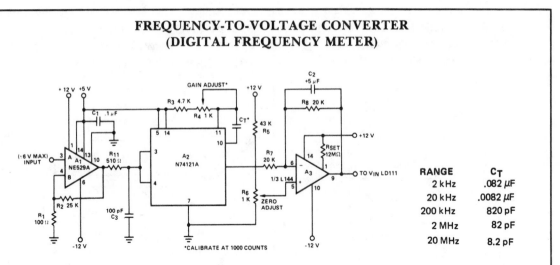

Fig. 37-2

RANGE	C_T
2 kHz	.082 μF
20 kHz	.0082 μF
200 kHz	820 pF
2 MHz	82 pF
20 MHz	8.2 pF

Circuit Notes

This circuit converts frequency to voltage by taking the average dc value of the pulses from the 74121 monostable multivibrator. The one shot is triggered by the positive-going ac signal at the input of the 529 comparator. The amplifier acts as a dc filter, and also provides zeroing. The accuracy is 2% over a 5 decade range. The input signal to the comparator should be greater than 0.1 volt peak-to-peak, and less than 12 volts peak-to-peak for proper operation.

ZENER REGULATED FREQUENCY-TO-VOLTAGE CONVERTER

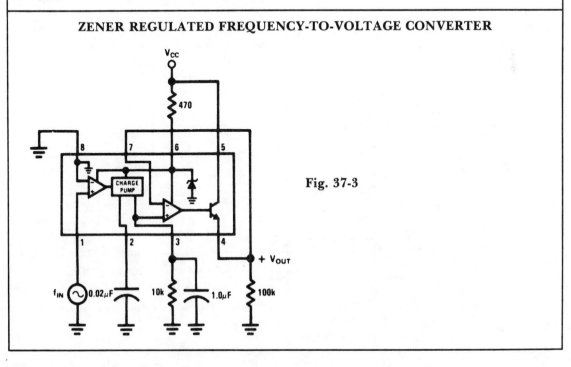

Fig. 37-3

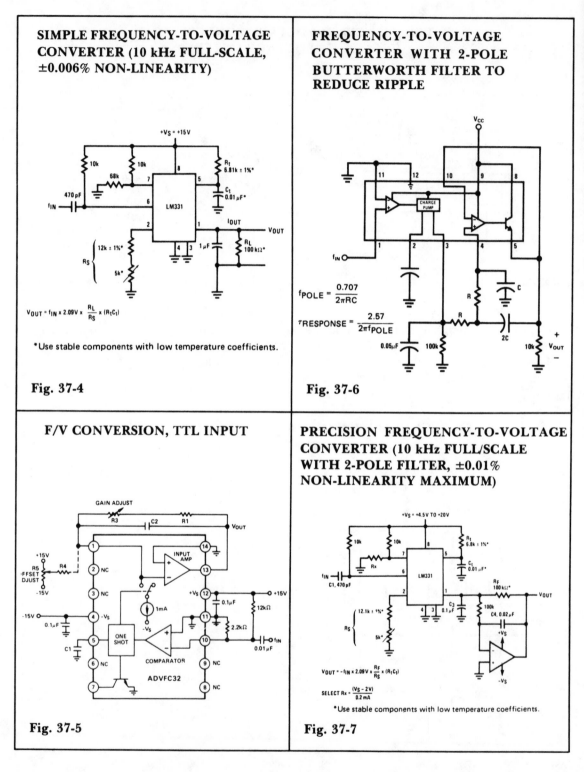

SIMPLE FREQUENCY-TO-VOLTAGE CONVERTER (10 kHz FULL-SCALE, ±0.006% NON-LINEARITY)

$+V_S = +15V$

10k 10k 8 R_t 6.81k ± 1%*

68k 7 5 C_t 0.01 μF*

470 pF 6 LM331

f_{IN}

2 1 I_{OUT} V_{OUT}

12k ± 1%* R_L 100 kΩ*

R_S

5k* 4 3 1 μF

$$V_{OUT} = f_{IN} \times 2.09V \times \frac{R_L}{R_S} \times (R_t C_t)$$

*Use stable components with low temperature coefficients.

Fig. 37-4

FREQUENCY-TO-VOLTAGE CONVERTER WITH 2-POLE BUTTERWORTH FILTER TO REDUCE RIPPLE

V_{CC}

11 12 10 9 8

CHARGE PUMP

f_{IN}

1 2 3 4 5

$$f_{POLE} = \frac{0.707}{2\pi RC}$$

$$\tau_{RESPONSE} = \frac{2.57}{2\pi f_{POLE}}$$

R C

R 2C

0.05μF 100k 10k V_{OUT} + −

Fig. 37-6

F/V CONVERSION, TTL INPUT

GAIN ADJUST

R3 C2 R1 V_{OUT}

+15V R5 FFSET DJUST R4

INPUT AMP

1 14

2 NC 13

3 NC +V_S 12 0.1μF +15V

-15V 12kΩ

-15V 4 -V_S 11 2.2kΩ

0.1μF -V_S

C1 5 ONE SHOT 10 f_{IN} 0.01μF

6 NC 9 NC

COMPARATOR

7 ADVFC32 8 NC

1mA

Fig. 37-5

PRECISION FREQUENCY-TO-VOLTAGE CONVERTER (10 kHz FULL/SCALE WITH 2-POLE FILTER, ±0.01% NON-LINEARITY MAXIMUM)

$+V_S = +4.5V$ TO $+20V$

10k 10k 8 R_t 6.8k ± 1%*

Rx 7 5 C_t 0.01 μF*

f_{IN} 6 LM331 R_F 100 kΩ*

C1, 470 pF 2 V_{OUT}

12.1k ± 1%* C_3 0.1 μF 100k C4, 0.02 μF

R_S

5k* 4 3 +V_S

−V_S

$$V_{OUT} = -f_{IN} \times 2.09V \times \frac{R_F}{R_S} \times (R_t C_t)$$

$$\text{SELECT } Rx = \frac{(V_S - 2V)}{0.2 \, mA}$$

*Use stable components with low temperature coefficients.

Fig. 37-7

38

Fuzz Circuits

The sources of the following circuits are contained in the Sources section beginning on page 730. The figure number contained in the box of each circuit correlates to the source entry in the Sources section.

FUZZ BOX 1

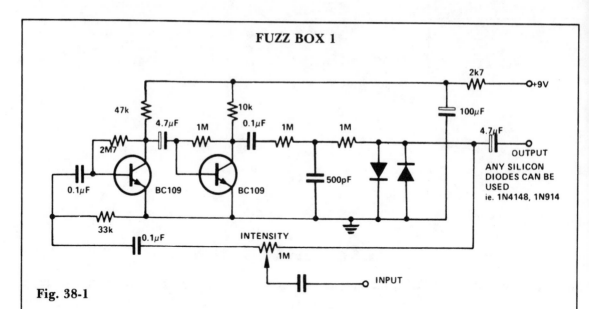

Fig. 38-1

Circuit Notes

The input signal is amplified by the transistors. The distorted output is then clipped by the two diodes and the high frequency noise is filtered from the circuit via the 500 pF capacitor. The 1 M pot adjusts the intensity of the fuzz from maximum to no fuzz (normal playing).

FUZZ BOX 2

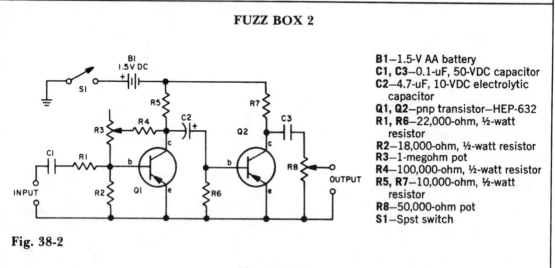

B1—1.5-V AA battery
C1, C3—0.1-uF, 50-VDC capacitor
C2—4.7-uF, 10-VDC electrolytic capacitor
Q1, Q2—pnp transistor—HEP-632
R1, R6—22,000-ohm, ½-watt resistor
R2—18,000-ohm, ½-watt resistor
R3—1-megohm pot
R4—100,000-ohm, ½-watt resistor
R5, R7—10,000-ohm, ½-watt resistor
R8—50,000-ohm pot
S1—Spst switch

Fig. 38-2

Circuit Notes

Potentiometer R3 sets the degree of fuzz, and R8 sets the output level. Since the fuzz effect cannot be completely eliminated by R3, fuzz-free sound requires a bypass switch from the input to output terminals.

FUZZ BOX 3

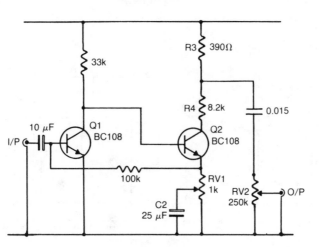

Fig. 38-3

Circuit Notes

Q1 and Q2 form a voltage amplifier which has sufficient gain to be overdriven by a relatively low input, such as an electric guitar. The result is that the output from Q2 is a Squared-Off verson of the input, giving the required fuzz sound. RV1 adjusts the amount of negative feedback inserted into the circuit by C2, and thus the amount of squaring of the signal. The purpose of R3 and R4 is to lower the output voltage to a suitable level, which is then adjusted as required with the volume control VR2.

FUZZ BOX 4

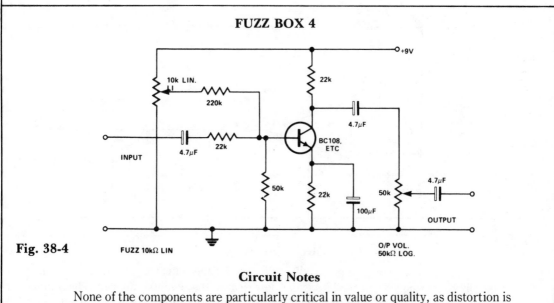

Fig. 38-4

Circuit Notes

None of the components are particularly critical in value or quality, as distortion is the sole object! The transistor could be BC107-8-9, 2N2926, etc.

FUZZ BOX 5

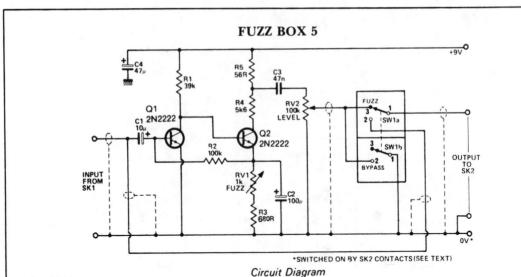

Circuit Diagram

Fig. 38-5

Circuit Notes

Transistors Q1 and Q2 amplify the incoming signal, and the gain is such that the input will overload when used with an electric guitar. RV1 adjusts the amount of feedback present, and hence voltage gain. The output is, therefore, a squared version of the input signal. The amount of squaring is varied by RV1.

GUITAR FUZZ

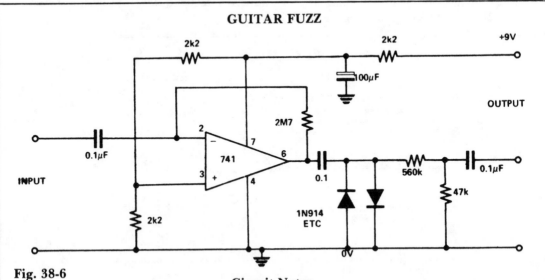

Fig. 38-6

Circuit Notes

The 741 has a maximum gain of 20,000, but the circuit is so designed that the IC's gain is 2,700,000 which then distorts the output. This distortion gives the fuzz effect. The two diodes clip the output to drop the level, also lowered by the potential divider. This circuit also sustains the notes, due to clipping, giving a totally new sound.

39

Games

The sources of the following circuits are contained in the Sources section beginning on page 730. The figure number contained in the box of each circuit correlates to the source entry in the Sources section.

Ready, Set, Go!
Electronic Dice
Game Roller or Chase Circuit
Toss-A-Coin Binary Box
Electronic Coin Tosser

Heads or Tails
Pot Shot
Low Cost Heads or Tails
Who Is First
Windicator

READY, SET, GO!

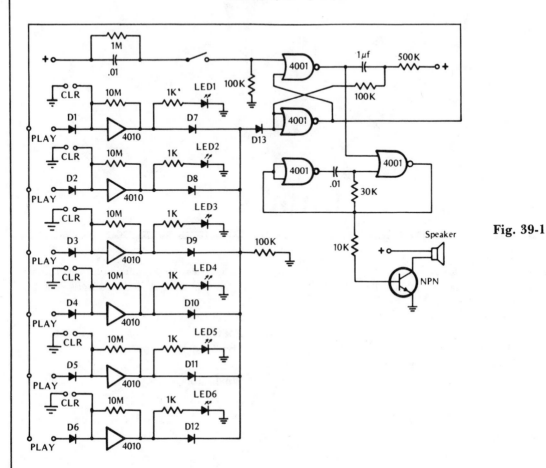

Fig. 39-1

Circuit Notes

This game tests a player's reaction time. It is activated by closing switch S1, which starts the tone generator and arms the circuit. The touchplate, labeled PLAY in the diagram, consists of two metal strips about 1/16th-inch apart. The first player to bridge the gap with his or her finger turns off the tone and lights the associated LED indicator. A second touchplate, labeled CLR in the diagram, clears the circuit, extinguishing the LED, when its gap is bridged by a fingertip.

ELECTRONIC DICE

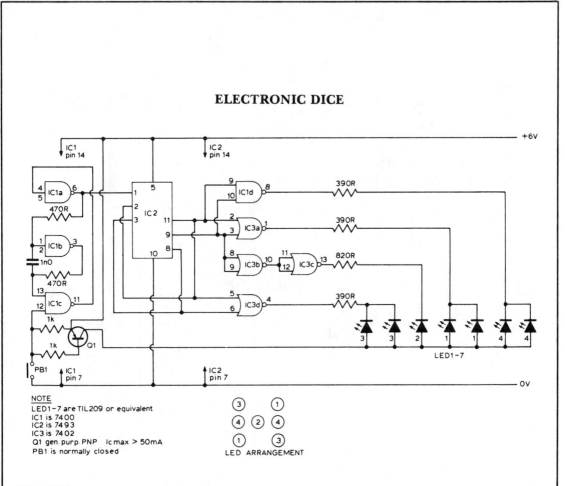

Fig. 39-2

Circuit Notes

Six LEDs are arranged to produce a display the same as the dots on a dice. When PBI is depressed, the display is blanked and the oscillator (IC1 a, b, c) clocks IC2 at about 1MHz.

IC2 counts from zero and resets on seven. When PBI is released, the display is enabled and a decoding system (IC3) produces the correct output on the LEDs.

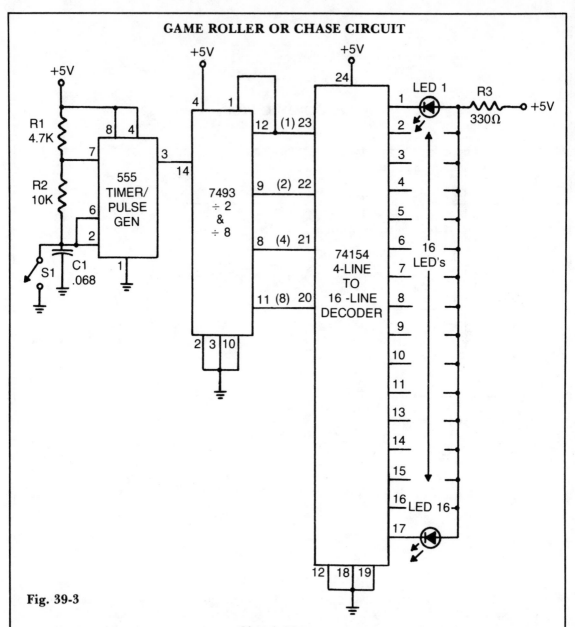

GAME ROLLER OR CHASE CIRCUIT

Fig. 39-3

Circuit Notes

The 555 timer produces a rapid series of pulses whenever switch S1 is open. These pulses are counted in groups of 16 and converted into binary form by the 7493 and applied to the 74154 (a 1-of-16 decoder/demultiplexer) wired so that each of its 16 output lines goes low sequentially and in step with the binary count delivered by the 7493. When the switch is closed, only one LED remains on. Only one current limiting resistor (R3) is used for all the LED's since only one is on at any one time.

TOSS-A-COIN BINARY BOX

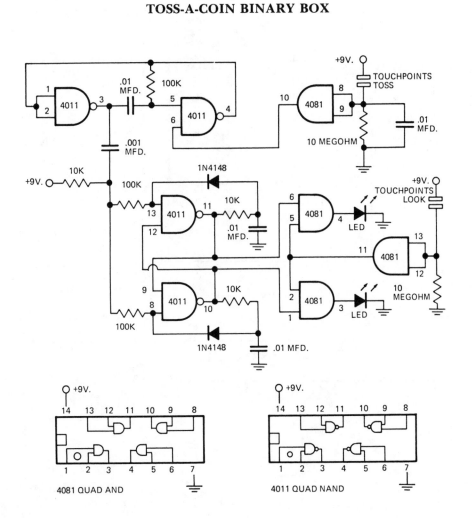

Fig. 39-4

Circuit Notes

Circuit uses an astable multivibrator to vary the heads-or-tails condition, and a flip-flop to store the condition given by the multivibrator. Consequently, the circuit is wired so that the flip-flop's state is changed once for each full cycle the multivibrator goes through to assure an absolutely even 50-50 chance of a heads or tails loss.

ELECTRONIC COIN TOSSER

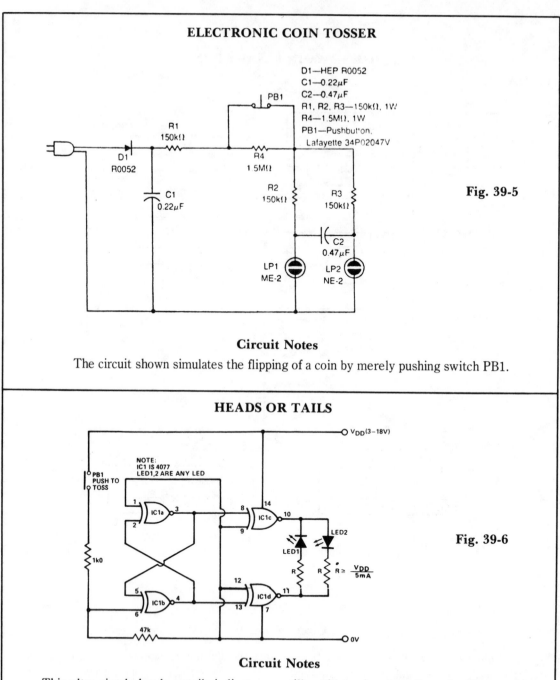

D1—HEP R0052
C1—0 22μF
C2—0.47μF
R1, R2, R3—150kΩ, 1W
R4—1.5MΩ, 1W
PB1—Pushbutton,
   Lafayette 34P02047V

Fig. 39-5

Circuit Notes

The circuit shown simulates the flipping of a coin by merely pushing switch PB1.

HEADS OR TAILS

NOTE:
IC1 IS 4077
LED1,2 ARE ANY LED

Fig. 39-6

Circuit Notes

This ultra-simple heads or tails indicator uses a single 4077 and no capacitor.

The circuit is normally in a latched bistable mode; when the switch is closed the circuit will oscillate, i.e. toss the coin. The astable frequency is approximately 5-10 MHz. PB1 is a normally closed switch.

POT SHOT

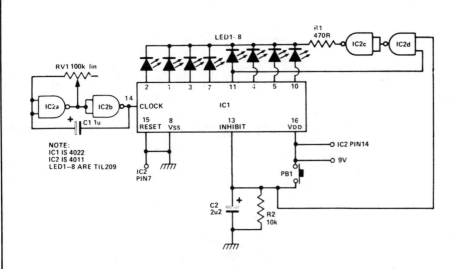

Fig. 39-7

NOTE:
IC1 IS 4022
IC2 IS 4011
LED1–8 ARE TIL209

Circuit Notes

This is a circuit for a game of the shooting gallery variety. IC2a and b form an astable multivibrator clocking IC1 which causes LEDs 1-8 to flash in turn LED 5 is the target LED and the object of the game is to depress PBI just as LED 5 comes on. If this is done, the whole display is blanked for a few seconds signifying a hit. Otherwise, the LED which was lit remains lit. When the push button is released, C2 discharges through R2 taking 8 pin 13 low again and the LEDs will start to flash again.

LOW COST "HEADS OR TAILS"

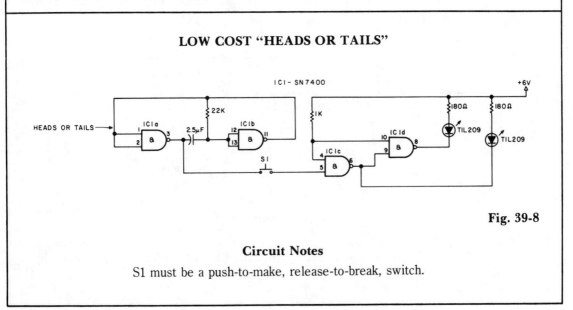

Fig. 39-8

Circuit Notes

S1 must be a push-to-make, release-to-break, switch.

WHO IS FIRST

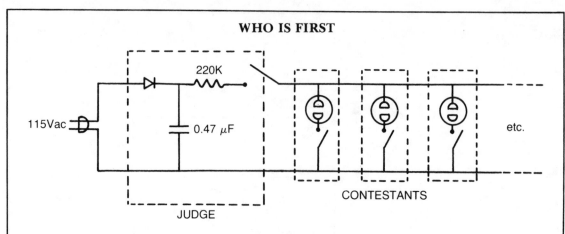

Fig. 39-9

Circuit Notes

Here is a circuit for any question-and-answer party game. The first button pushed ionizes the neon bulb dropping the dc voltage on the parallel neons (the other contestants) below the ionization level: determining unequivocally the first person to press the button.

WINDICATOR

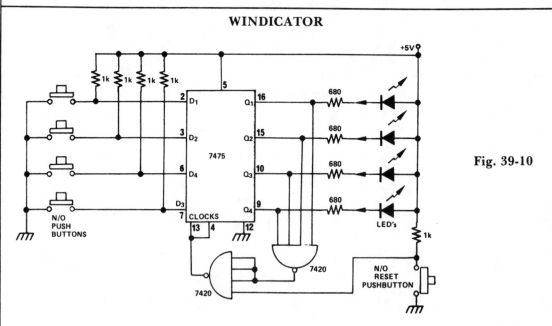

Fig. 39-10

Circuit Notes

Two TTL ICs and a handful of other components are all that is needed for a circuit that will indicate which of four buttons was pressed first, as well as lock out all other entries. A logic 0 at one of the Q outputs, lights the appropriate LED and locks out other entries by taking the clock input low.

40

Gas/Vapor Detectors

The sources of the following circuits are contained in the Sources section beginning on page 730. The figure number contained in the box of each circuit correlates to the source entry in the Sources section.

Gas and Smoke Detector   Ionization Chamber Smoke Detector
Ionization Chamber Smoke Detector

GAS AND SMOKE DETECTOR

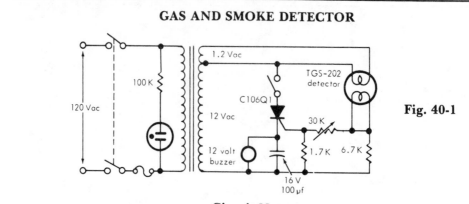

Fig. 40-1

Circuit Notes

This circuit can detect smoke and a number of gases (CO, CO$_2$, methane, coal gas and others) with a 10 ppm sensitivity. It uses a heated surface semiconductor sensor. Detection occurs when the gas concentration increase causes a decrease of the sensor element internal resistance. The switch in series with the SCR is used for resetting the alarm.

IONIZATION CHAMBER SMOKE DETECTOR

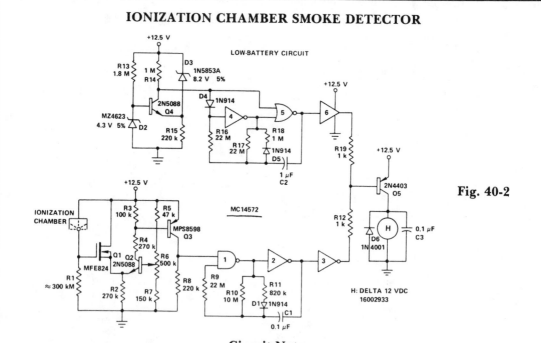

Fig. 40-2

Circuit Notes

Battery-operated, ionization chamber smoke detector includes a circuit to generate a unique alarm when the battery reaches the end of its useful life. The circuit uses the MCMOS MC14572 for two alarm oscillators (smoke and low battery). This circuit additionally uses five discrete transistors as buffers and comparators.

IONIZATION CHAMBER SMOKE DETECTOR

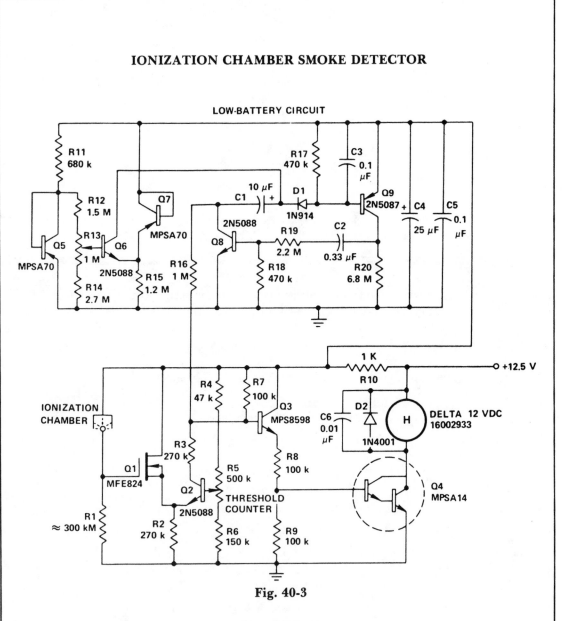

Fig. 40-3

Circuit Notes

If the smoke alarm signal must be a continuous one rather than pulsating, then the slightly less expensive, all discrete transistor version of the MC14572 may be used.

41

Indicators

The sources of the following circuits are contained in the Sources section beginning on page 730. The figure number contained in the box of each circuit correlates to the source entry in the Sources section.

Ten-Step Voltage-Level Indicator
Beat Frequency Indicator
Three-Step Level Indicator
Indicator and Alarm

Five-Step Voltage-Level Indicator
Visible Voltage Indicator
Voltage Level Detector
Zero Center Indicator for FM Receivers

Visual Zero-Beat Indicator

TEN-STEP VOLTAGE-LEVEL INDICATOR

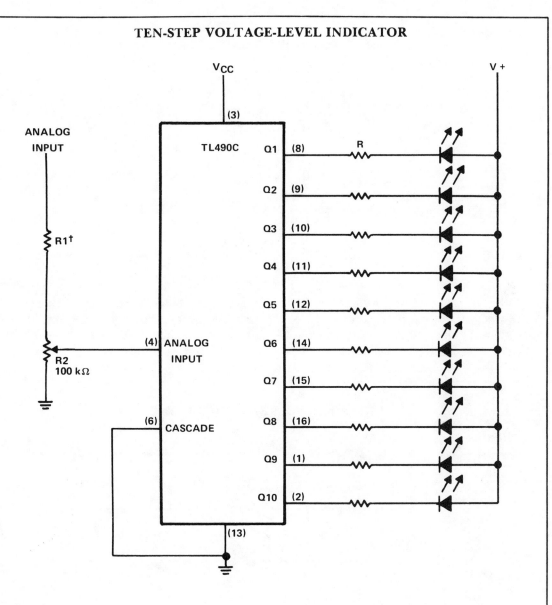

Fig. 41-1

Circuit Notes

This ten-step adjustable analog level detector is capable of sinking up to 40 milliamperes at each output. The voltage range at the input pin should range from 0 to 2 volts. Circuits of this type are useful as liquid-level indicators, pressure indicators, and temperature indicators. They may also be used with a set of active filters to provide a visual indication of harmonic content of audio signals.

BEAT FREQUENCY INDICATOR

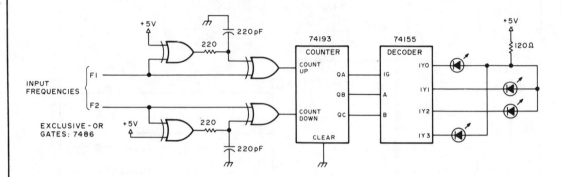

Fig. 41-2

Circuit Notes

This circuit uses LEDs to display the beat frequency of two-tone oscillators. Only one LED is on at a time, and the apparent rotation of the dot is an exact indication of the best frequency. When f1 is greater than f2, a dot of light rotates clockwise; when f1 is less than f2, the dot rotates counterclockwise; and when f1 equals f2, there is no rotation.

THREE-STEP LEVEL INDICATOR

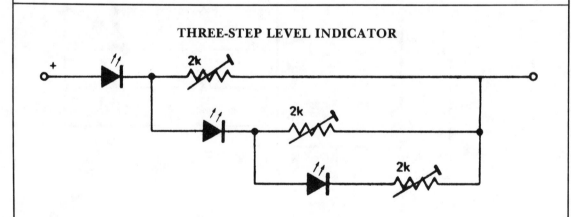

Fig. 41-3

Circuit Notes

This circuit makes a very compact level indicator where a meter would be impractical or not justified due to cost. Resistor values will depend on type of LED used. For MV50 LEDs the resistors are 2 K for steps of approx 2 V and current drain with all three LEDs on of 5 mA. The chain can be extended but current drain increases rapidly and the first LED carries all the current drawn from the supply.

INDICATOR AND ALARM

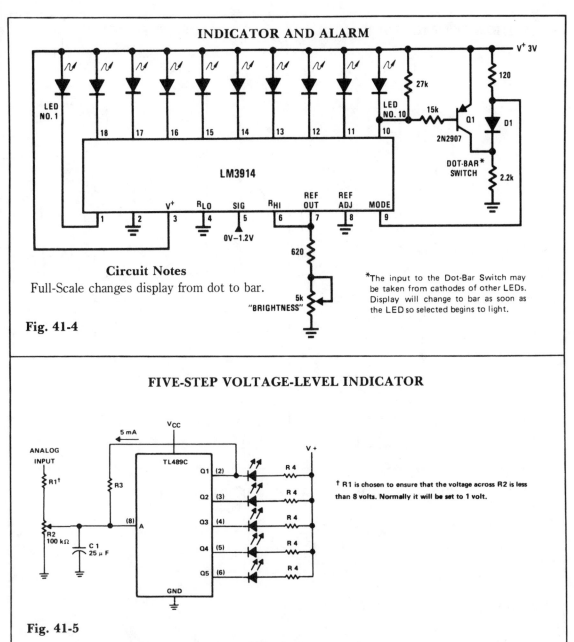

LM3914

V⁺ 3V

27k

120

LED NO. 10

15k

Q1

2N2907

D1

DOT-BAR* SWITCH

2.2k

LED NO. 1

18 17 16 15 14 13 12 11 10

V⁺ R_LO SIG R_HI REF OUT REF ADJ MODE

1 2 3 4 5 6 7 8 9

0V–1.2V

620

5k
"BRIGHTNESS"

Circuit Notes
Full-Scale changes display from dot to bar.

*The input to the Dot-Bar Switch may be taken from cathodes of other LEDs. Display will change to bar as soon as the LED so selected begins to light.

Fig. 41-4

FIVE-STEP VOLTAGE-LEVEL INDICATOR

V_CC

5 mA

ANALOG INPUT

R1†

R3

TL489C

Q1 (2)   R 4
Q2 (3)   R 4
Q3 (4)   R 4
Q4 (5)   R 4
Q5 (6)   R 4

V +

(8) A

R2
100 kΩ

C 1
25 µF

GND

† R1 is chosen to ensure that the voltage across R2 is less than 8 volts. Normally it will be set to 1 volt.

Fig. 41-5

Circuit Notes

This circuit provides a visual indication of the input analog voltage level. It has a high input impedance at pin 8 and open-collector outputs capable of sinking up to 40 milliamperes. It is suitable for driving a linear array of 5 LEDs to indicate the level is 5 steps. The voltage at the analog input should be in the range of zero to approximately one volt and should never exceed eight volts.

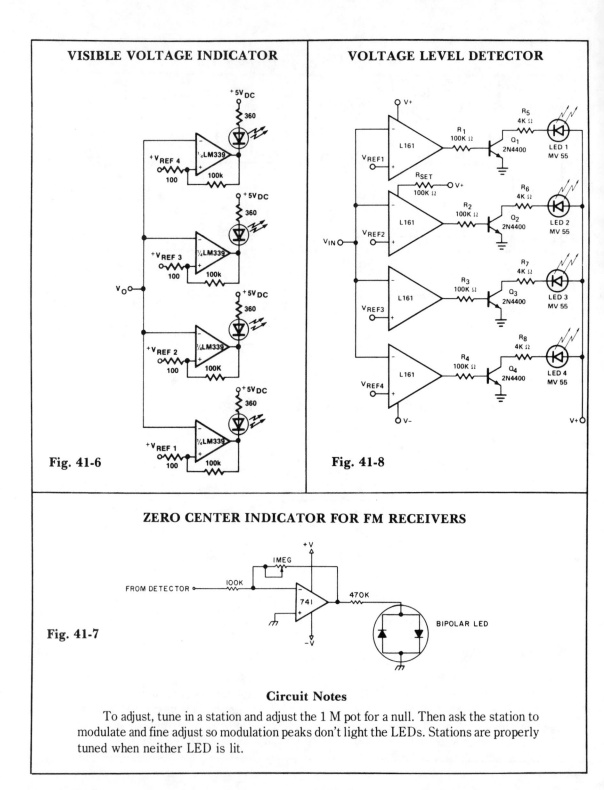

VISIBLE VOLTAGE INDICATOR

Fig. 41-6

VOLTAGE LEVEL DETECTOR

Fig. 41-8

ZERO CENTER INDICATOR FOR FM RECEIVERS

Fig. 41-7

Circuit Notes

To adjust, tune in a station and adjust the 1 M pot for a null. Then ask the station to modulate and fine adjust so modulation peaks don't light the LEDs. Stations are properly tuned when neither LED is lit.

338

VISUAL ZERO-BEAT INDICATOR

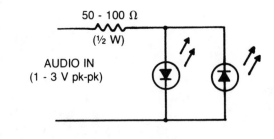

50 - 100 Ω
(½ W)

AUDIO IN
(1 - 3 V pk-pk)

LEDs: FAIRCHILD FLV-100 RED,
     OR MONSANTO MV-5094 RED/RED,
     OR MONSANTO MV-5491 RED/GREEN

Fig. 41-9

Circuit Notes

Light-emitting diodes connected with reverse polarity provide a visual indication of zero-beat frequency. Each LED is on for only half a cycle of the input. When the input frequency is more than 1 kilohertz away from the zero-beat frequency, both LEDs appear to be on all the time. As the input frequency comes within about 20 hertz of zero beat, the LEDs will flicker until zero beat is reached. Both LEDs glow or flicker until zero beat is reached, when they go out.

42

Infrared Circuits

The sources of the following circuits are contained in the Sources section beginning on page 730. The figure number contained in the box of each circuit correlates to the source entry in the Sources section.

IR Type Data Link
IR Remote Control Transmitter/Receiver
Compact IR Receiver

IR Transmitter
Remote Loudspeaker Via IR Link
Proximity Detector

IR TYPE DATA LINK

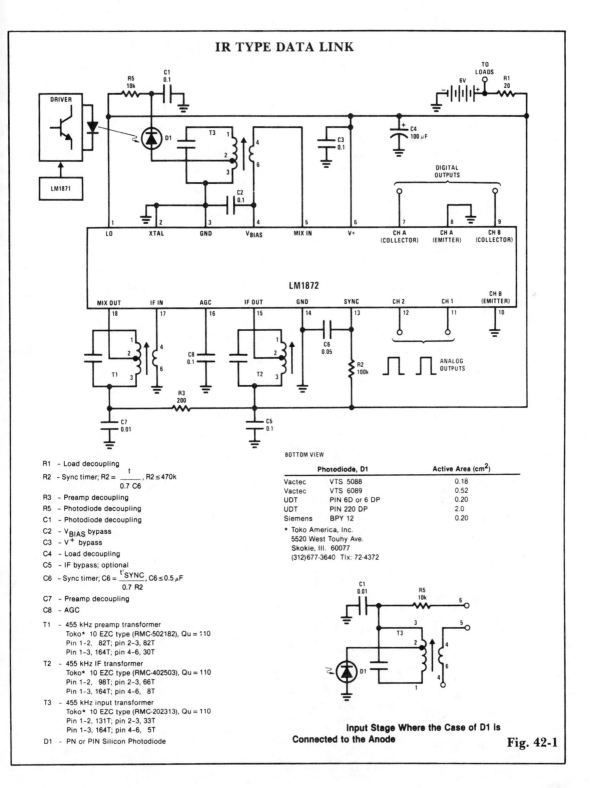

R1 - Load decoupling

R2 - Sync timer; $R2 = \dfrac{t}{0.7\ C6}$, $R2 \le 470k$

R3 - Preamp decoupling

R5 - Photodiode decoupling

C1 - Photodiode decoupling

C2 - V_{BIAS} bypass

C3 - V^+ bypass

C4 - Load decoupling

C5 - IF bypass; optional

C6 - Sync timer; $C6 = \dfrac{t'_{SYNC}}{0.7\ R2}$, $C6 \le 0.5\ \mu F$

C7 - Preamp decoupling

C8 - AGC

T1 - 455 kHz preamp transformer
Toko* 10 EZC type (RMC-502182), Qu = 110
Pin 1-2, 82T; pin 2-3, 82T
Pin 1-3, 164T; pin 4-6, 30T

T2 - 455 kHz IF transformer
Toko* 10 EZC type (RMC-402503), Qu = 110
Pin 1-2, 98T; pin 2-3, 66T
Pin 1-3, 164T; pin 4-6, 8T

T3 - 455 kHz input transformer
Toko* 10 EZC type (RMC-202313), Qu = 110
Pin 1-2, 131T; pin 2-3, 33T
Pin 1-3, 164T; pin 4-6, 5T

D1 - PN or PIN Silicon Photodiode

BOTTOM VIEW

Photodiode, D1		Active Area (cm²)
Vactec	VTS 5088	0.18
Vactec	VTS 6089	0.52
UDT	PIN 6D or 6 DP	0.20
UDT	PIN 220 DP	2.0
Siemens	BPY 12	0.20

* Toko America, Inc.
5520 West Touhy Ave.
Skokie, Ill. 60077
(312)677-3640 Tlx: 72-4372

**Input Stage Where the Case of D1 is
Connected to the Anode**

Fig. 42-1

341

IR REMOTE CONTROL TRANSMITTER/RECEIVER

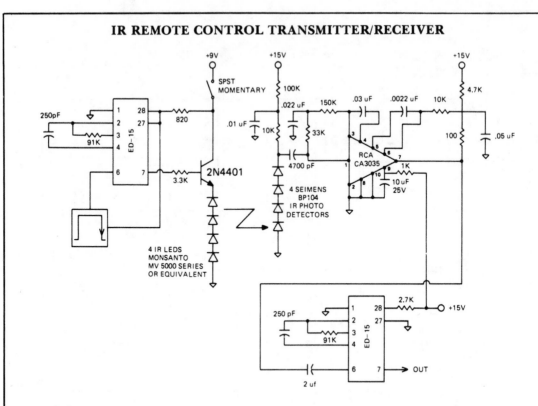

Fig. 42-2

Circuit Notes

The circuit is designed to operate at 25 kHz. The data stream turns the 2N4401 hard on or off depending upon the coded state. This in turn switches the series infrared LEDs on and off. The receiver circuit consists of a three stage amplifier with photo diodes arrayed for maximum coverage of the reception area. The range of this set-up should be about 10 meters.

COMPACT IR RECEIVER

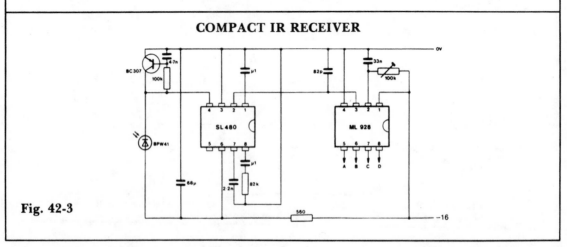

Fig. 42-3

IR TRANSMITTER

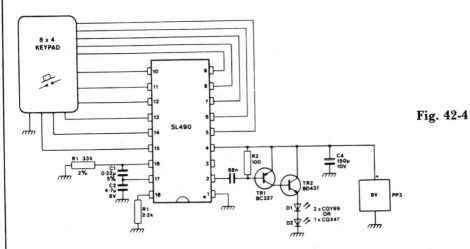

Fig. 42-4

Circuit Notes

This simple infra-red transmitter, where the PPM output from pin 2 of the SL490 is fed to the base of the PNP trasmitter TR1, pro- duces an amplified current pulse about 15 μsec wide. This pulse is further amplified by TR2 and applied to the infra-red diodes D1 and D2.

REMOTE LOUDSPEAKER VIA IR LINK

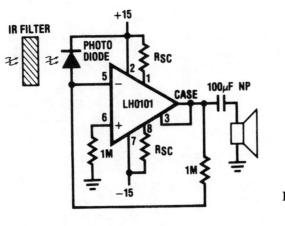

Fig. 42-5

343

PROXIMITY DETECTOR

NOTE:
IC1 IS CA3240
Q1 IS 2N3819
Q2,4 ARE BC184L
Q3 IS BD140

D1 IS PHOTODIODE
D2 IS 1N4148
ZD1 IS 2V7 400mW ZENER
LED1 IS 3mm RED LED
LED2 IS IS INFRA-RED LED

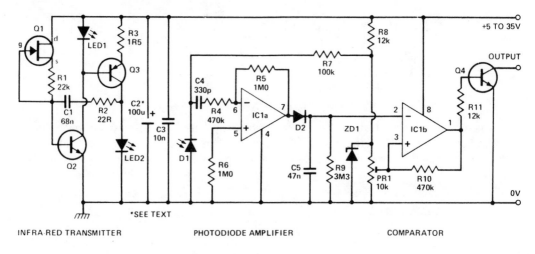

INFRA-RED TRANSMITTER   PHOTODIODE AMPLIFIER   COMPARATOR

Fig. 42-6

Circuit Notes

This circuit provides a means of detecting the presence of anything by the reflection of infra-red light and provides a direct digital output of object detection. By the use of modulation and high power bursts of infra-red at a very low duty cycle, a detection range of over a foot is achieved. Works on the principle of transmitting a beam of modulated infra-red light from the emitter diode LED2, and receiving reflections from objects passing in front of the beam with a photodiode detector D1. The circuit consists of an infra-red transmitter, photodiode amplifier, and a variable threshold comparator.

43

Instrumentation Amplifiers

The sources of the following circuits are contained in the Sources section beginning on page 730. The figure number contained in the box of each circuit correlates to the source entry in the Sources section.

INSTRUMENTATION AMPLIFIER

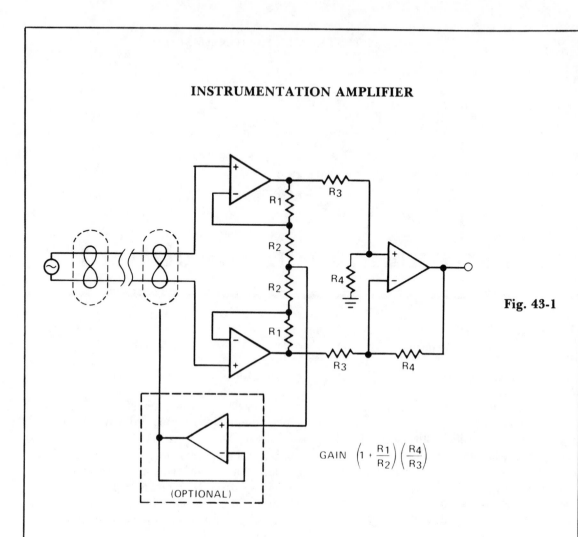

Fig. 43-1

$$\text{GAIN } \left(1 + \frac{R_1}{R_2}\right)\left(\frac{R_4}{R_3}\right)$$

Circuit Notes

Instrumentation amplifiers (differential amplifiers) are specifically designed to extract and amplify small differential signals from much larger common mode voltages. To serve as building blocks in instrumentation amplifiers, op amps must have very low offset voltage drift, high gain and wide bandwidth. The HA-4620/5604 is suited for this application. The optional circuitry makes use of the fourth amplifier section as a shield driver which enhances the ac common mode rejection by nullifying the effects of capacitance-to-ground mismatch between input conductors.

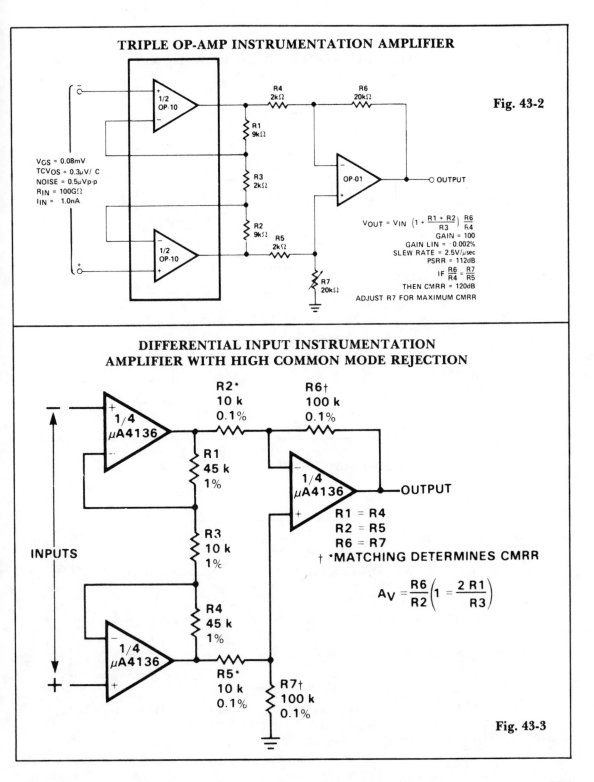

TRIPLE OP-AMP INSTRUMENTATION AMPLIFIER

Fig. 43-2

$V_{GS} = 0.08mV$
$TCV_{OS} = 0.3\mu V/ C$
$NOISE = 0.5\mu Vp\text{-}p$
$R_{IN} = 100G\Omega$
$I_{IN} = 1.0nA$

R4 2kΩ
R6 20kΩ
R1 9kΩ
R3 2kΩ
R2 9kΩ
R5 2kΩ
R7 20kΩ

1/2 OP-10
1/2 OP-10
OP-01
OUTPUT

$$V_{OUT} = V_{IN} \left(1 + \frac{R1 + R2}{R3}\right) \frac{R6}{R4}$$
GAIN = 100
GAIN LIN = ·0.002%
SLEW RATE = 2.5V/μsec
PSRR = 112dB
IF $\frac{R6}{R4} = \frac{R7}{R5}$
THEN CMRR = 120dB
ADJUST R7 FOR MAXIMUM CMRR

DIFFERENTIAL INPUT INSTRUMENTATION AMPLIFIER WITH HIGH COMMON MODE REJECTION

R2* 10 k 0.1%
R6† 100 k 0.1%
R1 45 k 1%
R3 10 k 1%
R4 45 k 1%
R5* 10 k 0.1%
R7† 100 k 0.1%

1/4 μA4136
1/4 μA4136
1/4 μA4136

INPUTS

OUTPUT

R1 = R4
R2 = R5
R6 = R7
† *MATCHING DETERMINES CMRR

$$A_V = \frac{R6}{R2}\left(1 = \frac{2\,R1}{R3}\right)$$

Fig. 43-3

347

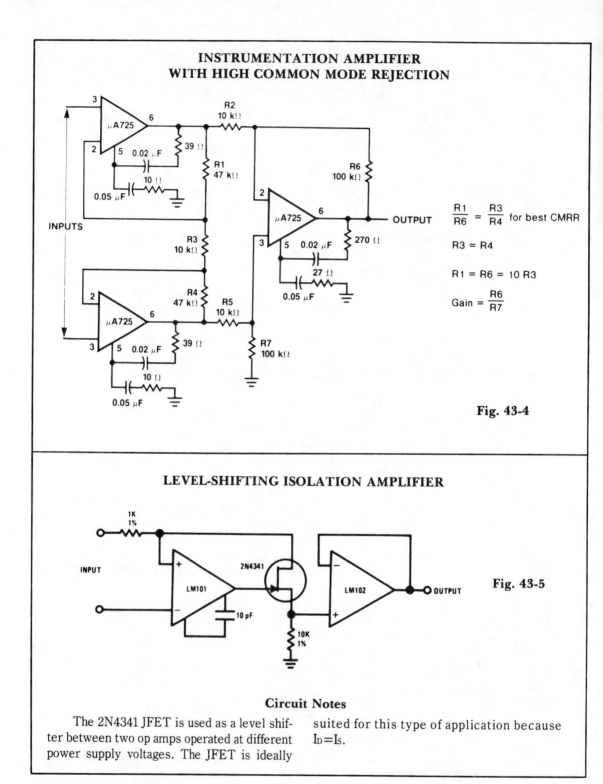

INSTRUMENTATION AMPLIFIER
WITH HIGH COMMON MODE REJECTION

$$\frac{R1}{R6} = \frac{R3}{R4} \text{ for best CMRR}$$

R3 = R4

R1 = R6 = 10 R3

$$\text{Gain} = \frac{R6}{R7}$$

Fig. 43-4

LEVEL-SHIFTING ISOLATION AMPLIFIER

Fig. 43-5

Circuit Notes

The 2N4341 JFET is used as a level shifter between two op amps operated at different power supply voltages. The JFET is ideally suited for this type of application because $I_D = I_S$.

348

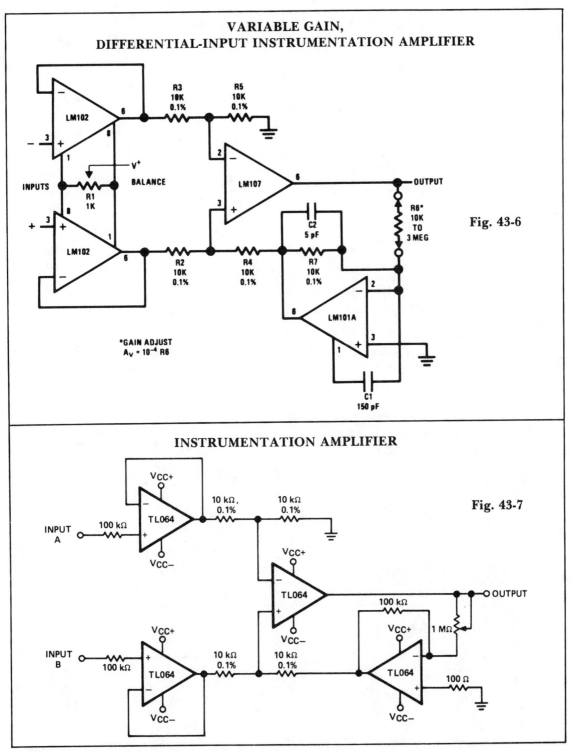

VARIABLE GAIN,
DIFFERENTIAL-INPUT INSTRUMENTATION AMPLIFIER

Fig. 43-6

INSTRUMENTATION AMPLIFIER

Fig. 43-7

349

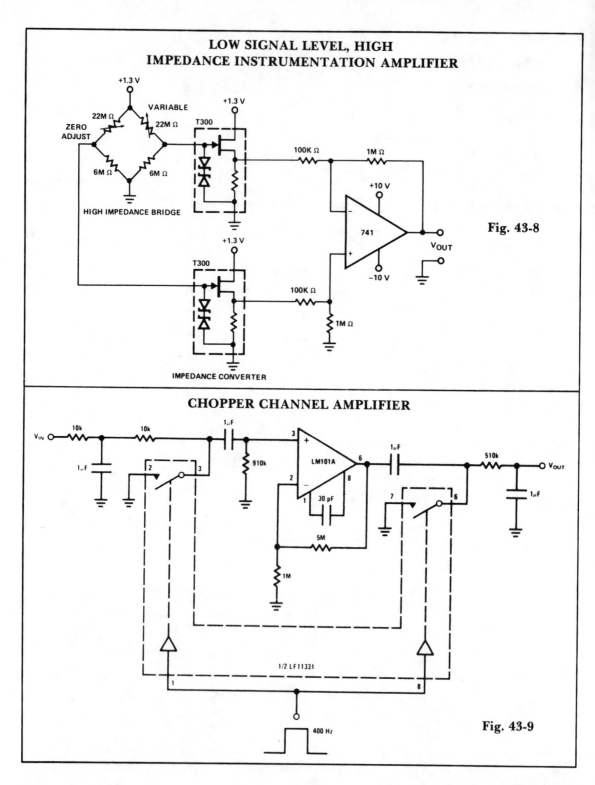

LOW SIGNAL LEVEL, HIGH
IMPEDANCE INSTRUMENTATION AMPLIFIER

+1.3 V

VARIABLE

22M Ω

ZERO
ADJUST

22M Ω

T300

+1.3 V

100K Ω

1M Ω

6M Ω

6M Ω

+10 V

−

741

+

V_OUT

−10 V

Fig. 43-8

HIGH IMPEDANCE BRIDGE

+1.3 V

T300

100K Ω

1M Ω

IMPEDANCE CONVERTER

CHOPPER CHANNEL AMPLIFIER

V_IN

10k

10k

1 μF

1 μF

2

3

910k

3

+

LM101A

6

1 μF

510k

V_OUT

1 μF

2

−

1

8

30 pF

7

6

5M

1M

1/2 LF11331

1

8

400 Hz

Fig. 43-9

350

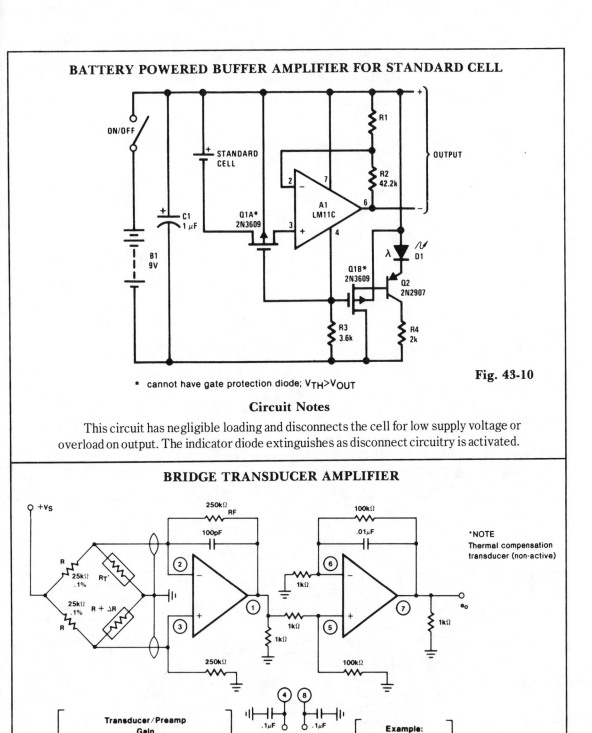

BATTERY POWERED BUFFER AMPLIFIER FOR STANDARD CELL

ON/OFF

STANDARD CELL

R1

R2
42.2k

OUTPUT

2 7

A1
LM11C

6

Q1A*
2N3609

3

+ 4

C1
1 μF

B1
9V

λ D1

Q1B*
2N3609

Q2
2N2907

R3
3.6k

R4
2k

* cannot have gate protection diode; $V_{TH} > V_{OUT}$

Fig. 43-10

Circuit Notes

This circuit has negligible loading and disconnects the cell for low supply voltage or overload on output. The indicator diode extinguishes as disconnect circuitry is activated.

BRIDGE TRANSDUCER AMPLIFIER

+v_S

250kΩ
RF

100pF

100kΩ

.01μF

*NOTE
Thermal compensation
transducer (non-active)

R
25kΩ
.1%

R_T*

2 −

1

R
25kΩ
.1%

$R + \Delta R$

3 +

1kΩ

6 −

1kΩ

7

e_o

1kΩ

5 +

1kΩ

250kΩ

100kΩ

1kΩ

4 8

.1μF   .1μF

−15   +15

$$
\begin{array}{l}
\text{Transducer/Preamp} \\
\text{Gain} \\
e_o \simeq \dfrac{R_F}{R} \cdot \dfrac{\Delta}{1 + \Delta} \cdot \dfrac{V_S}{(2 + \Delta)/[(1 + \Delta) + R/R_F]} \\
\Delta = \dfrac{\Delta R}{R}
\end{array}
$$

IC = NE/SE5512

Example:
$\Delta R = 5\Omega$ $\Delta e_o \approx 1.2V$
$V_S = +10V$

Fig. 43-11

INSTRUMENTATION AMPLIFIER

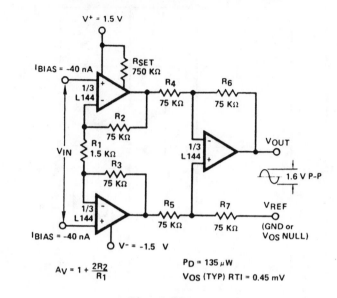

Fig. 43-12

$$A_V = 1 + \frac{2R2}{R1}$$

$P_D = 135\,\mu W$

V_{OS} (TYP) RTI = 0.45 mV

Circuit Notes

Three-amplifier circuit consumes only 135 μW of power from a ±1.5 V power supply. With a gain of 101, the instrumentation amplifier is ideal in sensor interface and biomedical preamplifier applications. The first stage provides all of the gain while the second stage is used to provide common mode rejection and double-ended to single-ended conversion.

ISOLATION AMPLIFIER FOR MEDICAL TELEMETRY

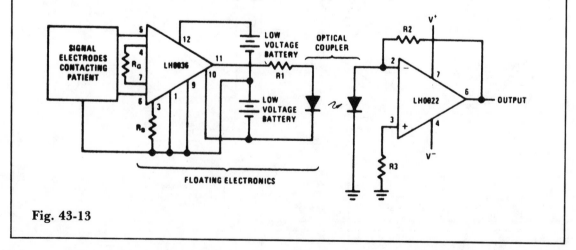

Fig. 43-13

HIGH GAIN DIFFERENTIAL INSTRUMENTATION AMPLIFIER

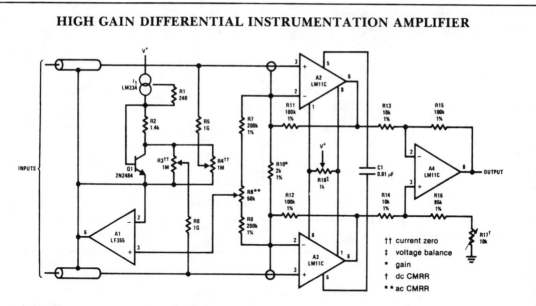

Fig. 43-14

Circuit Notes

This circuit includes input guarding, cable bootstrapping, and bias current compensation. Differential bandwidth is reduced by C1 which also makes common-mode rejection less dependent on matching of input amplifiers.

HIGH IMPEDANCE BRIDGE AMPLIFIER

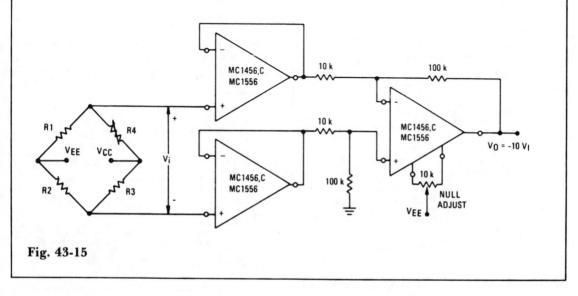

Fig. 43-15

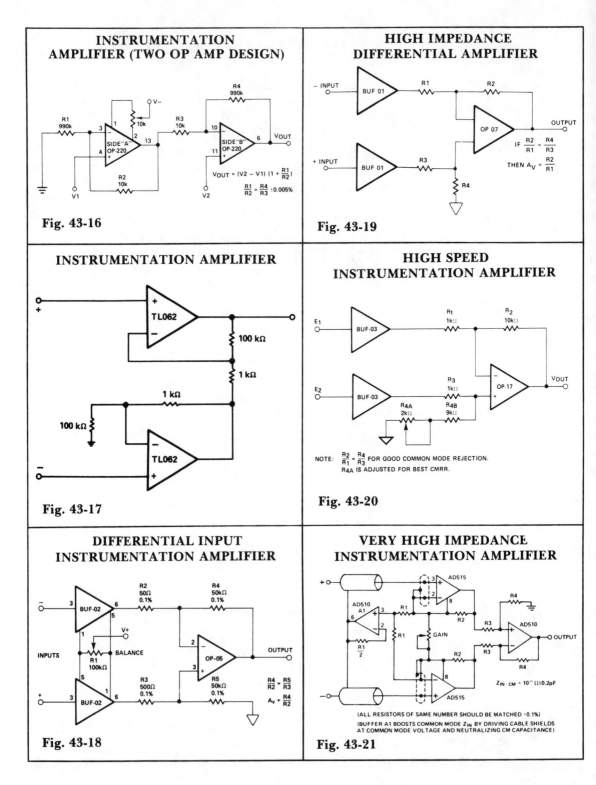

INSTRUMENTATION AMPLIFIER (TWO OP AMP DESIGN)

$V_{OUT} = (V2 - V1)(1 + \frac{R1}{R2})$

$\frac{R1}{R2} = \frac{R4}{R3} \pm 0.005\%$

Fig. 43-16

INSTRUMENTATION AMPLIFIER

Fig. 43-17

DIFFERENTIAL INPUT INSTRUMENTATION AMPLIFIER

$\frac{R4}{R2} = \frac{R5}{R3}$

$A_V = \frac{R4}{R2}$

Fig. 43-18

HIGH IMPEDANCE DIFFERENTIAL AMPLIFIER

IF $\frac{R2}{R1} = \frac{R4}{R3}$

THEN $A_V = \frac{R2}{R1}$

Fig. 43-19

HIGH SPEED INSTRUMENTATION AMPLIFIER

NOTE: $\frac{R2}{R1} = \frac{R4}{R3}$ FOR GOOD COMMON MODE REJECTION.
R4A IS ADJUSTED FOR BEST CMRR.

Fig. 43-20

VERY HIGH IMPEDANCE INSTRUMENTATION AMPLIFIER

$Z_{IN\cdot CM} = 10^{11} \, \Omega \| 0.2pF$

(ALL RESISTORS OF SAME NUMBER SHOULD BE MATCHED ±0.1%)
(BUFFER A1 BOOSTS COMMON MODE Z_{IN} BY DRIVING CABLE SHIELDS
AT COMMON MODE VOLTAGE AND NEUTRALIZING CM CAPACITANCE)

Fig. 43-21

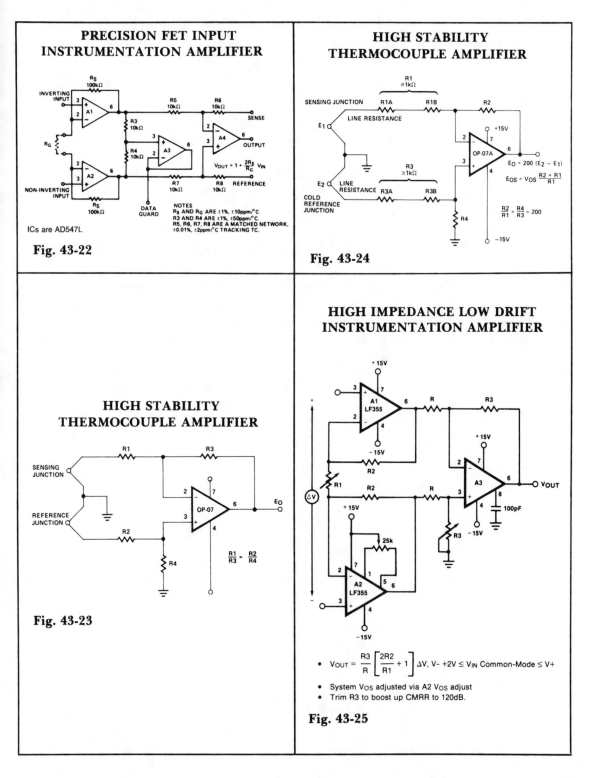

PRECISION FET INPUT INSTRUMENTATION AMPLIFIER

$V_{OUT} = 1 + \frac{2R_S}{R_G} \; V_{IN}$

NOTES
R_S AND R_G ARE ±1%, ±10ppm/°C
R3 AND R4 ARE ±1%, ±50ppm/°C.
R5, R6, R7, R8 ARE A MATCHED NETWORK,
±0.01%, ±2ppm/°C TRACKING TC.

ICs are AD547L.

Fig. 43-22

HIGH STABILITY THERMOCOUPLE AMPLIFIER

$E_O = 200 \, (E_2 - E_1)$

$E_{OS} = V_{OS} \frac{R2 + R1}{R1}$

$\frac{R2}{R1} = \frac{R4}{R3} = 200$

Fig. 43-24

HIGH STABILITY THERMOCOUPLE AMPLIFIER

$\frac{R1}{R3} = \frac{R2}{R4}$

Fig. 43-23

HIGH IMPEDANCE LOW DRIFT INSTRUMENTATION AMPLIFIER

- $V_{OUT} = \frac{R3}{R} \left[\frac{2R2}{R1} + 1 \right] \Delta V$, V- +2V ≤ V_{IN} Common-Mode ≤ V+
- System V_{OS} adjusted via A2 V_{OS} adjust
- Trim R3 to boost up CMRR to 120dB.

Fig. 43-25

355

44

Light Activated Circuits

The sources of the following circuits are contained in the Sources section beginning on page 730. The figure number contained in the box of each circuit correlates to the source entry in the Sources section.

PULSE GENERATION BY INTERRUPTING A LIGHT BEAM

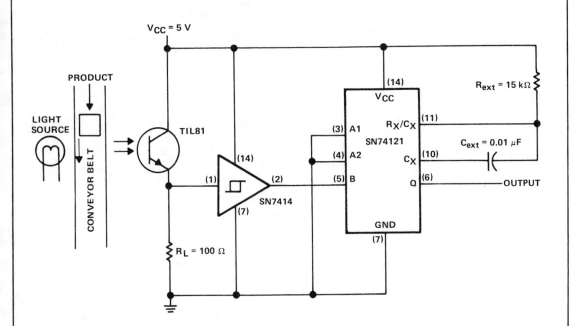

Fig. 44-1

Circuit Notes

This circuit puts out a pulse when an object on the conveyor belt blocks the light source. The light source keeps the phototransistor turned on. This produces a high-logic-level voltage at the Schmitt-trigger inverter and a TTL-compatible low logic level at pin 5 of the monostable. When an object blocks the light, TIL81 turns off the Schmitt-trigger inverter to triggers the one shot.

357

OPTICAL COMMUNICATION SYSTEM

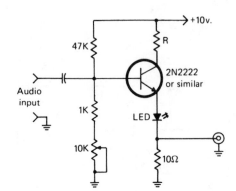

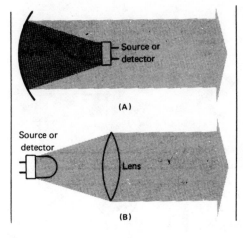

(A)

(B)

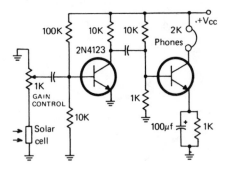

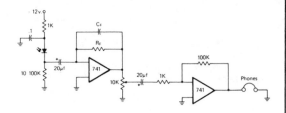

Circuit Notes

The simple modulator stage will accommodate most common LEDs. By adjusting the potentiometer, the bias of the transistor is varied until the LED is at its half output point. Then, audio will cause it to vary above and below this point. The purpose of R1 is to limit the current through the LED to a safe level and the purpose of the 10 ohm resistor is to allow a portion of the modulating signal to be observed on a scope.

Fig. 44-2

FOUR QUADRANT PHOTO-CONDUCTIVE DETECTOR AMPLIFIER

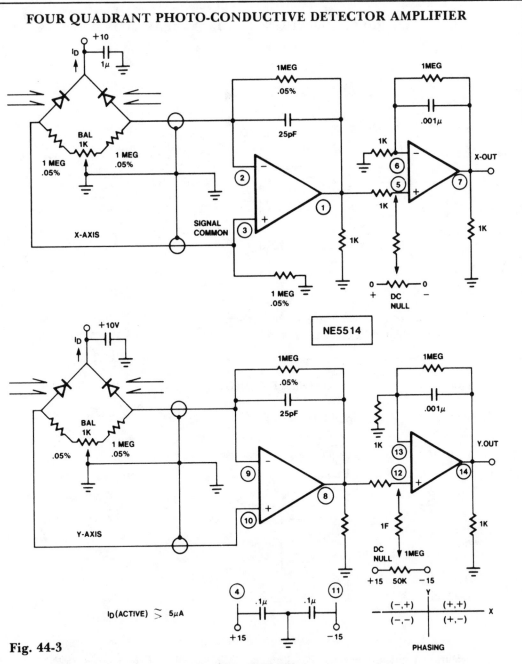

Fig. 44-3

Circuit Notes

Use this circuit to sense four quadrant motion of a light source. By proper summing of the signals from the X and Y axes, four quadrant output may be fed to an X-Y plotter, oscilloscope, or computer for simulation. IC = NE/SE5514

PRECISION PHOTODIODE COMPARATOR

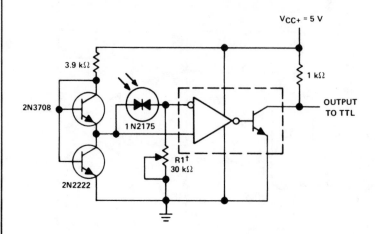

Fig. 44-4

Circuit Notes

R1 sets the comparison level. At comparison, the photodiode has less than 5 mV across it, decreasing dark current by an order of magnitude. IC = LM 111/211/311.

AUTOMATIC NIGHT LIGHT

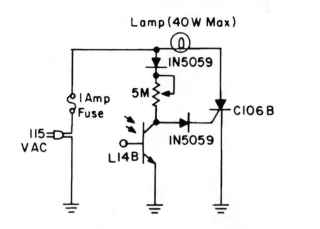

Fig. 44-5

Circuit Notes

During daylight hours, the L14B photo-Darlington (JEDEC registered as 2N5777 through 2N5780) shunts all gate current to ground. At night, the L14B effectively provides a high resistance, diverting the current into the gate of the C106B and turning on the lamp.

RECEIVER FOR 50 kHz FM OPTICAL TRANSMITTER

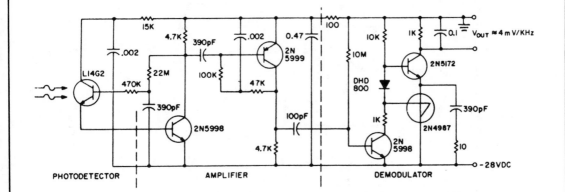

Fig. 44-6

Circuit Notes

This circuit consists of a L14G2 detector, two stages of gain, and a FM demodulator. Better sensitivity can be obtained using more stages of stabilized gain with AGC.

PHOTODIODE AMPLIFIER

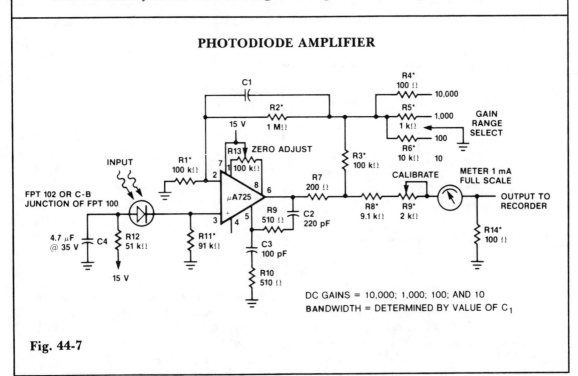

DC GAINS = 10,000; 1,000; 100; AND 10
BANDWIDTH = DETERMINED BY VALUE OF C_1

Fig. 44-7

OPTICAL SCHMITT TRIGGER

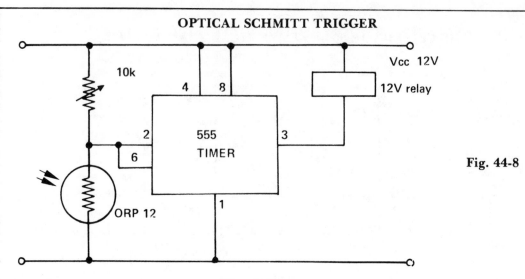

Fig. 44-8

Circuit Notes

This circuit shows a 555 with its trigger and threshold inputs connected together used to energize a relay when the light level on a photoconductive cell falls below a preset value.

Circuit can be used in other applications where a high input impedance and low output impedance are required with the minimum component count.

ADJUSTABLE LIGHT DETECTION SWITCH

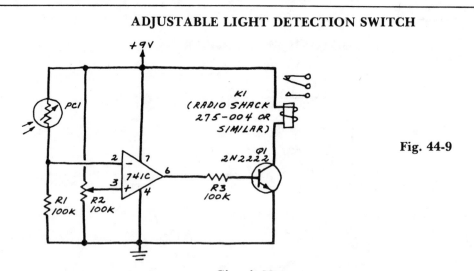

Fig. 44-9

Circuit Notes

R2 sets the circuit's threshold. When the light intensity at PCI's surface is decreased, the resistance of PC1 a cadmium-sulfide photoresistor is increased. This decreases the voltage at the inverting input of the 741. When the

reference voltage at the 741's noninverting input is properly adjusted via R2, the comparator will switch from low to high when PC1 is darkened. This turns on Q1 which, in turn, pulls in relay K1.

PHOTOCELL MEMORY SWITCH FOR AC POWER CONTROL

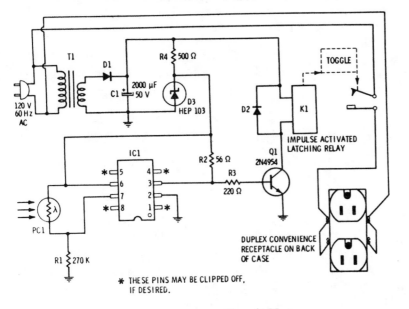

Fig. 44-10

Circuit Notes

Provides remote control for ac-powered devices by using the beam of a flashlight as a magic wand. The important aspect of this gadget is that it remembers. Activate it once to apply power to a device and it stays on. Acti- vate it a second time and power goes off and stays off. It consists of a combination of a high- sensitivity photocell, a high-gain IC Schmitt trigger, and an impulse-actuated latching relay.

OPTICAL TRANSMITTER

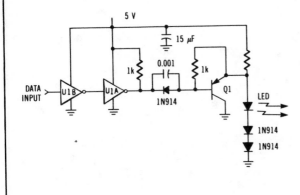

Circuit Notes

Driver circuit uses an MC74LS04 and one discrete transistor. The circuit can drive the LED (MFOE1200) at up to 1 Mbps data rate.

Fig. 44-11

LIGHT INTERRUPTION DETECTOR

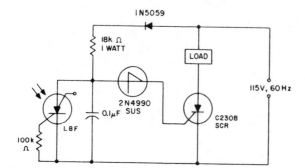

Fig. 44-12

Circuit Notes

When the light incident on the LASCR is interrupted, the voltage at the anode to the 2N4990 unilateral switch goes positive on the next positive cycle of the power which in turn triggers the switch and the C230 SCR when the switching voltage of the unilateral switch is reached. This will cause the load to be energized for as long as light is not incident on the LASCR.

OPTICAL RECEIVER

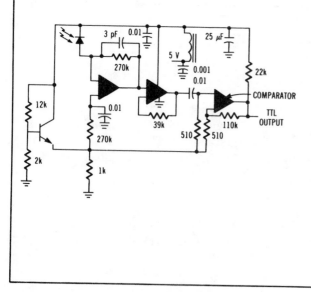

Circuit Notes

The MFOD1100 PIN diode requires shielding from emi.

Fig. 44-13

LIGHT ISOLATED SOLID STATE POWER RELAY CIRCUITS

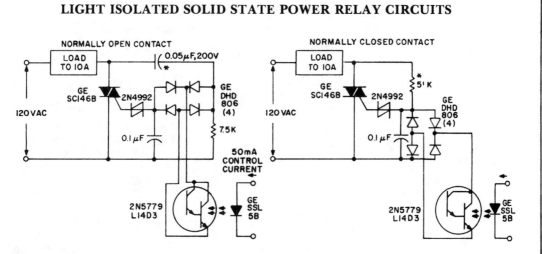

Fig. 44-14

Circuit Notes

Both circuits use the G.E. SC146B, 200 V, 10 A Triac as load current contacts. These triacs are triggered by normal SBS (2N4992) trigger circuits, which are controlled by the photo-Darlington, acting through the DA806 bridge as an ac photo switch. To operate the relays at other line voltages the asterisked (*) components are scaled to supply identical current. Ratings must be changed as required. Incandescent lamps may be used in place of the light emitting diodes, if desired.

PRECISION PHOTODIODE LEVEL DETECTOR

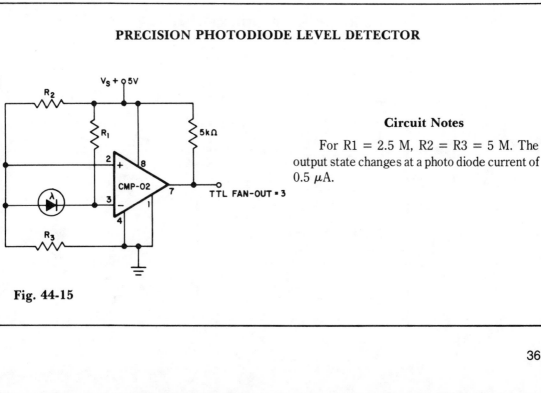

Circuit Notes

For R1 = 2.5 M, R2 = R3 = 5 M. The output state changes at a photo diode current of 0.5 μA.

Fig. 44-15

LIGHT BEAM OPERATED ON-OFF RELAY

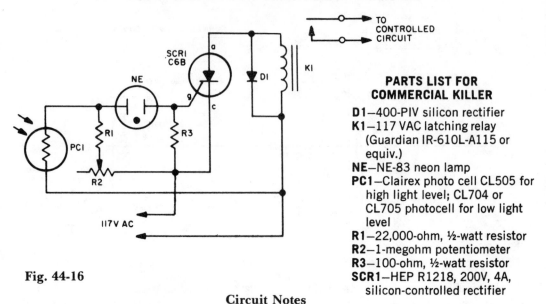

PARTS LIST FOR COMMERCIAL KILLER

D1—400-PIV silicon rectifier
K1—117 VAC latching relay (Guardian IR-610L-A115 or equiv.)
NE—NE-83 neon lamp
PC1—Clairex photo cell CL505 for high light level; CL704 or CL705 photocell for low light level
R1—22,000-ohm, ½-watt resistor
R2—1-megohm potentiometer
R3—100-ohm, ½-watt resistor
SCR1—HEP R1218, 200V, 4A, silicon-controlled rectifier

Fig. 44-16

Circuit Notes

When a beam of light strikes the photocell, the voltage across neon lamp NE-1 rises sharply. NE-1 turns on and fires the SCR. K1 is an impulse relay whose contacts stay in position even after coil current is removed. The first impulse opens K1's contacts, the second impulse closes them, etc.

LOGARITHMIC LIGHT SENSOR

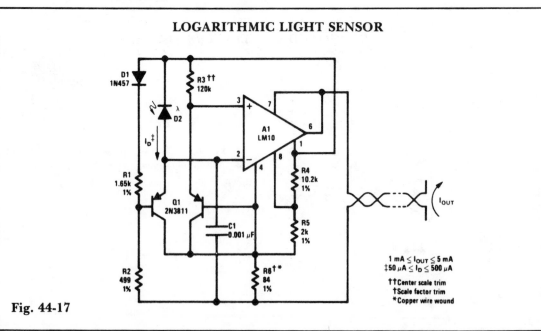

Fig. 44-17

FM (PRM) OPTICAL TRANSMITTER

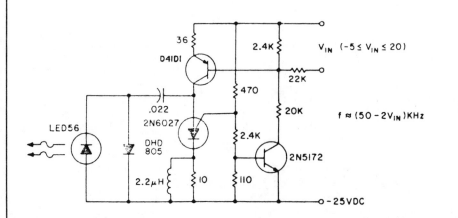

Fig. 44-18

V_{IN} ($-5 \le V_{IN} \le 20$)

$f \approx (50 - 2V_{IN})$ KHz

Circuit Notes

The basic circuit can be operated at 80 kHz and is limited by the PUT capacitor combination. 60 kHz is the maximum modulation frequency. The pulse repetition rate is a linear function of V_{IN}, the modulating voltage. Lenses or reflectors minimizes stray light noise effects. Greater output can be obtained by using a larger capacitor, which also gives a lower operating frequency, or using a higher power output IRED such as the F5D1. Average power consumption of the transmitter circuit is less than 3 watts.

LIGHT LEVEL SENSOR

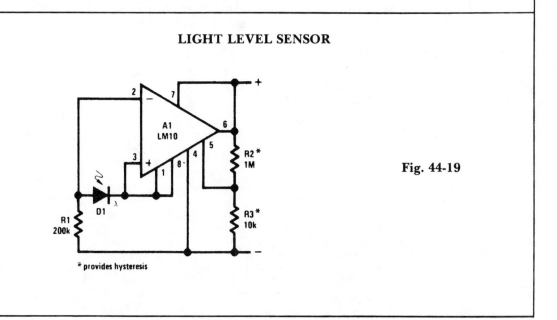

Fig. 44-19

* provides hysteresis

45

Light Controls

The sources of the following circuits are contained in the Sources section beginning on page 730. The figure number contained in the box of each circuit correlates to the source entry in the Sources section.

LIGHT DIMMERS

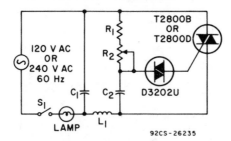

(a) Single-time-constant light-dimmer circuit.

Parts List

120-Volt, 60-Hz Operation

C_1, C_2 = 0.1 μF, 200 V
L_1 = 100 μH
R_1 = 3300 ohms, 0.5 watt
R_2 = light control, potentiometer, 0.25 megohm,
0.5 watt

240-Volt, 50/60 Hz Opera-tion

C_1 = 0.1 μF, 400 V

C_2 = 0.05 μF, 400 V
L_1 = 200 μH
R_1 = 4700 ohms, 0.5 watt
R_2 = light control, potentiometer, 0.25 megohm,
1 watt

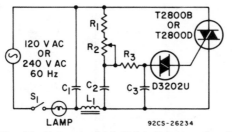

(b) Double-time-constant light-dimmer circuit.

Parts List

120-Volt, 60-Hz Operation

C_1, C_2 = 0.1 μF, 200 V
C_3 = 0.1 μF, 100 V
L_1 = 100 μH
R_1 = 1000 ohms, 0.5 watt
R_2 = light control, poten-

tiometer, 0.1 megohm,
0.5 watt

240-Volt, 60-Hz Operation

C_1 = 0.1 μF, 400 V
C_2 = 0.05 μF, 400 V

C_3 = 0.1 μF, 100 V
L_1 = 100 μH
R_1 = 7500 ohms, 2 watts
R_2 = light control, potentiometer, 0.2 megohm,
1 watt
R_3 = 7500 ohms, 2 watts

Fig. 45-1

Circuit Notes

The two lamp-dimmer circuits differ in that (a) employs a single-time-constant trigger network and (b) uses a double-time-constant trigger circuit that reduces hysteresis effects and thereby extends the effective range of the light-control potentiometer. (Hysteresis refers to a difference in the control potentiometer setting at which the lamp turns on and the setting at which the light is extin-guished.) The additional capacitor C2 in (b) reduces hysteresis by charging to a higher voltage than capacitor C3. During gate triggering, C3 discharges to form the gate current pulse. Capacitor C2, however, has a longer discharge time constant and this capacitor restores some of the charge removed from C3 by the gate current pulse.

REMOTE CONTROL FOR LAMP OR APPLIANCE

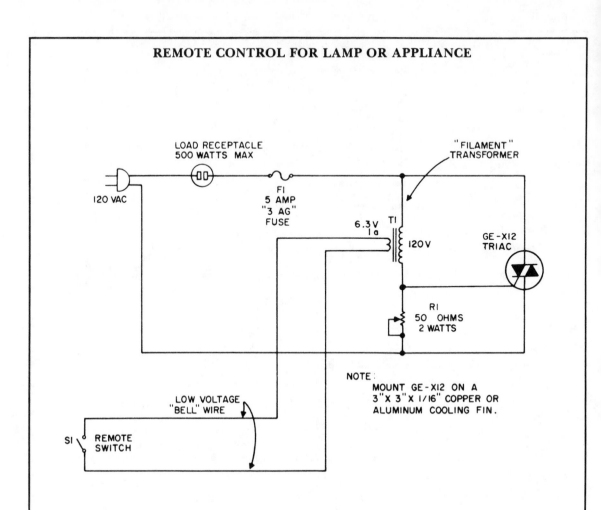

Fig. 45-2

Circuit Notes

The circuit uses the primary current of a small 6.3 volt filament transformer to actuate a triac and energize the load. When switch S1, in the six-volt secondary, of the transformer is open, a small "magnetizing" current flows through the primary winding. This magnetizing current may be large enough to trigger the triac. Therefore, a shunting resistor, R1, is required to prevent such triggering. R1, is ad-justed for the highest resistance that will not cause the triac to trigger with S1 open. When single-pole remote switch, S1, closes, the secondary of the transformer is shorted and a high current flows through the 120-volt primary. This triggers the triac and energizes the load. When the triac conducts, current through the primary stops and thus prevents burning out the transformer.

HIGH POWER CONTROL FOR SENSITIVE CONTACTS

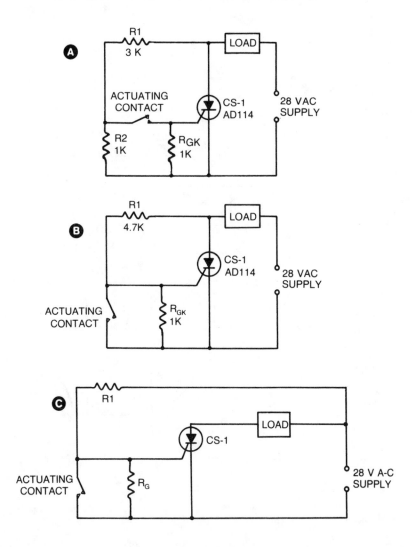

Fig. 45-3

Circuit Notes

Two simple arrangements for resistive loads are shown in A & B. The circuit in A will provide load power when the actuating contact is closed, and no power when the contact is open. B provides the reverse of this action—power being supplied to the load when the contact is open with no load power when the contact is closed. If desired, both circuits can be made to latch by operating with dc instead of the indicated ac supply. In both of these circuits, voltage across the sensitive contacts is under 5 volts, and contact current is below 5 mA. For inductive loads, R1 would normally be returned to the opposite side of the load as shown in C.

371

COMPLEMENTARY LIGHTING CONTROL

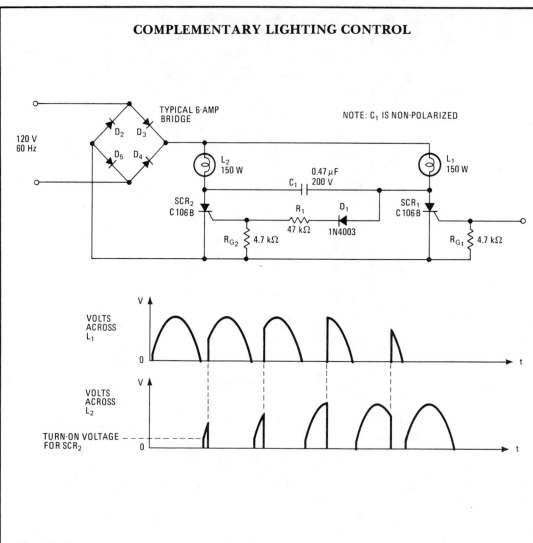

Fig. 45-4

Circuit Notes

This lighting-control unit will fade out one lamp while simultaneously increasing the light output of another. The two loads track each other accurately without adjustments. The gate of SCR1, a silicon-controlled rectifier, is driven from a standard phase-control circuit, based, for example, on a unijunction transistor or a diac. It controls the brightness of lamp L1 directly. Whenever SCR1 is not on, a small current flows through L1, D1, and R1, permitting SCR2 to fire. When SCR1 turns on, current flow ceases through D1 and R1; the energy stored in C1 produces a negative spike that turns SCR2 off.

FLOODLAMP POWER CONTROL

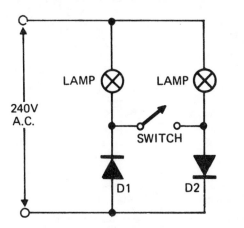

Fig. 45-5

Circuit Notes

When setting up photographic floodlamps, it is sometimes desirable to operate the lamps at lower power levels until actually ready to take the photograph. The circuit allows the lamps to operate on half cycle power when the switch is open, and full power, when the switch is closed. The diodes D1 and D2 should have a 400 volt PIV rating at 5 amps.

HYSTERESIS-FREE PHASE CONTROL CIRCUIT

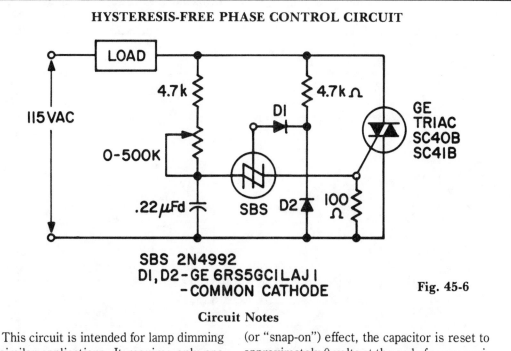

SBS 2N4992
DI, D2-GE 6RS5GCILAJI
 -COMMON CATHODE

Fig. 45-6

Circuit Notes

This circuit is intended for lamp dimming and similar applications. It requires only one RC phase lag network. To avoid the hysteresis (or "snap-on") effect, the capacitor is reset to approximately 0 volts at the end of every positive half cycle using the gate lead.

LOW COST LAMP DIMMER

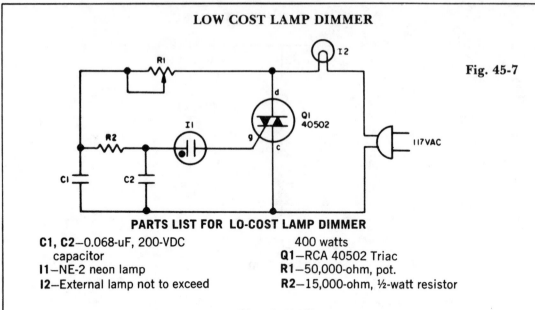

Fig. 45-7

PARTS LIST FOR LO-COST LAMP DIMMER

C1, C2—0.068-uF, 200-VDC capacitor
I1—NE-2 neon lamp
I2—External lamp not to exceed

400 watts
Q1—RCA 40502 Triac
R1—50,000-ohm, pot.
R2—15,000-ohm, ½-watt resistor

Circuit Notes

Without a heatsink, Triac Q1 handles up to a 400-watt lamp. The neon lamp does not trip the gate until it conducts so the lamp turns on a medium brilliance. The lamp can then be backed off to a soft glow.

ZERO-POINT SWITCH

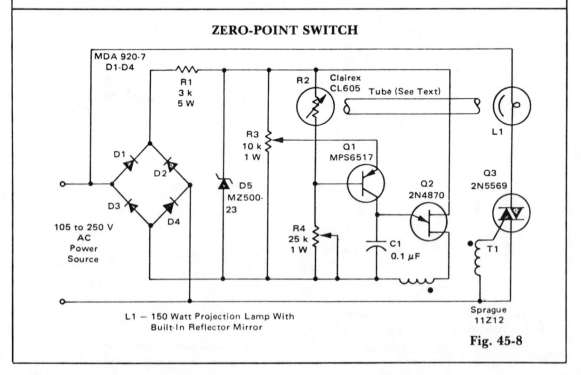

L1 — 150 Watt Projection Lamp With
Built-In Reflector Mirror

Fig. 45-8

800 W TRIAC LIGHT DIMMER

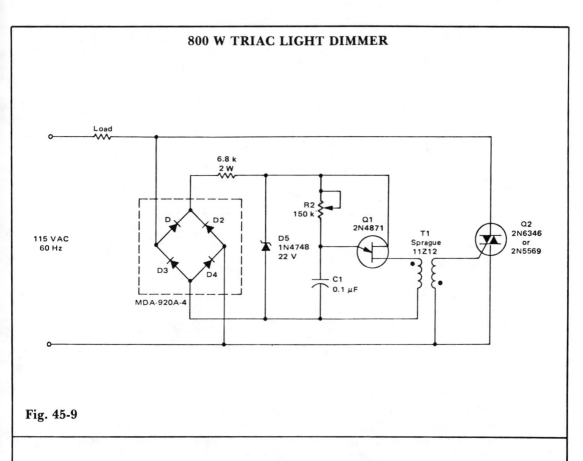

Fig. 45-9

FULL-WAVE SCR CONTROL

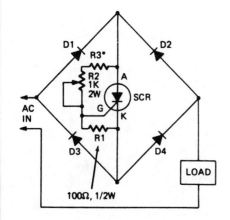

Circuit Notes

This circuit enables a single SCR to provide fullwave control of resistive loads. Resistor R3 should be chosen so that when potentiometer R2 is at its minimum setting, the current in the load is at the required minimum level. Diodes should have same current and voltage rating as the SCR.

Fig. 45-10

860 WATT LIMITED-RANGE LOW COST PRECISION LIGHT CONTROL

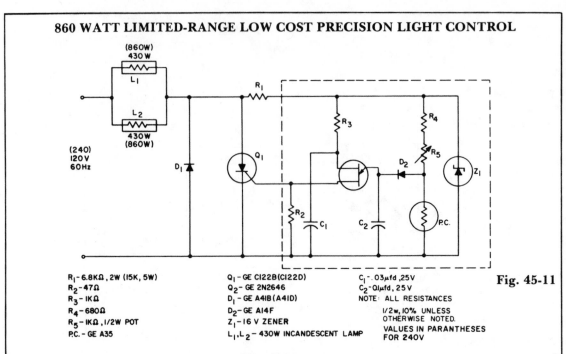

R₁ - 6.8KΩ, 2W (15K, 5W)
R₂ - 47Ω
R₃ - 1KΩ
R₄ - 680Ω
R₅ - 1KΩ, 1/2W POT
P.C. - GE A35

Q₁ - GE C122B(C122D)
Q₂ - GE 2N2646
D₁ - GE A41B(A41D)
D₂ - GE A14F
Z₁ - 16 V ZENER
L₁,L₂ - 430W INCANDESCENT LAMP

C₁ - .03μfd, 25V
C₂ - 01μfd, 25 V
NOTE: ALL RESISTANCES
1/2w, 10% UNLESS
OTHERWISE NOTED.
VALUES IN PARANTHESES
FOR 240V

Fig. 45-11

Circuit Notes

The system is designed to regulate an 860 watt lamp load from half to full power. This is achieved by the controlled-half-plus-fixed-half-wave phase control method. Half power applied to an incandescent lamp results in 30% of the full light output. Consequently the circuit is designed to control the light output of the lamp from 30% to 100% of maximum.

800 W SOFT-START LIGHT DIMMER

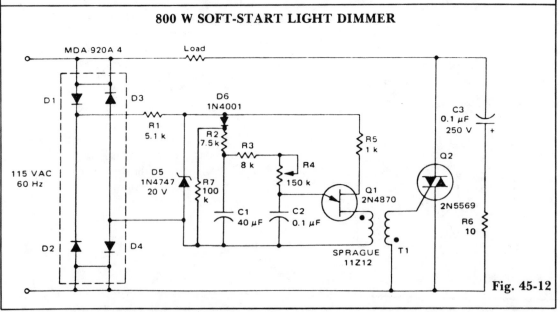

Fig. 45-12

LOW LOSS BRIGHTNESS CONTROL

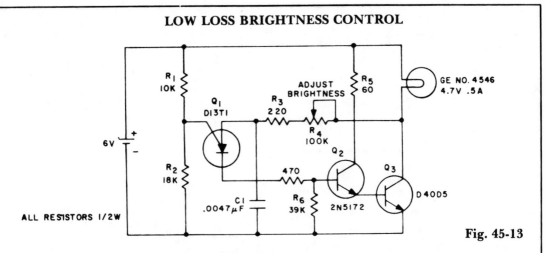

Fig. 45-13

ALL RESISTORS 1/2W

Circuit Notes

This circuit changes the average value of the dc supply voltage because of the high switching frequency. The tungsten lamp will have an almost continuous adjustable light output between 0 and 100%. If a light emitting diode is used as the emitting device, the irradiance will be in phase with the applied current pulses and will decrease to zero when the supply current is zero.

HALF WAVE AC PHASE-CONTROLLED CIRCUIT

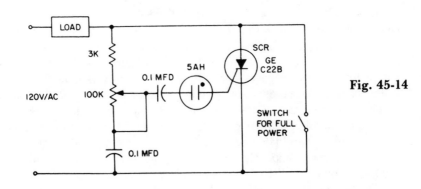

Fig. 45-14

Circuit Notes

The 5AH will trigger when the voltage across the two 0.1 μF capacitors reaches the breakdown voltage of the lamp. Control can be obtained full off to 95% of the half wave RMS output voltage. Full power can be obtained with the addition of the switch across the SCR.

377

EMERGENCY LIGHT

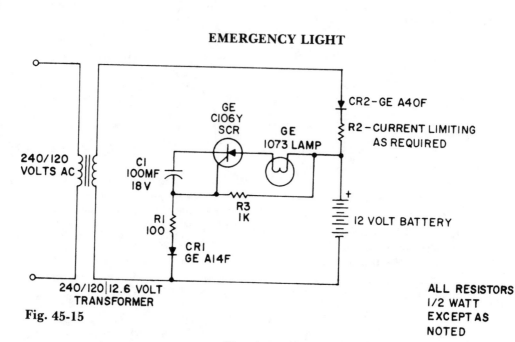

Fig. 45-15

240/120|12.6 VOLT TRANSFORMER

ALL RESISTORS 1/2 WATT EXCEPT AS NOTED

Circuit Notes

This simple circuit provides battery operated emergency lighting instantaneously upon failure of the regular ac service. When line power is restored, the emergency light turns off and the battery recharges automatically. The circuit is ideal for use in elevator cars, corridors and similar places where loss of light due to power failure would be undesirable. Completely static in operation, the circuit requires no maintenance. With ac power on, capacitor C1 charges through rectifier CR1 and resistor R1 to develop a negative voltage at the gate of the C106Y SCR. By this means, the SCR is prevented from being triggered, and the emergency light stays off. At the same time, the battery is kept fully charged by rectifier CR2 and resistor R2. Should the ac power fail, C1 discharges and the SCR is triggered on by battery power through resistor R3. The SCR then energizes the emergency light. Reset is automatic when ac is restored, because the peak ac line voltage biases the SCR and turns it off.

NEON LAMP DRIVER

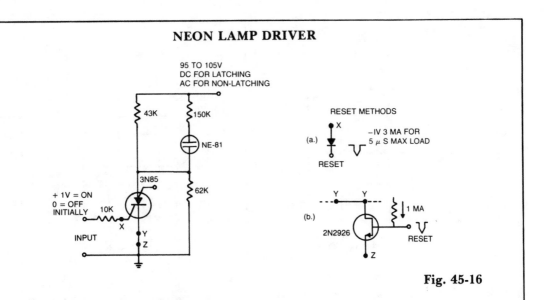

Fig. 45-16

COMPLEMENTARY AC POWER SWITCHING

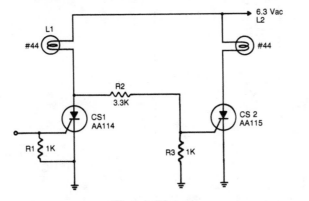

Fig. 45-17

Circuit Notes

An input signal of less than 1 mA and 1 V is required to switch on CS1. As long as this input signal is maintained, CS1 will conduct during each positive half cycle of anode voltage, thereby energizing load L1 with half-wave rectified dc. L2 remains de-energized, since the anode of CS1 will not go more positive than 1.5 volts, and voltage divider R2 - R3 cannot provide enough voltage to trigger CS2. Upon removal of the input signal, CS1 will drop out. L1 will be de-energized, except for a small amount of ac current through R2 and R3. CS2 will be triggered on at the beginning of each positive half-cycle, when CS1 anode voltage reaches 2 to 3 volts. CS2 will conduct for nearly the entire positive half-cycle energizing L2. It should be noted that the 6.3 volt lamps used will operate at ⅓ the rated brilliance because of the controlled switch half-wave rectifying action and will extend the operating lamp life by several orders of magnitude. Should full brilliance be desired, the anode supply voltage level should be raised to 9 volts ac.

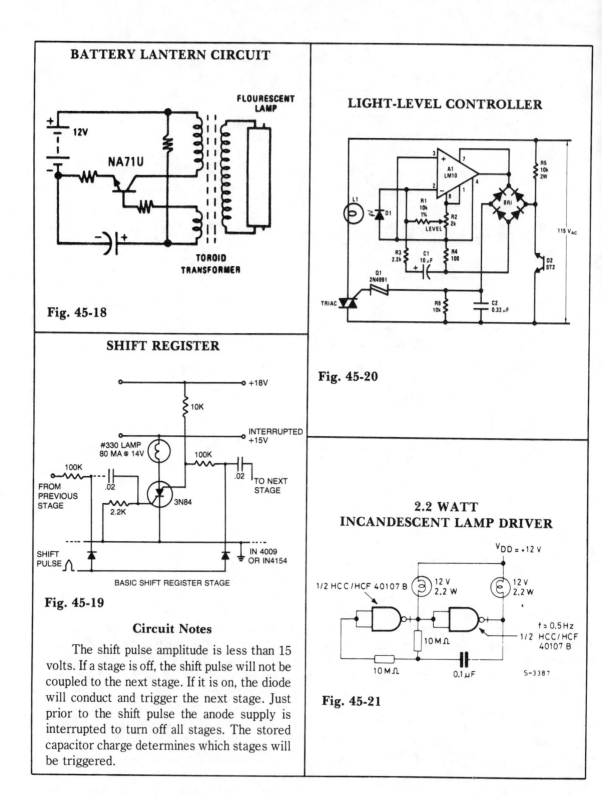

BATTERY LANTERN CIRCUIT

12V

NA71U

FLOURESCENT
LAMP

TOROID
TRANSFORMER

Fig. 45-18

LIGHT-LEVEL CONTROLLER

L1

D1

A1
LM10

R1
10k
1%

R2
2k

LEVEL

R5
10k
2W

BRI

115 V_{AC}

R3
2.2k

C1
10 μF

R4
100

Q1
2N4991

D2
ST2

TRIAC

R6
10k

C2
0.33 μF

Fig. 45-20

SHIFT REGISTER

+18V

10K

INTERRUPTED
+15V

#330 LAMP
80 MA @ 14V

100K

100K

.02

FROM
PREVIOUS
STAGE

.02

TO NEXT
STAGE

2.2K

3N84

SHIFT
PULSE

IN 4009
OR IN4154

BASIC SHIFT REGISTER STAGE

Fig. 45-19

Circuit Notes

The shift pulse amplitude is less than 15 volts. If a stage is off, the shift pulse will not be coupled to the next stage. If it is on, the diode will conduct and trigger the next stage. Just prior to the shift pulse the anode supply is interrupted to turn off all stages. The stored capacitor charge determines which stages will be triggered.

2.2 WATT
INCANDESCENT LAMP DRIVER

V_{DD} = +12 V

1/2 HCC/HCF 40107 B

12 V
2.2 W

12 V
2.2 W

f ≅ 0.5 Hz
1/2 HCC/HCF
40107 B

10 M Ω

10 M Ω

0.1 μF

S-3387

Fig. 45-21

46

Light Measuring Circuits

The sources of the following circuits are contained in the Sources section beginning on page 730. The figure number contained in the box of each circuit correlates to the source entry in the Sources section.

LINEAR LIGHT-METER CIRCUIT

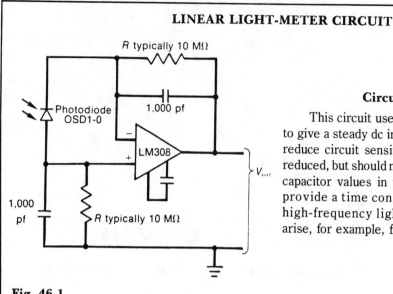

Circuit Notes

This circuit uses a low-input-bias op amp to give a steady dc indication of light level. To reduce circuit sensitivity to light, R1 can be reduced, but should not be less than 100 K. The capacitor values in the circuit are chosen to provide a time constant sufficient to filter high-frequency light variations that might arise, for example, from fluorescent lights.

Fig. 46-1

LOGARITHMIC LIGHT-METER CIRCUIT

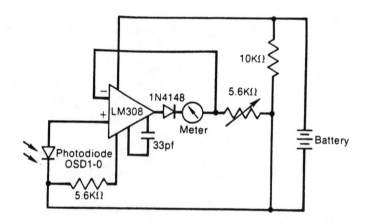

Fig. 46-2

Circuit Notes

The meter reading is directly proportional to the logarithm of the input light power. The logarithmic circuit behavior arises from the nonlinear diode pn junction current/voltage relationship. The diode in the amplifier output prevents output voltage from becoming negative (thereby pegging the meter), which may happen at low light levels due to amplifier bias currents. R1 adjusts the meter full-scale deflection, enabling the meter to be calibrated.

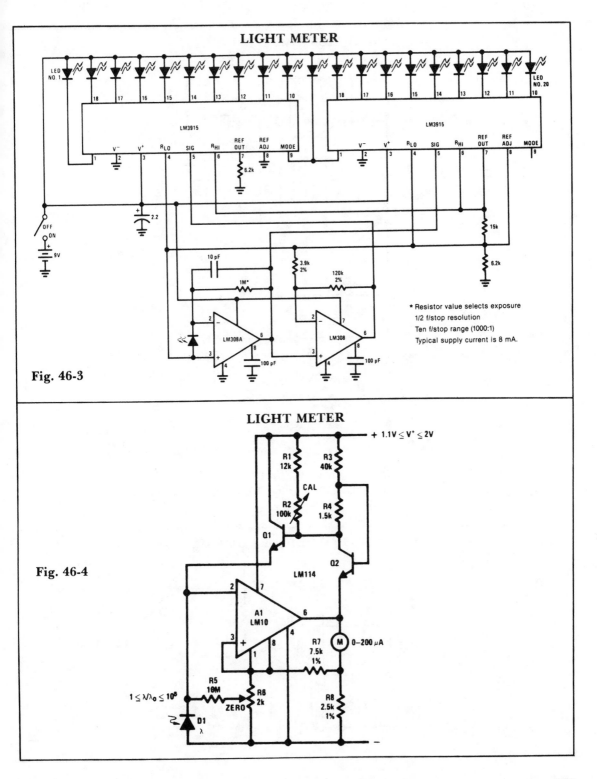

LIGHT METER

Fig. 46-3

LED NO. 1

LED NO. 20

LM3915

LM3915

V⁻ V⁺ R_LO SIG R_HI REF OUT REF ADJ MODE

6.2k

OFF ON 9V

2.2

10 pF

1M*

3.9k 2%

120k 2%

15k

6.2k

LM308A

6

LM308

6

100 pF

100 pF

* Resistor value selects exposure
1/2 f/stop resolution
Ten f/stop range (1000:1)
Typical supply current is 8 mA.

LIGHT METER

Fig. 46-4

+ 1.1V ≤ V⁺ ≤ 2V

R1 12k
R3 40k
CAL
R2 100k
R4 1.5k

Q1

Q2

LM114

A1 LM10

6

M 0–200 μA

R7 7.5k 1%

R5 10M

R6 2k
ZERO

R8 2.5k 1%

$1 \le \lambda/\lambda_0 \le 10^5$

D1 λ

LIGHT METER

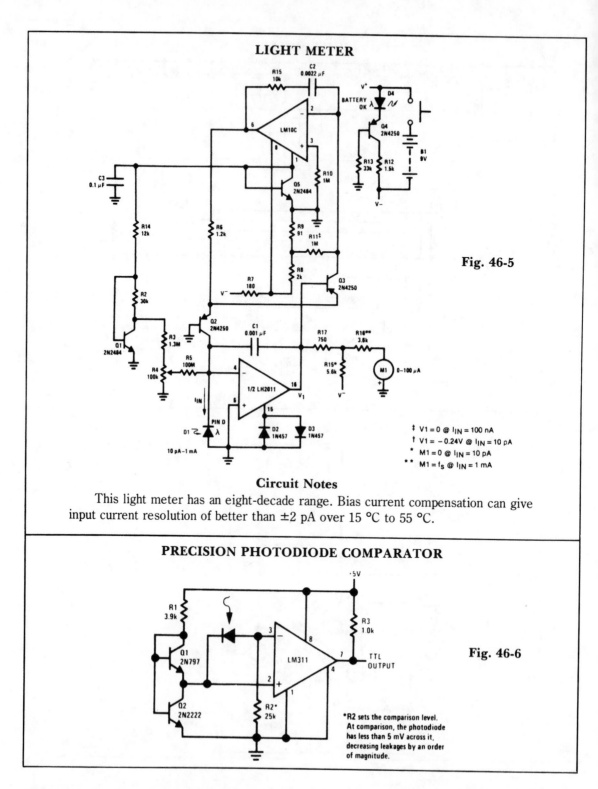

Fig. 46-5

‡ V1 = 0 @ I_IN = 100 nA
† V1 = −0.24V @ I_IN = 10 pA
* M1 = 0 @ I_IN = 10 pA
** M1 = f_S @ I_IN = 1 mA

Circuit Notes

This light meter has an eight-decade range. Bias current compensation can give input current resolution of better than ±2 pA over 15 °C to 55 °C.

PRECISION PHOTODIODE COMPARATOR

Fig. 46-6

*R2 sets the comparison level.
At comparison, the photodiode
has less than 5 mV across it,
decreasing leakages by an order
of magnitude.

47

Liquid Level Detectors

The sources of the following circuits are contained in the Sources section beginning on page 730. The figure number contained in the box of each circuit correlates to the source entry in the Sources section.

Level Sensor for Cryogenic Fluids
Fluid Level Controller
High Level Warning Device
Liquid Level Control
Liquid Level Detector Latching

Water Level Alarm
Water-Level Sensing Control Circuit
Flood Alarm
Liquid Level Detector
Low-Level Warning with Audio Output

LEVEL SENSOR FOR CRYOGENIC FLUIDS

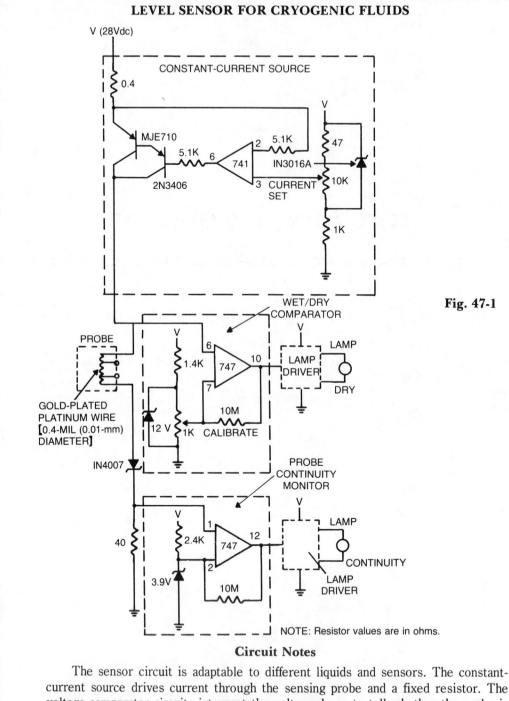

Fig. 47-1

NOTE: Resistor values are in ohms.

Circuit Notes

The sensor circuit is adaptable to different liquids and sensors. The constant-current source drives current through the sensing probe and a fixed resistor. The voltage-comparator circuits interpret the voltage drops to tell whether the probe is immersed in liquid and whether there is current in the probe.

FLUID LEVEL CONTROLLER

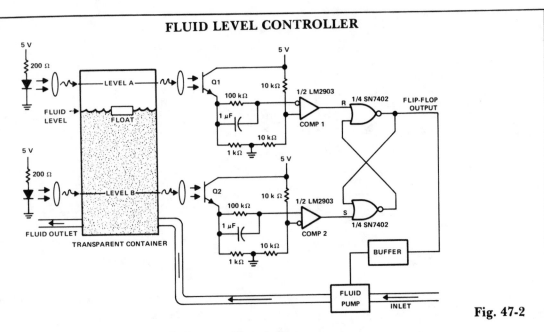

Fig. 47-2

Circuit Notes

This circuit can be used to maintain fluid between two levels. Variations on this control circuit can be made to keep something that moves within certain boundary conditions.

HIGH LEVEL WARNING DEVICE

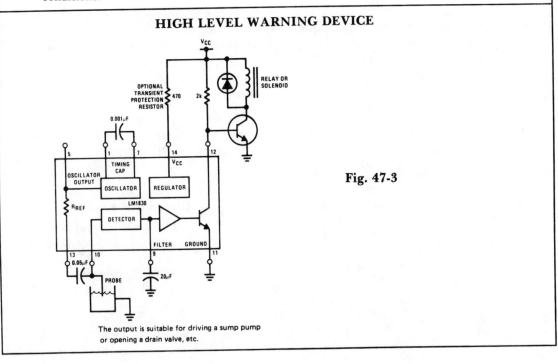

Fig. 47-3

The output is suitable for driving a sump pump or opening a drain valve, etc.

LIQUID LEVEL CONTROL

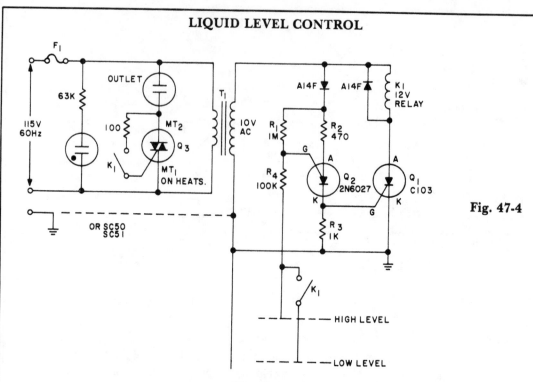

Fig. 47-4

Circuit Notes

Use this circuit to keep the fluid level of a liquid between two fixed points. Two modes, for filling or emptying are possible by simple reversing the contact connections of K1. The loads can be either electric motors or solenoid operated valves, operating from ac power. Liquid level detection is accomplished by two metal probes, one measuring the high level and the other the low level. An inversion of the logic (keeping the container filled) can be accomplished by replacing the normally open contact on the gate of Q3 with a normally closed contact.

LIQUID LEVEL DETECTOR (LATCHING)

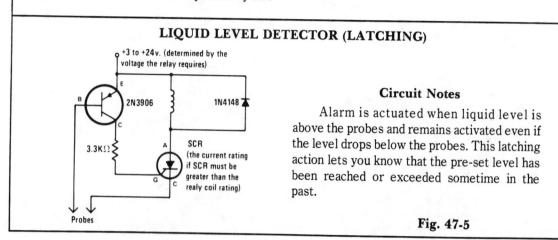

Circuit Notes

Alarm is actuated when liquid level is above the probes and remains activated even if the level drops below the probes. This latching action lets you know that the pre-set level has been reached or exceeded sometime in the past.

Fig. 47-5

WATER LEVEL ALARM

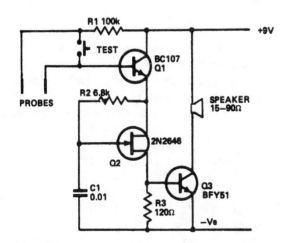

Fig. 47-6

Circuit Notes

The circuit draws so little current that the shelf-line of the battery is the limiting factor. The only current drawn is the leakage of the transistor. The circuit is shown in the form of a water level alarm but by using different forms of probe can act as a rain alarm or shorting alarm; anything from zero to about 1 M between the probes will trigger it. Q1 acts as a switch which applies current to the unijunction relaxation oscillator Q2. Alarm signal frequency is controlled by values and ratios of C1/R2. Pulses switch Q3 on and off, applying a signal to the speaker. Almost any NPN silicon transistor can be used for Q1 and Q3 and almost any unijunction for Q2.

WATER-LEVEL SENSING CONTROL CIRCUIT

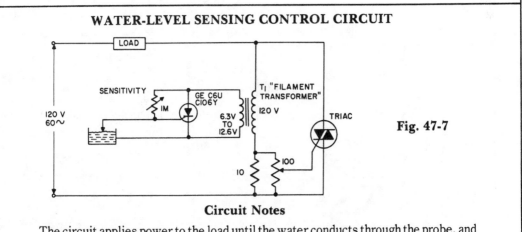

Fig. 47-7

Circuit Notes

The circuit applies power to the load until the water conducts through the probe, and bypasses gate current from the low current SCR. This gives an isolated low voltage probe to satisfy safety requirements.

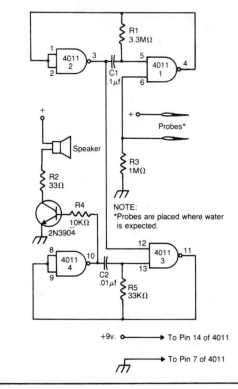

Circuit Notes

The alarm is built around two audio oscillators, each using two NAND gates. The detection oscillator is gated on by a pair of remote probes. One of the probes is connected to the battery supply, the other to the input of one of the gates. When water flows between the probes, the detection oscillator is gated on. The alarm oscillator is gated on by the output of the detection oscillator. The values given produce an audio tone of about 3000 Hz. The detection oscillator gates this audio tone at a rate of about 3 Hz. The result is a unique pulsating note. Use any 8 ohm speaker to sound the alarm. The 2N3904 can be replaced by any similar NPN transistor. The circuit will work from any six to 12-volt supply.

Fig. 47-8

LIQUID LEVEL DETECTOR

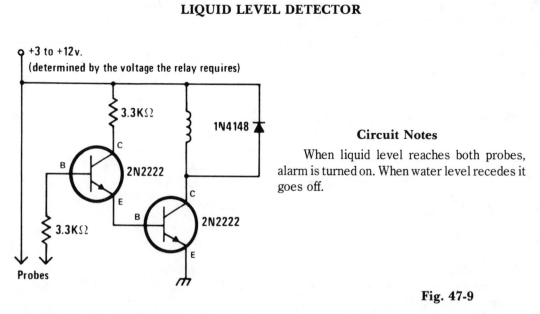

Circuit Notes

When liquid level reaches both probes, alarm is turned on. When water level recedes it goes off.

Fig. 47-9

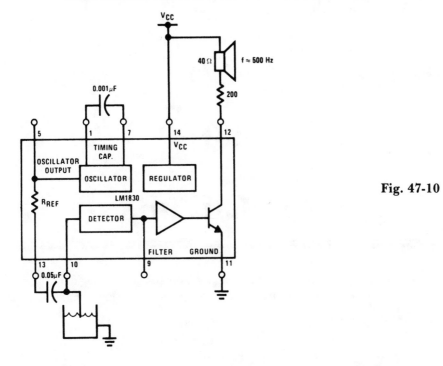

LOW-LEVEL WARNING WITH AUDIO OUTPUT

Fig. 47-10

48

Logic Circuits

The sources of the following circuits are contained in the Sources section beginning on page 730. The figure number contained in the box of each circuit correlates to the source entry in the Sources section.

LIGHT ACTIVATED LOGIC CIRCUITS

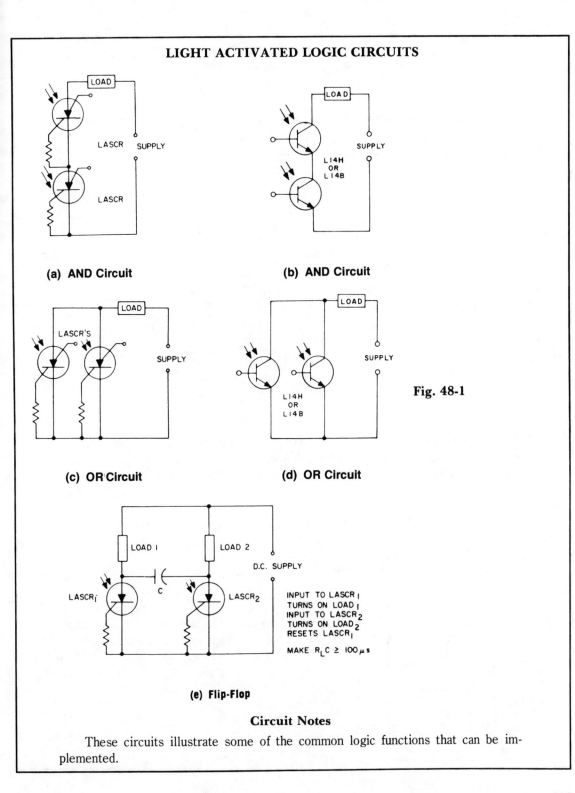

(a) AND Circuit

(b) AND Circuit

Fig. 48-1

(c) OR Circuit

(d) OR Circuit

(e) Flip-Flop

INPUT TO LASCR₁
TURNS ON LOAD₁
INPUT TO LASCR₂
TURNS ON LOAD₂
RESETS LASCR₁

MAKE $R_L C \geq 100 \mu s$

Circuit Notes

These circuits illustrate some of the common logic functions that can be implemented.

PROGRAMMABLE GATE

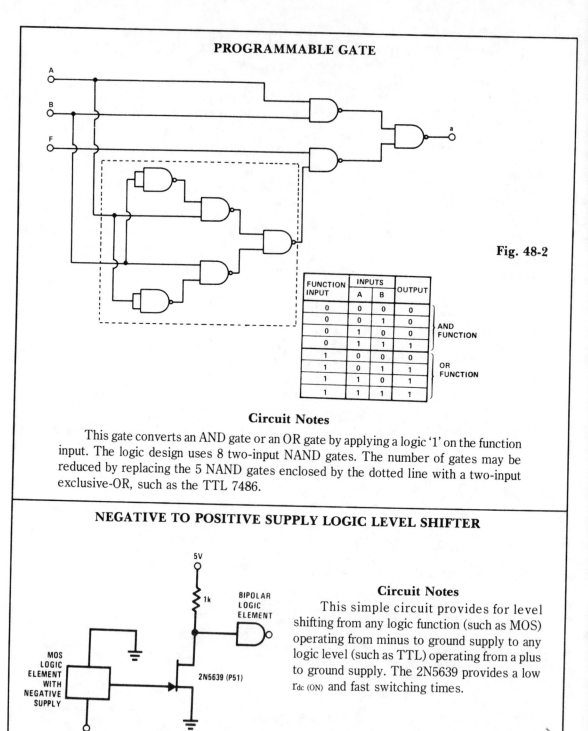

Fig. 48-2

FUNCTION INPUT	INPUTS		OUTPUT	
	A	B		
0	0	0	0	
0	0	1	0	AND
0	1	0	0	FUNCTION
0	1	1	1	
1	0	0	0	
1	0	1	1	OR
1	1	0	1	FUNCTION
1	1	1	1	

Circuit Notes

This gate converts an AND gate or an OR gate by applying a logic '1' on the function input. The logic design uses 8 two-input NAND gates. The number of gates may be reduced by replacing the 5 NAND gates enclosed by the dotted line with a two-input exclusive-OR, such as the TTL 7486.

NEGATIVE TO POSITIVE SUPPLY LOGIC LEVEL SHIFTER

Circuit Notes

This simple circuit provides for level shifting from any logic function (such as MOS) operating from minus to ground supply to any logic level (such as TTL) operating from a plus to ground supply. The 2N5639 provides a low $r_{dc\ (ON)}$ and fast switching times.

Fig. 48-3

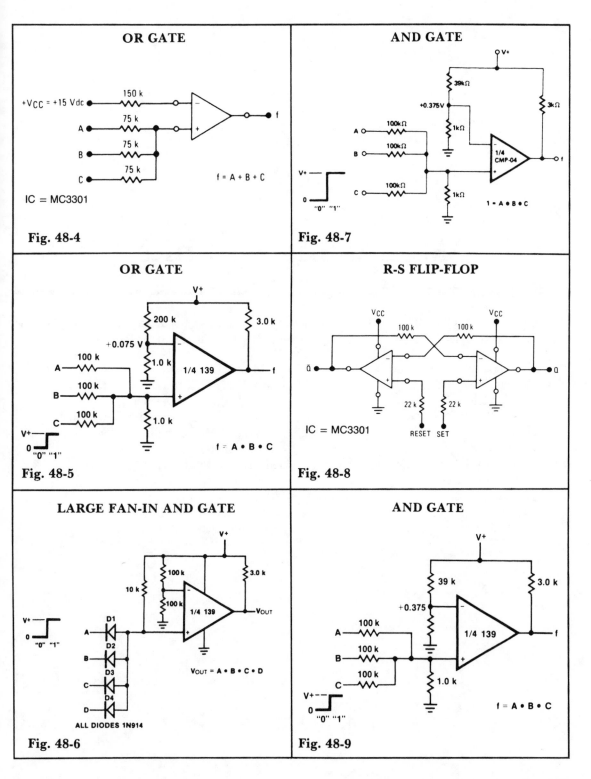

OR GATE

+V_CC = +15 Vdc
150 k
A 75 k
B 75 k
C 75 k

f = A + B + C

IC = MC3301

Fig. 48-4

AND GATE

V+
39kΩ
+0.375V
1kΩ
A 100kΩ
B 100kΩ
3kΩ
1/4 CMP-04
f
C 100kΩ
1kΩ

V+ ⌐
0
"0" "1"

1 = A • B • C

Fig. 48-7

OR GATE

V+
200 k
3.0 k
+0.075 V
1.0 k
1/4 139
f
A 100 k
B 100 k
1.0 k
C 100 k

V+ ⌐
0
"0" "1"

f = A • B • C

Fig. 48-5

R-S FLIP-FLOP

V_CC
100 k
100 k
V_CC
Q̄
Q
22 k
22 k
RESET SET

IC = MC3301

Fig. 48-8

LARGE FAN-IN AND GATE

V+
100 k
10 k
100 k
3.0 k
1/4 139
V_OUT
V+ ⌐
0
"0" "1"
D1
A
D2
B
D3
C
D4
D

ALL DIODES 1N914

V_OUT = A • B • C • D

Fig. 48-6

AND GATE

V+
39 k
3.0 k
+0.375
1/4 139
f
A 100 k
B 100 k
1.0 k
C 100 k

V+ ⌐
0
"0" "1"

f = A • B • C

Fig. 48-9

49

Measuring Circuits

The sources of the following circuits are contained in the Sources section beginning on page 730. The figure number contained in the box of each circuit correlates to the source entry in the Sources section.

FET Curve Tracer
Digital Weight Scale
Low Cost pH Meter
pH Probe Amplifier/Temperature
   Compensator
Capacitance Meter
Zener Tester
Transistor Sorter/Tester
Go/No-Go Diode Tester
Diode Tester
Peak Level Indicator

Sound Level Monitor
Linear Variable Differential Transformer
   (LVDT) Driver Demodulator
Linear Variable Differential Transformer
   (LVDT) Measuring Gauge
Vibration Meter
Sensitive RF Voltmeter
Minimum Component Tachometer
Phase Meter
Precision Calibration Standard
Zener Diode Checker

FET CURVE TRACER

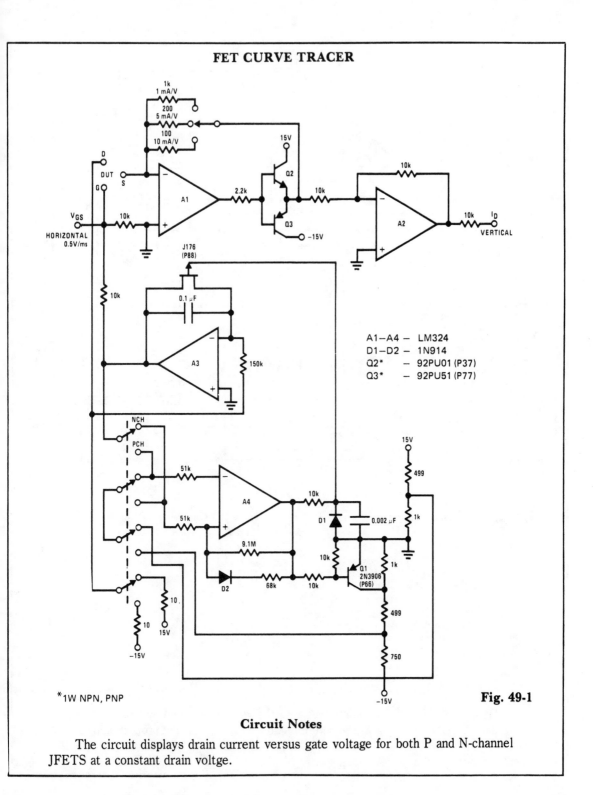

A1–A4 — LM324
D1–D2 — 1N914
Q2* — 92PU01 (P37)
Q3* — 92PU51 (P77)

*1W NPN, PNP

Fig. 49-1

Circuit Notes

The circuit displays drain current versus gate voltage for both P and N-channel JFETS at a constant drain voltge.

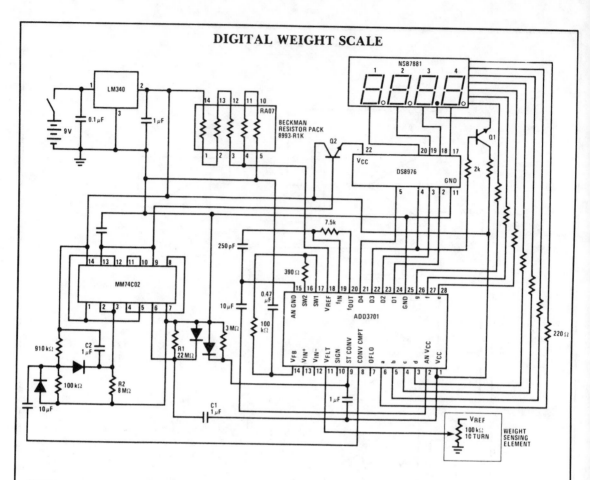

DIGITAL WEIGHT SCALE

Notes:

1. R1, C1 defines POWER ON display blanking interval. R2, C2 defines display ON time.

2. All V_{CC} connections should use a single V_{CC} point and all ground/analog ground connections should use a single ground/analog ground point.

3. Display sequence for Rev A ckt implementation:

t = 0 sec	• power ON
t = 0 → 5 sec	• display blanked
	• system converging
t = 5 → 10 sec	• conversion complete
	• display ENABLE
t ≥ 10 sec	• display blanked
	• wait for new POWER UP cycle

Fig. 49-2

Circuit Notes

This circuit employs a potentiometer as the weight sensing element. An object placed upon the scale displaces the potentiometer wiper, an amount proportional to its weight. Conversion of the wiper voltage to digital information is performed, decoded, and interfaced to the numeric display.

LOW COST pH METER

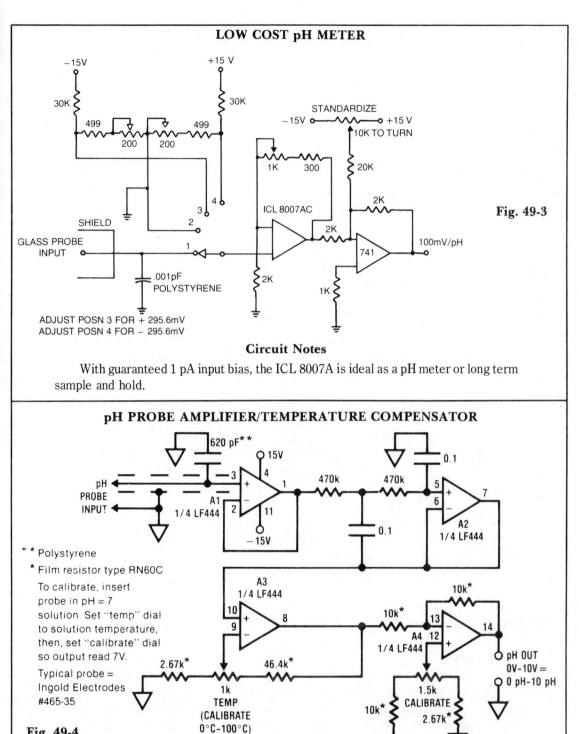

Fig. 49-3

ADJUST POSN 3 FOR + 295.6mV
ADJUST POSN 4 FOR − 295.6mV

Circuit Notes

With guaranteed 1 pA input bias, the ICL 8007A is ideal as a pH meter or long term sample and hold.

pH PROBE AMPLIFIER/TEMPERATURE COMPENSATOR

** Polystyrene

* Film resistor type RN60C

To calibrate, insert probe in pH = 7 solution. Set "temp" dial to solution temperature, then, set "calibrate" dial so output read 7V.

Typical probe = Ingold Electrodes #465-35

Fig. 49-4

399

CAPACITANCE METER

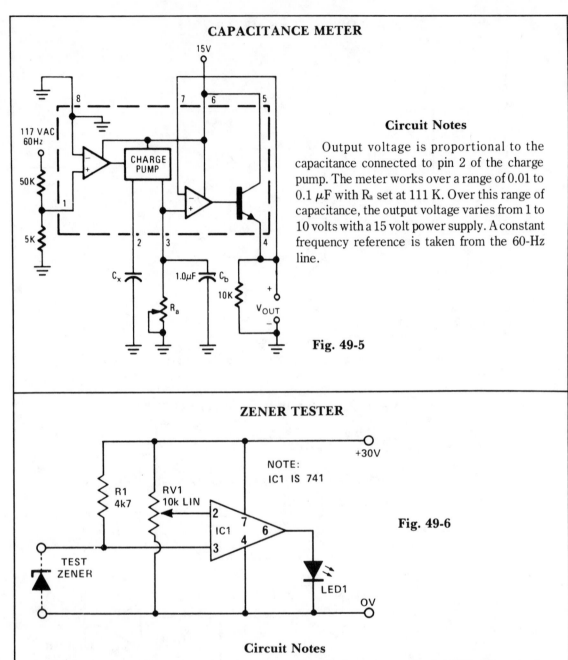

Circuit Notes

Output voltage is proportional to the capacitance connected to pin 2 of the charge pump. The meter works over a range of 0.01 to 0.1 μF with R_a set at 111 K. Over this range of capacitance, the output voltage varies from 1 to 10 volts with a 15 volt power supply. A constant frequency reference is taken from the 60-Hz line.

Fig. 49-5

ZENER TESTER

NOTE:
IC1 IS 741

Fig. 49-6

Circuit Notes

This circuit provides a low cost and reliable method of testing zener diodes. RV1 can be calibrated in volts, so that when LED 1 just lights, the voltage on pins 2 and 3 are nearly equal. Hence, the zener voltage can be read directly from the setting of RV1. The supply need only be as high a value as the zener itself. For a more accurate measurement, a precision pot could be added and calibrated.

TRANSISTOR SORTER/TESTER

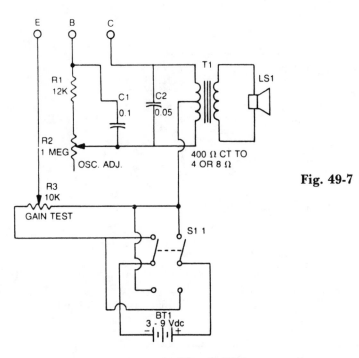

Fig. 49-7

Circuit Notes

This tester checks transistor for polarity (PNP or NPN). An audible signal will give an indication of gain. Tester can also be used as a GO/NO GO tester to match unmarked devices.

GO/NO-GO DIODE TESTER

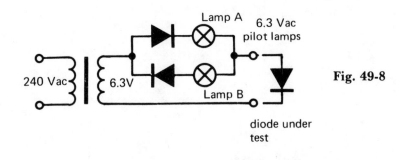

Fig. 49-8

Circuit Notes

If lamp A or B is illuminated, the diode is serviceable. If both light, the diode is short circuited. If neither light, diode is an open circuit.

DIODE TESTER

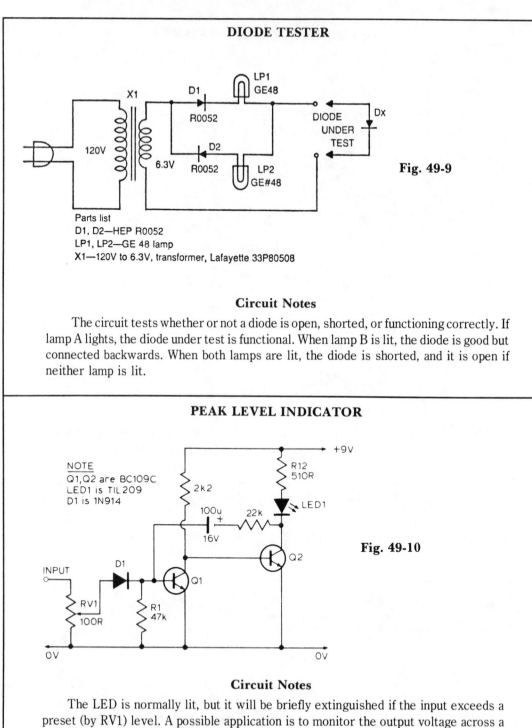

Fig. 49-9

Parts list
D1, D2—HEP R0052
LP1, LP2—GE 48 lamp
X1—120V to 6.3V, transformer, Lafayette 33P80508

Circuit Notes

The circuit tests whether or not a diode is open, shorted, or functioning correctly. If lamp A lights, the diode under test is functional. When lamp B is lit, the diode is good but connected backwards. When both lamps are lit, the diode is shorted, and it is open if neither lamp is lit.

PEAK LEVEL INDICATOR

NOTE
Q1,Q2 are BC109C
LED1 is TIL209
D1 is 1N914

Fig. 49-10

Circuit Notes

The LED is normally lit, but it will be briefly extinguished if the input exceeds a preset (by RV1) level. A possible application is to monitor the output voltage across a loudspeaker; the LED will flicker with large signals.

SOUND LEVEL MONITOR

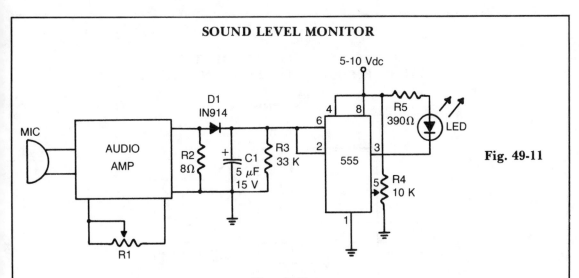

Fig. 49-11

Circuit Notes

Loudness detector consists of a 555 IC wired as a Schmitt trigger. The output changes state—from high to low—whenever the input crosses a certain voltage. That threshold voltage is established by the setting of R4.

LINEAR VARIABLE DIFFFERENTIAL TRANSFORMER (LVDT) DRIVER DEMODULATOR

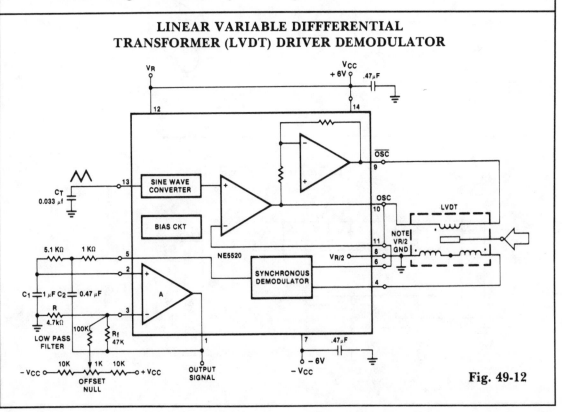

Fig. 49-12

LINEAR VARIABLE DIFFERENTIAL TRANSFORMER (LVDT) MEASURING GAUGE

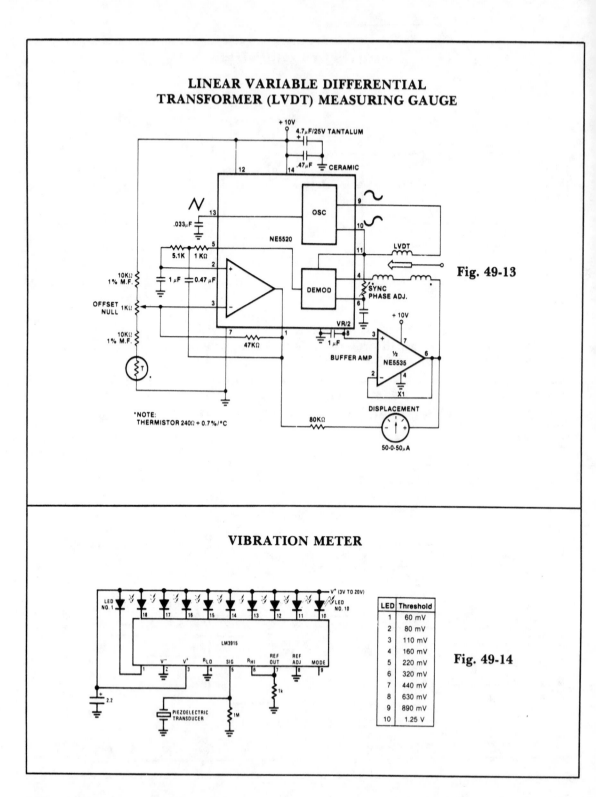

Fig. 49-13

VIBRATION METER

LED	Threshold
1	60 mV
2	80 mV
3	110 mV
4	160 mV
5	220 mV
6	320 mV
7	440 mV
8	630 mV
9	890 mV
10	1.25 V

Fig. 49-14

SENSITIVE RF VOLTMETER

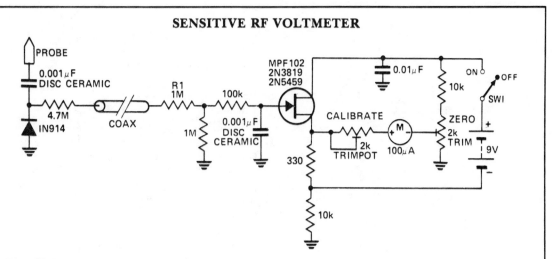

Fig. 49-15

Circuit Notes

This circuit measures RF voltages beyond 200 MHz and up to about 5 V. The diode should be mounted in a remote probe, close to the probe tip. Sensitivity is excellent and voltages less than 1 V peak can be easily measured. The unit can be calibrated by connecting the input to a known level of RF voltage, such as a calibrated signal generator, and setting the calibrate control.

MINIMUM COMPONENT TACHOMETER

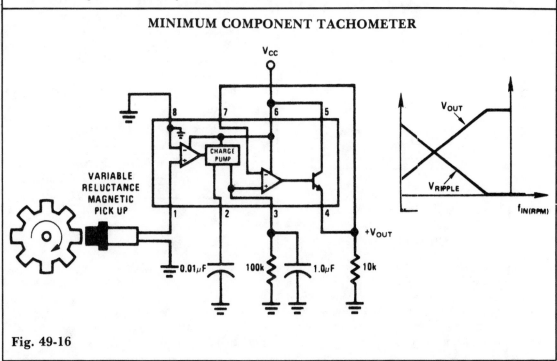

Fig. 49-16

PHASE METER

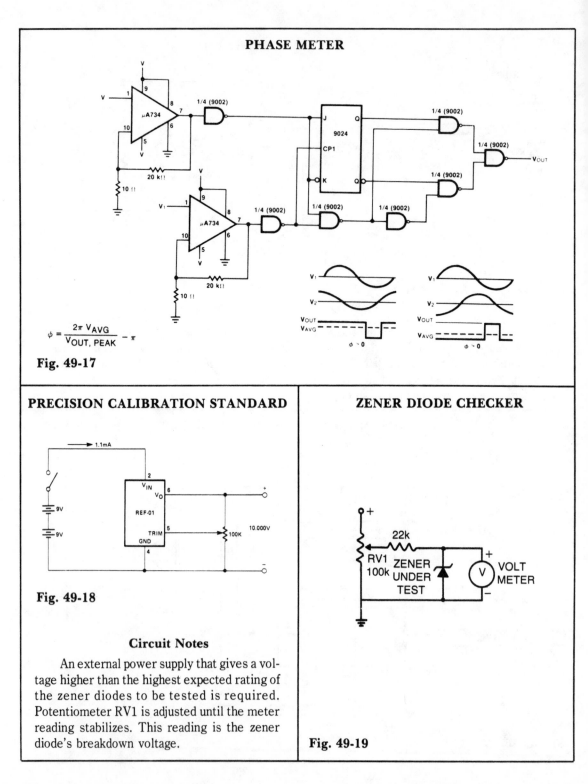

$$\phi = \frac{2\pi \, V_{AVG}}{V_{OUT, \, PEAK}} - \pi$$

Fig. 49-17

PRECISION CALIBRATION STANDARD

Fig. 49-18

Circuit Notes

An external power supply that gives a voltage higher than the highest expected rating of the zener diodes to be tested is required. Potentiometer RV1 is adjusted until the meter reading stabilizes. This reading is the zener diode's breakdown voltage.

ZENER DIODE CHECKER

Fig. 49-19

50

Metal Detectors

The sources of the following circuits are contained in the Sources section beginning on page 730. The figure number contained in the box of each circuit correlates to the source entry in the Sources section.

Micropower Metal Detector   Lo-Parts Treasure Locator

MICROPOWER METAL DETECTOR

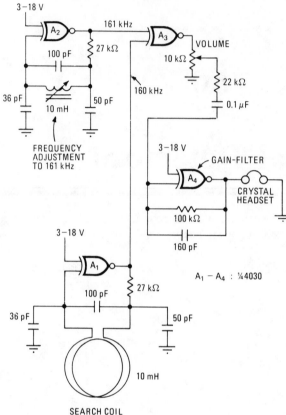

Fig. 50-1

A₁ – A₄ : ¼ 4030

Circuit Notes

This battery-powered metal detector uses four exclusive-OR gates contained in the 4030 CMOS integrated circuit. The gates are wired as a twin-oscillators and a search coil serves as the inductance element in one of the oscillators. When the coil is brought near metal, the resultant change in its effective inductance changes the oscillator's frequency. Gates A1 and A2 form the two oscillators which are tuned to 160 and 161 kilohertz respectively. The pulses produced by each oscillator are mixed in A3, its output contains sum and difference frequencies at 1 and 321 kHz. The 321 kHz signal is filtered out by the 10 kHz low-pass filter at A4, leaving the 1 kHz signal to be amplified for the crystal headset connected at the output. The device's sensitivity is sufficient to detect coin-sized objects a foot away.

LO-PARTS TREASURE LOCATOR

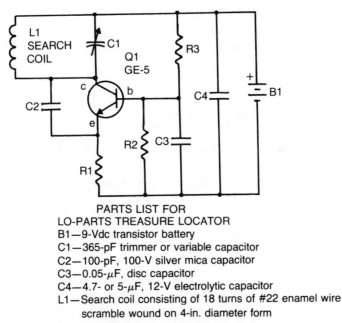

Fig. 50-2

PARTS LIST FOR
LO-PARTS TREASURE LOCATOR
B1—9-Vdc transistor battery
C1—365-pF trimmer or variable capacitor
C2—100-pF, 100-V silver mica capacitor
C3—0.05-μF, disc capacitor
C4—4.7- or 5-μF, 12-V electrolytic capacitor
L1—Search coil consisting of 18 turns of #22 enamel wire
    scramble wound on 4-in. diameter form
Q1—RCA SK3011 npn transistor or equiv.
R1—680-ohm, ½-watt resistor
R2—10,000-ohm, ½-watt resistor
R3—47,000-ohm, ½-watt resistor

Circuit Notes

Locator uses a transistor radio as the detector. With the radio tuned to a weak station, adjust C1 so the locator oscillator beats against the received signal. When the search head passes over metal, the inductance of L1 changes thereby changing the locator oscillator's frequency and changing the beat tone in the radio.

The search coil consists of 18 turns of #22 enameled wire scramble wound on a 4-in. diameter form. After the coil is wound and checked for proper operation, saturate the coil with RTV adhesive for stable operation of the locator.

51

Metronomes

The sources of the following circuits are contained in the Sources section beginning on page 730. The figure number contained in the box of each circuit correlates to the source entry in the Sources section.

Accentuated Beat Metronome

Sight N' Sound Metronome

Micrometronome

ACCENTUATED BEAT METRONOME

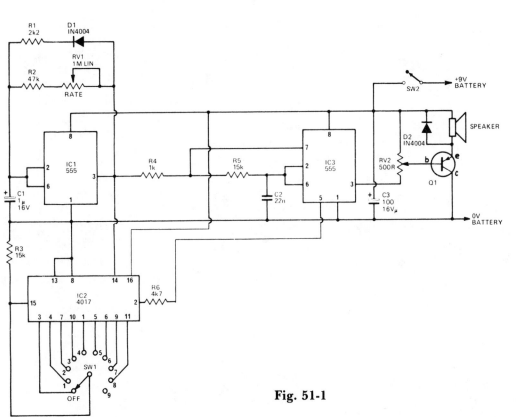

Fig. 51-1

Circuit Notes

IC3 acts as an oscillator which operates if the output of IC1 is high. With the values used the two frequencies produced are about 800 Hz and 2500 Hz. The output is buffered by Q1 which drives the speaker. The first IC is used to generate the tone duration and the time interval between beats. The interval is adjustable by RV1 while the tone duration is set by R1. The output of IC1 also clocks IC2, a decade counter with 10 decoded outputs. Each of these outputs go high in sequence on each clock. The second output of IC2 is connected to the control input of IC3 and is used to change the frequency. Therefore the first tone will be high frequency, the second low and the third to tenth will be high again. This gives the 9-1 beat. If for example the 5th output is connected to the reset, the first tone will be high, the second low, and the third and fourth high, then when the 5th output goes to a high it resets it back to the first which is a high tone. We then have 3 high and one low tones or a 3-1.

SIGHT N' SOUND METRONOME

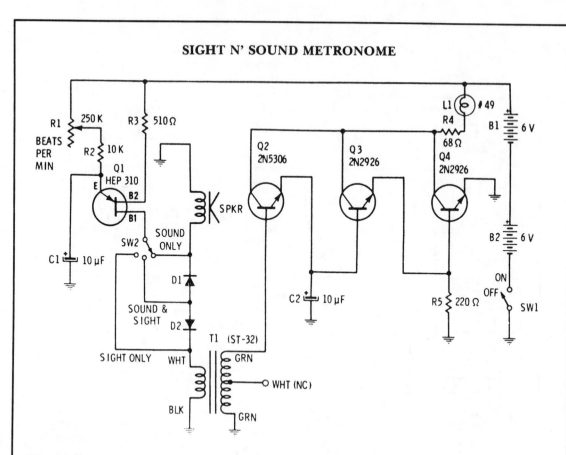

Fig. 51-2

Circuit Notes

Precise, adjustable control of beats per minute from a largo of 18 to a frenzied, high presto of 500. These beats are produced acoustically through a speaker. A light flashes at the same rate. When SW1 is closed, C1 begins to charge through R1 and R2. C1 will eventually reach a voltage at which the emitter of unijunction transistor is switched on, "dumping" the energy stored in C1 into an 8 ohm speaker. To produce a distinct "plop", brief pulses across T2 secondary drive Q2 into conduction. The extra gain of Q3 and Q4 are sufficient to briefly switch L1 on, then off, as the pulse wave passes. Capacitor C2 "stretches" the pulse slightly to overcome the thermal inertia of the lamp, so that a bright flash occurs.

MICROMETRONOME

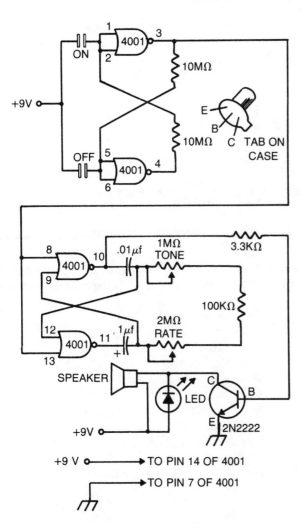

Fig. 51-3

Circuit Notes

This compact metronome will run for years on a single nine-volt transistor battery. Has both tone and pulse rate controls, and uses touch plates to start and stop, can be built in a case no larger than a pack of cigarettes. The touch plates consist of two strips of metal about 1/16-inch apart mounted on, but insulated from, the case. Bridging the gap closes the switch.

52

Miscellaneous Circuits

The sources of the following circuits are contained in the Sources section beginning on page 730. The figure number contained in the box of each circuit correlates to the source entry in the Sources section.

Intercom
Musical Organ
Laser Diode Pulser
Capacitance Multiplier
Simulated Inductor
Active Inductor
Positive Input/Negative Output Charge
    Pump
Shift Register Driver
Tape Recorder
Negative-Edge Differentiator
Stylus Organ

Positive-Edge Differentiator
Four Channel Data Acquisition System
Triac Trigger
Precision Rectifiers
Voltage Control Resistor
Fast Inverter Circuit
Inverse Scaler
5.0 V Square Wave Calibrator
Low Drift Integrator and Low-Leakage
    Guarded Reset
Differentiator with High Common Mode
    Noise Rejection

Digital Transmission Isolator

414

INTERCOM

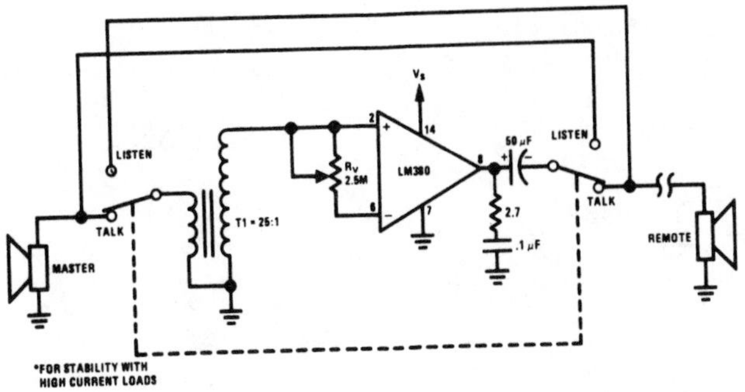

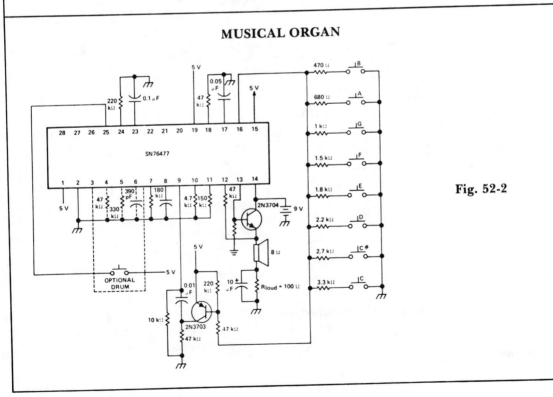

Fig. 52-1

Circuit Notes

The circuit provides a minimum component intercom. With switch S1 in the talk position, the speaker of the master station acts as the microphone with the aid of step-up transformer T1. A turns ratio of 25 and a device gain of 50 allows a maximum loop gain of 1250. R$_v$ provides a common mode volume control. Switching S1 to the listen position reverses the role of the master and remote speakers.

MUSICAL ORGAN

Fig. 52-2

LASER DIODE PULSER

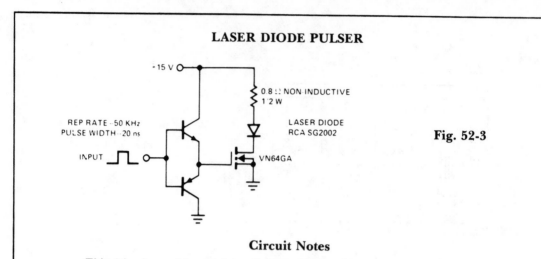

Fig. 52-3

Circuit Notes

This drive is capable of driving the laser diode with 10 ampere, 20 ns pulses. For a 0.1% duty cycle, the repetition rate will be 50 kHz. A complementary emitter-follower is used as a driver. Switching speed is determined by the fᴛ of the bipolar transistors used and the impedance of the drive source.

CAPACITANCE MULTIPLIER

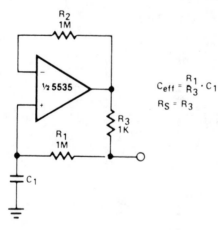

$$C_{eff} = \frac{R_1}{R_3} \cdot C_1$$

$$R_S = R_3$$

Fig. 52-4

All resistor values are in ohms

Circuit Notes

This circuit can be used to simulate large capacitances using small value components. With the values shown and $C = 10 \ \mu F$, an effective capacitance of 10,000 μF was obtained. The Q available is limited by the effective series resistance. So R1 should be as large as practical.

SIMULATED INDUCTOR

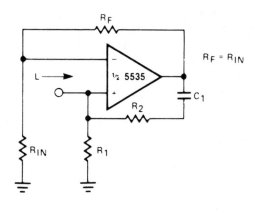

$R_F = R_{IN}$

Fig. 52-5

Circuit Notes

With a constant current excitation, the voltage dropped across an inductance increases with frequency. Thus, an active device whose output increases with frequency can be characterized as an inductance. The circuit yields such a response with the effective inductance being equal to: $L = R1R2C$. The Q of this inductance depends upon R1 being equal to R2. At the same time, however, the positive and negative feedback paths of the amplifier are equal leading to the distinct possibility of instability at high frequencies. R1 should, therefore, always be slightly smaller than R2 to assure stable operation.

ACTIVE INDUCTOR

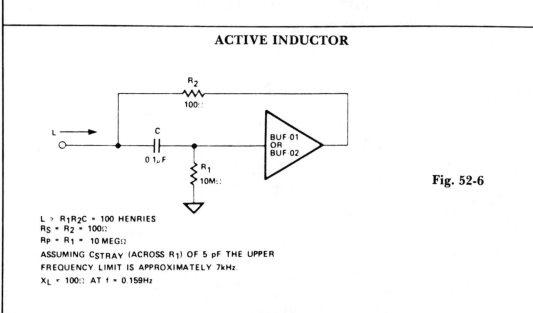

Fig. 52-6

$L = R_1R_2C = 100$ HENRIES
$R_S = R_2 = 100\Omega$
$R_P = R_1 = 10$ MEGΩ
ASSUMING C_{STRAY} (ACROSS R_1) OF 5 pF THE UPPER
FREQUENCY LIMIT IS APPROXIMATELY 7kHz.
$X_L = 100\Omega$ AT $f = 0.159$Hz

Circuit Notes

An active inductor is realized with an eight-lead IC, two carbon resistors, and a small capacitor. A commercial inductor of 50 henries may occupy up to five cubic inches.

POSITIVE INPUT/NEGATIVE OUTPUT CHARGE PUMP

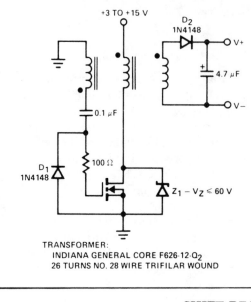

TRANSFORMER:
INDIANA GENERAL CORE F626-12-Q₂
26 TURNS NO. 28 WIRE TRIFILAR WOUND

Circuit Notes

A simple means of generating a low-power voltage supply of opposite polarity from the main supply. Self oscillating driver produces pulses at a repetition frequency of 100 kHz. When the VMOS device is off, capacitor C is charged to the positive supply. When the VMOS transistor switches on, C delivers a negative voltage through the series diode to the output. The zener serves as a dissipative regulator.

Fig. 52-7

SHIFT REGISTER DRIVER

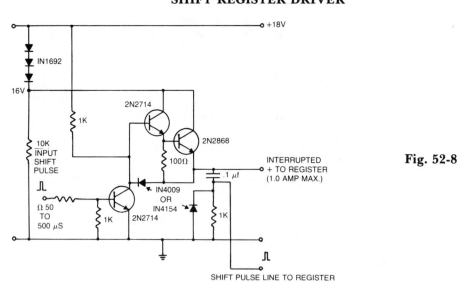

SHIFT PULSE LINE TO REGISTER

Fig. 52-8

Circuit Notes

A 16 V power supply can be synthesized as shown using IN1692 rectifiers. A shift pulse input saturates the 2N2714 depriving the Darlington combination (2N2714 and 2N2868) of base drive. The negative pulse so generated on the 15 V line is differentiated to produce a positive trigger pulse at its trailing edge.

418

TAPE RECORDER

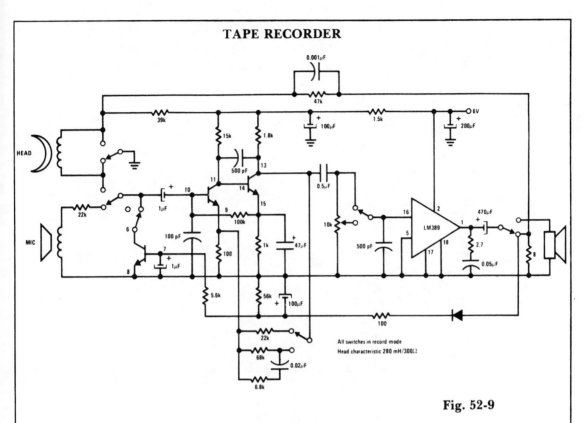

Fig. 52-9

Circuit Notes

Complete record/playback cassette tape machine amplifier. Two of the transistors act as signal amplifiers, with the third used for automatic level control during the record mode.

NEGATIVE-EDGE DIFFERENTIATOR

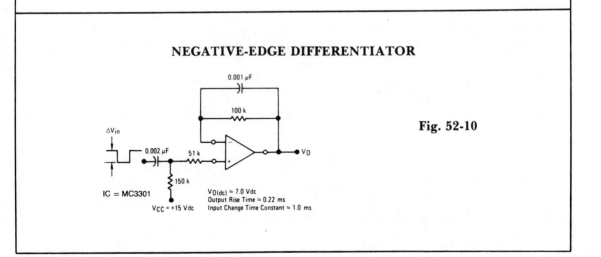

Fig. 52-10

$V_{O(dc)} \approx 7.0 \text{ Vdc}$
Output Rise Time ≈ 0.22 ms
Input Change Time Constant ≈ 1.0 ms

IC = MC3301

$V_{CC} = +15$ Vdc

419

STYLUS ORGAN

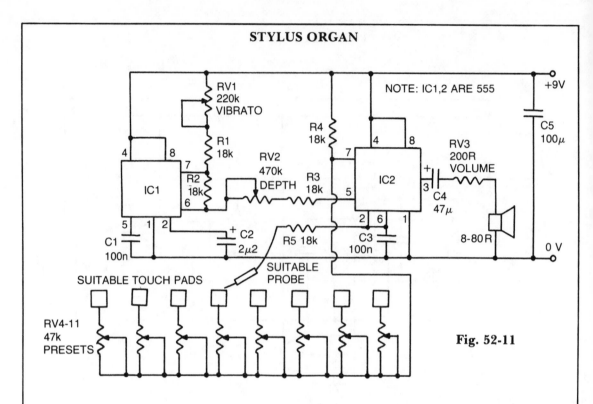

NOTE: IC1,2 ARE 555

Fig. 52-11

Circuit Notes

IC2 is an audio frequency oscillator. Its frequency is primarily controlled by the resistance between pins 2 and 7. RV4-11 control the oscillator frequency and by touching a stylus (connected via limiting resistor R5 to pin 2) to each preset, different notes can be played. IC1 is a low frequency oscillator (approximately 3-10Hz), the frequency of which is variable by RV1. The output of this oscillator is connected through depth control RV2 and limiting resistor R3 to the voltage control input of the audio frequency oscillator. Thus a vibrato effect occurs.

POSITIVE-EDGE DIFFERENTIATOR

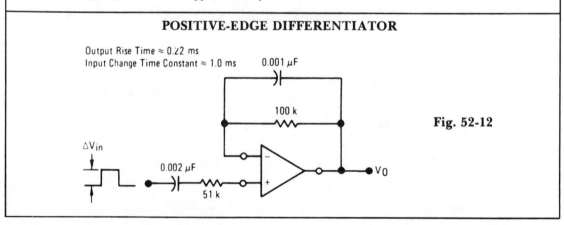

Output Rise Time ≈ 0.22 ms
Input Change Time Constant ≈ 1.0 ms

Fig. 52-12

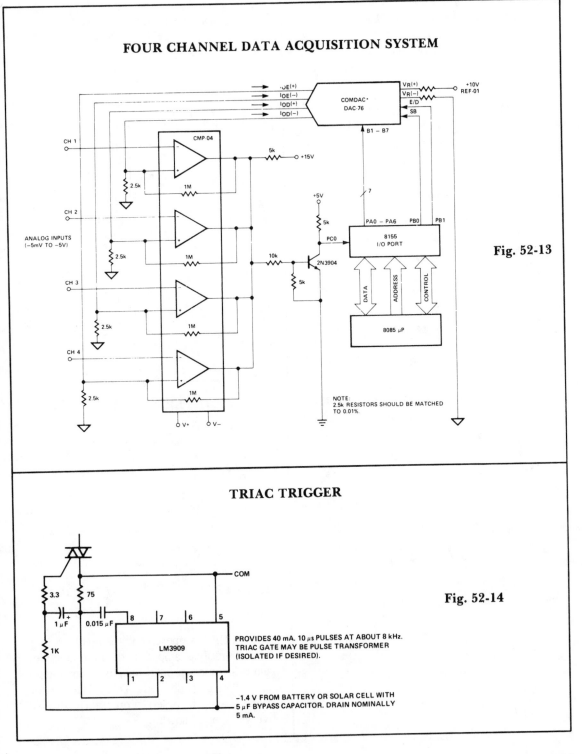

FOUR CHANNEL DATA ACQUISITION SYSTEM

Fig. 52-13

NOTE:
2.5k RESISTORS SHOULD BE MATCHED TO 0.01%.

ANALOG INPUTS
(−5mV TO −5V)

TRIAC TRIGGER

Fig. 52-14

PROVIDES 40 mA, 10 μs PULSES AT ABOUT 8 kHz.
TRIAC GATE MAY BE PULSE TRANSFORMER
(ISOLATED IF DESIRED).

−1.4 V FROM BATTERY OR SOLAR CELL WITH
5 μF BYPASS CAPACITOR. DRAIN NOMINALLY
5 mA.

421

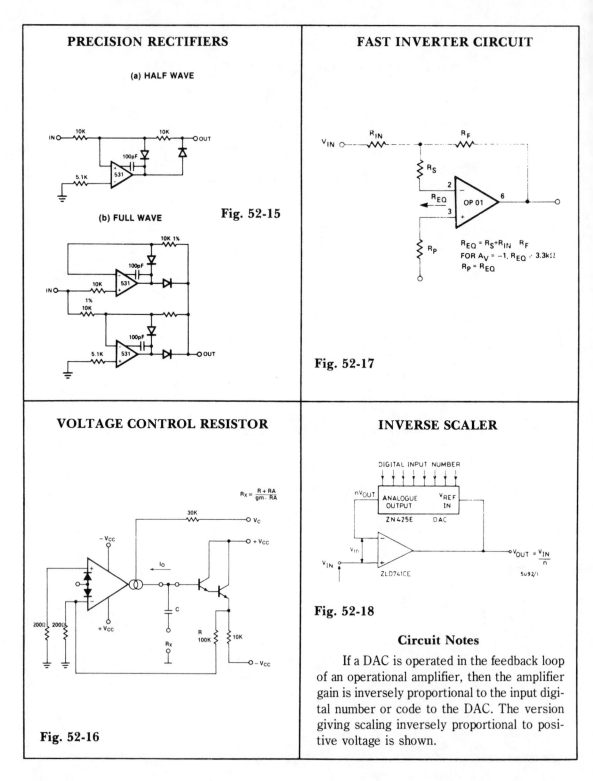

PRECISION RECTIFIERS

(a) HALF WAVE

Fig. 52-15

(b) FULL WAVE

Fig. 52-16

FAST INVERTER CIRCUIT

$R_{EQ} = R_S + R_{IN} \parallel R_F$
FOR $A_V = -1$, $R_{EQ} \approx 3.3k\Omega$
$R_P = R_{EQ}$

Fig. 52-17

VOLTAGE CONTROL RESISTOR

$R_X = \dfrac{R + RA}{gm \cdot RA}$

INVERSE SCALER

Fig. 52-18

Circuit Notes

If a DAC is operated in the feedback loop of an operational amplifier, then the amplifier gain is inversely proportional to the input digital number or code to the DAC. The version giving scaling inversely proportional to positive voltage is shown.

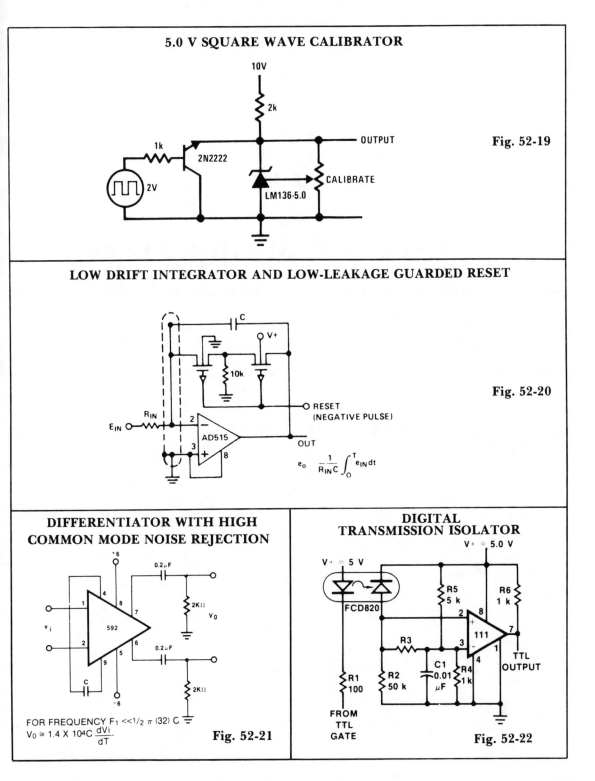

5.0 V SQUARE WAVE CALIBRATOR

Fig. 52-19

LOW DRIFT INTEGRATOR AND LOW-LEAKAGE GUARDED RESET

Fig. 52-20

$$e_o \quad \frac{1}{R_{IN}C}\int_0^T e_{IN}\,dt$$

DIFFERENTIATOR WITH HIGH COMMON MODE NOISE REJECTION

FOR FREQUENCY $F_1 \ll 1/2\,\pi\,(32)\,C$

$V_0 \cong 1.4 \times 10^4 C \frac{dV_i}{dT}$

Fig. 52-21

DIGITAL TRANSMISSION ISOLATOR

Fig. 52-22

423

53

Mixers and Multiplexers

The sources of the following circuits are contained in the Sources section beginning on page 730. The figure number contained in the box of each circuit correlates to the source entry in the Sources section.

DIFFERENTIAL MUX/DEMUX SYSTEM

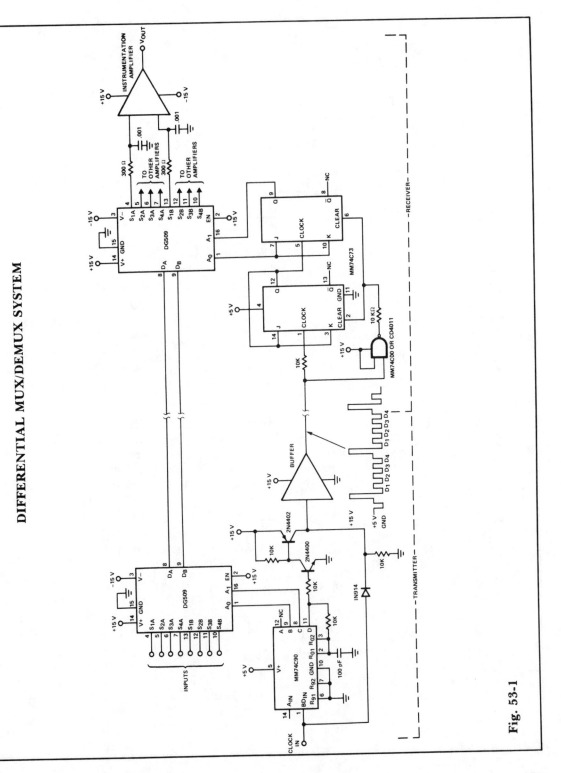

Fig. 53-1

EIGHT CHANNEL MUX/DEMUX SYSTEM

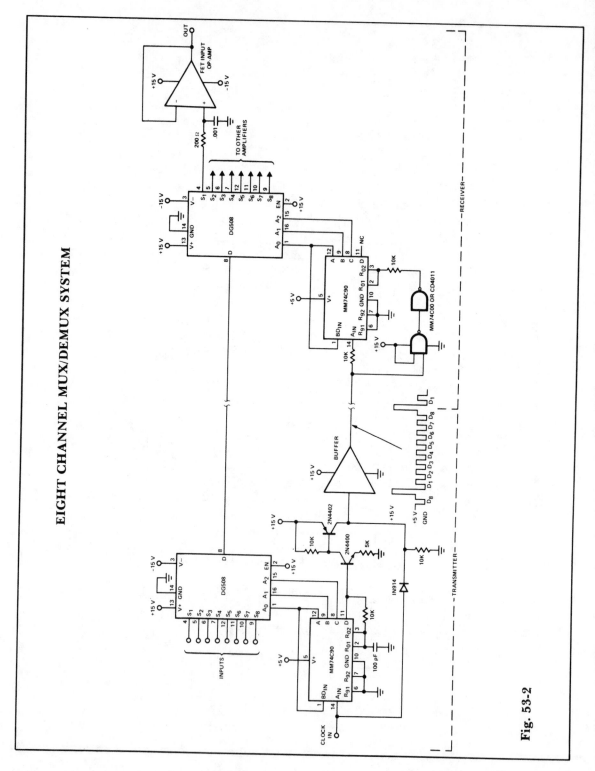

Fig. 53-2

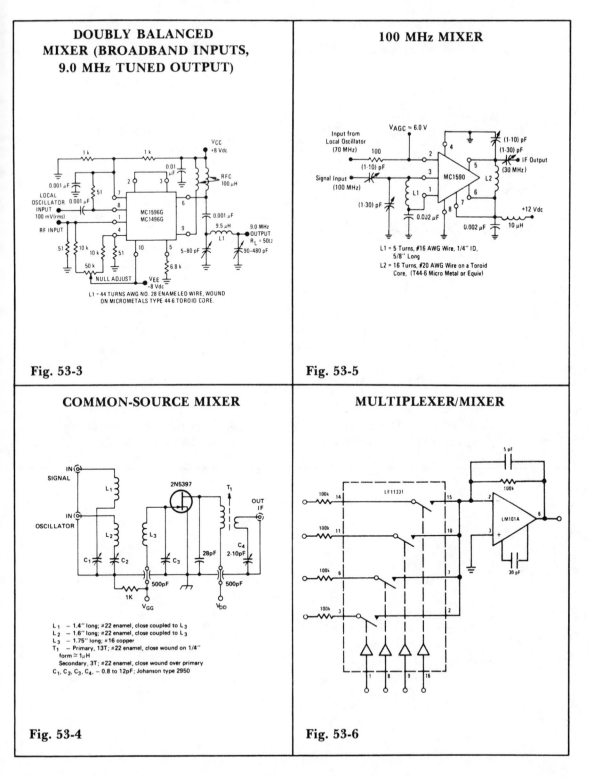

DOUBLY BALANCED MIXER (BROADBAND INPUTS, 9.0 MHz TUNED OUTPUT)

Fig. 53-3

100 MHz MIXER

Fig. 53-5

COMMON-SOURCE MIXER

Fig. 53-4

MULTIPLEXER/MIXER

Fig. 53-6

427

WIDE BAND DIFFERENTIAL MULTIPLEXER

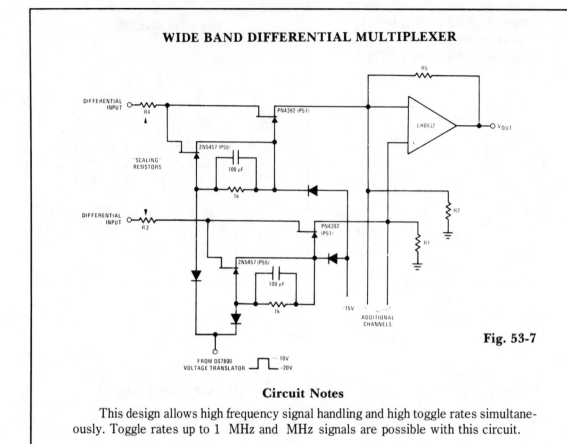

Fig. 53-7

Circuit Notes

This design allows high frequency signal handling and high toggle rates simultaneously. Toggle rates up to 1 MHz and MHz signals are possible with this circuit.

54

Modulation Monitors

The sources of the following circuits are contained in the Sources section beginning on page 730. The figure number contained in the box of each circuit correlates to the source entry in the Sources section.

Modulation Monitor   Visual Modulation Indicator

CB Modulation Monitor

MODULATION MONITOR

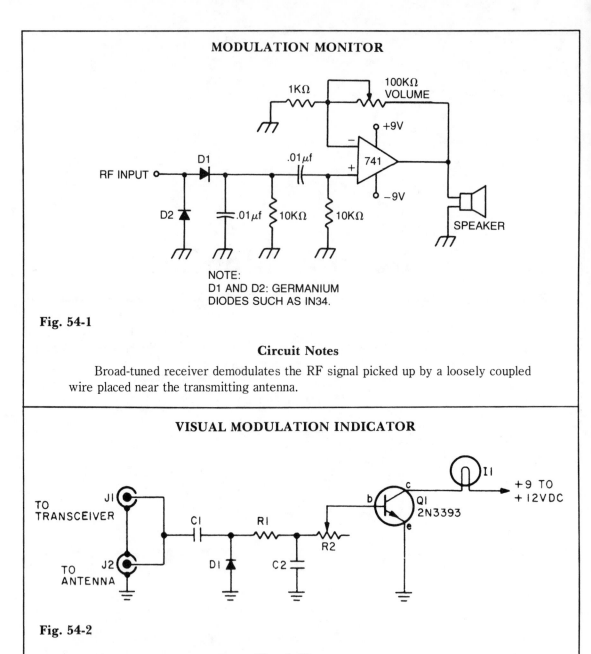

Fig. 54-1

NOTE:
D1 AND D2: GERMANIUM
DIODES SUCH AS IN34.

Circuit Notes

Broad-tuned receiver demodulates the RF signal picked up by a loosely coupled wire placed near the transmitting antenna.

VISUAL MODULATION INDICATOR

Fig. 54-2

Circuit Notes

Indicator lamp brightness varies in step with modulated RF signal. Adjust R2 with transmitter on (modulated) until the lamp flashes in step with modulation. C1 = 5 pf, C2 = 100 pF, D1 = 1N60 or 1N34 (Germanium), R3 = 10 K pot, I1 = 6-8 V, 30-60 mA incandescent bulb, Q1 = 2N3393 (for increased sensitivity use 2N3392 or other high-gain transistor).

CB MODULATION MONITOR

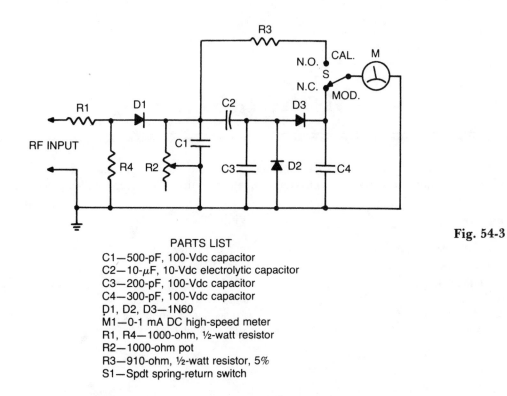

Fig. 54-3

PARTS LIST
C1—500-pF, 100-Vdc capacitor
C2—10-μF, 10-Vdc electrolytic capacitor
C3—200-pF, 100-Vdc capacitor
C4—300-pF, 100-Vdc capacitor
D1, D2, D3—1N60
M1—0-1 mA DC high-speed meter
R1, R4—1000-ohm, ½-watt resistor
R2—1000-ohm pot
R3—910-ohm, ½-watt resistor, 5%
S1—Spdt spring-return switch

Circuit Notes

Connect this circuit to a transceiver with a coaxial T connector in the transmission line. Key the transmitter (unmodulated), set S1 to CAL, and adjust R2 for a full scale reading. Return S1 to MOD position. The meter will read % modulation with 10% accuracy.

55

Modulators

The sources of the following circuits are contained in the Sources section beginning on page 730. The figure number contained in the box of each circuit correlates to the source entry in the Sources section.

TV Modulator

TV Modulator

Pulse-Position Modulator

Pulse-Width Modulator

Pulse-Width Modulator

RF Modulator

Linear Pulse-Width Modulator

Balanced Modulator

Video Modulator

Modulator

Pulse-Width Modulator

AM Modulator

TV Modulator Using a Motorola MC1374

Pulse-Width Modulator

Pulse-Width Modulator

VHF Modulator

TV MODULATOR

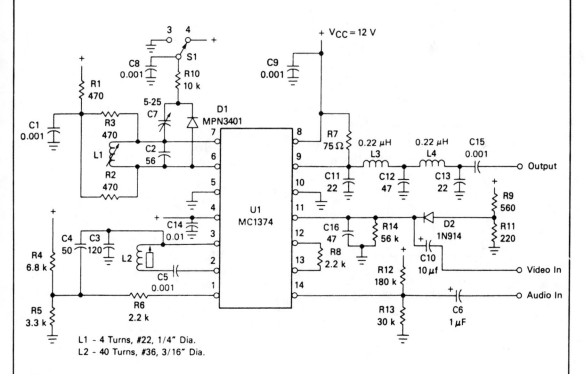

Fig. 55-1

Circuit Notes

The FM oscillator/modulator is a voltage-controlled oscillator, which exhibits a nearly linear output frequency versus input voltage characteristic for a wide deviation. It provides a good FM source with a few inexpensive external parts. It has a frequency range of 1.4 to 14 MHz and can typically produce a ±25 kHz modulated 4.5 MHz signal with about 0.6% total harmonic distortion.

TV MODULATOR

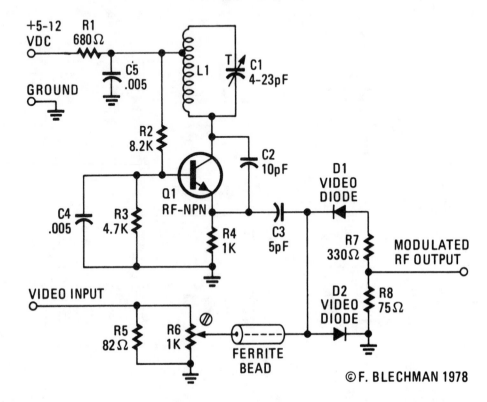

Fig. 55-2

Circuit Notes

The VHF frequency is generated by a tuned Hartley oscillator circuit. Resistors R2, R3, and R4 bias the transistor, with tapped inductor L1 and trimmer capacitor C1 forming the tank circuit. Adjusting C1 determines the frequency. Capacitor C2 provides positive feedback from the tank circuit to the emitter at Q1. Capacitor C4 provides an RF ground for the base of Q1. Bypass capacitor C5 and resistor R1 filter out the radio frequencies generated in the tank circuit to prevent radiation from the power-supply lines. The video signal enters the parallel combination of resistors R5 and R6; this combination closely matches the 75 ohm impedance of most video cables. Resistor R6 is a small screwdriver-adjusted potentiometer that is used to control the video input level to mixer diodes D1 and D2.

PULSE-POSITION MODULATOR

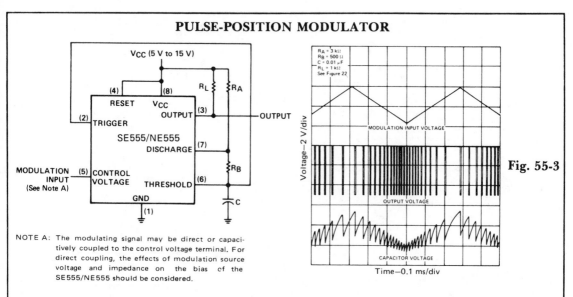

NOTE A: The modulating signal may be direct or capacitively coupled to the control voltage terminal. For direct coupling, the effects of modulation source voltage and impedance on the bias of the SE555/NE555 should be considered.

Fig. 55-3

Circuit Notes

The threshold voltage, and thereby the time delay, of a free-running oscillator is shown modulated with a triangular-wave modulation signal; however, any modulating wave-shape could be used.

PULSE-WIDTH MODULATOR

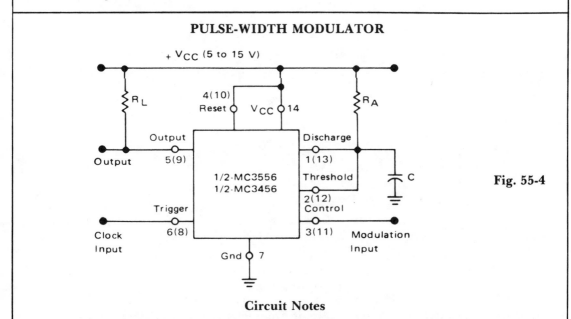

Fig. 55-4

Circuit Notes

If the timer is triggered with a continuous pulse train in the monostable mode of operation, the charge time of the capacitor can be varied by changing the control voltage at pin 3.

In this manner, the output pulse width can be modulated by applying a modulating signal that controls the threshold voltage.

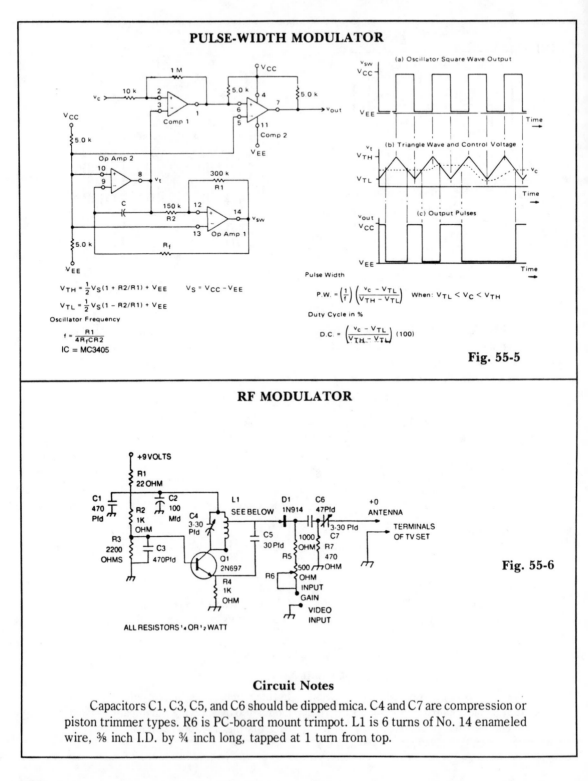

PULSE-WIDTH MODULATOR

$$V_{TH} = \frac{1}{2}V_S(1 + R2/R1) + V_{EE} \qquad V_S = V_{CC} - V_{EE}$$

$$V_{TL} = \frac{1}{2}V_S(1 - R2/R1) + V_{EE}$$

Oscillator Frequency

$$f = \frac{R1}{4R_fCR2}$$

IC = MC3405

Pulse Width

$$P.W. = \left(\frac{1}{f}\right)\left(\frac{v_c - V_{TL}}{V_{TH} - V_{TL}}\right) \quad \text{When: } V_{TL} < V_C < V_{TH}$$

Duty Cycle in %

$$D.C. = \left(\frac{v_c - V_{TL}}{V_{TH} - V_{TL}}\right)(100)$$

Fig. 55-5

(a) Oscillator Square Wave Output

(b) Triangle Wave and Control Voltage

(c) Output Pulses

RF MODULATOR

Fig. 55-6

ALL RESISTORS ¼ OR ½ WATT

Circuit Notes

Capacitors C1, C3, C5, and C6 should be dipped mica. C4 and C7 are compression or piston trimmer types. R6 is PC-board mount trimpot. L1 is 6 turns of No. 14 enameled wire, ⅜ inch I.D. by ¾ inch long, tapped at 1 turn from top.

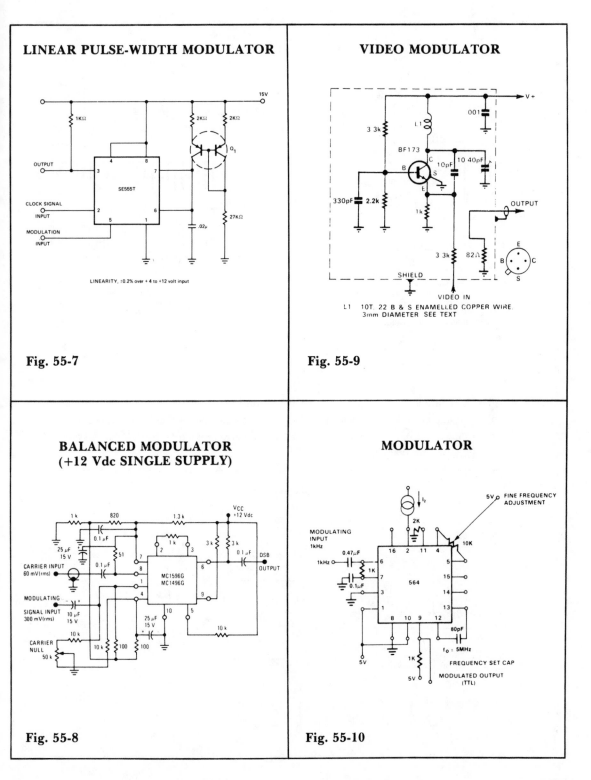

LINEAR PULSE-WIDTH MODULATOR

15V

1KΩ

2KΩ   2KΩ

Q₁

OUTPUT

4   8

3   SE555T   7

CLOCK SIGNAL
INPUT

2   6

MODULATION
INPUT

5   1

27KΩ

.02μ

LINEARITY, ±0.2% over + 4 to +12 volt input

Fig. 55-7

VIDEO MODULATOR

V +

3.3k   L1   001

BF173

B   C   10pF   10 40pF
S
E

330pF   2.2k

1k   OUTPUT

3 3k   82Ω

E
B   C
S

SHIELD

VIDEO IN

L1   10T. 22 B & S ENAMELLED COPPER WIRE.
3mm DIAMETER SEE TEXT

Fig. 55-9

BALANCED MODULATOR
(+12 Vdc SINGLE SUPPLY)

1 k   820   1.3 k   Vcc
+12 Vdc

0.1 μF

25 μF
15 V   51   7   2   1 k   3   3 k   3 k

0.1 C   DSB
OUTPUT

CARRIER INPUT
60 mV(rms)   0.1 μF   8   6

1
MC1596G
MODULATING   MC1496G
4
SIGNAL INPUT   10   9
10 μF
15 V   25 μF   10   5
15 V

CARRIER
NULL   10 k   10 k   100   100   10 k

50 k

Fig. 55-8

MODULATOR

I₂

5V   FINE FREQUENCY
ADJUSTMENT

2K

MODULATING
INPUT
1kHz   10K

0.47μF   16   2   11   4

1kHz   6

1K
7   15
564

0.1μF   3   14

1   8   10   9   12   13

80pF

5V   1K   f₀ = 5MHz

5V   FREQUENCY SET CAP

MODULATED OUTPUT
(TTL)

Fig. 55-10

437

PULSE-WIDTH MODULATOR

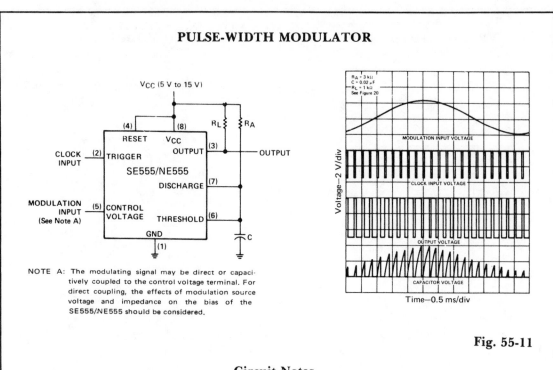

NOTE A: The modulating signal may be direct or capacitively coupled to the control voltage terminal. For direct coupling, the effects of modulation source voltage and impedance on the bias of the SE555/NE555 should be considered.

Fig. 55-11

Circuit Notes

The monostable circuit is triggered by a continuous input pulse train and the threshold voltage is modulated by a control signal. The resultant effect is a modulation of the output pulse width, as shown. A sine-wave modulation signal is illustrated, but any wave-shape could be used.

AM MODULATOR

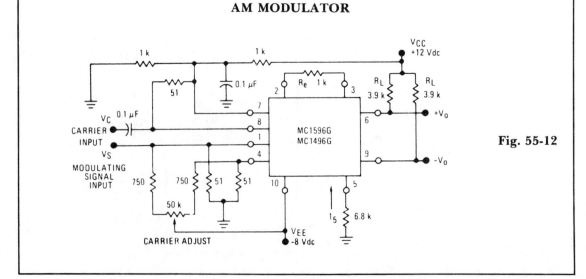

Fig. 55-12

TV MODULATOR USING A MOTOROLA MC1374

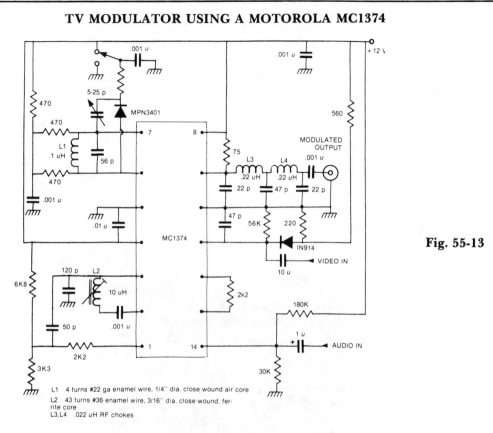

Fig. 55-13

L1 4 turns #22 ga enamel wire, 1/4'' dia. close wound air core
L2 43 turns #36 enamel wire, 3/16'' dia. close wound, ferrite core
L3, L4 .022 uH RF chokes

Circuit Notes

This one-chip modulator requires some outboard circuitry and a shielded box.

PULSE-WIDTH MODULATOR

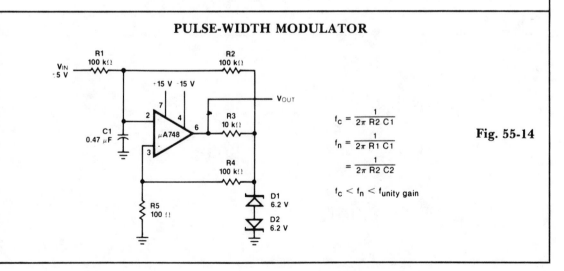

$$f_c = \frac{1}{2\pi\, R2\, C1}$$

$$f_n = \frac{1}{2\pi\, R1\, C1}$$

$$= \frac{1}{2\pi\, R2\, C2}$$

$$f_c < f_n < f_{unity\ gain}$$

Fig. 55-14

PULSE-WIDTH MODULATOR

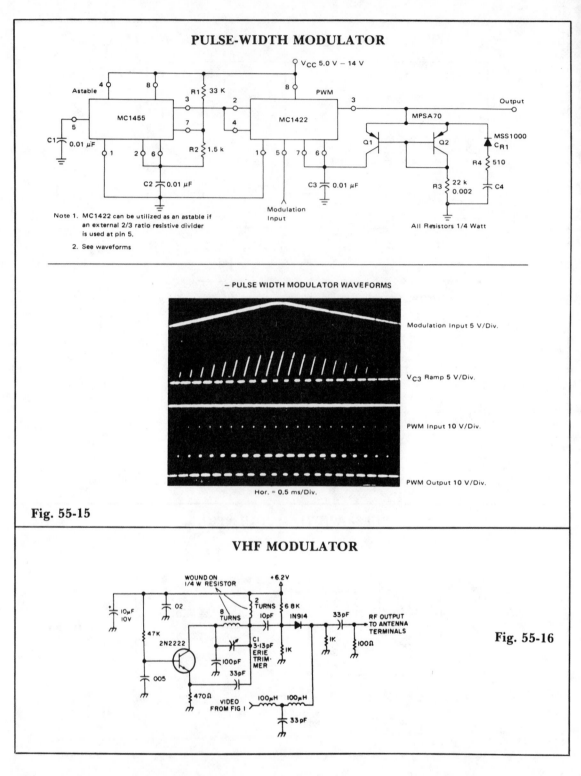

Note 1. MC1422 can be utilized as an astable if
      an external 2/3 ratio resistive divider
      is used at pin 5.

2. See waveforms

– PULSE WIDTH MODULATOR WAVEFORMS

Modulation Input 5 V/Div.

V_{C3} Ramp 5 V/Div.

PWM Input 10 V/Div.

PWM Output 10 V/Div.

Hor. = 0.5 ms/Div.

Fig. 55-15

VHF MODULATOR

Fig. 55-16

56

Moisture and Rain Detectors

The sources of the following circuits are contained in the Sources section beginning on page 730. The figure number contained in the box of each circuit correlates to the source entry in the Sources section.

Rain Alarm
Moisture Detector

Automatic Plant Waterer
Rain Alarm/Door Bell

RAIN ALARM

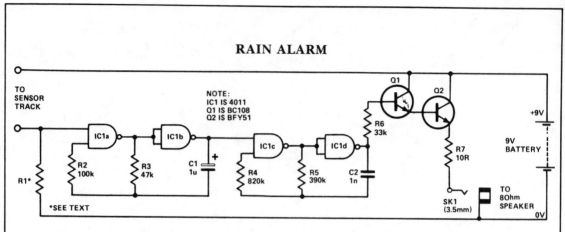

Fig. 56-1

Circuit Notes

The circuit uses four NAND gates of a 4011 package. In each oscillator, while one gate is configured as a straightforward inverter, the other has one input that can act as a control input. Oscillator action is inhibited if this input is held low. The first oscillator (IC1a and IC1b) has this input tied low via a high value resistor (R1) that acts as a sensitivity control. Thus this oscillator will be disabled until the control input is taken high. Any moisture bridging the sensor track will so enable the output which is a square wave at about 10 Hz. This in turn will gate on and off the 500 Hz oscillator formed by IC1c and IC1d. This latter oscillator drives the loudspeaker via R6, the Darlington pair formed by Q1 and Q2, and resistor R7.

MOISTURE DETECTOR

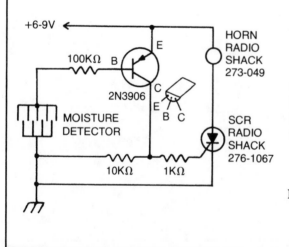

Circuit Notes

The detector is made of fine wires spaced about one or two inches apart. When the area between a pair of wires becomes moistened, the horn will sound. To turn it off, dc power must be disconnected.

Fig. 56-2

AUTOMATIC PLANT WATERER

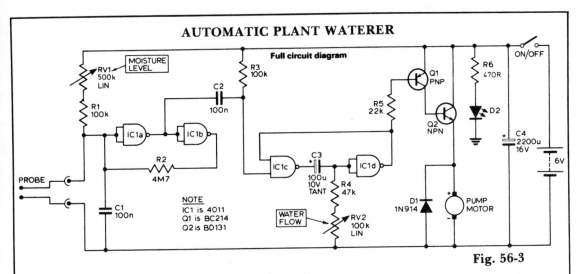

Fig. 56-3

Circuit Notes

The unit consists of a sensor, timer, and electric water pump. The sensor is embedded in the soil, and when dry, the electronics operate the water pump for a preset time. The circuit is composed of a level sensitive Schmitt trigger, variable time monostable, and output driver. When the resistance across the probe increases beyond a set value (i.e., the soil dries), the Schmitt is triggered. C2 feeds a negative going pulse to the monostable when the Schmitt triggers and R2 acts as feedback, to ensure a fast switching action.

RAIN ALARM/DOOR BELL

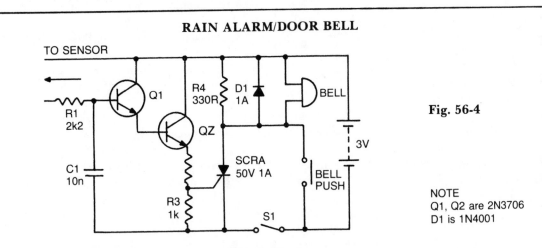

Fig. 56-4

NOTE
Q1, Q2 are 2N3706
D1 is 1N4001

Circuit Notes

With S1 open the circuit functions as a doorbell. With S1 closed, rain falling on the sensor will turn on Q1, triggering Q2 and the thyristor and activating the bell, R4 provides the holding for the thyristor while D1 prevents any damage to the thyristor from back EMF in the bell coil. The sensor can be made from 3 square inches of copper clad board with a razor cut down the center. C1 prevents any mains pickup in the sensor leads.

57

Motor Controls

The sources of the following circuits are contained in the Sources section beginning on page 730. The figure number contained in the box of each circuit correlates to the source entry in the Sources section.

MOTOR SPEED CONTROL

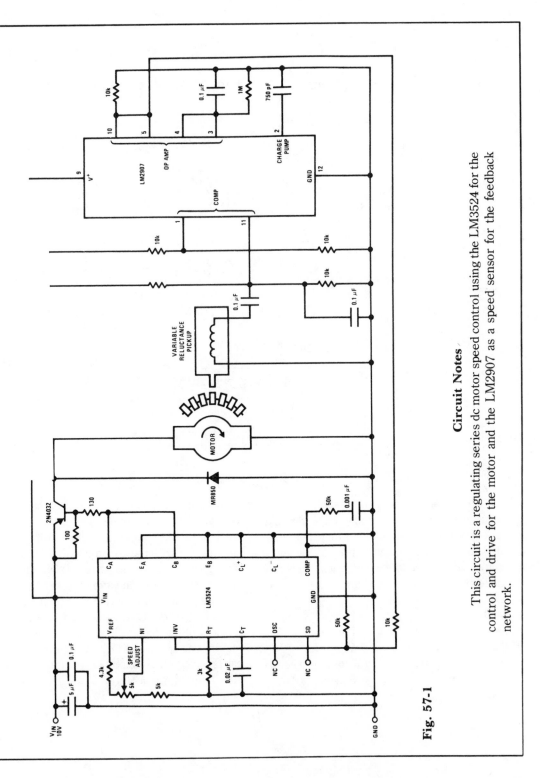

Fig. 57-1

Circuit Notes

This circuit is a regulating series dc motor speed control using the LM3524 for the control and drive for the motor and the LM2907 as a speed sensor for the feedback network.

PLUG-IN SPEED CONTROL FOR TOOLS OR APPLIANCES

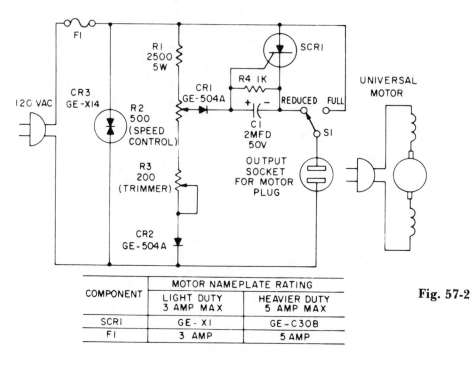

COMPONENT	MOTOR NAMEPLATE RATING	
	LIGHT DUTY 3 AMP MAX	HEAVIER DUTY 5 AMP MAX
SCRI	GE-XI	GE-C30B
FI	3 AMP	5 AMP

Fig. 57-2

Circuit Notes

Most standard household appliances and portable hand tools can be adapted to variable-speed operation by use of this simple half-wave SCR phase control. It can be used as the speed control unit for the following typical loads providee they use series universal (brush type) motors.

Drills	Fans
Sewing Machines	Lathes
Saber saws	Vibrators
Portable band saws	Movie projectors
Food mixers	Sanders
Food blenders	

During the positive half cycle of the supply voltage, the arm on potentiometer R2 taps off a traction of the sine wave supply voltage and compares it with the counter emf of the motor through the gate of the SCR. When the pot voltage rises above the armature voltage, current flows through CR1 into the gate of the SCR, triggering it, and thus applying the remainder of that half cycle supply voltage to the motor. The speed at which the motor operates can be selected by R2. Stable operation is possible over approximately a 3-to-1 speed range.

MOTOR SPEED CONTROL WITH FEEDBACK

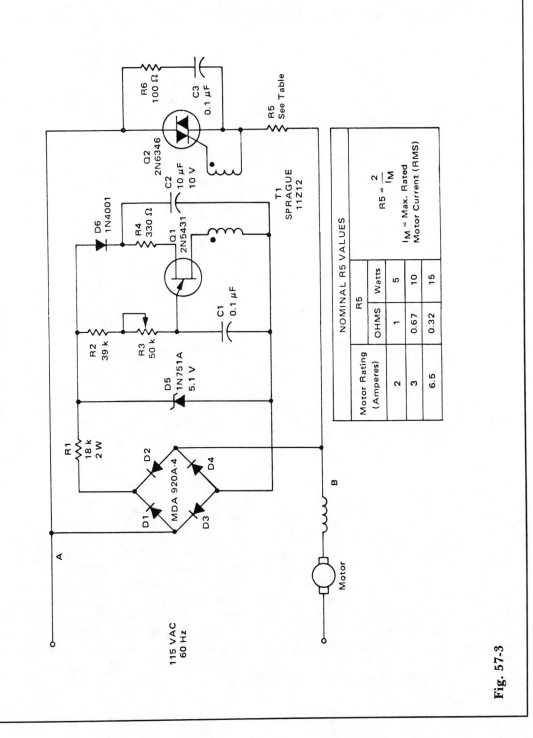

NOMINAL R5 VALUES			
Motor Rating (Amperes)	R5		
	OHMS	Watts	
2	1	5	
3	0.67	10	
6.5	0.32	15	

$$R5 = \frac{2}{I_M}$$

I_M = Max. Rated Motor Current (RMS)

Fig. 57-3

447

DIRECTION AND SPEED CONTROL FOR SERIES-WOUND MOTORS

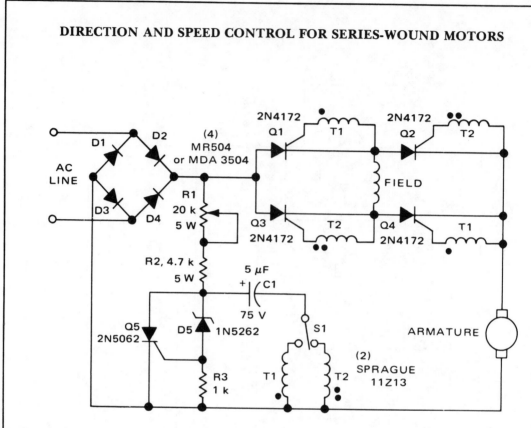

Fig. 57-4

Circuit Notes

The circuit shown here can be used to control the speed and direction of rotation of a series-wound dc motor. Silicon controlled rectifiers Q1-Q4, which are connected in a bridge arrangement, are triggered in diagonal pairs. Which pair is turned on is controlled by switch S1 since it connects either coupling transformer T1 or coupling transformer T2 to a pulsing circuit. The current in the field can be reversed by selecting either SCRs Q2 and Q3 for conduction, or SCRs Q1 and Q4 for conduction. Since the armature current is always in the same direction, the field current reverses in relation to the armature current, thus reversing the direction of rotation of the motor. A pulse circuit is used to drive the SCRs through either transformer T1 or T2. The pulse required to fire the SCR is obtained from energy stored in capacitor C1.

HIGH-TORQUE MOTOR SPEED CONTROL

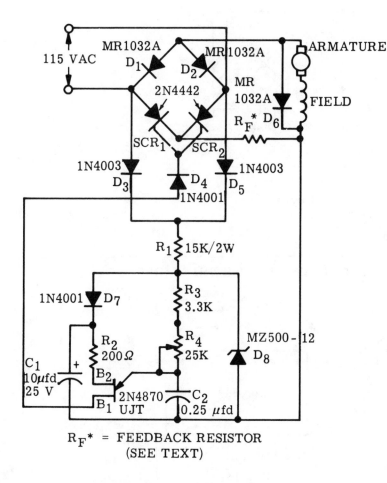

R_F^* = FEEDBACK RESISTOR
(SEE TEXT)

Fig. 57-5

Circuit Notes

A bridge circuit consisting of two SCRs and two silicon rectifiers furnishes full-wave power to the motor. Diodes, D3 and D5, supply dc to the trigger circuit through dropping resistors, R1. Phase delay of SCR firing is obtained by charging C2 through resistors R3 and R4 from the voltage level established by the zener diode, D8. When C2 charges to the firing voltage of the unijunction transistor, the UJT fires, triggering the SCR that has a positive voltage on its anode. When C2 discharges sufficiently, the unijunction transistor drops out of conduction. The value of RF is dependent upon the size of the motor and on the amount of feedback desired. A typical value for RF can be calculated from: $R_F = \dfrac{2}{I_M}$ where S_{IM} is the max rated load current (rms).

MOTOR SPEED CONTROL

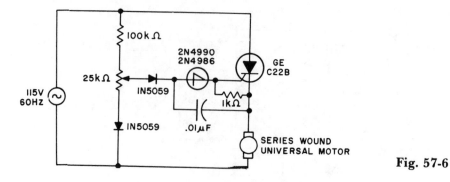

Fig. 57-6

Circuit Notes

Switching action of the 2N4990 allows smaller capacitors to be used while achieving reliable thyristor triggering.

CONSTANT CURRENT MOTOR DRIVE CIRCUIT

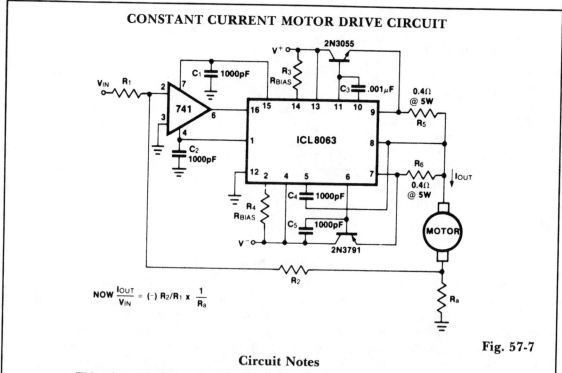

$$NOW \frac{I_{OUT}}{V_{IN}} = (-) R_2/R_1 \times \frac{1}{R_a}$$

Fig. 57-7

Circuit Notes

This minimum device circuit can be used to drive dc motors where there is some likelihood of stalling or lock up; if the motor locks, the current drive remains constant and the system does not destroy itself.

AC MOTOR BRAKE

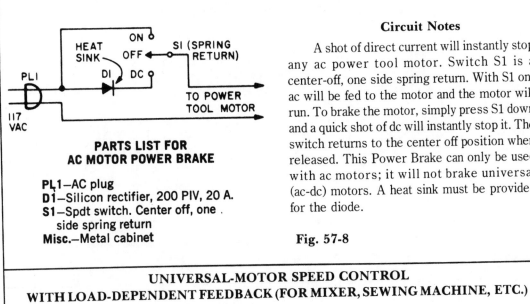

Circuit Notes

A shot of direct current will instantly stop any ac power tool motor. Switch S1 is a center-off, one side spring return. With S1 on, ac will be fed to the motor and the motor will run. To brake the motor, simply press S1 down and a quick shot of dc will instantly stop it. The switch returns to the center off position when released. This Power Brake can only be used with ac motors; it will not brake universal (ac-dc) motors. A heat sink must be provided for the diode.

PARTS LIST FOR
AC MOTOR POWER BRAKE

PL1—AC plug
D1—Silicon rectifier, 200 PIV, 20 A.
S1—Spdt switch. Center off, one side spring return
Misc.—Metal cabinet

Fig. 57-8

UNIVERSAL-MOTOR SPEED CONTROL
WITH LOAD-DEPENDENT FEEDBACK (FOR MIXER, SEWING MACHINE, ETC.)

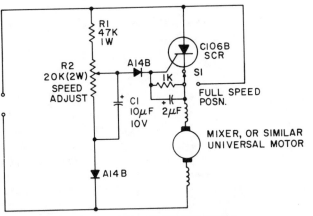

NOTE: RESISTORS 1/2 WATT EXCEPT AS NOTED

Fig. 57-9

Circuit Notes

Simple half-wave motor speed control is effective for use with small universal (ac/dc) motors. Maximum current capability 2.0 amps RMS. Because speed-dependent feedback is provided, the control gives excellent torque characteristics to the motor, even at low rotational speeds. Normal operation at maximum speed can be achieved by closing switch S1, thus bypassing the SCR.

DC MOTOR SPEED/DIRECTION CONTROL CIRCUIT

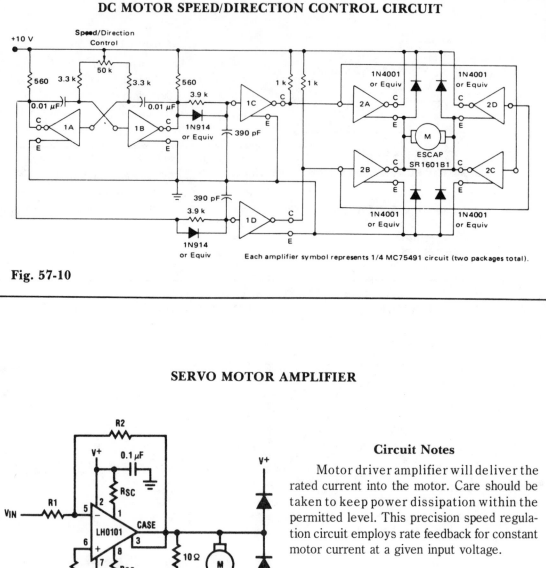

Fig. 57-10

Each amplifier symbol represents 1/4 MC75491 circuit (two packages total).

SERVO MOTOR AMPLIFIER

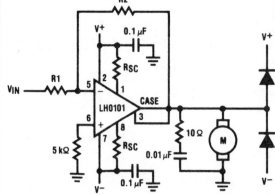

Fig. 57-11

Circuit Notes

Motor driver amplifier will deliver the rated current into the motor. Care should be taken to keep power dissipation within the permitted level. This precision speed regulation circuit employs rate feedback for constant motor current at a given input voltage.

MOTOR SPEED CONTROL

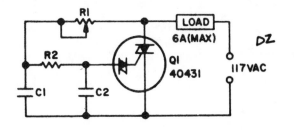

C1, C2–0.1-uF, 200-VDC capacitor
Q1–RCA 40431 Triac-Diac
R1–100,000-ohm linear taper
    potentiometer
R2–10,000-ohm, 1-watt resistor

Fig. 57-12

Circuit Notes

Universal motors and shaded-pole induction motors can be easily controlled with a full-wave Triac speed controller. Q1 combines both the triac and diac trigger diodes in the same case. The motor used for the load must be limited to 6 amperes maximum. Triac Q1 must be provided with a heat sink. With the component values shown, the Triac controls motor speed from full off to full on.

MODEL TRAIN SPEED CONTROL

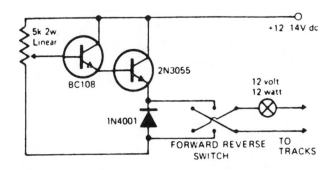

Fig. 57-13

Circuit Notes

Virtually any NPN small signal transistor may be used in place of the BC 108 shown. Likewise any suitable NPN power transistor can be used in place of the 2N3055. The output transistor must be mounted on a suitable heat-sink. Short circuit protection may be provided by wiring a 12 volt 12 watt bulb in series with the output. This will glow in event of a short circuit and thus effectively current-limit the output, it also acts as a visual short-circuit alarm.

INDUCTION-MOTOR CONTROL

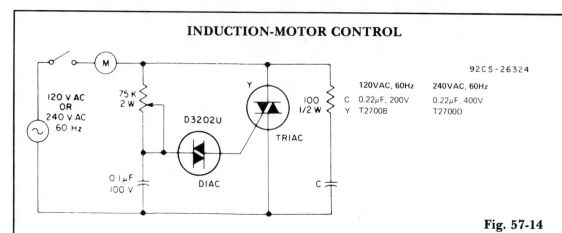

92CS-26324

	120VAC, 60Hz	240VAC, 60Hz
C	0.22µF, 200V	0.22µF, 400V
Y	T2700B	T2700D

Fig. 57-14

Circuit Notes

This single time-constant circuit can be used as proportional speed control for induction motors such as shaded pole or permanent split-capacitor motors when the load is fixed.

The circuit is best suited to applications which require speed control in the medium to full-power range.

DC MOTOR SPEED CONTROL

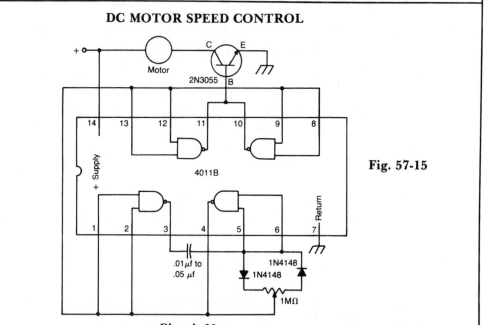

Fig. 57-15

Circuit Notes

The circuit uses a 4011 CMOS NAND gate, a pair of diodes and an NPN power transistor to provide a variable duty-cycle dc source. Adjusting the speed control varies the average voltage applied to the motor. The peak voltage, however, is not changed. This pulse power is effective at very low speeds, constantly kicking the motor along. At higher speeds, the motor behaves in a nearly normal manner.

UNIVERSAL MOTOR CONTROL WITH BUILT-IN SELF TIMER

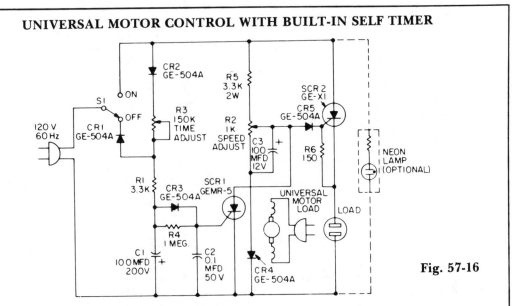

Fig. 57-16

Circuit Notes

When the time delay expires, SCR1 conducts and removes the gate signal from SCR2, which stops the motor. Both the time delay and motor speed are adjustable by potentiometers R2 and R3. If heavier motor loads are antici-pated, use the larger C30B SCR in place of the GE-X1 for SCR2. Also, the capacitance of C1 can be increased to lengthen the time delay, if desired.

SPEED CONTROL FOR MODEL TRAINS OR CARS

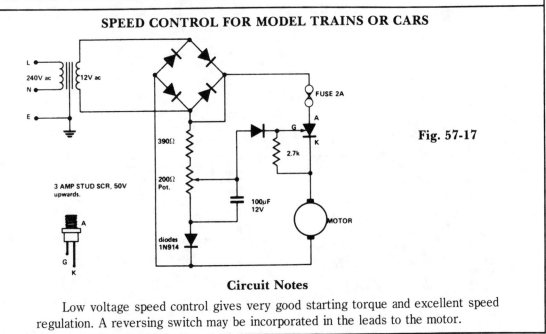

Fig. 57-17

Circuit Notes

Low voltage speed control gives very good starting torque and excellent speed regulation. A reversing switch may be incorporated in the leads to the motor.

DIRECTION AND SPEED CONTROL FOR SHUNT-WOUND MOTORS

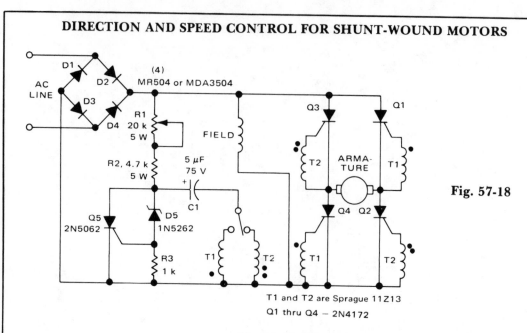

Fig. 57-18

T1 and T2 are Sprague 11Z13
Q1 thru Q4 — 2N4172

Circuit Notes

This circuit operates like the one shown in Fig. 57-4. The only differences are that the field is placed across the rectified supply and the armature is placed in the SCR bridge. Thus the field current is unidirectional but armature current is reversible; consequently the motor's direction of rotation is reversible. Potentiometer R1 controls the speed.

TWO-PHASE MOTOR DRIVE

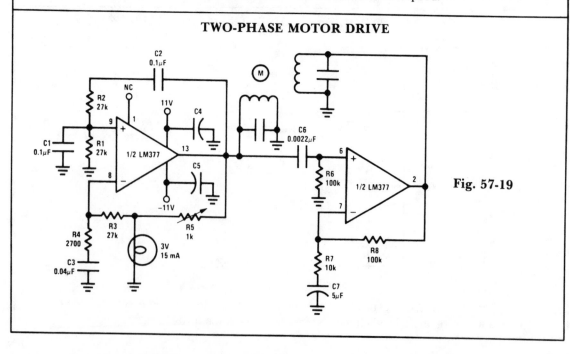

Fig. 57-19

DC SERVO AMPLIFIER

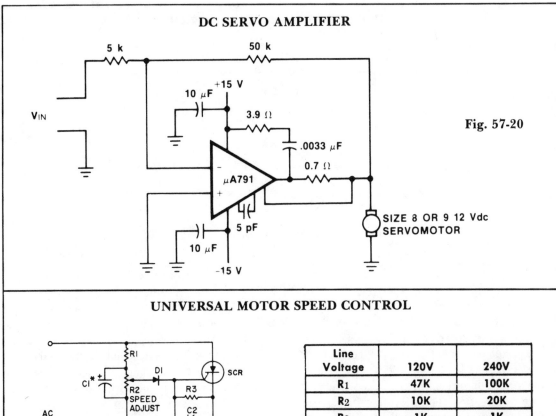

Fig. 57-20

UNIVERSAL MOTOR SPEED CONTROL

Line Voltage	120V	240V
R1	47K	100K
R2	10K	20K
R3	1K	1K
C1	1µF, 50V	1µF, 100V
C2	0.1µF, 50V	0.1µF, 50V
D1	1N5059	1N5060
D2	1N5059	1N5060
SCR	C106B1	C106D1

Fig. 57-21

Circuit Notes

The resistor capacitor network R1-R2-C1 provides a ramp-type reference voltage superimposed on top of a dc voltage adjustable with the speed-setting potentiometer R2. This reference voltage appearing at the wiper of R2 is balanced against the residual counter emf of the motor through the SCR gate. As the motor slows down due to heavy loading, its counter emf falls, and the reference ramp triggers the SCR earlier in the ac cycle. More voltage is thereby applied to the motor causing it to pick up speed again. Performance with the C106 SCR is particularly good because the low trigger current requirements of this device allow use of a flat top reference voltage, which provides good feedback gain and close speed regulation.

457

POWER TOOL TORQUE CONTROL

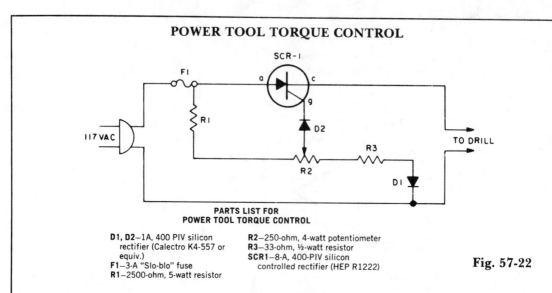

PARTS LIST FOR
POWER TOOL TORQUE CONTROL

D1, D2—1A, 400 PIV silicon
   rectifier (Calectro K4-557 or
   equiv.)
F1—3-A "Slo-blo" fuse
R1—2500-ohm, 5-watt resistor

R2—250-ohm, 4-watt potentiometer
R3—33-ohm, ½-watt resistor
SCR1—8-A, 400-PIV silicon
   controlled rectifier (HEP R1222)

Fig. 57-22

Circuit Notes

As the speed of an electric drill is decreased by loading, its torque also drops. A compensating speed control like this one puts the oomph back into the motor. When the drill slows down, a back voltage developed across the motor—in series with the SCR cathode and gate—decreases. The SCR gate voltage therefore increases relatively as the back voltage is reduced. The extra gate voltage causes the SCR to conduct over a larger angle and more current is driven into the drill, even as speed falls under load. The SCR should be mounted in ¼-in. thick block of aluminum or copper at least 1-in. square. If the circuit is used for extended periods use a 2 inch square piece.

AC SERVO AMPLIFIER—BRIDGE TYPE

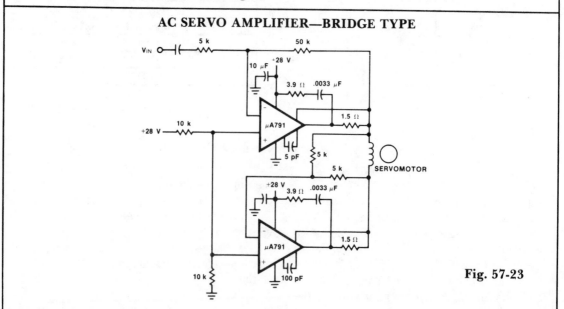

Fig. 57-23

58

Multivibrators

The sources of the following circuits are contained in the Sources section beginning on page 730. The figure number contained in the box of each circuit correlates to the source entry in the Sources section.

Monostable Circuit
Astable Multivibrator
Astable Oscillator
Digitally Controlled Astable Multivibrator
Dual Astable Multivibrator
UJT Monostable
Monostable Multivibrator with Input
    Lock-Out

TTL Monostable
Monostable Circuit
One-Shot Multivibrator
Monostable Multivibrator
Bistable Multivibrator
100 kHz Free-Running Multivibrator

MONOSTABLE CIRCUIT

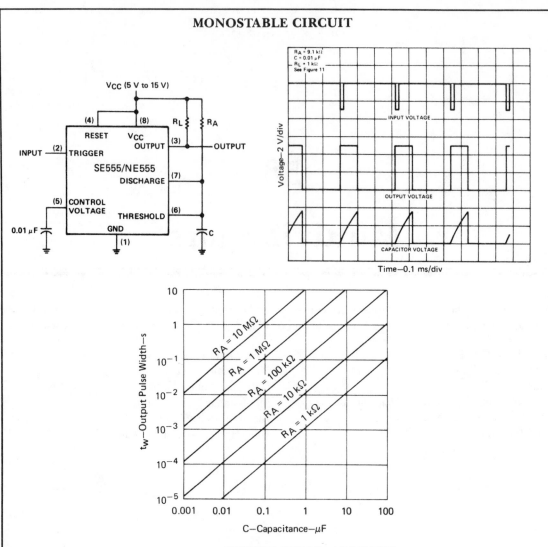

Fig. 58-1

OUTPUT PULSE WIDTH vs CAPACITANCE

Circuit Notes

If the output is low, application of a negative-going pulse to the trigger input sets the flip-flop (Q goes low), drives the output high, and turns off 1. Capacitor C is then charged through R_A until the voltage across the capacitor reaches the threshold voltage of the threshold input. If the trigger input has returned to a high level, the output of the threshold comparator will reset the flip-flop (Q goes high), drive the output low, and discharge C through Q1. Monostable operations is initiated when the trigger input voltage falls below the trigger threshold. Once initiated, the sequence will complete only if the trigger input is high at the end of the timing interval.

ASTABLE MULTIVIBRATOR

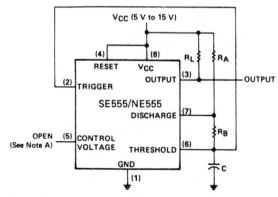

NOTE A: Decoupling the control voltage input (pin 5) to ground with a capacitor may improve operation. This should be evaluated for individual applications.

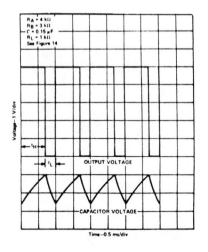

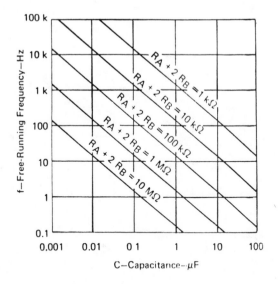

Fig. 58-2

Circuit Notes

The capacitor C will charge through R_A and R_B, and then discharge through R_B only. The duty cycle may be controlled by the values of R_A and R_B.

ASTABLE OSCILLATOR

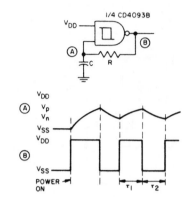

Circuit Notes

Before power is applied, the input and output are at ground potential and capacitor C is discharged. On power-on, the output goes high (V_{DD}) and C charges through R until V is reached; the output then goes low (V_{SS}). C is now discharged through R until V_n is reached. The output then goes high and charges C towards V_p through R. Thus input A alternately swings between V_p and V_n as the output goes high and low. This circuit is self-starting at power-on.

Fig. 58-3

DIGITALLY CONTROLLED ASTABLE MULTIVIBRATOR

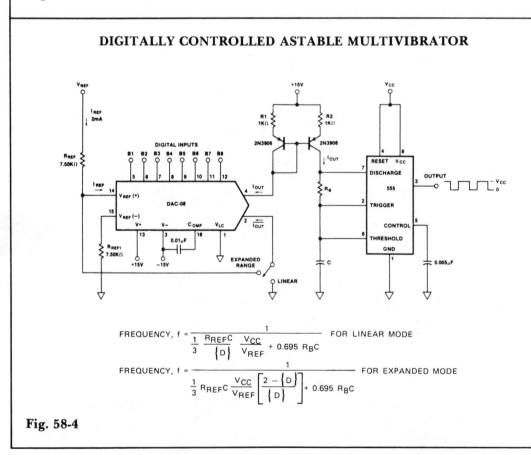

$$\text{FREQUENCY, } f = \cfrac{1}{\cfrac{1}{3}\ \cfrac{R_{REF}C}{\{D\}}\ \cfrac{V_{CC}}{V_{REF}} + 0.695\ R_B C} \quad \text{FOR LINEAR MODE}$$

$$\text{FREQUENCY, } f = \cfrac{1}{\cfrac{1}{3}\ R_{REF}C\ \cfrac{V_{CC}}{V_{REF}} \left[\cfrac{2 - \{D\}}{\{D\}}\right] + 0.695\ R_B C} \quad \text{FOR EXPANDED MODE}$$

Fig. 58-4

DUAL ASTABLE MULTIVIBRATOR

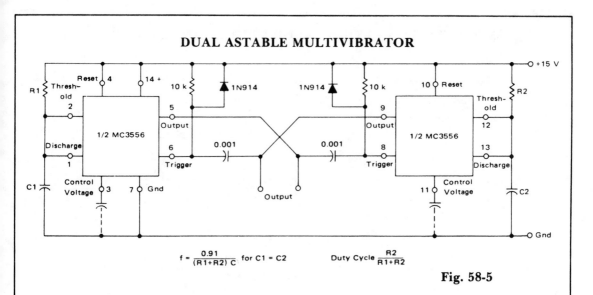

$$f = \frac{0.91}{(R1+R2)\,C} \text{ for C1 = C2}$$

Duty Cycle $\frac{R2}{R1+R2}$

Fig. 58-5

Circuit Notes

This dual astable multivibrator provides versatility not available with single timer circuits. The duty cycle can be adjusted from 5% to 95%. The two outputs provide two phase clock signals often required in digital systems. It can also be inhibited by use of either reset terminal.

UJT MONOSTABLE

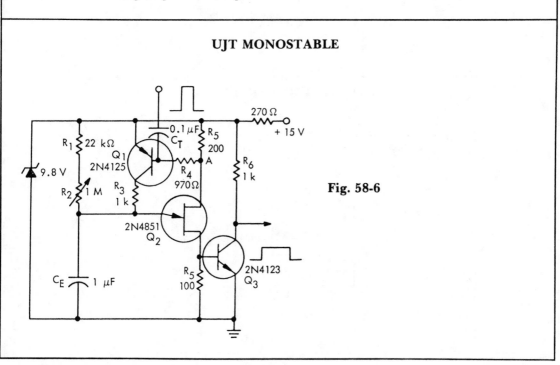

Fig. 58-6

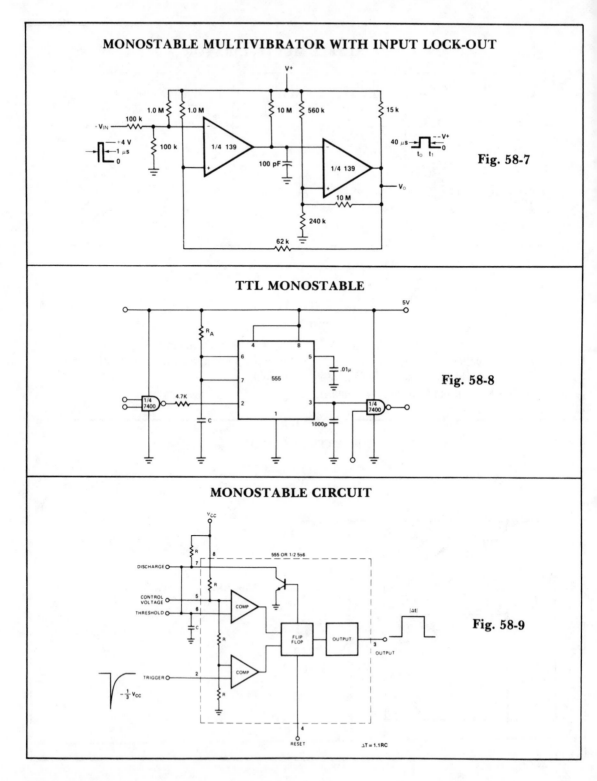

MONOSTABLE MULTIVIBRATOR WITH INPUT LOCK-OUT

Fig. 58-7

TTL MONOSTABLE

Fig. 58-8

MONOSTABLE CIRCUIT

Fig. 58-9

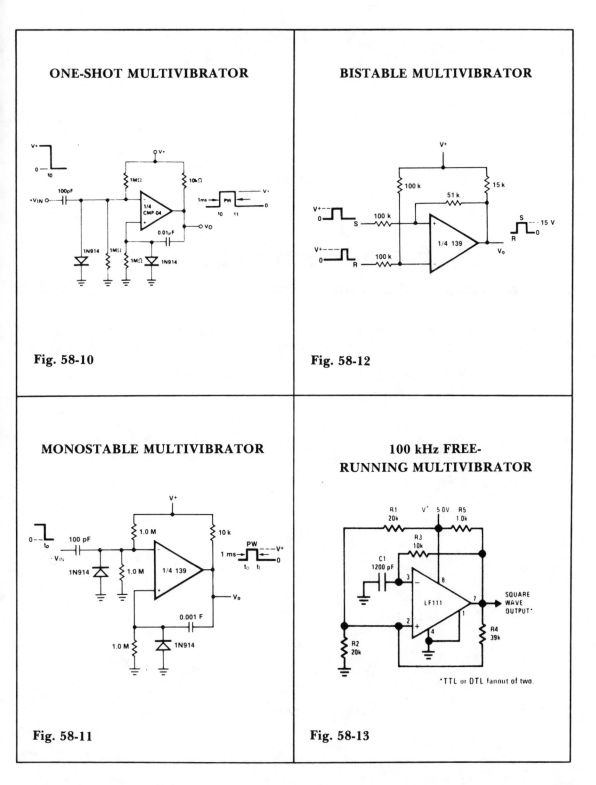

ONE-SHOT MULTIVIBRATOR

Fig. 58-10

BISTABLE MULTIVIBRATOR

Fig. 58-12

MONOSTABLE MULTIVIBRATOR

Fig. 58-11

100 kHz FREE-
RUNNING MULTIVIBRATOR

Fig. 58-13

465

59

Noise Generators

The sources of the following circuits are contained in the Sources section beginning on page 730. The figure number contained in the box of each circuit correlates to the source entry in the Sources section.

Audio Noise Generator   Noise Generator
Pink Noise Generator   Wideband Noise Generator

Noise Generator Circuit

AUDIO NOISE GENERATOR

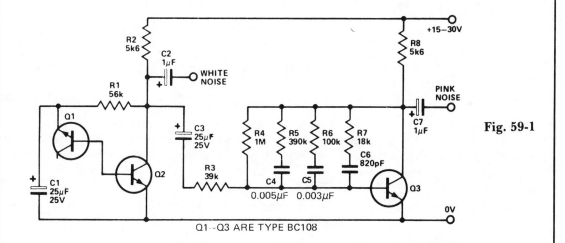

Fig. 59-1

Q1—Q3 ARE TYPE BC108

Circuit Notes

This simple circuit generates both white and pin noise. Transistor Q1 is used as a zener diode. The normal base-emitter junction is reverse-biased and goes into zener break-down at about 7 to 8 volts. The zener noise current from Q1 flows into the base of Q2 such that an output of about 150 millivolts of white noise is available. To convert the white noise to pink, a filter is required which provides a 3 dB cut per octave as the frequency increases.

Since such a filter attenuates the noise considerably an amplifier is used to restore the output level. Transistor Q3 is this amplifier and the pink noise filter is connected as a feedback network. between collector and base in order to obtain the required characteristic by controlling the gain-versus-frequency of the transistor. The output of transistor Q3 is thus the pink noise required and is fed to the relevant output socket.

PINK NOISE GENERATOR

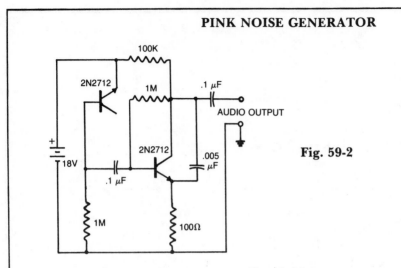

Fig. 59-2

Circuit Notes

A reverse-biased pn junction of a 2N2712 transistor is used as a noise generator. The second 2N2712 is an audio amplifier. The 0.005 µF capacitor across the amplifier output removes some high-frequency components to simulate pink noise more closely. The audio output may be connected to high-impedance earphones or to a driver amplifier for speaker listening.

NOISE GENERATOR

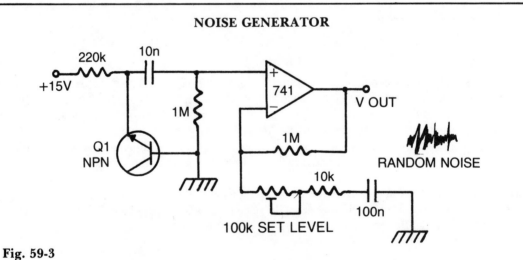

Fig. 59-3

Circuit Notes

The zener breakdown of a transistor junction is used as a noise generator. The breakdown mechanism is random and this voltage has a high source impedance. By using the op amp as a high input impedance, high ac gain amplifier, a low impedance, large signal noise source is obtained. The 100K potentiometer is used to set the noise level by varying the gain from 40 to 20 dB.

WIDEBAND NOISE GENERATOR

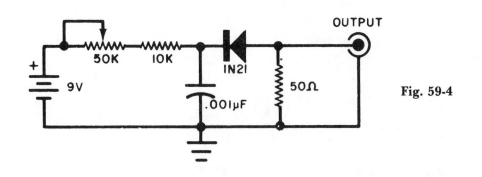

Fig. 59-4

Circuit Notes

This circuit will produce wideband rf noise. It uses a reverse-biased diode and has a low-impedance output. Can be used to align receivers for optimum performance.

NOISE GENERATOR CIRCUIT

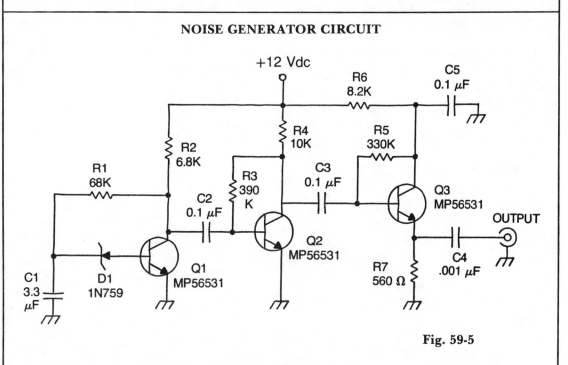

Fig. 59-5

Circuit Notes

The zener diode is an avalanche rectifier in the reverse bias mode connected to the input circuit of a wideband rf amplifier. The noise is amplified and applied to the cascade wideband amplifier, transistors Q2 and Q3.

60

Oscilloscope Circuits

The sources of the following circuits are contained in the Sources section beginning on page 730. The figure number contained in the box of each circuit correlates to the source entry in the Sources section.

OSCILLOSCOPE CONVERTER PROVIDES FOUR-CHANNEL DISPLAYS

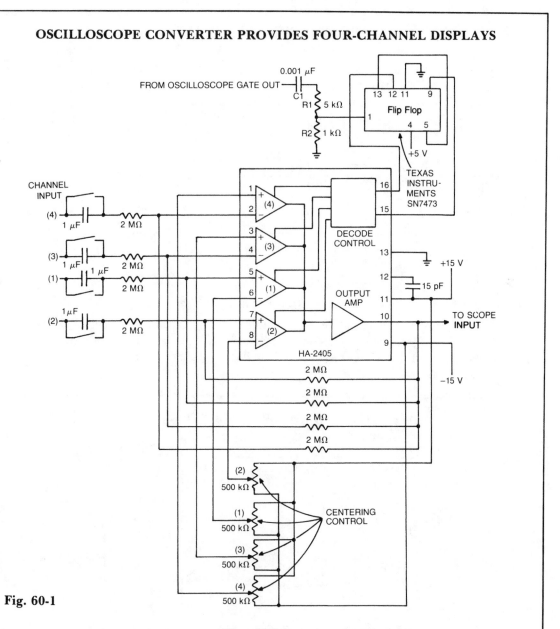

Fig. 60-1

Circuit Notes

The monolithic quad operational amplifier provides an inexpensive way to increase display capability of a standard oscilloscope. Binary inputs drive the IC op amp; a dual flip-flop divides the scope's gate output to obtain channel selection signals. All channels have centering controls for nulling offset voltage. A negative-going scope gate signal selects the next channel after each trace. The circuit operates out to 5 MHz.

ADD-ON TRIGGERED SWEEP

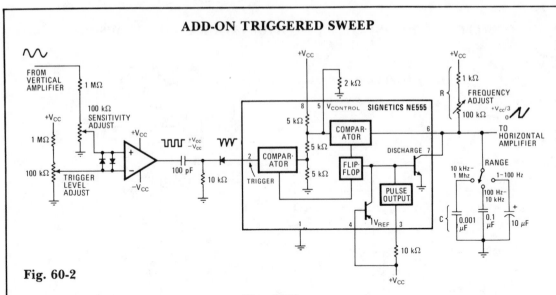

Fig. 60-2

Circuit Notes

The circuit's input op amp triggers the timer, setting its flip-flop and cutting off its discharge transistor so that capacitor C can charge. When the capacitor voltage reaches the timer's control voltage (0.33Vcc), the flip-flop resets and the transistor conducts, discharging the capacitor.

10.7 MHz SWEEP GENERATOR

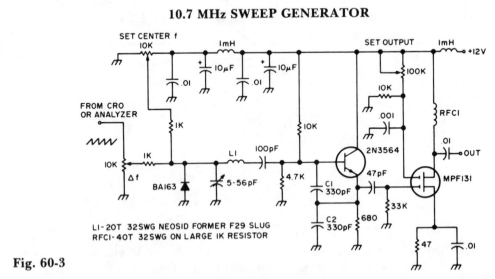

Fig. 60-3

Circuit Notes

This circuit is used to observe the response of an if amp or a filter. It can be used with an oscilloscope or, for more dynamic range, with a spectrum analyzer.

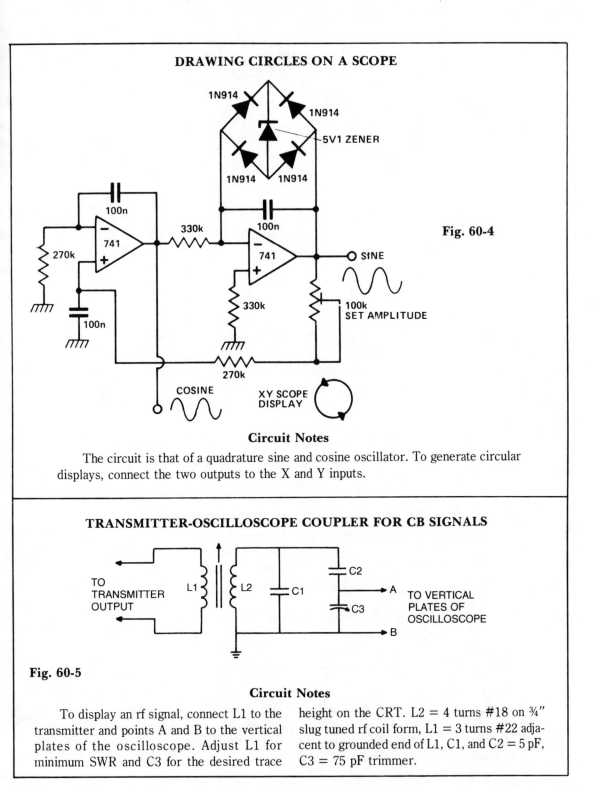

DRAWING CIRCLES ON A SCOPE

Fig. 60-4

SINE

COSINE

XY SCOPE
DISPLAY

Circuit Notes

The circuit is that of a quadrature sine and cosine oscillator. To generate circular displays, connect the two outputs to the X and Y inputs.

TRANSMITTER-OSCILLOSCOPE COUPLER FOR CB SIGNALS

Fig. 60-5

Circuit Notes

To display an rf signal, connect L1 to the transmitter and points A and B to the vertical plates of the oscilloscope. Adjust L1 for minimum SWR and C3 for the desired trace height on the CRT. L2 = 4 turns #18 on ¾" slug tuned rf coil form, L1 = 3 turns #22 adjacent to grounded end of L1, C1, and C2 = 5 pF, C3 = 75 pF trimmer.

OSCILLOSCOPE MONITOR

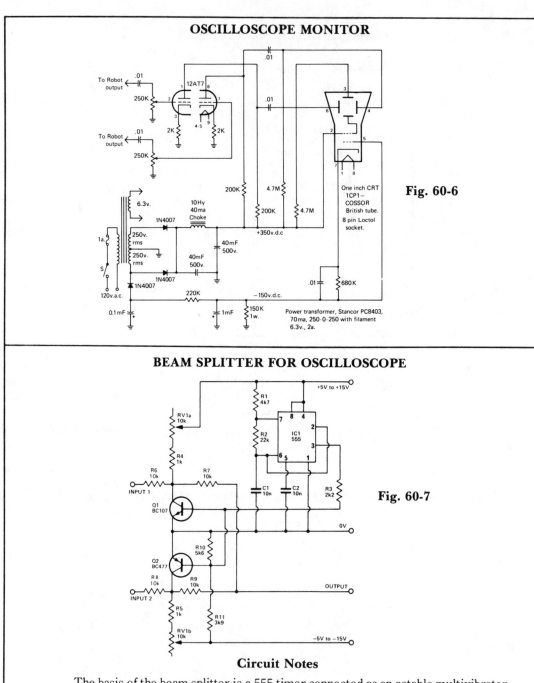

Fig. 60-6

One inch CRT 1CP1— COSSOR British tube. 8 pin Loctol socket.

Power transformer, Stancor PC8403, 70ma, 250·0·250 with filament 6·3v., 2a.

BEAM SPLITTER FOR OSCILLOSCOPE

Fig. 60-7

Circuit Notes

The basis of the beam splitter is a 555 timer connected as an astable multivibrator. Signals at the two inputs are alternately displayed on the oscilloscope with a clear separation between them. The output is controlled by the tandem potentiometer RV1a/b which also varies the amplitude of the traces.

61

Phase Sequence
and Phase Shift Circuits

The sources of the following circuits are contained in the Sources section beginning on page 730. The figure number contained in the box of each circuit correlates to the source entry in the Sources section.

PHASE SEQUENCE INDICATOR

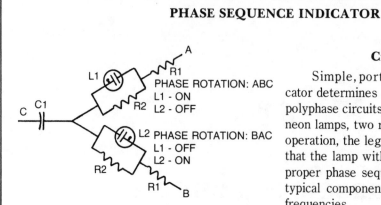

PHASE ROTATION: ABC
L1 - ON
L2 - OFF

PHASE ROTATION: BAC
L1 - OFF
L2 - ON

Fig. 61-1

Circuit Notes

Simple, portable phase-sequence indicator determines the proper phase rotation in polyphase circuits. Major components are two neon lamps, two resistors, and a capacitor. In operation, the leg voltages are unbalanced, so that the lamp with the maximum voltage—or proper phase sequence—lights. Table shows typical component values for various circuit frequencies.

SINGLE TRANSISTOR PHASE SHIFTER

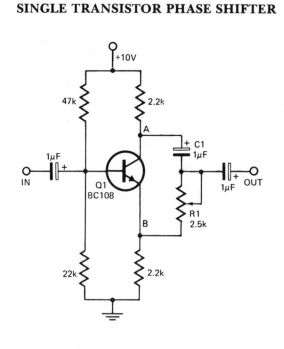

Fig. 61-2

Circuit Notes

This circuit provides a simple means of obtaining phase shifts between zero and 170°. The transistor operates as a phase splitter, the output at point A being 180° out of phase with the input. Point B is in phase with the input phase. Adjusting R1 provides the sum of various proportions of these and hence a continuously variable phase shift is provided. The circuit operates well in the 600 Hz to 4 kHz range.

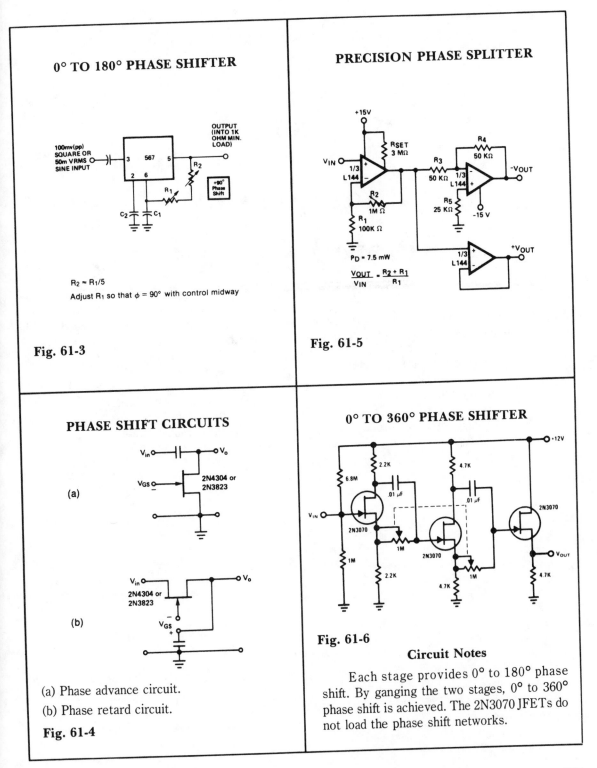

0° TO 180° PHASE SHIFTER

100mv(pp) SQUARE OR 50m VRMS SINE INPUT

567

OUTPUT (INTO 1K OHM MIN. LOAD)

R_2

+90° Phase Shift

R_1

C_2 C_1

$R_2 \approx R_1/5$

Adjust R_1 so that $\phi = 90°$ with control midway

Fig. 61-3

PRECISION PHASE SPLITTER

+15V

RSET 3 MΩ

V_{IN}

1/3 L144

R_2 1MΩ

R_1 100K Ω

R_3 50 KΩ

1/3 L144

R_4 50 KΩ

–VOUT

R_5 25 KΩ

–15 V

P_D = 7.5 mW

$$\frac{V_{OUT}}{V_{IN}} = \frac{R_2 + R_1}{R_1}$$

1/3 L144

+VOUT

Fig. 61-5

PHASE SHIFT CIRCUITS

V_{in} V_o

2N4304 or 2N3823

(a)

V_{GS}

V_{in} V_o

2N4304 or 2N3823

(b)

V_{GS}

(a) Phase advance circuit.

(b) Phase retard circuit.

Fig. 61-4

0° TO 360° PHASE SHIFTER

+12V

6.8M

2.2K

.01 μF

4.7K

.01 μF

2N3070

V_{IN}

2N3070

1M

1M

2N3070

V_{OUT}

1M

2.2K

4.7K

4.7K

1M

4.7K

Fig. 61-6

Circuit Notes

Each stage provides 0° to 180° phase shift. By ganging the two stages, 0° to 360° phase shift is achieved. The 2N3070 JFETs do not load the phase shift networks.

477

62

Photography
Related Circuits

The sources of the following circuits are contained in the Sources section beginning on page 730. The figure number contained in the box of each circuit correlates to the source entry in the Sources section.

Automatic Contrast Meter
Darkroom Timer
Photo Stop Action
Sound Light-Flash Trigger
Sound Activated Strobe Trip

Flash Slave Driver
Remote Flash Trigger
Flash Exposure Meter
Shutter Tester
Photographic Timer

AUTOMATIC CONTRAST METER

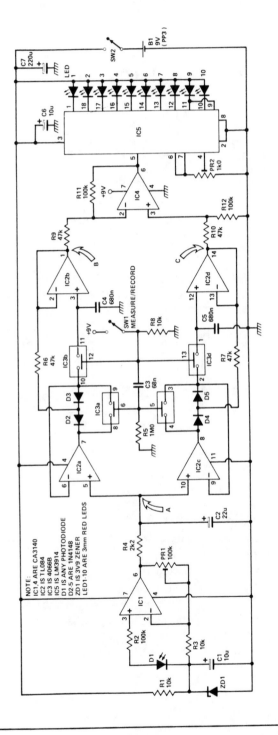

Fig. 62-1

Circuit Notes

The circuit arrangement consists of a photo-amplifier which feeds a voltage derived from varying light levels in an enlarger to a pair of peak detectors. One follows the peak positive voltage and the other the peak negative voltage. The capacitors used for storing the voltage peaks in the followers also form part of sample and hold circuits which are then switched to hold after the measurement. Their outputs represent the maximum and minimum values of light intensity. A differential amplifier then computes the ratio of these values, and the result is displayed on an LED bargraph meter.

DARKROOM TIMER

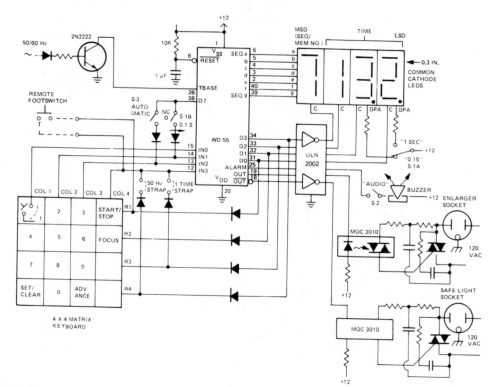

Fig. 62-2

Circuit Notes

The darkroom timer/controller uses few external components: a display, a digit driver, keyboard, and output switching devices. A 4-digit common-cathode LED display is desirable for dark room environments. The time base is provided by shaping up the 50/60 Hz ac line. A DPDT switch (S1) is used to select a resolution of .1 or 1 seconds and to simultane-ously move the decimal point. Timer/controller has two switched ac outlets, one for the enlarger and one for the safe light. They are the complements of each other in that the safe light is on when the enlarger is not active and is off when the enlarger is printing. The buzzer is of the self-contained oscillator variety and operates with dc drive.

PHOTO STOP ACTION

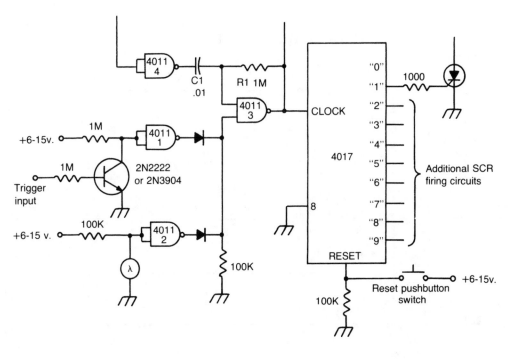

Fig. 62-3

Bulb firing system SCR

Circuit Notes

This circuit gives multiple "stop-action" photographic effects like showing a bouncing ball in up to nine locations in a single photograph. The circuit will automatically fire the bulbs sequentially with the time between each firing variable. The circuit is functionally complete except for the actual firing system. In many cases, a simple SCR will work, as shown. The firing can be initiated in one of two ways. A trigger pulse can be applied to the trigger input terminal through a capacitor, or can operate the unit as a slave. Light from a camera-mounted flash will activate the circuit through its built-in photocell pickup. The time period between each successive flash is determined by C1 and R1, which is variable. After firing the circuit, it must be reset by momentarily depressing the reset button.

SOUND LIGHT-FLASH TRIGGER

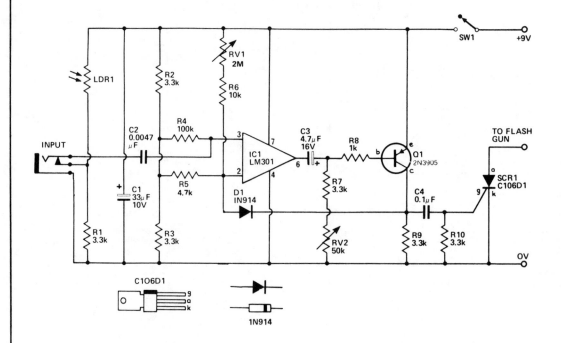

Fig. 62-4

Circuit Notes

Sound input to the microphone triggers the IC monostable circuit which subsequently triggers an SCR, and hence the flash, after a time delay. This delay is adjustable—by varying the monostable on-time—from from 5 milliseconds to 200 milliseconds.

SOUND ACTIVATED STROBE TRIP

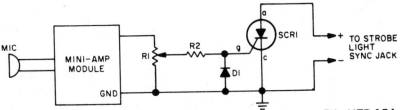

Fig. 62-5

D1—HEP-154 silicon rectifier
R1—5000-ohm potentiometer
R2—2700-ohm, ½-watt resistor
SCR1— silicon- controlled
rectifier
MIC.—Ceramic microphone

Circuit Notes

Take strobe-flash pictures the instant a pin pricks a balloon, a hammer breaks a lamp bulb or a bullet leaves a gun. Use a transistor amplifier of 1-watt rating or less. (It must have an output transformer.) The amplifier is terminated with a resistor on its highest output impedance, preferably 16 ohms. To test, darken room lights, open camera shutter, and break a lamp bulb with a hammer. The sound of the hammer striking the lamp will trigger the flash, and the picture will have been taken at that instant.

FLASH SLAVE DRIVER

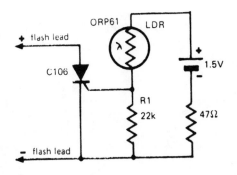

Fig. 62-6

Circuit Notes

In photography, a separate flash, triggered by the light of a master flash light, is often required to provide more light, fill-in shadows etc. The sensitivity of this circuit depends on the proximity of the master flash and the value of R1. Increasing R1 gives increased sensitivity.

REMOTE FLASH TRIGGER

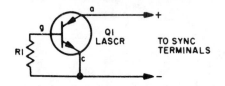

Q1—300-V light-activated silicon-
controlled rectifier (LASCR)
R1—47,000-ohm, ½-watt resistor

Fig. 62-7

Circuit Notes

Transistor Q1 is a light-activated silicon-controlled rectifier (LASCR). The gate is tripped by light entering a small lens built into the top cap. To operate, provide a 6-in. length of stiff wire for the anode and cathode connections and terminate the wires in a polarized power plug that matches the sync terminals on your electronic flashgun (strobelight). Make certain the anode lead connects to the positive sync terminal. When using the device, bend the connecting wires so the LASCR lens faces the main flash. This will fire the remote unit.

FLASH EXPOSURE METER

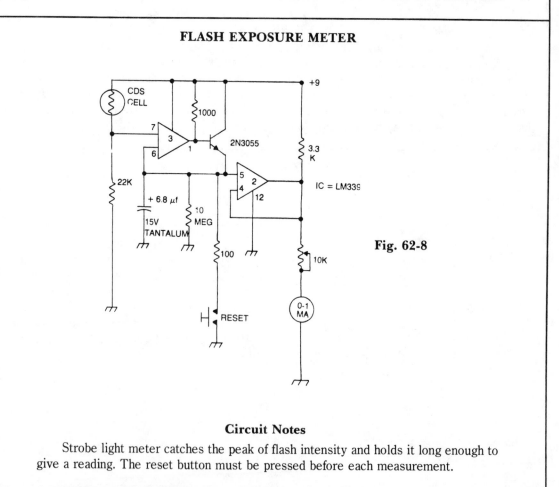

Fig. 62-8

Circuit Notes

Strobe light meter catches the peak of flash intensity and holds it long enough to give a reading. The reset button must be pressed before each measurement.

SHUTTER TESTER

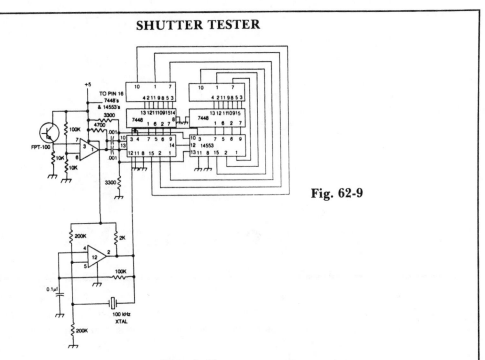

Fig. 62-9

Circuit Notes

Shutter speed tester combines frequency counter, crystal oscillator, and photo-transistor-operated gate generator. Oscillator pulses are counted as long as the shutter is open. Reset is automatic at the instant the shutter opens.

PHOTOGRAPHIC TIMER

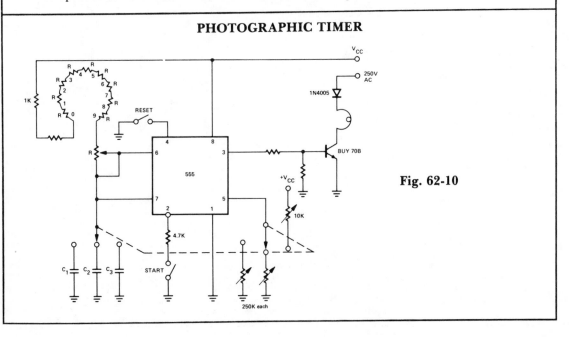

Fig. 62-10

485

63

Power Measuring Circuits

The sources of the following circuits are contained in the Sources section beginning on page 730. The figure number contained in the box of each circuit correlates to the source entry in the Sources section.

Extended Range VU Meter (Dot Mode)
Audio Power Meter

Audio Power Meter
Power Meter (1 kW Full Scale)

60 MHz Power Gain Test Circuit

EXTENDED RANGE VU METER (DOT MODE)

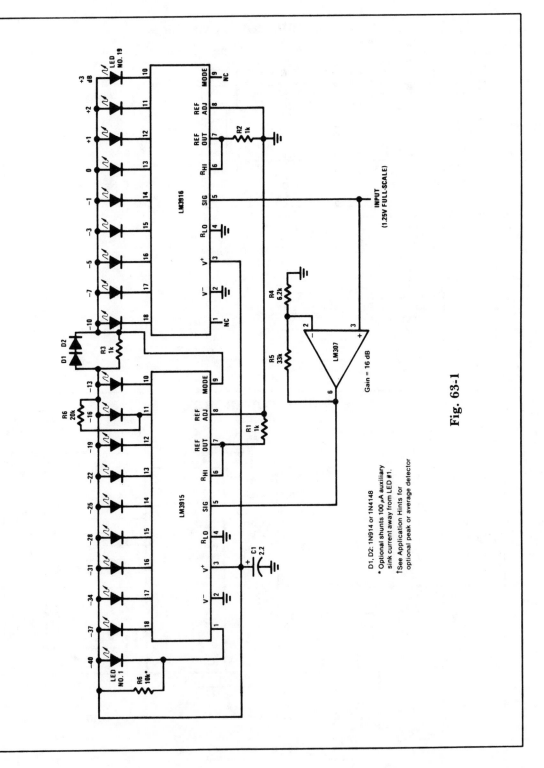

Fig. 63-1

D1, D2: 1N914 or 1N4148

* Optional shunts 100 μA auxiliary sink current away from LED #1.

†See Application Hints for optional peak or average detector

AUDIO POWER METER

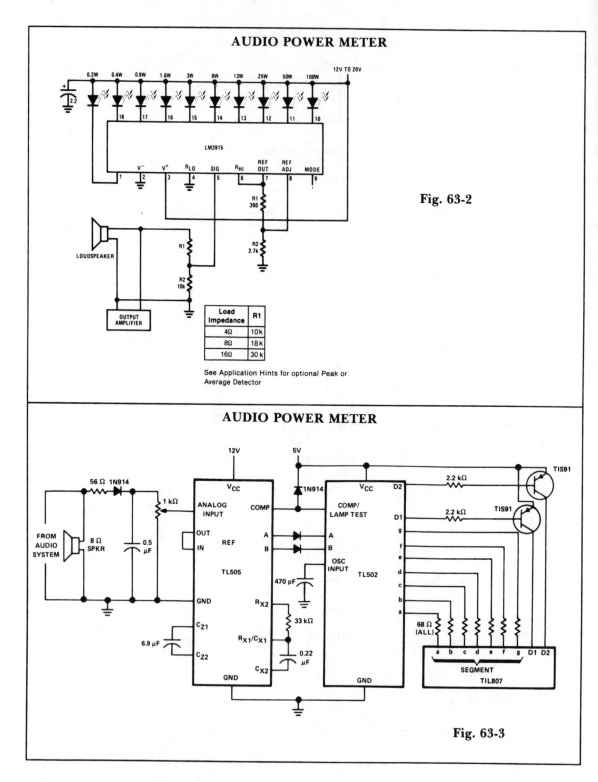

Fig. 63-2

Load Impedance	R1
4Ω	10k
8Ω	18k
16Ω	30k

See Application Hints for optional Peak or Average Detector

AUDIO POWER METER

Fig. 63-3

POWER METER (1 kW FULL SCALE)

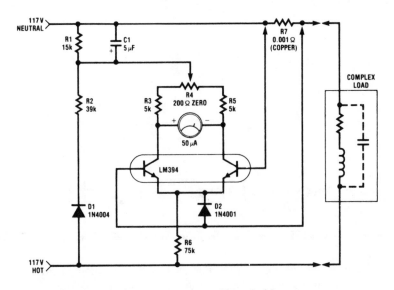

Fig. 63-4

Circuit Notes

The circuit is intended for 117 Vac ± 50 Vac operation, but can be easily modified for higher or lower voltages. It measures true (nonreactive) power being delivered to the load and requires no external power supply. Idling power drain is only 0.5 W. Load current sensing voltage is only 10 mV, keeping load voltage loss to 0.01%. Rejection of reactive load currents is better than 100:1 for linear loads. Nonlinearity is about 1% full scale when using a 50 μA meter movement.

60 MHz POWER GAIN TEST CIRCUIT

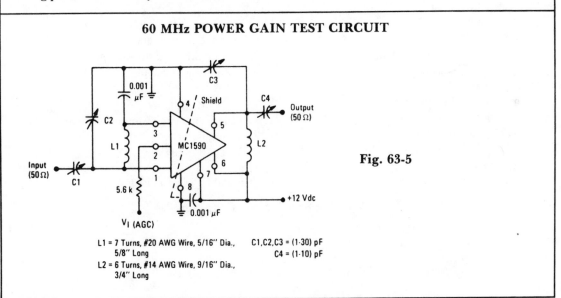

Fig. 63-5

L1 = 7 Turns, #20 AWG Wire, 5/16" Dia., 5/8" Long

L2 = 6 Turns, #14 AWG Wire, 9/16" Dia., 3/4" Long

C1,C2,C3 = (1-30) pF
C4 = (1-10) pF

64

Power Supplies (Fixed)

The sources of the following circuits are contained in the Sources section beginning on page 730. The figure number contained in the box of each circuit correlates to the source entry in the Sources section.

SWITCHING REGULATOR OPERATING AT 200 kHz

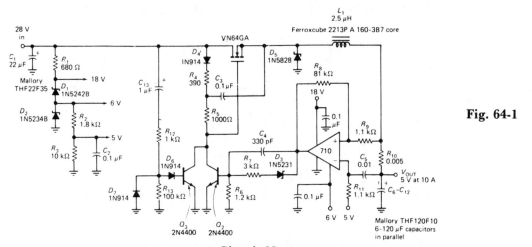

Fig. 64-1

Circuit Notes

This circuit provides a regulated dc with less than 100 mV of ripple for microprocessor applications. Necessary operating voltages are taken from the bleeder resistor network connected across the unregulated 28 V supply. The output of the LM710 comparator (actually an oscillator running at 200 kHz) is fed through a level-shifting circuit to the base of bipolar transistor Q2. This transistor is part of a bootstrap circuit necessary to turn the power MOSFET full on in totem-pole MOSFET arrays.

5 V, 0.5 A POWER SUPPLY

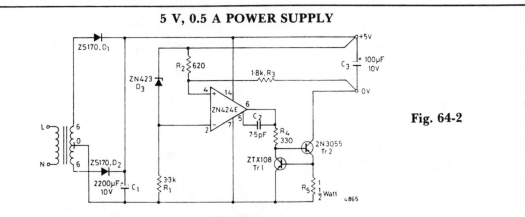

Fig. 64-2

Circuit Notes

The circuit is essentially a constant source modified by the feedback components R2 and R3 to give a constant voltage output. The output of the ZN424E need only be 2 volts above the negative rail, by placing the load in the collector of the output transistor Tr2. The current circuit is achieved by Tr1 and R5. This simple circuit has the following performance characteristics: Output noise and ripple (full load) = 1 mV rms. Load regulation (0 to 0.5 A) = 0.1%. Temperature coefficient = ± 100 ppm/°C. Current limit = 0.65 A.

3 W SWITCHING REGULATOR APPLICATION CIRCUIT

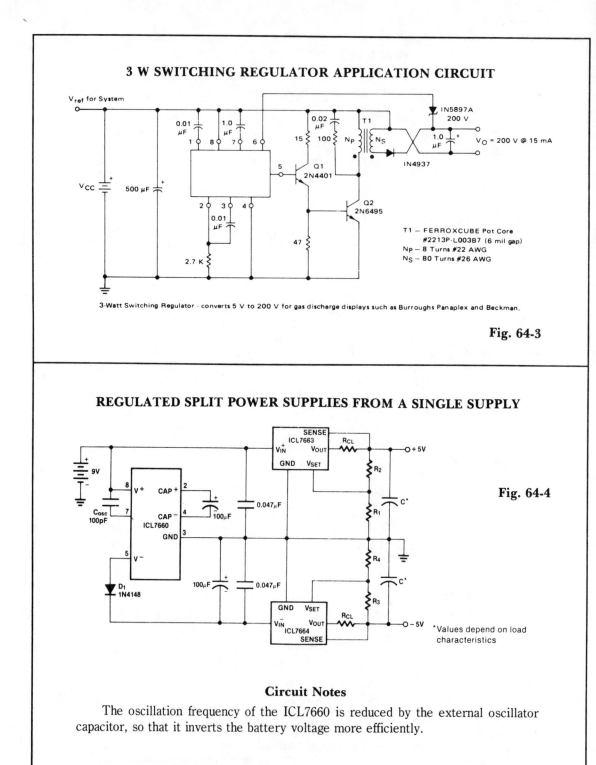

3-Watt Switching Regulator - converts 5 V to 200 V for gas discharge displays such as Burroughs Panaplex and Beckman.

Fig. 64-3

REGULATED SPLIT POWER SUPPLIES FROM A SINGLE SUPPLY

Fig. 64-4

*Values depend on load characteristics

Circuit Notes

The oscillation frequency of the ICL7660 is reduced by the external oscillator capacitor, so that it inverts the battery voltage more efficiently.

SWITCHING STEP-DOWN REGULATOR

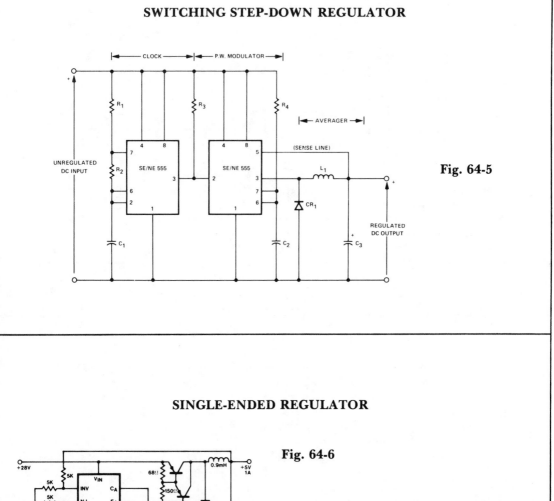

Fig. 64-5

SINGLE-ENDED REGULATOR

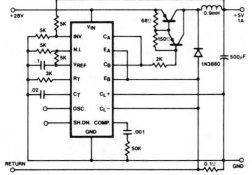

Fig. 64-6

Circuit Notes

In this conventional single-ended regulator circuit, the two outputs of the SG1524 are connected in parallel for effective 0-90% duty-cycle modulation. The use of an output inductor requires an RC phase compensation network for loop stability.

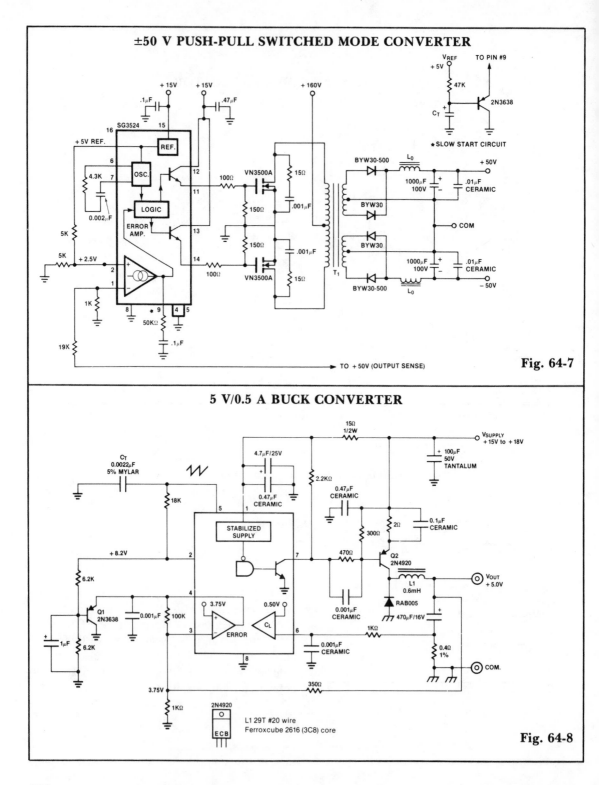

±50 V PUSH-PULL SWITCHED MODE CONVERTER

Fig. 64-7

5 V/0.5 A BUCK CONVERTER

Fig. 64-8

±50 V FEED FORWARD SWITCH MODE CONVERTER

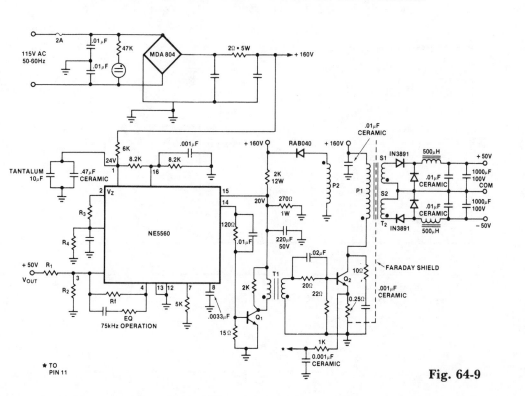

Fig. 64-9

TRAVELLER'S SHAVER ADAPTER

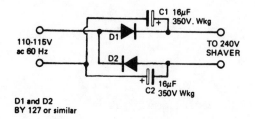

D1 and D2
BY 127 or similar

Fig. 64-10

Circuit Notes

Many countries have 115 volts mains supplies. This can be a problem if your electric shaver is designed for 220/240 volts only. This simple rectifier voltage doubler enables motor driven 240 volt shavers to be operated at full speed from a 115 volt supply. As the output voltage is dc, the circuit can only be used to drive small ac/dc motors. It cannot be used, for example, to operate vibrator-type shavers, or radios unless the latter are ac/dc operated.

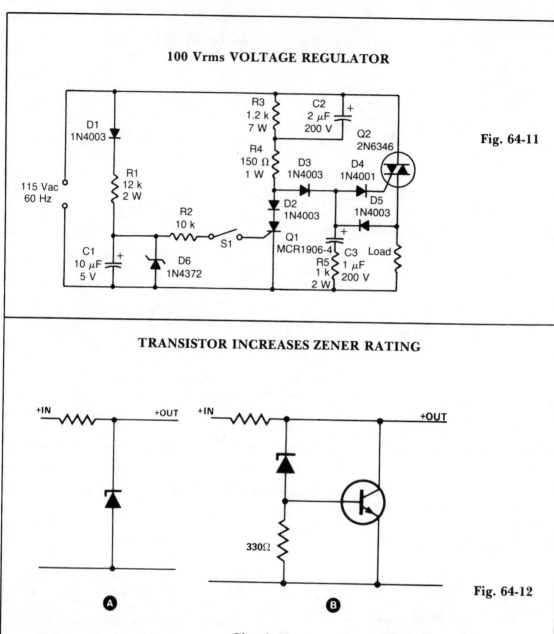

100 Vrms VOLTAGE REGULATOR

Fig. 64-11

TRANSISTOR INCREASES ZENER RATING

A

B

Fig. 64-12

Circuit Notes

The simple zener shunt in A may not handle sufficient current if the zener available is of low wattage. A power transistor will do most of the work for the zener as shown in B. Once the zener starts conducting, a bias voltage develops across the resistor (330 Ω to 1 K), turning on the transistor. The output voltage is 0.7 V greater than the zener voltage.

DUAL POLARITY POWER SUPPLY

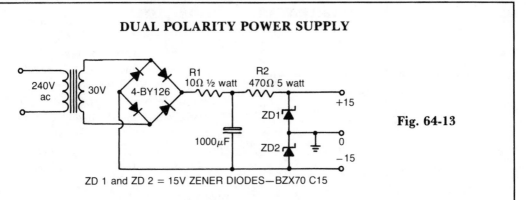

ZD 1 and ZD 2 = 15V ZENER DIODES—BZX70 C15

Fig. 64-13

Circuit Notes

This simple circuit gives a positive and negative supply from a single transformer winding and one full-wave bridge. Two zener diodes in series provide the voltage division and their centerpoint is grounded. (The filter capacitor must not be grounded via its case).

5.0 V/6.0 A 25 kHz SWITCHING REGULATOR WITH SEPARATE ULTRA-STABLE REFERENCE

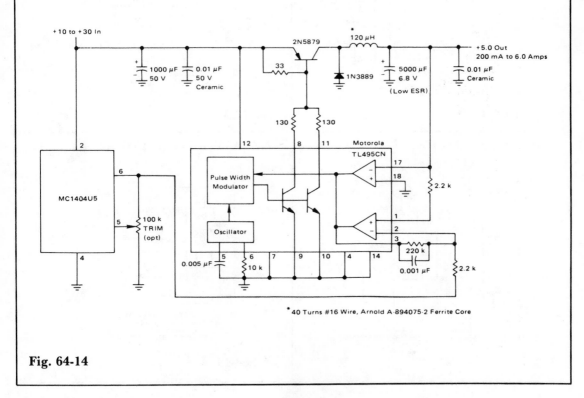

* 40 Turns #16 Wire, Arnold A-894075-2 Ferrite Core

Fig. 64-14

497

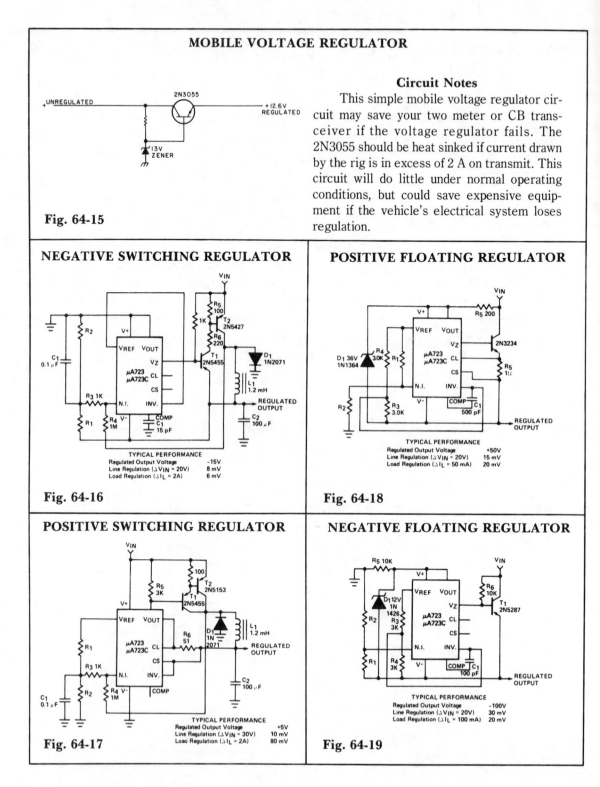

MOBILE VOLTAGE REGULATOR

Fig. 64-15

Circuit Notes

This simple mobile voltage regulator circuit may save your two meter or CB transceiver if the voltage regulator fails. The 2N3055 should be heat sinked if current drawn by the rig is in excess of 2 A on transmit. This circuit will do little under normal operating conditions, but could save expensive equipment if the vehicle's electrical system loses regulation.

NEGATIVE SWITCHING REGULATOR

TYPICAL PERFORMANCE

Regulated Output Voltage	-15V
Line Regulation (ΔV_{IN} = 20V)	8 mV
Load Regulation (ΔI_L = 2A)	6 mV

Fig. 64-16

POSITIVE FLOATING REGULATOR

TYPICAL PERFORMANCE

Regulated Output Voltage	+50V
Line Regulation (ΔV_{IN} = 20V)	15 mV
Load Regulation (ΔI_L = 50 mA)	20 mV

Fig. 64-18

POSITIVE SWITCHING REGULATOR

TYPICAL PERFORMANCE

Regulated Output Voltage	+5V
Line Regulation (ΔV_{IN} = 30V)	10 mV
Load Regulation (ΔI_L = 2A)	80 mV

Fig. 64-17

NEGATIVE FLOATING REGULATOR

TYPICAL PERFORMANCE

Regulated Output Voltage	-100V
Line Regulation (ΔV_{IN} = 20V)	30 mV
Load Regulation (ΔI_L = 100 mA)	20 mV

Fig. 64-19

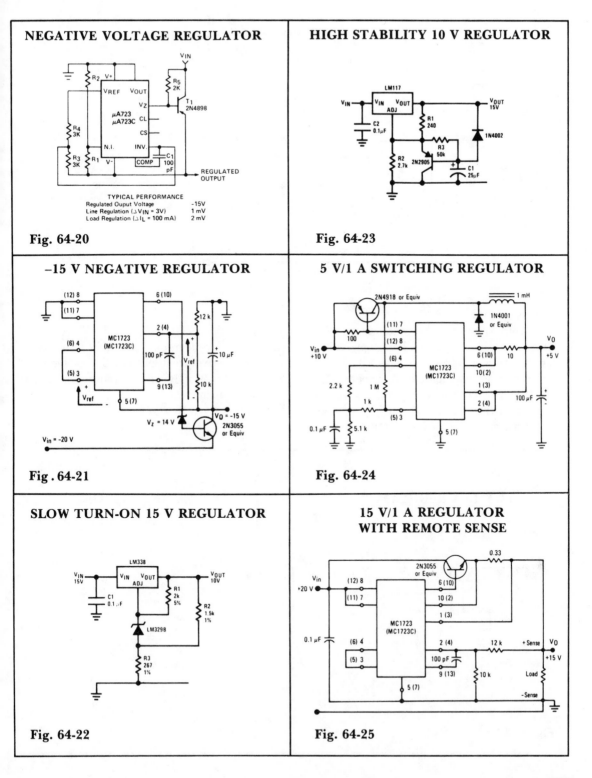

NEGATIVE VOLTAGE REGULATOR

TYPICAL PERFORMANCE

Regulated Ouput Voltage	-15V
Line Regulation (ΔV_{IN} = 3V)	1 mV
Load Regulation (ΔI_L = 100 mA)	2 mV

Fig. 64-20

HIGH STABILITY 10 V REGULATOR

Fig. 64-23

–15 V NEGATIVE REGULATOR

Fig . 64-21

5 V/1 A SWITCHING REGULATOR

Fig. 64-24

SLOW TURN-ON 15 V REGULATOR

Fig. 64-22

15 V/1 A REGULATOR WITH REMOTE SENSE

Fig. 64-25

499

LOW RIPPLE POWER SUPPLY

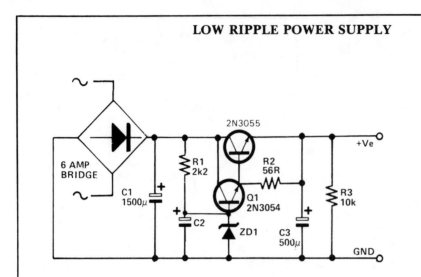

Fig. 64-26

Circuit Notes

This circuit may be used where a high current is required with a low ripple voltage (such as in a high powered class AB amplifier when high quality reproduction is necessary). Q1, Q2, and R2 may be regarded as a power darlington transistor. ZD1 and R1 provide a reference voltage at the base of Q1. ZD1 should be chosen thus: $ZD1 = V_{out} - 1.2$. C2 can be chosen for the degree of smoothness as its value is effectively multiplied by the combined gains of Q1/Q2, if 100 μF is chosen for C2, assuming minimum hfe for Q1 and Q2, C = 100 × 15(Q1) × 25(Q2) = 37,000 μF.

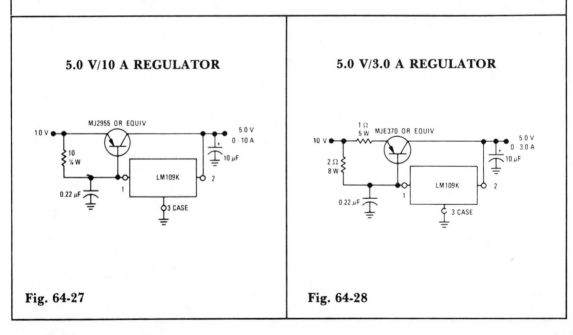

5.0 V/10 A REGULATOR

5.0 V/3.0 A REGULATOR

Fig. 64-27

Fig. 64-28

100 V/10.25 A SWITCH MODE CONVERTER

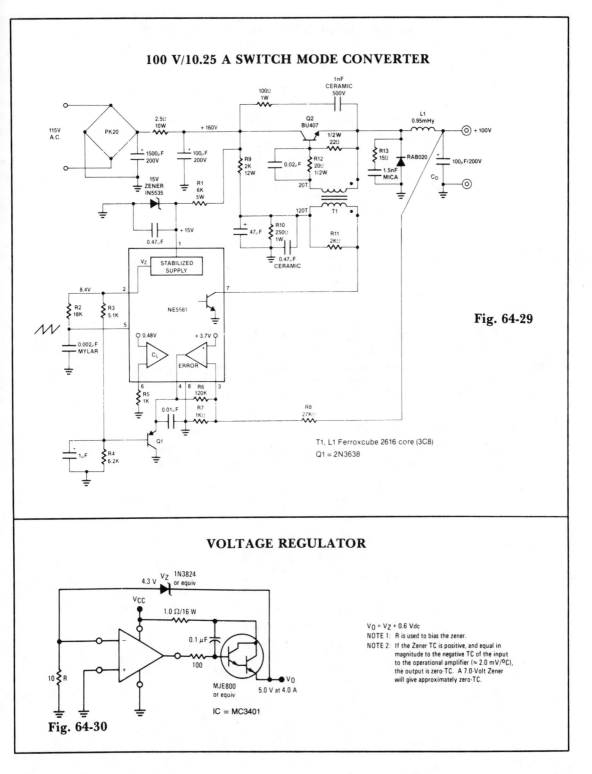

Fig. 64-29

T1, L1 Ferroxcube 2616 core (3C8)
Q1 = 2N3638

VOLTAGE REGULATOR

$V_O = V_Z + 0.6$ Vdc
NOTE 1: R is used to bias the zener.
NOTE 2: If the Zener TC is positive, and equal in
magnitude to the negative TC of the input
to the operational amplifier (≈ 2.0 mV/ºC),
the output is zero-TC. A 7.0-Volt Zener
will give approximately zero-TC.

Fig. 64-30

LOW VOLTAGE REGULATORS WITH SHORT CIRCUIT PROTECTION

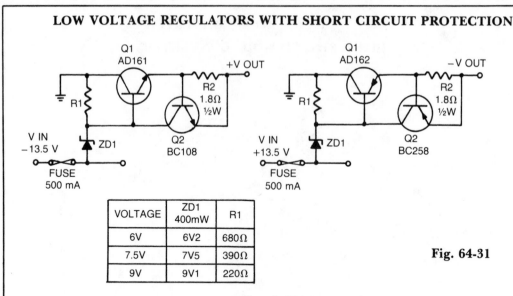

VOLTAGE	ZD1 400mW	R1
6V	6V2	680Ω
7.5V	7V5	390Ω
9V	9V1	220Ω

Fig. 64-31

Circuit Notes

These short-circuit protected regulators give 6, 7.5, and 9 V from an automobile battery supply of 13.5 V nominal; however, they will function just as well if connected to a smoothed dc output from a transformer/rectifier circuit. Two types are shown for both positive and negative ground systems. The power transistors can be mounted on the heatsink without a mica insulating spacer thus allowing for greater cooling efficiency. Both circuits are protected against overload or short-circuits. The current cannot exceed 330 mA. Under normal operating conditions the voltage across R2 does not rise above the 500 mV necessary to turn Q2 on and the circuit behaves as if there was only Q1 present. If excessive current is drawn, Q2 turns on and cuts off Q1, protecting the regulating transistor. The table gives the values of R1 for different zener voltages.

HIGH STABILITY 1 A REGULATOR

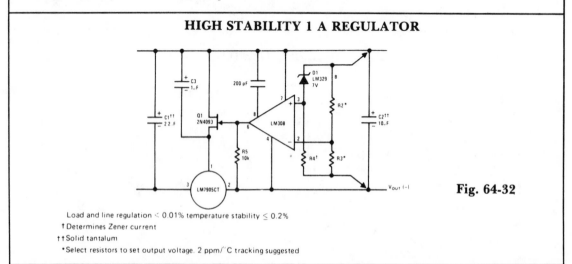

Fig. 64-32

Load and line regulation < 0.01% temperature stability ≤ 0.2%

† Determines Zener current

†† Solid tantalum

*Select resistors to set output voltage. 2 ppm/°C tracking suggested

100 V/0.25 A SWITCH MODE CONVERTER

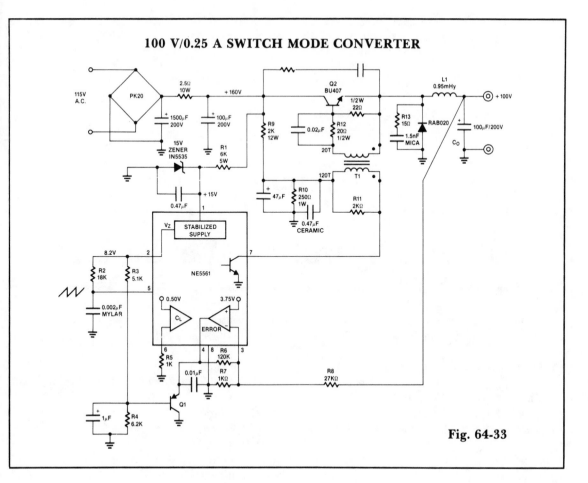

Fig. 64-33

503

65

Power Supplies (Variable)

The sources of the following circuits are contained in the Sources section beginning on page 730. The figure number contained in the box of each circuit correlates to the source entry in the Sources section.

DUAL OUTPUT BENCH POWER SUPPLY

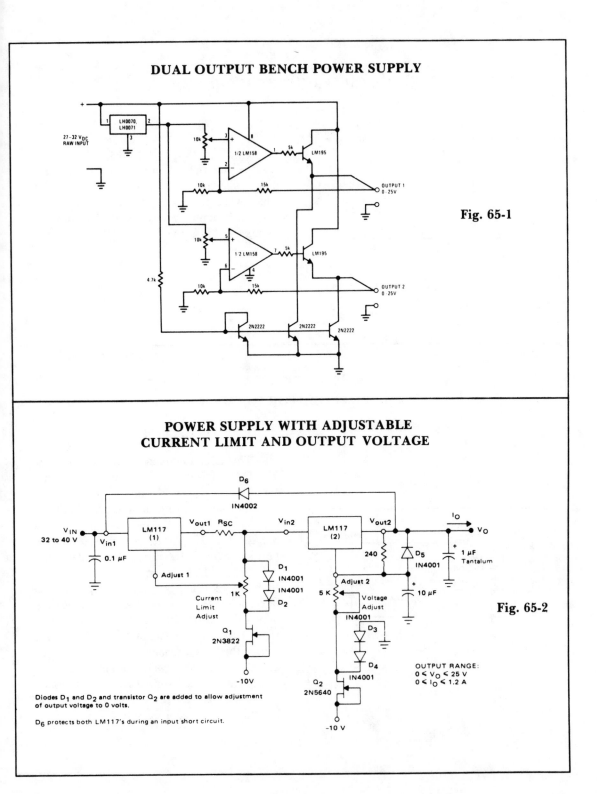

Fig. 65-1

POWER SUPPLY WITH ADJUSTABLE
CURRENT LIMIT AND OUTPUT VOLTAGE

Fig. 65-2

Diodes D₁ and D₂ and transistor Q₂ are added to allow adjustment of output voltage to 0 volts.

D₆ protects both LM117's during an input short circuit.

ADJUSTABLE OUTPUT REGULATOR

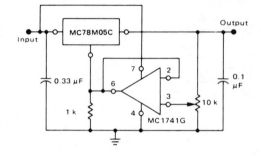

V_O, 7.0 V to 20 V
V_{IN} $V_O \geqslant 2.0$ V

Fig. 65-3

Circuit Notes

The addition of an operational amplifier allows adjustment to higher or intermediate values while retaining regulation characteristics. The minimum voltage obtainable with this arrangement is 2.0 volts greater than the regulator voltage.

RF PROBE FOR VTVM

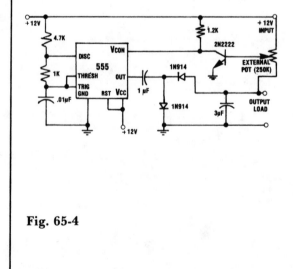

Fig. 65-4

Circuit Notes

This circuit combines a 555 timer with a 2N2222 transistor and an external potentiometer. The pot adjusts the output voltage to the desired value. To regulate the output voltage, the 2N2222 varies the control voltage of the 555 IC, increasing or decreasing the pulse repetition rate. A 1.2 K resistor is used as a collector load. The transistor base is driven from the external pot. If the output voltage becomes less negative, the control voltage moves closer to ground, causing the repetition rate of the 555 to increase, which, in turn, causes the 3 μF capacitor to charge more frequently. Output voltage for the circuit is 0 to 10 V, adjusted by the external pot. Output regulation is less than five percent for 0 to 10 mA and less than .05 percent for 0 to 0.2 mA.

REGULATED VOLTAGE DIVIDER

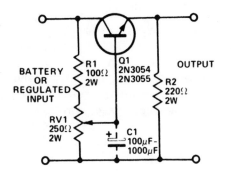

Fig. 65-5

Circuit Notes

ICs requiring 3.6 or 6 volts can be run from a battery or fixed regulated supply of a higher voltage by using the circuit shown. The transistor should be mounted on a heatsink as considerable power will be dissipated by its collector. Additional filtering can be obtained by fitting a capacitor (C1) as shown. The capacitance is effectively multiplied by the gain of the transistor. A ripple of 200 mV (peak to peak) at the input can be reduced to 2 mV in this fashion. Maximum output current depends on the supply rating and transistor type (with heatsink) used.

VARIABLE ZENER DIODE

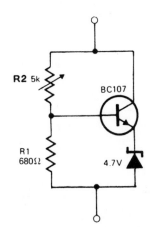

Fig. 65-6

Circuit Notes

The circuit behaves like a zener diode over a large range of voltages. The current passing through the voltage divider R1-R2 is substantially larger than the transistor base current and is in the region of 8 mA. The stabilizing voltage is adjustable over the range 5-45 V by changing the value of R2. The total current drawn by the circuit is variable over the range 15 mA to 50 mA. This value is determined by the maximum dissipation of the zener diode. In the case of a 250 mW device, this is of the order of 50 mA.

12 V TO 9, 7.5 or 6 V CONVERTER

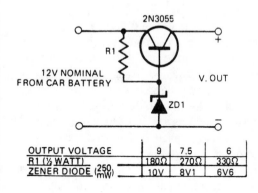

Circuit Notes

This circuit enables transistorized items such as radio, cassettes, and other electrical devices to be operated from a car's electrical supply. The table gives values for resistors and specified diode types for different voltage. Should more than one voltage be required a switching arrangement could be incorporated. For high currents, the transistor should be mounted on a heatsink.

OUTPUT VOLTAGE	9	7.5	6
R1 (½ WATT)	180Ω	270Ω	330Ω
ZENER DIODE (250 mW)	10V	8V1	6V6

Fig. 65-7

5 A CONSTANT VOLTAGE/CONSTANT CURRENT REGULATOR

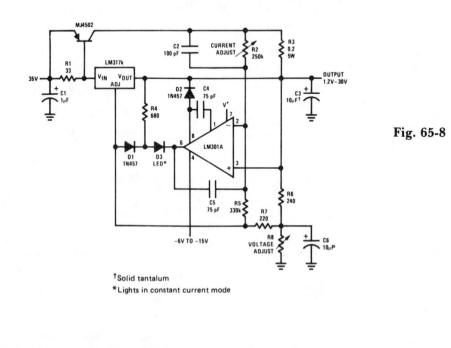

Fig. 65-8

†Solid tantalum
*Lights in constant current mode

POWER PACK FOR BATTERY-POWERED CALCULATORS, RADIOS, OR CASSETTE PLAYERS

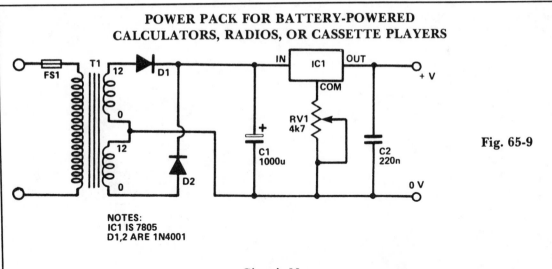

Fig. 65-9

NOTES:
IC1 IS 7805
D1,2 ARE 1N4001

Circuit Notes

This circuit gives a regulated output of between 5 V and 15 Vdc, adjusted and set by a preset resistor. Current output up to about 350 mA. An integrated circuit regulates the output voltage and although this IC (the 7805) is normally used in a fixed-voltage (5 Vdc) supply it is for a variable output voltage.

PRECISION HIGH VOLTAGE REGULATOR

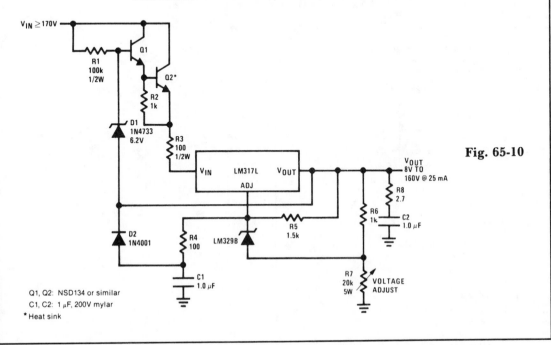

Fig. 65-10

Q1, Q2: NSD134 or similar
C1, C2: 1 μF, 200V mylar
* Heat sink

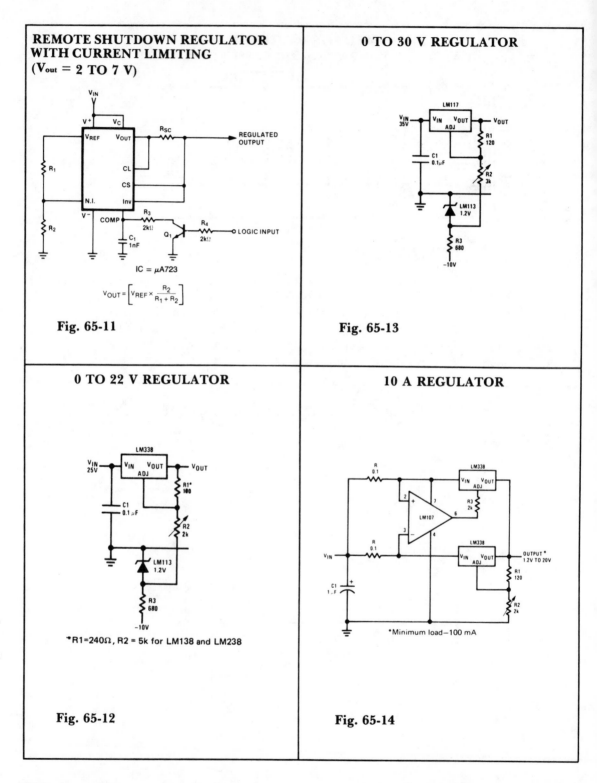

REMOTE SHUTDOWN REGULATOR WITH CURRENT LIMITING
(V_{out} = 2 TO 7 V)

IC = μA723

$$V_{OUT} = \left[V_{REF} \times \frac{R_2}{R_1 + R_2} \right]$$

Fig. 65-11

0 TO 30 V REGULATOR

Fig. 65-13

0 TO 22 V REGULATOR

*R1=240Ω, R2 = 5k for LM138 and LM238

Fig. 65-12

10 A REGULATOR

*Minimum load—100 mA

Fig. 65-14

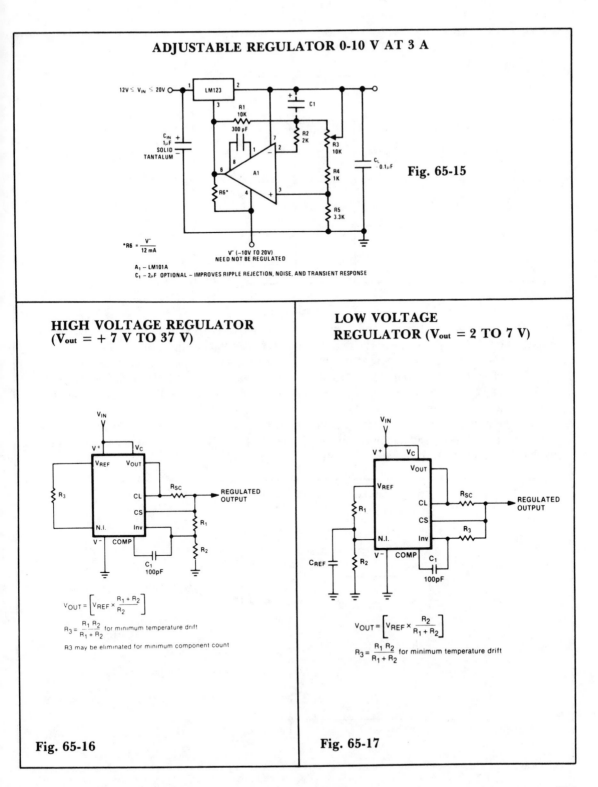

ADJUSTABLE REGULATOR 0-10 V AT 3 A

$12V \le V_{IN} \le 20V$

LM123

C1

R1 10K

300 pF

C_{IN} 1μF SOLID TANTALUM

R2 2K

R3 10K

C_L 0.1μF

A1

R4 1K

R5 3.3K

R6*

$*R6 = \dfrac{V^-}{12\,mA}$

V^- (−10V TO 20V) NEED NOT BE REGULATED

A₁ – LM101A

C₁ – 2μF OPTIONAL – IMPROVES RIPPLE REJECTION, NOISE, AND TRANSIENT RESPONSE

Fig. 65-15

HIGH VOLTAGE REGULATOR
($V_{out} = +7$ V TO 37 V)

V_{IN}

V^+ V_C

V_{REF} V_{OUT}

R_3

CL R_{SC} REGULATED OUTPUT

CS

N.I.   Inv R_1

V^- COMP

C_1 100pF R_2

$$V_{OUT} = \left[V_{REF} \times \frac{R_1 + R_2}{R_2} \right]$$

$R_3 = \dfrac{R_1 R_2}{R_1 + R_2}$ for minimum temperature drift

R3 may be eliminated for minimum component count

Fig. 65-16

LOW VOLTAGE REGULATOR ($V_{out} = 2$ TO 7 V)

V_{IN}

V^+ V_C

V_{OUT}

V_{REF}

CL R_{SC} REGULATED OUTPUT

R_1 CS

N.I.   Inv R_3

C_{REF} R_2 V^- COMP C_1 100pF

$$V_{OUT} = \left[V_{REF} \times \frac{R_2}{R_1 + R_2} \right]$$

$R_3 = \dfrac{R_1 R_2}{R_1 + R_2}$ for minimum temperature drift

Fig. 65-17

SIMPLE SPLIT POWER SUPPLY

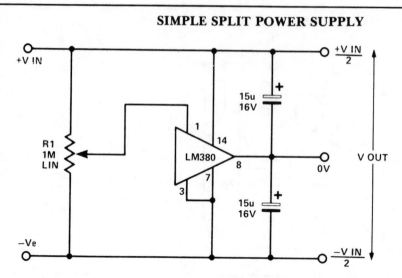

Fig. 65-18

Circuit Notes

This circuit utilizes the quasi-complementary output stage of the popular LM380 audio power IC. The device is internally biased so that with no input the output is held midway between the supply rails. R1, which should be initially set to mid-travel, is used to nullify any inbalance in the output. Regulation of V_{out} depends upon the circuit feeding the LM380, but positive and negative outputs will track accurately irrespective of input regulation and unbalanced loads. The free-air dissipation is a little over 1 watt, and so extra cooling may be required. The device is fully protected and will go into thermal shutdown if its rated dissipation is exceeded. Current limiting occurs if the output current exceeds 1.3 A. The input voltage should not exceed 20 V.

ADJUSTABLE OUTPUT REGULATOR

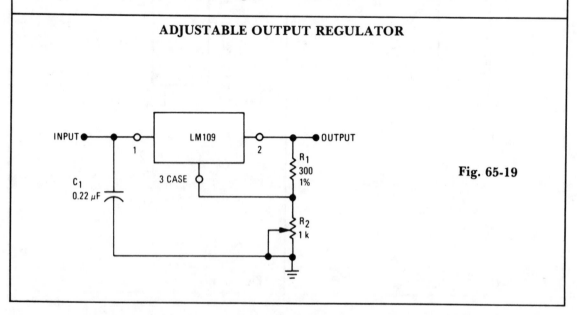

Fig. 65-19

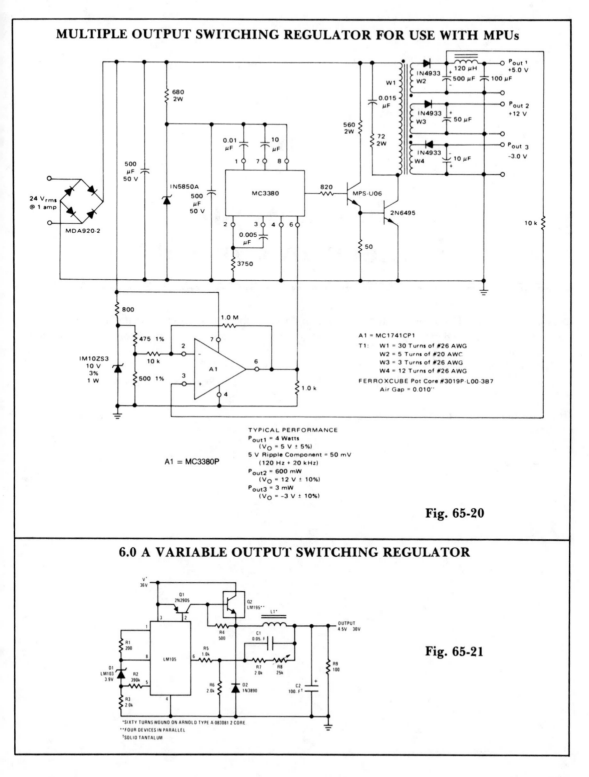

MULTIPLE OUTPUT SWITCHING REGULATOR FOR USE WITH MPUs

A1 = MC1741CP1
T1: W1 = 30 Turns of #26 AWG
     W2 = 5 Turns of #20 AWG
     W3 = 3 Turns of #26 AWG
     W4 = 12 Turns of #26 AWG
FERROXCUBE Pot Core #3019P-L00-3B7
     Air Gap = 0.010"

A1 = MC3380P

TYPICAL PERFORMANCE
P_{out1} = 4 Watts
     (V_O = 5 V ± 5%)
5 V Ripple Component = 50 mV
     (120 Hz + 20 kHz)
P_{out2} = 600 mW
     (V_O = 12 V ± 10%)
P_{out3} = 3 mW
     (V_O = –3 V ± 10%)

Fig. 65-20

6.0 A VARIABLE OUTPUT SWITCHING REGULATOR

*SIXTY TURNS WOUND ON ARNOLD TYPE A-083081-2 CORE
**FOUR DEVICES IN PARALLEL
†SOLID TANTALUM

Fig. 65-21

513

66

Power Supply Protection Circuits

The sources of the following circuits are contained in the Sources section beginning on page 730. The figure number contained in the box of each circuit correlates to the source entry in the Sources section.

ELECTRONIC CROWBAR FOR AC OR DC LINES

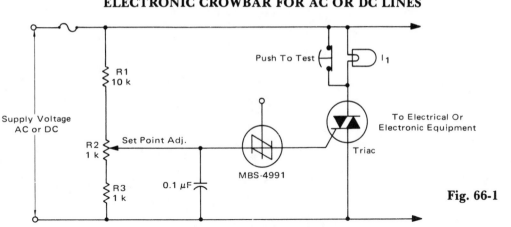

Fig. 66-1

Circuit Notes

For positive protection of electrical or electronic equipment, use this against excessive supply voltage. Due to improper switching, wiring, short circuits, or failure of regulators, an electronic crowbar circuit can quickly place a short circuit across the power lines, thereby dropping the voltage across the protected device to near zero and blowing a fuse. The triac and SBS are both bilateral devices, the circuit is equally useful on ac or dc supply lines. With the values shown for R1, R2, and R3, the crowbar operating point can be adjusted over the range of 60 to 120 volts dc or 42 to 84 volts ac. The resistor values can be changed to cover a different range of supply voltages. The voltage rating of the triac must be greater than the highest operating point as set by R2. I1 is a low power incandescent lamp with a voltage rating equal to the supply voltage. It may be used to check the set point and operation of the unit by opening the test switch and adjusting the input or set point to fire the SBS. An alarm unit such as the Mallory Sonalert may be connected across the fuse to provide an audible indication of crowbar action. (This circuit may not act on short, infrequent power line transients).

POWER PROTECTION CIRCUIT

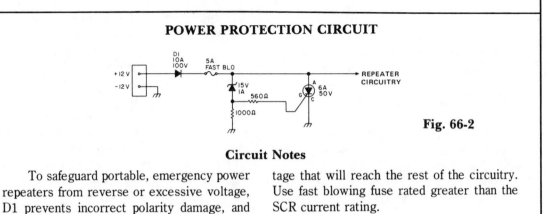

Fig. 66-2

Circuit Notes

To safeguard portable, emergency power repeaters from reverse or excessive voltage, D1 prevents incorrect polarity damage, and zener voltage determines the maximum voltage that will reach the rest of the circuitry. Use fast blowing fuse rated greater than the SCR current rating.

SIMPLE CROWBAR

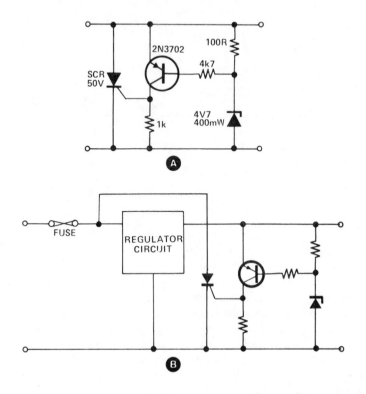

Fig. 66-3

Circuit Notes

These circuits provide overvoltage protection in case of voltage regulator failure or application of an external voltage. Intended to be used with a supply offering some form of short circuit protection, either foldback, current limiting, or a simple fuse. The most likely application is a 5 V logic supply, since TTL is easily damaged by excess voltage. The values chosen in A are for a 5 V supply, although any supply up to about 25 V can be protected by simply choosing the appropriate zener diode. When the supply voltage exceeds the zener voltage +0.7 V, the transistor turns on and fires the thyristor. This shorts out the supply, and prevents the voltage rising any further. In the case of a supply with only fuse protection, it is better to connect the thyristor the regulator circuit when the crowbar operates. The thyristor should have a current rating about twice the expected short circuit current and a maximum voltage greater than the supply voltage. The circuit can be reset by either switching off the supply, or by breaking the thyristor circuit with a switch.

OVERVOLTAGE PROTECTION WITH AUTOMATIC RESET

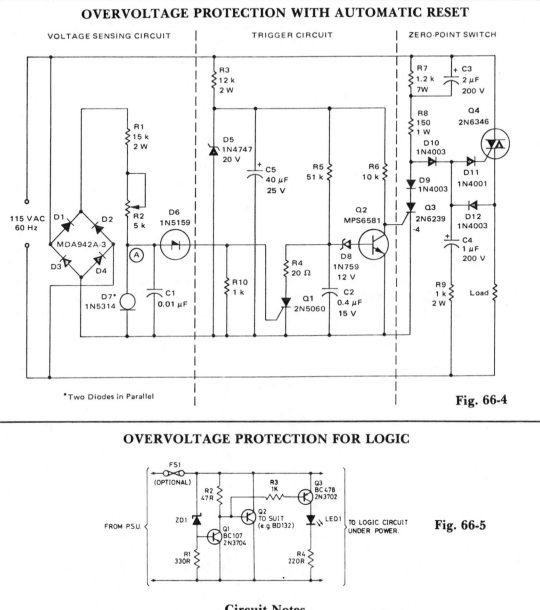

VOLTAGE SENSING CIRCUIT | TRIGGER CIRCUIT | ZERO-POINT SWITCH

*Two Diodes in Parallel

Fig. 66-4

OVERVOLTAGE PROTECTION FOR LOGIC

Fig. 66-5

Circuit Notes

Zener diode ZD1 senses the supply, and should the supply rise above 6 V, Q1 will turn on. In turn, Q2 conducts clamping the rail. Subsequent events depend on the source supply. It will either shut down, go into current limit or blow its supply fuse. None of these will damage the TTL chips. The rating of Q2 depends on the source supply, and whether it will be required to operate continuously in the event of failure. Its current rating has to be in excess of the source supply.

FAST ACTING POWER SUPPLY PROTECTION

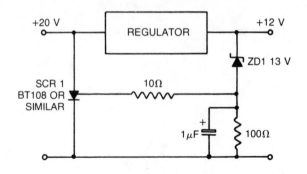

Fig. 66-6

Circuit Notes

When using a regulated power supply to reduce a supply voltage, there is always the danger that component failure in the power supply might lead to a severe overvoltage condition across the load. To cope with overvoltage situations, the circuit is designed to protect the load under overvoltage conditions. Component values given are for a 20 V supply with regulated output at 12 V. The zener diode can be changed according to whatever voltage is to be the maximum. If the voltage at the regulator output rises to 13 V or above, the zener diode breaks down and triggers the thyristor which shorts out the supply line and blows the main fuse.

5 V CROWBAR

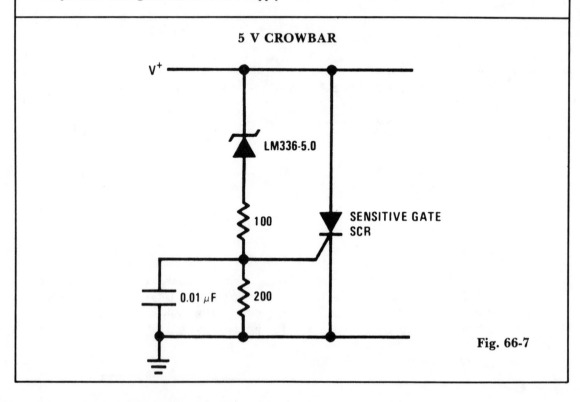

Fig. 66-7

518

67

Probes

The sources of the following circuits are contained in the Sources section beginning on page 730. The figure number contained in the box of each circuit correlates to the source entry in the Sources section.

Logic Probe Yields Three Discrete States
Signal Injector/Tracer
Injector/Tracer
CMOS Logic Probe
RF Probe for VOM
100 K Megohm DC Probe

Audible TTL Probe
Logic Probe
Logic Test Probe with Memory
Logic Probe
Simple Logic Probe
Audio-RF Signal Tracer Probe

TTL Logic Tester

LOGIC PROBE YIELDS THREE DISCRETE STATES

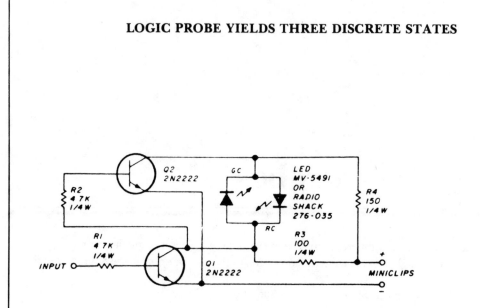

Fig. 67-1

Circuit Notes

The circuit uses a dual LED. When power is applied to the probe through the power leads, and the input is touched to a low level or ground, Q1 is cut off. This will cause Q2 to conduct since the base is positive with respect to the emitter. With Q1 cut off and Q2 conducting, the green diode of the dual LED will be forward biased, yielding a green output. Touching the probe tip to a high level will cause Q1 and Q2 to complement, and the red diode will be forward biased, yielding a red output from the LED. An alternating signal will cause alternating conduction of the red and green diodes and will yield an indication approximately amber. In this manner, both static and dynamic signals can be traced with the logic probe.

SIGNAL INJECTOR/TRACER

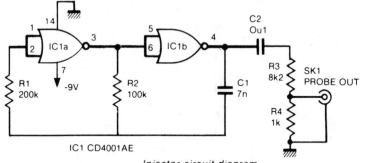

Injector circuit diagram.

Fig. 67-2

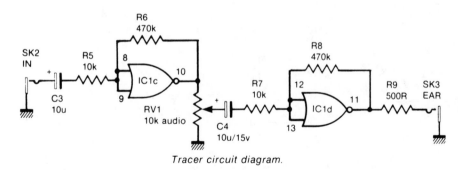

Tracer circuit diagram.

Circuit Notes

The injector is a CMOS oscillator with period approximately equal to $1.4 \times C1 \times R2$ seconds. The values are given for 1 kHz operation. Resistors R3 and R4 divide the output to 1 V. Whereas the oscillator employs the gates in their digital mode, the tracer used them in a linear fashion by applying negative feedback from output to input. They are used in much the same way as op amps. The circuit uses positive ground. It offers an advantage at the earphone output because one side of the earphone must be connected to ground via the case. Use of a positive ground allows the phone to be driven by the two N-channel transistors inside the CD4001 which are arranged in parallel and are thus able to handle more current for better volume.

INJECTOR/TRACER

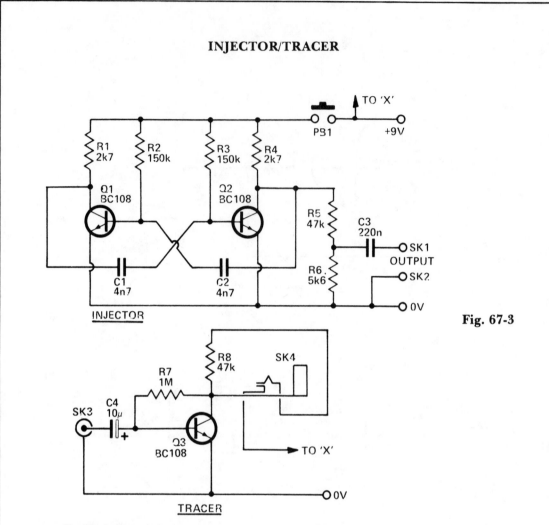

Fig. 67-3

The circuit diagrams for both parts of the injector/tracer. Note that SK4 is used to apply power to the amplifier section.

Circuit Notes

The unit has a separate amplifier and oscillator section allowing them to be used separately if need be. The injector is a multivibrator running at 1 kHz, with R5 and R6 dividing down the output to a suitable level ($\approx$1 V). The tracer is a single-stage amplifier that drives the high impedance earpiece. C4 decouples the input.

CMOS LOGIC PROBE

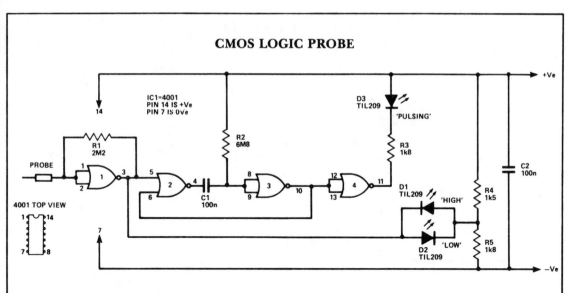

Fig. 67-4

Circuit Notes

The logic probe can indicate four input states, as follows: floating input—all LEDs off; logic 0 input—D2 switched on (D3 will briefly flash on); logic 1 input—D1 switched on; pulsing input—D3 switched on, or pulsing in the case of a low frequency input signal (one or both of the other indicators will switch on, showing if one input state predominates).

RF PROBE FOR VOM

PARTS LIST FOR RF PROBE FOR VOM

C1—500-pF, 400-VDC capacitor
C2—0.001-uF, disc capacitor
D1—1N4149 diode
R1—15,000-ohm, ½-watt resistor

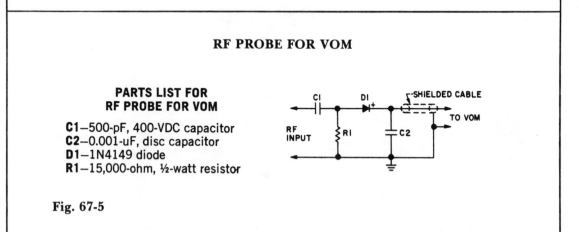

Fig. 67-5

Circuit Notes

This probe makes possible relative measurements of rf voltages to 200 MHz on a 20,000 ohms-per-volt multimeter. Rf voltage must not exceed the breakdown rating of the 1N4149—approximately 100 V.

100 K MEGOHM DC PROBE

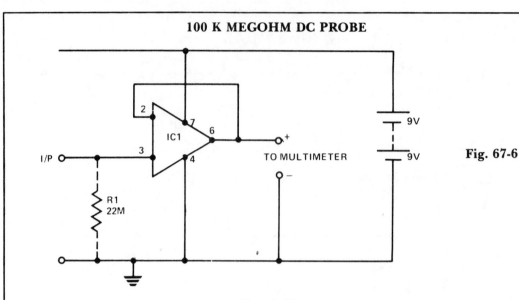

Fig. 67-6

Circuit Notes

A 741 op amp is used with 100% ac and dc feedback to provide a typical input impedance of 10^{11} ohm and unity gain. To avoid hum and rf pickup the input leads should be kept as short as possible and the circuit should be mounted in a small grounded case. Output leads may be long since the output impedance of the circuit is a fraction of an ohm. With no input the output level is indeterminate. Including R1 in the circuit through lowers the input impedance to 22 M.

AUDIBLE TTL PROBE

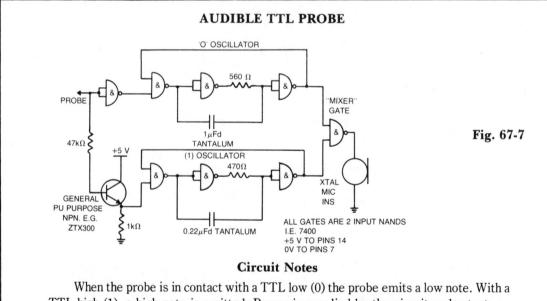

Fig. 67-7

Circuit Notes

When the probe is in contact with a TTL low (0) the probe emits a low note. With a TTL high (1), a high note is emitted. Power is supplied by the circuit under test.

LOGIC PROBE

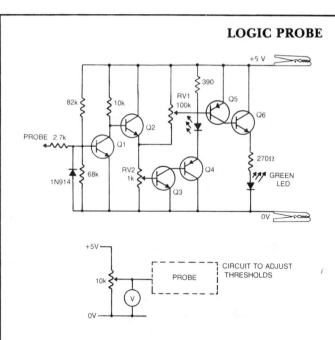

Fig. 67-8

Circuit Notes

Transistors Q1 and Q2 form a buffer, providing the probe with a reasonable input impedance. Q3 and Q4 form a level detecting circuit. As the voltage across the base-emitter junction of the Q3 rises above 0.6 V the transistor turns on thus turning on Q4 and lighting the red (high) LED. Q5 and Q6 perform the same function but for the green (low) LED. Q1, Q4, Q5 are all PNP general purpose silicon transistors (BC178 etc). Q2, Q3, Q6 are all PNP general purpose silicon transistors (BC 108 etc.) The threshold low is ≤ 0.8 V, and the threshold high is ≥ 2.4 V.

LOGIC TEST PROBE WITH MEMORY

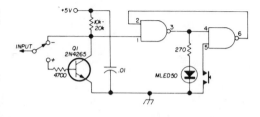

Fig. 67-9

Circuit Notes

There are two switches: a memory disable switch and a pulse polarity switch. Memory disable is a push-button that resets the memory to the low state when depressed. Pulse polarity is a toggle switch that selects whether the probe responds to a high-level or pulse (+5 V) or a low-level or pulse (ground). (Use IC logic of the same type as is being tested).

LOGIC PROBE

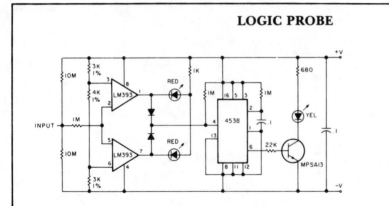

Fig. 67-10

Circuit Notes

The probe indicates a high or low at 70% and 30% of V+ (5 to 12 V). One section of the voltage comparator (LM393) senses V in over 70% of supply and the second section senses V in under 30%. These two sections direct-drive the appropriate LEDs. The pulse detector is a CMOS oneshot (MC14538) triggered on the rising edge of the LM393 outputs through 1N4148 diodes. With the RC values shown, it triggered reliably at greater than 30 kHz on both sine and square waves.

SIMPLE LOGIC PROBE

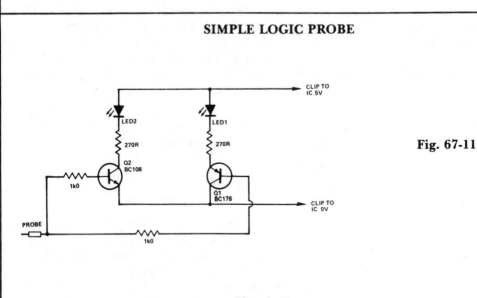

Fig. 67-11

Circuit Notes

If the probe is connected to logic 0, Q1 will be turned on lighting D1. At logic 1, Q2 will be turned on lighting D2. For Q1 and Q2 any NPN or PNP transistors will do. Similarly, D1 and D2 can be any LEDs.

AUDIO-RF SIGNAL TRACER PROBE

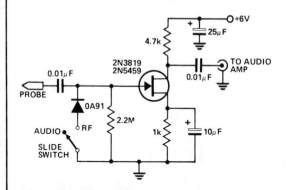

Circuit Notes

This economical signal tracer is useful for servicing and alignment work in receivers and low power transmitters. When switched to RF, the modulation on any signal is detected by the diode and amplified by the FET. A twin-core shielded lead can be used to connect the probe to an amplifier and to feed 6 volts to it.

Fig. 67-12

TTL LOGIC TESTER

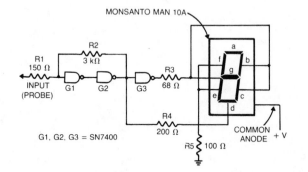

Fig. 67-13

Circuit Notes

Gates G1 and G2 together with resistors R1 and R2 form a simple voltage monitor that has a trip point of 1.4 volts. Gate G3 is simply an inverter. The display section of the tester consists of a common anode alphanumeric LED and current-limiting resistors. It indicates whether the input voltage is above or below 1.4 V, and displays a H or a L (for high or low logic-level) respectively.

68

Pulse Generators

The sources of the following circuits are contained in the Sources section beginning on page 730. The figure number contained in the box of each circuit correlates to the source entry in the Sources section.

Pulse Generator
Single Op Amp Oscillator
Programmable Pulse Generator
Unijunction Transistor Pulse Generators
Pulse Generator

Pulse Generator
Free-Running Oscillator
Pulse Generator with 25% Duty Cycle
Pulse Generator
555 Timer Oscillator

Versatile Two-Phase Pulse Generator

PULSE GENERATOR

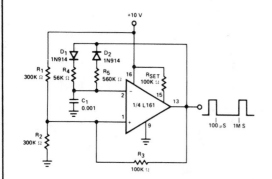

Fig. 68-1

Circuit Notes

The duty cycle of the output pulse is equal to R4/(R4 + R5) × 100%. For duty cycles of less than 50%, D1 can be eliminated and R2 raised according to the following formula:

$$R4(actual) = \frac{R5 \times R4(eff)}{R5 - R4(eff)}$$

R4(eff) is the effective value of R4 in the circuit and R4(actual) is the actual value used; R4(actual) will always be larger than R4(eff).

SINGLE OP AMP OSCILLATOR

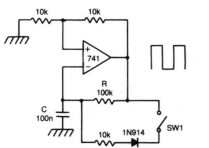

Circuit Notes

This circuit has a Schmitt trigger and integrator built around one op amp. Timing is controlled by the RC network. Voltage at the inverting input follows the RC charging exponential within the upper and lower hysteresis levels. By closing the switch SW1, the discharge time of the capacitor becomes ten times as fast as the rise time. Thus a square wave with an 10:1 mark space ratio is generated.

Fig. 68-2

PROGRAMMABLE PULSE GENERATOR

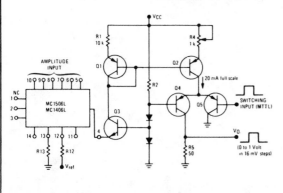

Circuit Notes

Fast rise and fall times require the use of high speed switching transistors for the differential pair, Q4 and Q5. Linear ramps and sine waves may be generated by the appropriate reference input.

Fig. 68-3

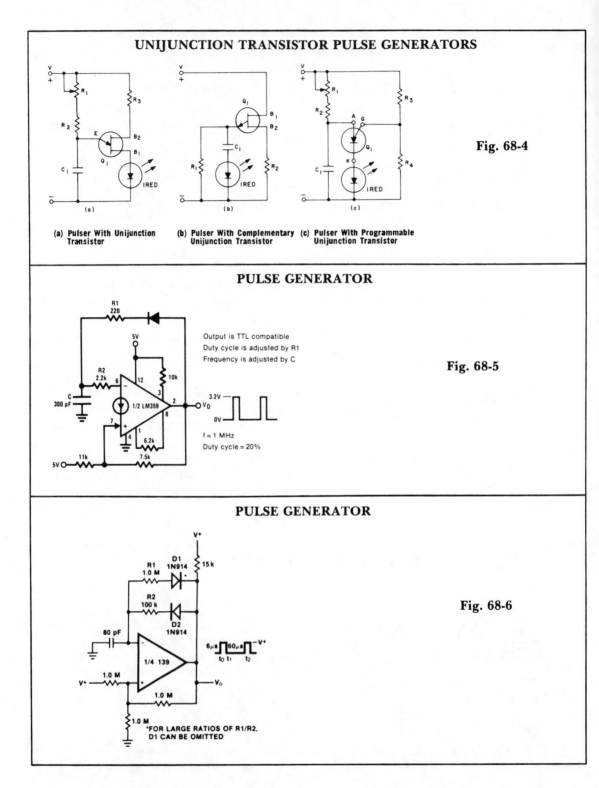

UNIJUNCTION TRANSISTOR PULSE GENERATORS

Fig. 68-4

(a) Pulser With Unijunction Transistor

(b) Pulser With Complementary Unijunction Transistor

(c) Pulser With Programmable Unijunction Transistor

PULSE GENERATOR

Output is TTL compatible
Duty cycle is adjusted by R1
Frequency is adjusted by C

f = 1 MHz
Duty cycle = 20%

Fig. 68-5

PULSE GENERATOR

*FOR LARGE RATIOS OF R1/R2, D1 CAN BE OMITTED

Fig. 68-6

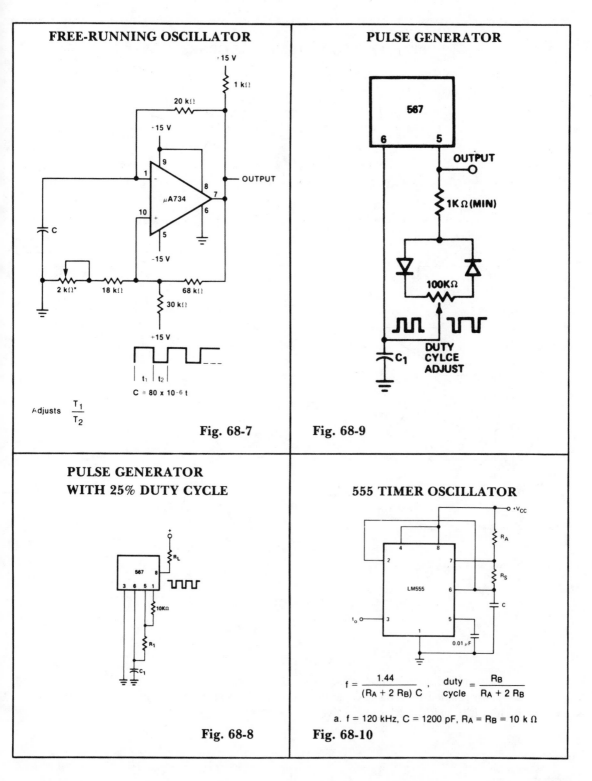

FREE-RUNNING OSCILLATOR

+15 V

1 kΩ

20 kΩ

+15 V

9

μA734

1

−

8

7

OUTPUT

10

+

6

5

−15 V

C

2 kΩ*

18 kΩ

68 kΩ

30 kΩ

+15 V

t₁ t₂

$C \cong 80 \times 10^{-6} t$

Adjusts $\dfrac{T_1}{T_2}$

Fig. 68-7

PULSE GENERATOR

567

6   5

OUTPUT

1KΩ (MIN)

100KΩ

C_1

**DUTY
CYLCE
ADJUST**

Fig. 68-9

PULSE GENERATOR
WITH 25% DUTY CYCLE

R_L

567

3 6 5 1   8

10KΩ

R₁

C₁

Fig. 68-8

555 TIMER OSCILLATOR

+V_CC

R_A

2   4   8

7

LM555

R_S

C

6

I_o

3

5

1

0.01 μF

$$f = \frac{1.44}{(R_A + 2 R_B)\,C} \quad , \quad \frac{duty}{cycle} = \frac{R_B}{R_A + 2 R_B}$$

a. f = 120 kHz, C = 1200 pF, $R_A = R_B = 10\ k\Omega$

Fig. 68-10

VERSATILE TWO-PHASE PULSE GENERATOR

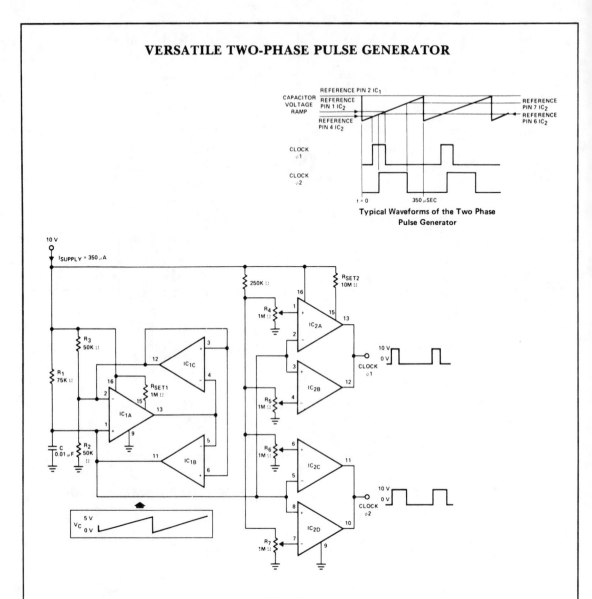

Fig. 68-11

Circuit Notes

Two-phase clock generator uses two L161s to generate pulses of adjustable widths and phase relationships. Ramp generator feeds two variable window comparators formed by IC_{2A}-IC_{2B} and IC_{2C}-IC_{2D} respectively.

69

Radiation Detectors

The sources of the following circuits are contained in the Sources section beginning on page 730. The figure number contained in the box of each circuit correlates to the source entry in the Sources section.

Dosage-Rate Meter
Wideband Radiation Monitor
Gamma Ray Pulse Integrator

Sensitive Geiger Counter
Geiger Counter
Nuclear Particle Detector

DOSAGE-RATE METER

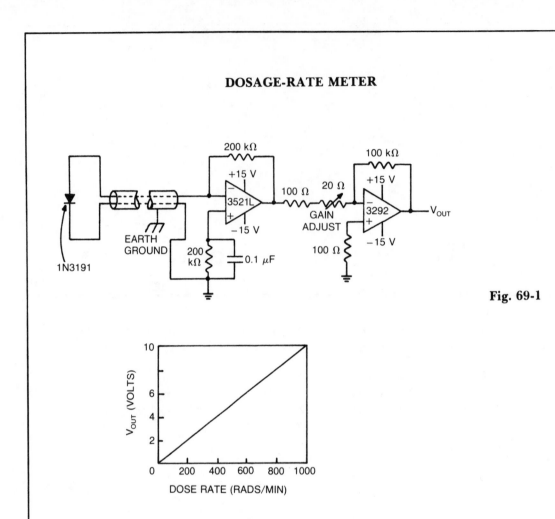

Fig. 69-1

Circuit Notes

A commercial diode is the detector in this highly accurate radiation monitor. The lowdrift FET-input op amp amplifies detector current to a usable level, and the chopper-stabilized amplifier then provides additional gain while minimizing any error caused by ambient-temperature fluctuations. Gain is adjusted so that the output voltage is 1% of incident radiation intensity in rads per minute; therefore voltage can be displayed on 3½ digit DVM for direct reading of dosage rate. Output voltage from the monitor is linearly proportional to radiation intensity at the diode.

WIDEBAND RADIATION MONITOR

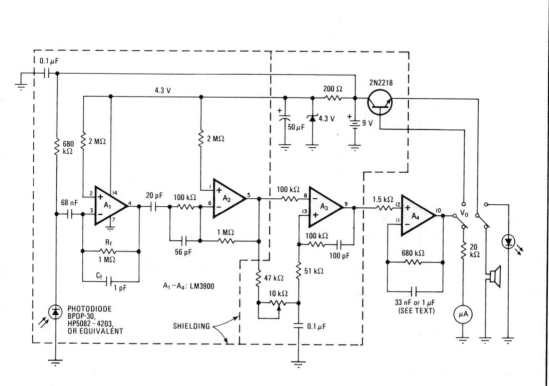

Fig. 69-2

Circuit Notes

A sensitive radiation monitor may be simply constructed with a large-area photodiode and a quad operational amplifier. Replacing the glass window of the diode with Mylar foil will shield it from light and infrared energy, enabling it to respond to such nuclear radiation as alpha and beta particles and gamma rays. A4 integrates the output of A3 in order to drive a microammeter. A 1 microfarad capacitor is used in the integrating network. A lower value, say, 33 nanofarads, will make it possible to drive a small loudspeaker (50-hertz output signal) or light-emitting diode.

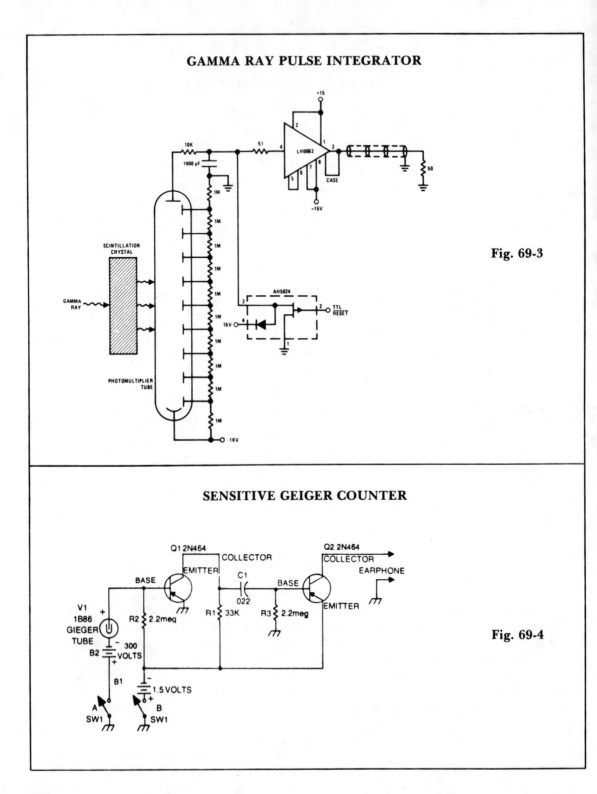

GAMMA RAY PULSE INTEGRATOR

Fig. 69-3

SENSITIVE GEIGER COUNTER

Fig. 69-4

GEIGER COUNTER

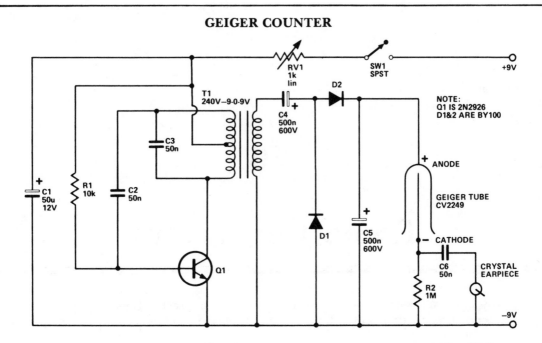

Fig. 69-5

Circuit Notes

The Geiger tube needs a high voltage supply which consists of Q1 and its associated components. The transformer is connected in reverse; the secondary is connected as a Hartley oscillator, and R1 provides base bias.

D1, D2, C4, and C5 comprise a voltage doubler. RV1 should be set so that each click heard is nice and clean because over a certain voltage range all that will be heard is a continuous buzz.

NUCLEAR PARTICLE DETECTOR

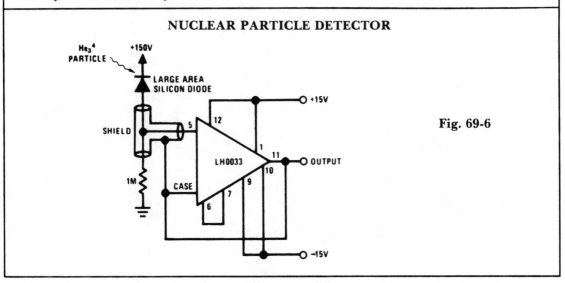

Fig. 69-6

70

Ramp Generators

The sources of the following circuits are contained in the Sources section beginning on page 730. The figure number contained in the box of each circuit correlates to the source entry in the Sources section.

Staircase Generator
Linear Voltage Ramp Generator

Precision Ramp Generator
Ramp Generator with Variable Reset Level

STAIRCASE GENERATOR

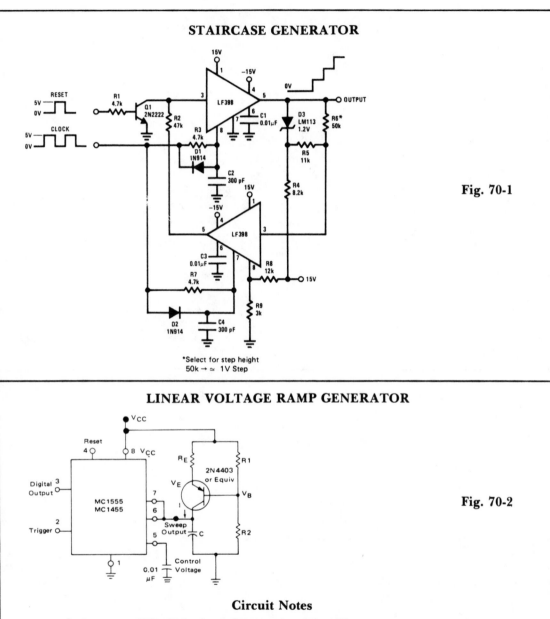

Fig. 70-1

*Select for step height
50k → ≃ 1V Step

LINEAR VOLTAGE RAMP GENERATOR

Fig. 70-2

Circuit Notes

In the monostable mode, the resistor can be replaced by a constant current source to provide a linear ramp voltage. The capacitor still charges from 0 to 2/3 Vcc. The linear ramp time is given by the following equation:

$$I = \frac{V_{CC} - V_B - V_{BE}}{R_E} \qquad t = \frac{2}{3} \frac{V_{CC}}{I}$$

If V_B is much larger than V_{BE}, then t can be made independent of V_{CC}.

PRECISION RAMP GENERATOR

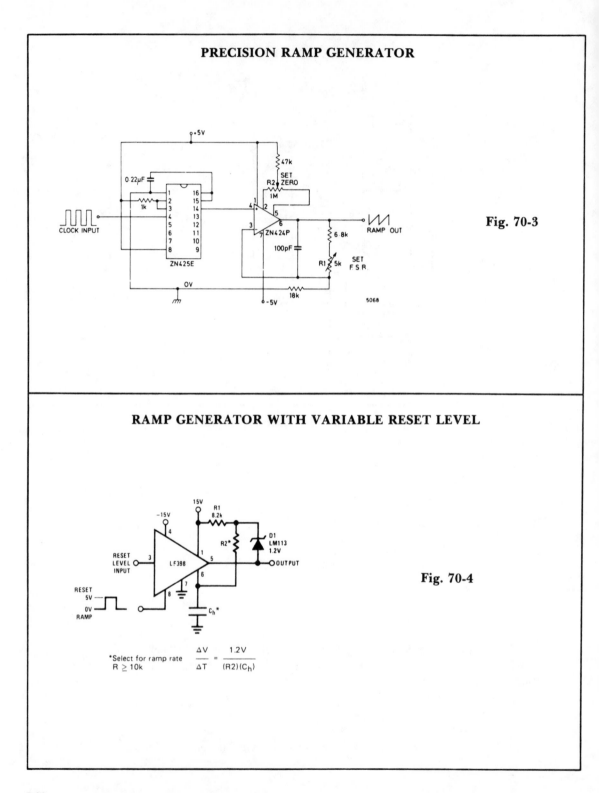

Fig. 70-3

RAMP GENERATOR WITH VARIABLE RESET LEVEL

Fig. 70-4

$$\frac{\Delta V}{\Delta T} = \frac{1.2V}{(R2)(C_h)}$$

*Select for ramp rate
R ≥ 10k

71

Receivers

The sources of the following circuits are contained in the Sources section beginning on page 730. The figure number contained in the box of each circuit correlates to the source entry in the Sources section.

Clock Radio
AM/FM Clock Radio
AM Radio
FM Stereo Demodulation System
Analog Receiver

FM Radio
Simple LF Converter
CMOS Line Receiver
Squelch Circuit for AM or FM
VLF Converter

CLOCK RADIO

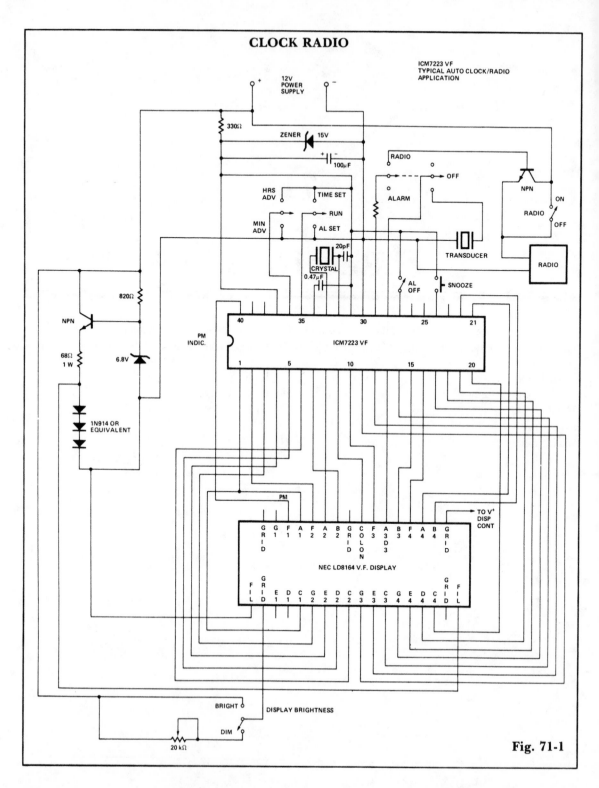

ICM7223 VF
TYPICAL AUTO CLOCK/RADIO
APPLICATION

Fig. 71-1

542

AM/FM CLOCK RADIO

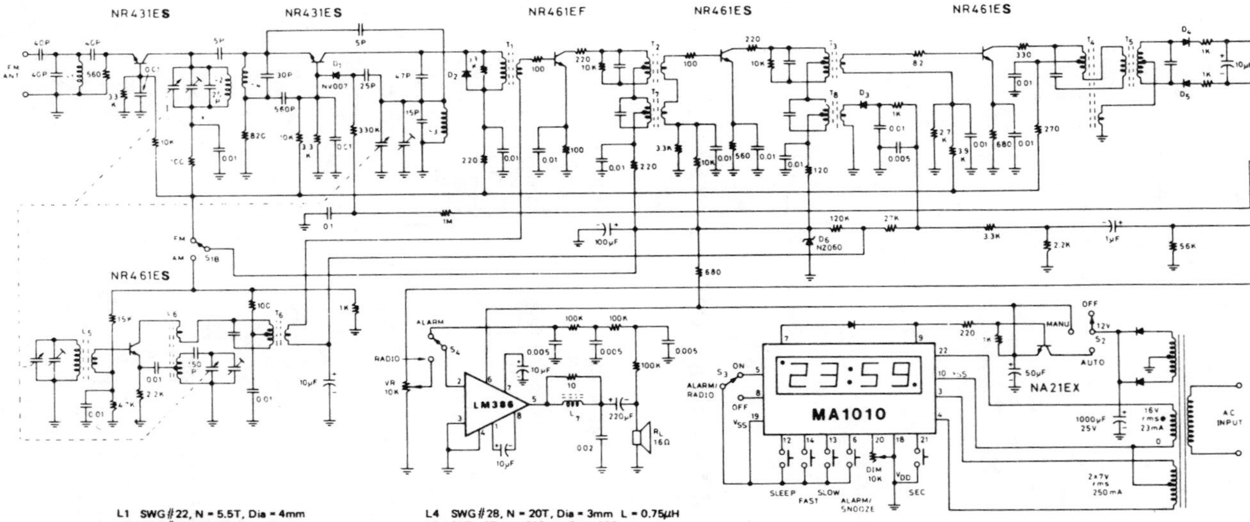

Fig. 71-2

L1 SWG#22, N = 5.5T, Dia = 4mm
L2 SWG#22, N = 4T, Dia = 4mm
L3 SWG#22, N = 4T, Dia = 3mm
L4 SWG#28, N = 20T, Dia = 3mm, L = 0.75μH
L5 9.5T : 8T, L = 600μH, Qu = 300
L7 N = 1.5T, PHILIPS #4312-020-34401

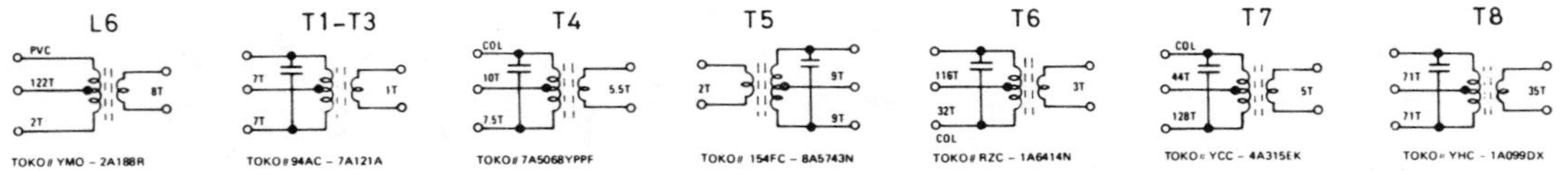

T1–T3   TOKO# 9MAC-7A121A
L6   TOKO# YMO-2A18BR
T4   TOKO# 7A506BYPPF
T5   TOKO# 154FC-8A5742N
T6   TOKO# RZC-1A9R14N
T7   TOKO# YCC-4A3156K
T8   TOKO# YHC-1A099DX

FM performance (88–108 MHz)

- 30dB quieting sensitivity: 5μV
- limiting sensitivity: 20μV
- AM rejection: 40dB
- AFC holding range: 800KHz
- Bandwidth: 180KHz

AM performance (525–1650 KHz)

- maximum sensitivity: 100μV/M
- 20dB quieting sensitivity: 280μV/M
- selectivity ± 10KHz: −28dB
- AGC figure of merit: 40dB
- overload distortion: 6%

AUDIO performance

- gain at 1 KHz: 200
- 10% THD output power: 900mW
- frequency response: 70Hz – 12KHz
- typical system dist: 0.8%
- alarm tone frequency: 600Hz

AM RADIO

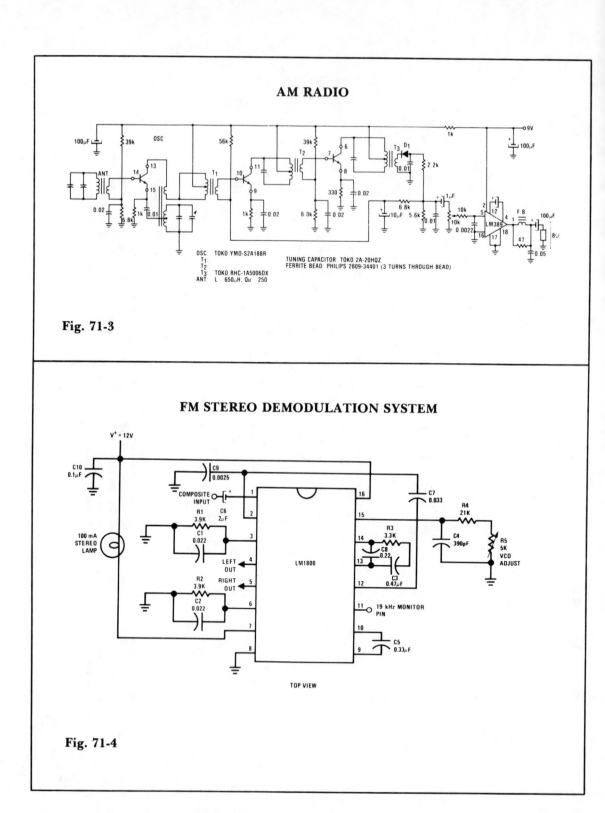

Fig. 71-3

FM STEREO DEMODULATION SYSTEM

Fig. 71-4

ANALOG RECEIVER (LOW TEMPERATURE DRIFT)

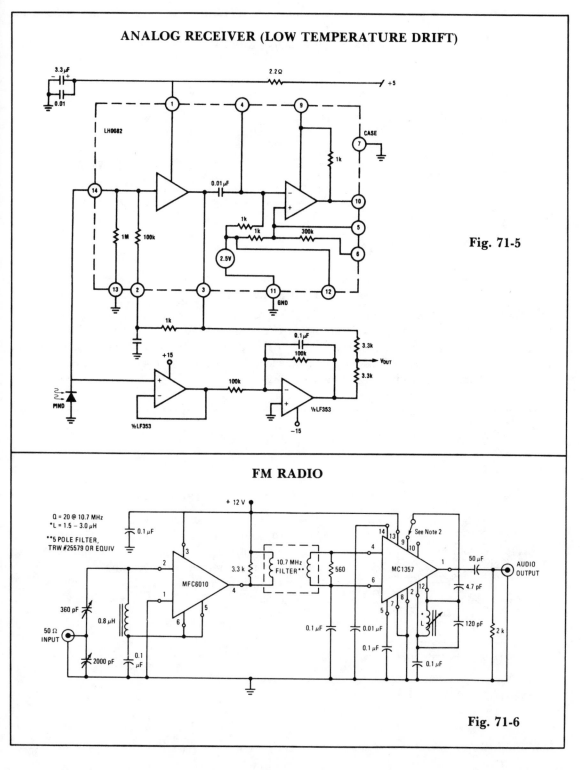

Fig. 71-5

FM RADIO

Fig. 71-6

SIMPLE LF CONVERTER

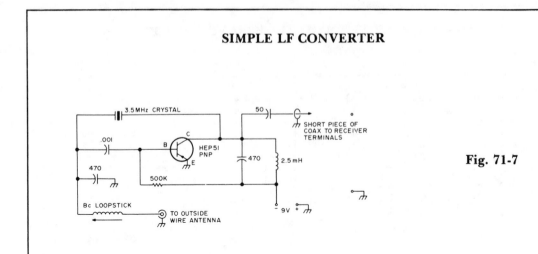

Fig. 71-7

Circuit Notes

This converter allows coverage from 25 kHz up to 500 kHz. Use short coax from the converter to receiver antenna input. Tune the receiver to 3.5 MHz, peak for loudest crystal calibrator and tune your receiver higher in frequency to 3.6 MHz and you're tuning the 100 kHz range. 3.7 MHz puts you at 200 kHz, 3.8 MHz equals 300 kHz, 3.9 MHz yields 500 kHz, and 4.0 MHz gives you 500 kHz.

CMOS LINE RECEIVER

Circuit Notes

The trip point is set half way between the supplies by R1 and R2; R3 provides over 200 mV of hysteresis to increase noise immunity. Maximum frequency of operation is about 300 kHz. If response to TTL levels is desired, change R2 to 39 K. The trip point is now centered at 1.4 V.

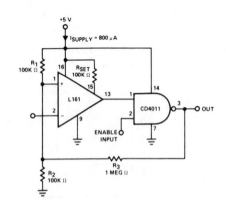

Fig. 71-8

SQUELCH CIRCUIT FOR AM OR FM

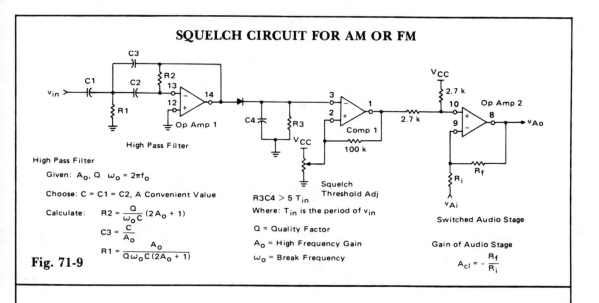

High Pass Filter

Given: A_o, Q, $\omega_o = 2\pi f_o$

Choose: $C = C_1 = C_2$, A Convenient Value

Calculate: $R_2 = \dfrac{Q}{\omega_o C}(2A_o + 1)$

$C_3 = \dfrac{C}{A_o}$

$R_1 = \dfrac{A_o}{Q\omega_o C(2A_o + 1)}$

Fig. 71-9

$R_3 C_4 > 5\,T_{in}$
Where: T_{in} is the period of v_{in}

Q = Quality Factor

A_o = High Frequency Gain

ω_o = Break Frequency

Gain of Audio Stage

$A_{cl} = -\dfrac{R_f}{R_i}$

VLF CONVERTER

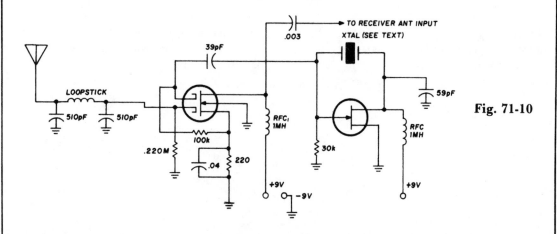

Fig. 71-10

Circuit Notes

This converter uses a low-pass filter instead of the usual tuned circuit so the only tuning required is with the receiver. The dual-gate MOSFET and FET used in the mixer and oscillator aren't critical. Any crystal having a frequency compatible with the receiver tuning range may be used. For example, with a 3500 kHz crystal, 3500 kHz on the receiver dial corresponds to zero kHz; 3600 to 100 kHz; 3700 to 200 kHz, etc. (At 3500 khz on the receiver all one can hear is the converter oscillator, and VLF signals start to come in about 20 kHz higher.)

72

Resistance and
Continuity Measuring Circuits

The sources of the following circuits are contained in the Sources section beginning on page 730. The figure number contained in the box of each circuit correlates to the source entry in the Sources section.

Linear Scale Ohmmeter
Ohmmeter
Low Parts Count Ratiometric Resistance
    Measurement

Audio Continuity Tester
Low Resistance Continuity Tester
"Buzz Box" Continuity and Coil Checker
Linear Scale Ohmmeter

Bridge Circuit

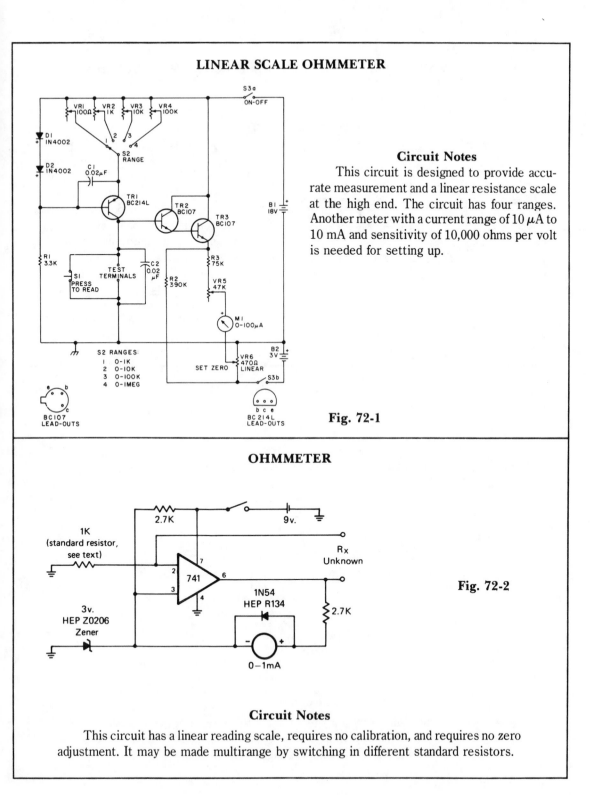

LINEAR SCALE OHMMETER

Circuit Notes

This circuit is designed to provide accurate measurement and a linear resistance scale at the high end. The circuit has four ranges. Another meter with a current range of 10 μA to 10 mA and sensitivity of 10,000 ohms per volt is needed for setting up.

Fig. 72-1

OHMMETER

Fig. 72-2

Circuit Notes

This circuit has a linear reading scale, requires no calibration, and requires no zero adjustment. It may be made multirange by switching in different standard resistors.

LOW PARTS COUNT RATIOMETRIC RESISTANCE MEASUREMENT

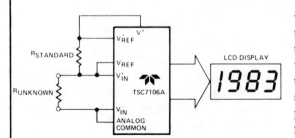

Fig. 72-3

Circuit Notes

The unknown resistance is put in series with a known standard and a current passed through the pair. The voltage developed across the unknown is applied to the input and the voltage across the known resistor applied to the reference input. If the unknown equals the standard, the display will read 1000. The displayed reading can be determined from the following expression:

$$\text{Displayed Reading} = \frac{R_{unknown}}{R_{standard}} \times 1000$$

The display will overrange for $R_{unknown} \geq 2 \times R_{standard}$.

AUDIO CONTINUITY TESTER

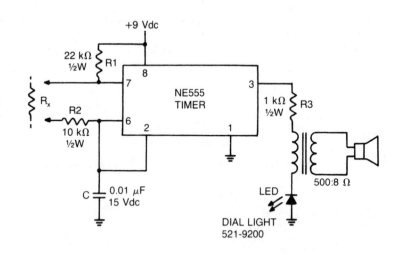

Fig. 72-4

Circuit Notes

This low-current audio continuity tester indicates the unknown resistance value by the frequency of audio tone. A high tone indicates a low resistance, and a tone of a few pulses per second indicates a resistance as high as 30 megohms.

LOW RESISTANCE CONTINUITY TESTER

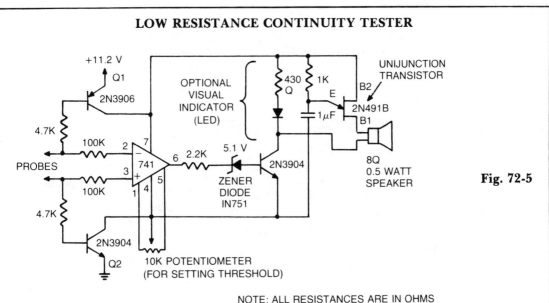

Fig. 72-5

NOTE: ALL RESISTANCES ARE IN OHMS
UNLESS OTHERWISE INDICATED.

Circuit Notes

This tester can be used to check IC printed circuit boards. Two 4.7 K resistors and the transistors connected to them prevent current flow through the operational amplifier until the probe circuit is completed. The zener diode in series with the operational amplifier output prevents audio oscillator operation until the positive output of the operational amplifier has sufficient amplitude.

"BUZZ BOX" CONTINUITY AND COIL CHECKER

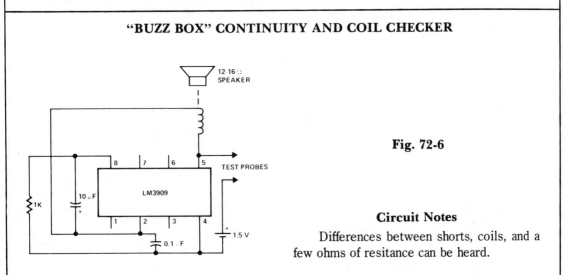

Fig. 72-6

Circuit Notes

Differences between shorts, coils, and a few ohms of resitance can be heard.

LINEAR SCALE OHMMETER

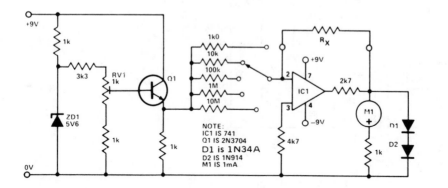

Fig. 72-7

NOTE:
IC1 IS 741
Q1 IS 2N3704
D1 is 1N34A
D2 IS 1N914
M1 IS 1mA

Circuit Notes

One preset resistor is used for all the ranges, simplifying the setting up. Diode clamping is included to prevent damage to the meter if the unknown resistor is higher than the range selected. When the meter has been as-sembled, a 10 K precision resistor is placed in the test position, R_x; the meter is set to the 10 K range and RV1 is adjusted for full scale deflection.

BRIDGE CIRCUIT

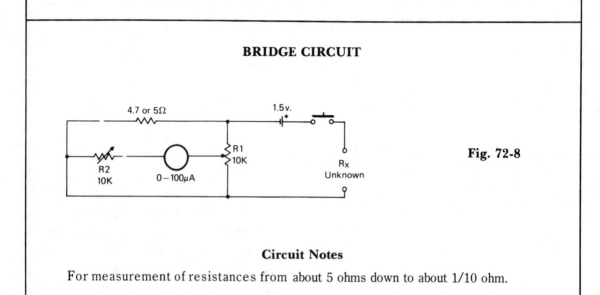

Fig. 72-8

Circuit Notes

For measurement of resistances from about 5 ohms down to about 1/10 ohm.

73

RF Amplifiers

The sources of the following circuits are contained in the Sources section beginning on page 730. The figure number contained in the box of each circuit correlates to the source entry in the Sources section.

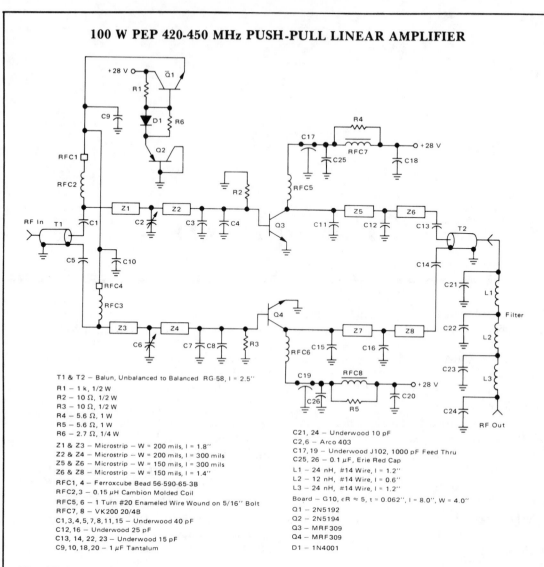

100 W PEP 420-450 MHz PUSH-PULL LINEAR AMPLIFIER

T1 & T2 — Balun, Unbalanced to Balanced RG-58, l = 2.5''

R1 — 1 k, 1/2 W
R2 — 10 Ω, 1/2 W
R3 — 10 Ω, 1/2 W
R4 — 5.6 Ω, 1 W
R5 — 5.6 Ω, 1 W
R6 — 2.7 Ω, 1/4 W

Z1 & Z3 — Microstrip — W = 200 mils, l = 1.8''
Z2 & Z4 — Microstrip — W = 200 mils, l = 300 mils
Z5 & Z6 — Microstrip — W = 150 mils, l = 300 mils
Z6 & Z8 — Microstrip — W = 150 mils, l = 1.4''

RFC1, 4 — Ferroxcube Bead 56-590-65-3B
RFC2,3 — 0.15 μH Cambion Molded Coil
RFC5, 6 — 1 Turn #20 Enameled Wire Wound on 5/16'' Bolt
RFC7, 8 — VK200 20/4B
C1,3,4,5,7,8,11,15 — Underwood 40 pF
C12, 16 — Underwood 25 pF
C13, 14, 22, 23 — Underwood 15 pF
C9, 10, 18, 20 — 1 μF Tantalum

C21, 24 — Underwood 10 pF
C2,6 — Arco 403
C17, 19 — Underwood J102, 1000 pF Feed Thru
C25, 26 — 0.1 μF, Erie Red Cap
L1 — 24 nH, #14 Wire, l = 1.2''
L2 — 12 nH, #14 Wire, l = 0.6''
L3 — 24 nH, #14 Wire, l = 1.2''
Board — G10, εR ≈ 5, t = 0.062'', l = 8.0'', W = 4.0''
Q1 — 2N5192
Q2 — 2N5194
Q3 — MRF309
Q4 — MRF309
D1 — 1N4001

Fig. 73-1

Circuit Notes

This 100 watt linear amplifier may be constructed using two MRF309 transistors in push-pull, requiring only 16 watts drive from 420 to 450 MHz. Operating from a 28 volt supply, eight dB of power gain is achieved along with excellent practical performance featuring: maximum input SWR of 2:1, harmonic suppression more than—63 dB below 100 watts output, efficiency greater than 40%, circuit stability with a 3:1 collector mismatch at all phase angles.

140 W (PEP) AMATEUR RADIO LINEAR AMPLIFIER (2-30 MHz)

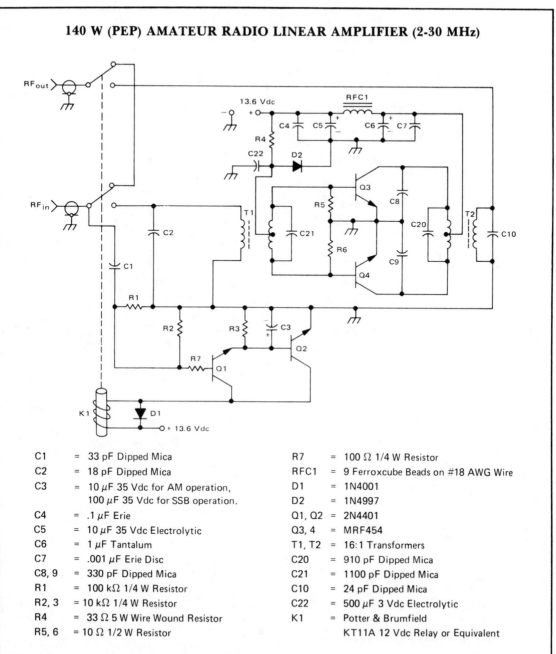

Fig. 73-2

C1	= 33 pF Dipped Mica		R7	= 100 Ω 1/4 W Resistor
C2	= 18 pF Dipped Mica		RFC1	= 9 Ferroxcube Beads on #18 AWG Wire
C3	= 10 μF 35 Vdc for AM operation,		D1	= 1N4001
	100 μF 35 Vdc for SSB operation.		D2	= 1N4997
C4	= .1 μF Erie		Q1, Q2	= 2N4401
C5	= 10 μF 35 Vdc Electrolytic		Q3, 4	= MRF454
C6	= 1 μF Tantalum		T1, T2	= 16:1 Transformers
C7	= .001 μF Erie Disc		C20	= 910 pF Dipped Mica
C8, 9	= 330 pF Dipped Mica		C21	= 1100 pF Dipped Mica
R1	= 100 kΩ 1/4 W Resistor		C10	= 24 pF Dipped Mica
R2, 3	= 10 kΩ 1/4 W Resistor		C22	= 500 μF 3 Vdc Electrolytic
R4	= 33 Ω 5 W Wire Wound Resistor		K1	= Potter & Brumfield
R5, 6	= 10 Ω 1/2 W Resistor			KT11A 12 Vdc Relay or Equivalent

Circuit Notes

This inexpensive, easy to construct amplifier uses two MRF454 devices. Specified at 80 W power output with 5 W of input drive, 30 MHz, and 12.5 Vdc.

160 W (PEP) BROADBAND LINEAR AMPLIFIER

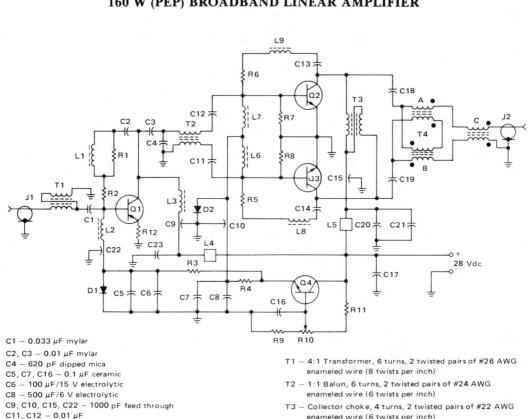

C1 — 0.033 μF mylar
C2, C3 — 0.01 μF mylar
C4 — 620 pF dipped mica
C5, C7, C16 — 0.1 μF ceramic
C6 — 100 μF/15 V electrolytic
C8 — 500 μF/6 V electrolytic
C9, C10, C15, C22 — 1000 pF feed through
C11, C12 — 0.01 μF
C13, C14 — 0.015 μF mylar
C17 — 10 μF/35 V electrolytic
C18, C19, C21 — Two 0.068 μF mylars in parallel
C20 — 0.1 μF disc ceramic
C23 — 0.1 μF disc ceramic
R1 — 220 Ω, 1/4 W carbon
R2 — 47 Ω, 1/2 W carbon
R3 — 820 Ω, 1 W wire W
R4 — 35 Ω, 5 W wire W
R5, R6 — Two 150 Ω, 1/2 W carbon in parallel
R7, R8 — 10 Ω, 1/2 W carbon
R9, R11 — 1 k, 1/2 W carbon
R10 — 1 k, 1/2 W potentiometer
R12 — 0.85 Ω (6 5.1 Ω or 4 3.3 Ω 1/4 W resistors in parallel, divided equally between both emitter leads)

T1 — 4:1 Transformer, 6 turns, 2 twisted pairs of #26 AWG enameled wire (8 twists per inch)

T2 — 1:1 Balun, 6 turns, 2 twisted pairs of #24 AWG enameled wire (6 twists per inch)

T3 — Collector choke, 4 turns, 2 twisted pairs of #22 AWG enameled wire (6 twists per inch)

T4 — 1:4 Transformer Balun, A&B — 5 turns, 2 twisted pairs of #24, C — 8 turns, 1 twisted pair of #24 AWG enameled wire (All windings 6 twists per inch). (T4 — Indiana General F624-19Q1, — All others are Indiana General F627-8Q1 ferrite toroids or equivalent.)

PARTS LIST

L1 — .33 μH, molded choke	Q1 — 2N6370
L2, L6, L7 — 10 μH, molded choke	Q2, Q3 — 2N5942
L3 — 1.8 μH (Ohmite 2-144)	Q4 — 2N5190
L4, L5 — 3 ferrite beads each	
L8, L9 — .22 μH, molded choke	D1 — 1N4001
	D2 — 1N4997
	J1, J2 — BNC connectors

Fig. 73-3

80 W (PEP) BROADBAND/LINEAR AMPLIFIER

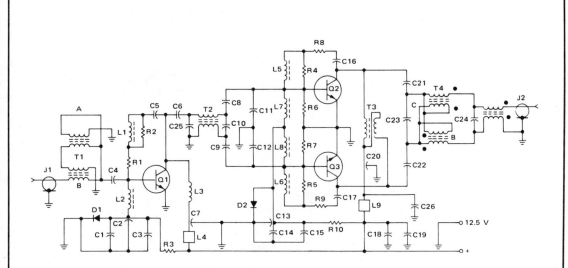

C1, C14, C18 — 0.1 µF ceramic.
C2, C7, C13, C20 — 0.001 µF feed through.
C3 — 100 µF/3V.
C4, C6 — 0.033 µF mylar
C5 — 0.0047 µF mylar.
C8, C9 — 0.015 and 0.033 µF mylars in parallel.
C10 — 470 pF mica.
C11, C12 — 560 pF mica.
C15 — 1000 µF/3 V
C16, C17 — 0.015 µF mylar
C19 — 10 pF 15 V
C21, C22 — two 0.068 µF mylars in parallel.
C23 — 330 pF mica
C24 — 39 pF mica
C25 — 680 pF mica
C26 — .01 µF ceramic

R1, R6, R7 — 10 Ω, 1/2 W carbon.
R2 — 51 Ω, 1/2 W carbon
R3 — 240 Ω, 1 wire W
R4, R5 — 18 Ω, 1 W carbon
R8, R9 — 27 Ω, 2 W carbon
R10 — 33 Ω, 6 W wire W

L1 — 0.22 µh molded choke
L2, L7, L8 — 10 µh molded choke
L5, L6 — 0.15 µh
L3 — 25 t, #26 wire, wound on a 100 Ω, 2 W resistor. (1.0 µh)
L4, L9 — 3 ferrite beads each.

T1 — 2 twisted pairs of #26 wire, 8 twists per inch. A = 4 turns,
      B = 8 turns. Core- -Stackpole 57-9322-11, Indiana General
      F627-8Q1 or equivalent

T2 — 2 twisted pairs of #24 wire, 8 twists per inch, 6 turns.
      (Core as above.)

T3 — 2 twisted pairs of #20 wire, 6 twists per inch, 4 turns.
      (Core as above.)

T4 — A and B = 2 twisted pairs of #24 wire, 8 twists per inch.
      5 turns each. C = 1 twisted pair of #24 wire, 8 turns.
      Core - -Stackpole 57-9074-11, Indiana General F624-19Q1
      or equivalent.

Q1 — 2N6367

Q2, Q3 — 2N6368

D1 — 1N4001
D2 — 1N4997

J1, J2 — BNC connectors

Fig. 73-4

SINGLE-DEVICE, 80 W, 50 Ohm VHF AMPLIFIER

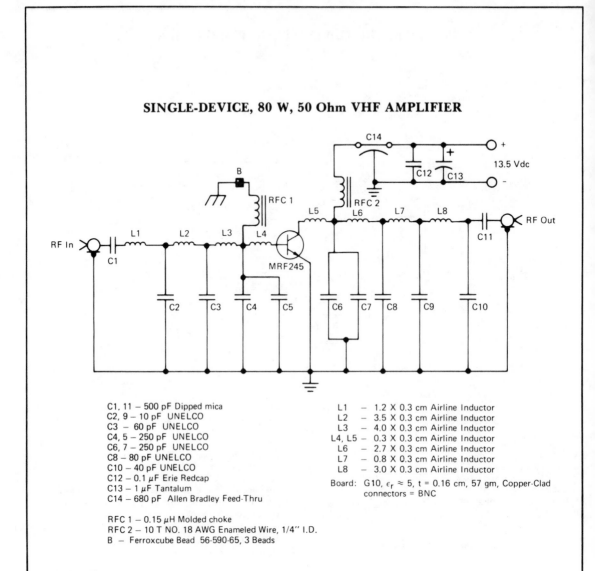

Fig. 73-5

C1, 11 – 500 pF Dipped mica
C2, 9 – 10 pF UNELCO
C3 – 60 pF UNELCO
C4, 5 – 250 pF UNELCO
C6, 7 – 250 pF UNELCO
C8 – 80 pF UNELCO
C10 – 40 pF UNELCO
C12 – 0.1 μF Erie Redcap
C13 – 1 μF Tantalum
C14 – 680 pF Allen Bradley Feed-Thru

RFC 1 – 0.15 μH Molded choke
RFC 2 – 10 T NO. 18 AWG Enameled Wire, 1/4" I.D.
B – Ferroxcube Bead 56-590-65, 3 Beads

L1   – 1.2 X 0.3 cm Airline Inductor
L2   – 3.5 X 0.3 cm Airline Inductor
L3   – 4.0 X 0.3 cm Airline Inductor
L4, L5 – 0.3 X 0.3 cm Airline Inductor
L6   – 2.7 X 0.3 cm Airline Inductor
L7   – 0.8 X 0.3 cm Airline Inductor
L8   – 3.0 X 0.3 cm Airline Inductor

Board: G10, $\epsilon_r \approx 5$, t = 0.16 cm, 57 gm, Copper-Clad connectors = BNC

Circuit Notes

The amplifier uses a single MRF245 and provides 80 W with 9.4 dB gain across the 143 to 156 MHz band.

600 W RF POWER AMPLIFIER

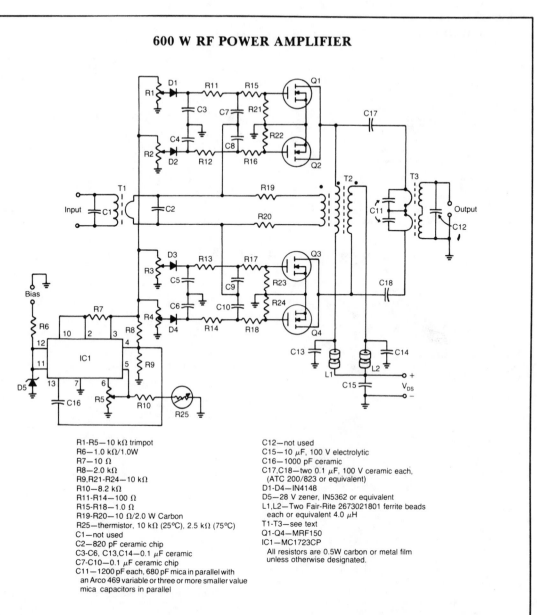

Fig. 73-6

R1-R5—10 kΩ trimpot
R6—1.0 kΩ/1.0W
R7—10 Ω
R8—2.0 kΩ
R9,R21-R24—10 kΩ
R10—8.2 kΩ
R11-R14—100 Ω
R15-R18—1.0 Ω
R19-R20—10 Ω/2.0 W Carbon
R25—thermistor, 10 kΩ (25°C), 2.5 kΩ (75°C)
C1—not used
C2—820 pF ceramic chip
C3-C6, C13,C14—0.1 μF ceramic
C7-C10—0.1 μF ceramic chip
C11—1200 pF each, 680 pF mica in parallel with
 an Arco 469 variable or three or more smaller value
 mica capacitors in parallel

C12—not used
C15—10 μF, 100 V electrolytic
C16—1000 pF ceramic
C17,C18—two 0.1 μF, 100 V ceramic each,
 (ATC 200/823 or equivalent)
D1-D4—IN4148
D5—28 V zener, IN5362 or equivalent
L1,L2—Two Fair-Rite 2673021801 ferrite beads
 each or equivalent 4.0 μH
T1-T3—see text
Q1-Q4—MRF150
IC1—MC1723CP
 All resistors are 0.5W carbon or metal film
 unless otherwise designated.

Circuit Notes

A unique push-pull parallel circuit. It uses four MRF150 RF power FETs paralleled at relatively high power levels. Supply voltages of 40 to 50 Vdc can be used, depending on linearity requirements. The bias for each device is independently adjustable; therefore, no matching is required for the gate threshold voltages.

WIDEBAND UHF AMPLIFIER WITH HIGH-PERFORMANCE FETs

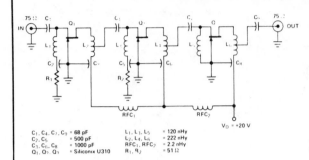

C₁, C₄, C₇, C₉ = 68 pF
C₂, C₅ = 500 pF
C₃, C₆, C₈ = 1000 pF
Q₁, Q₅, Q₇ = Siliconix U310

L₁, L₃, L₅ = 120 nHy
L₂, L₄, L₆ = 222 nHy
RFC₁, RFC₂ = 2.2 nHy
R₁, R₂ = 51 Ω

Circuit Notes

The amplifier circuit is designed for 225 MHz center frequency, 1 dB bandwidth of 50 MHz, low input VSWR in a 75-ohm system, and 24 dB gain. Three stages of U310 FETs are used in a straight forward design.

Fig. 73-7

---

10 MHz COAXIAL LINE DRIVER

B.W = 10MHz
V_O = ±2V
I_O = ±40mA

Circuit Notes

The circuit will find excellent usage in high frequency line driving systems that require wide-power bandwidths at high output current levels. (IC=HA2530) The bandwidth of the circuit is limited only by the single pole response of the feedback components; namely $f(-3 dB) = \frac{1}{2} \pi R_f C_f$. As such, the response is flat with no peaking and yields minimum distortion.

Fig. 73-8

VHF PREAMPLIFIER

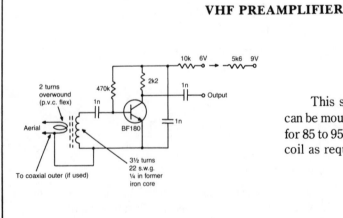

Circuit Notes

This simple circuit gives 15 dB gain and can be mounted on 1 in²PCB. Coil data is given for 85 to 95 MHz. For other frequencies modify coil as required.

Fig. 73-9

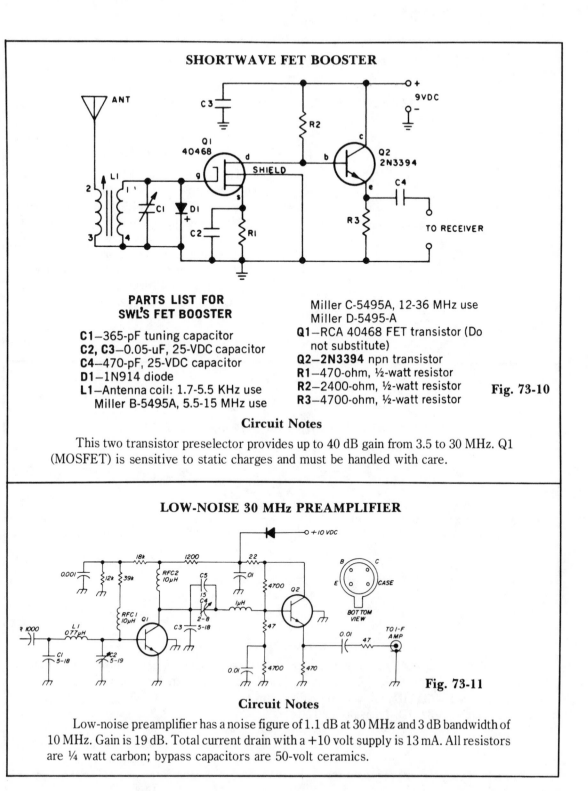

SHORTWAVE FET BOOSTER

PARTS LIST FOR SWL'S FET BOOSTER

C1–365-pF tuning capacitor
C2, C3–0.05-uF, 25-VDC capacitor
C4–470-pF, 25-VDC capacitor
D1–1N914 diode
L1–Antenna coil: 1.7-5.5 KHz use
  Miller B-5495A, 5.5-15 MHz use

Miller C-5495A, 12-36 MHz use
  Miller D-5495-A
Q1–RCA 40468 FET transistor (Do
  not substitute)
Q2–2N3394 npn transistor
R1–470-ohm, ½-watt resistor
R2–2400-ohm, ½-watt resistor
R3–4700-ohm, ½-watt resistor

Fig. 73-10

Circuit Notes

This two transistor preselector provides up to 40 dB gain from 3.5 to 30 MHz. Q1 (MOSFET) is sensitive to static charges and must be handled with care.

LOW-NOISE 30 MHz PREAMPLIFIER

Fig. 73-11

Circuit Notes

Low-noise preamplifier has a noise figure of 1.1 dB at 30 MHz and 3 dB bandwidth of 10 MHz. Gain is 19 dB. Total current drain with a +10 volt supply is 13 mA. All resistors are ¼ watt carbon; bypass capacitors are 50-volt ceramics.

LOW-NOISE BROADBAND AMPLIFIER

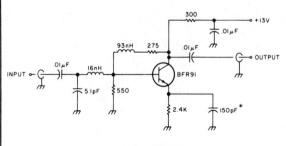

Fig. 73-12

Circuit Notes

The amplifier provides 10 dB of gain from 10-600 MHz and has a 1.5-to-1 match at 50 ohms. The BFR91 has a 1.5 dB noise figures at 500 MHz. The circuit requires 13.5 Vdc at about 13 mA. Keep the leads on the 150 pF emitter bypass capacitor as short as possible. The 16 nH coil is 2.5 turns of #26 enamel wire on the shank of a #40 drill. The 93 nH inductor is 10 turns of the same material.

TWO-METER 10 WATT POWER AMPLIFIER

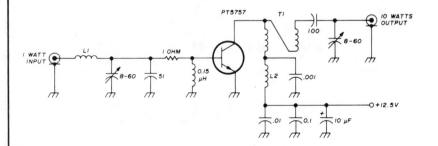

Fig. 73-13

Circuit Notes

This 10-watt, 144-MHz power amplifier uses a TRW PT5757 transistor. L1 is 4 turns of no. 20 enameled, 3/32″ ID; L2 is 10 turns of no. 20 enameled, 3/32″ ID. Transformer T1 is a 4:1 transmission-line transformer made from a 3″ length of twisted pair of no. 20 enameled wire.

TWO-STAGE 60 MHz IF AMPLIFIER
(POWER GAIN ≈ 80 dB, BW ≈ 1.5 MHz)

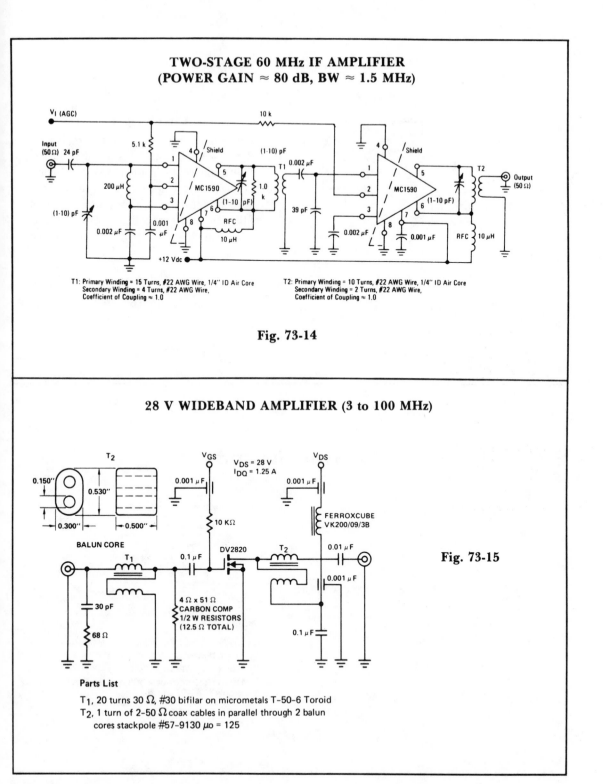

T1: Primary Winding = 15 Turns, #22 AWG Wire, 1/4" ID Air Core
Secondary Winding = 4 Turns, #22 AWG Wire,
Coefficient of Coupling ≈ 1.0

T2: Primary Winding = 10 Turns, #22 AWG Wire, 1/4" ID Air Core
Secondary Winding = 2 Turns, #22 AWG Wire,
Coefficient of Coupling ≈ 1.0

Fig. 73-14

28 V WIDEBAND AMPLIFIER (3 to 100 MHz)

Fig. 73-15

Parts List

T_1, 20 turns 30 Ω, #30 bifilar on micrometals T-50-6 Toroid
T_2, 1 turn of 2-50 Ω coax cables in parallel through 2 balun
    cores stackpole #57-9130 μ_o = 125

563

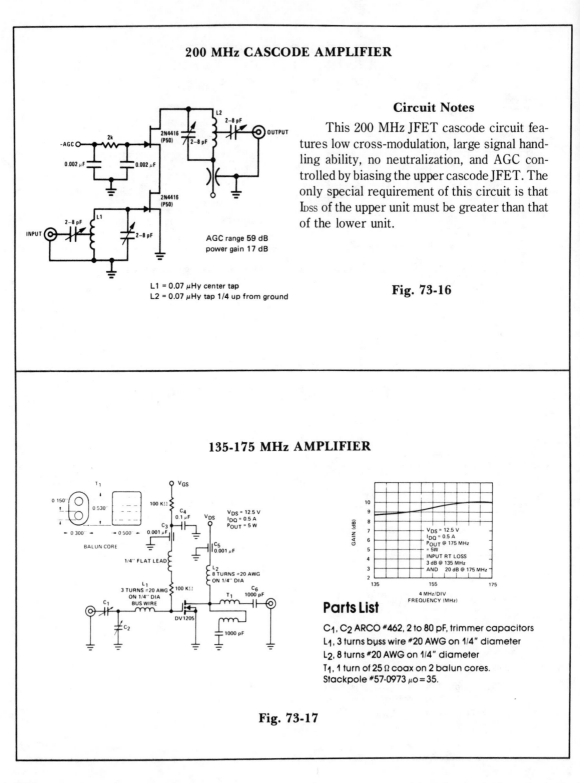

200 MHz CASCODE AMPLIFIER

Circuit Notes

This 200 MHz JFET cascode circuit features low cross-modulation, large signal handling ability, no neutralization, and AGC controlled by biasing the upper cascode JFET. The only special requirement of this circuit is that I$_{DSS}$ of the upper unit must be greater than that of the lower unit.

AGC range 59 dB
power gain 17 dB

L1 = 0.07 μHy center tap
L2 = 0.07 μHy tap 1/4 up from ground

Fig. 73-16

135-175 MHz AMPLIFIER

Parts List

C$_1$, C$_2$ ARCO #462, 2 to 80 pF, trimmer capacitors
L$_1$, 3 turns buss wire #20 AWG on 1/4" diameter
L$_2$, 8 turns #20 AWG on 1/4" diameter
T$_1$, 1 turn of 25 Ω coax on 2 balun cores.
Stackpole #57-0973 μo = 35.

Fig. 73-17

200 MHz CASCODE AMPLIFIER

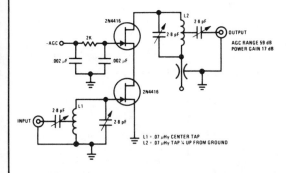

Circuit Notes

This 200 MHz JFET cascode circuit features low cross-modulation, large signal handling ability, no neutralization, and AGC controlled by biasing the upper cascode JFET. The only special requirement of this circuit is that I_{DSS} of the upper unit must be greater than that of the lower unit.

Fig. 73-18

100 MHz AND 400 MHz NEUTRALIZED COMMON SOURCE AMPLIFIER

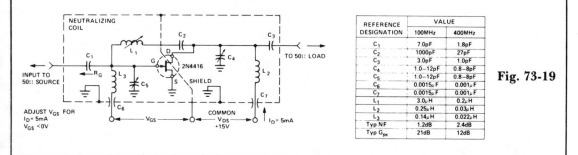

REFERENCE DESIGNATION	VALUE	
	100MHz	400MHz
C_1	7.0pF	1.8pF
C_2	1000pF	27pF
C_3	3.0pF	1.0pF
C_4	1.0–12pF	0.8–8pF
C_5	1.0–12pF	0.8–8pF
C_6	0.0015μF	0.001μF
C_7	0.0015μF	0.001μF
L_1	3.0μH	0.2μH
L_2	0.25μH	0.03μH
L_3	0.14μH	0.022μH
Typ NF	1.2dB	2.4dB
Typ G_{ps}	21dB	12dB

Fig. 73-19

ULTRA HIGH FREQUENCY AMPLIFIER

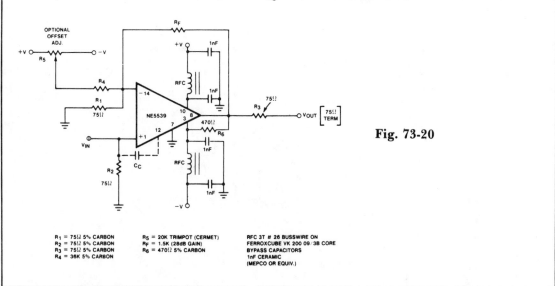

Fig. 73-20

R_1 = 75Ω 5% CARBON
R_2 = 75Ω 5% CARBON
R_3 = 75Ω 5% CARBON
R_4 = 36K 5% CARBON

R_5 = 20K TRIMPOT (CERMET)
R_F = 1.5K (28dB GAIN)
R_6 = 470Ω 5% CARBON

RFC 3T # 26 BUSSWIRE ON
FERROXCUBE VK 200 09/3B CORE
BYPASS CAPACITORS
1nF CERAMIC
(MEPCO OR EQUIV.)

UHF AMPLIFIER WITH INVERTING GAIN OF 2 AND LAG-LEAD COMPENSATION (GAIN BANDWIDTH PRODUCT 350 MHz)

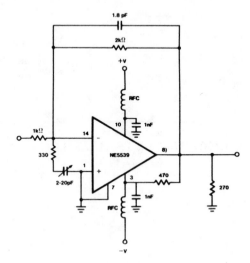

NOTE
Resistors—1/4 watt carbon.
RFC-3T #26 bus wire on Ferroxcube VK200 09/3B
wideband threaded core.

Fig. 73-21

TRANSISTORIZED Q-MULTIPLIER
FOR USE WITH IFS IN THE 1400 kHz RANGE

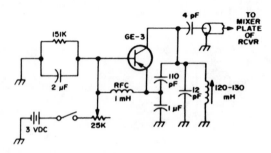

Fig. 73-22

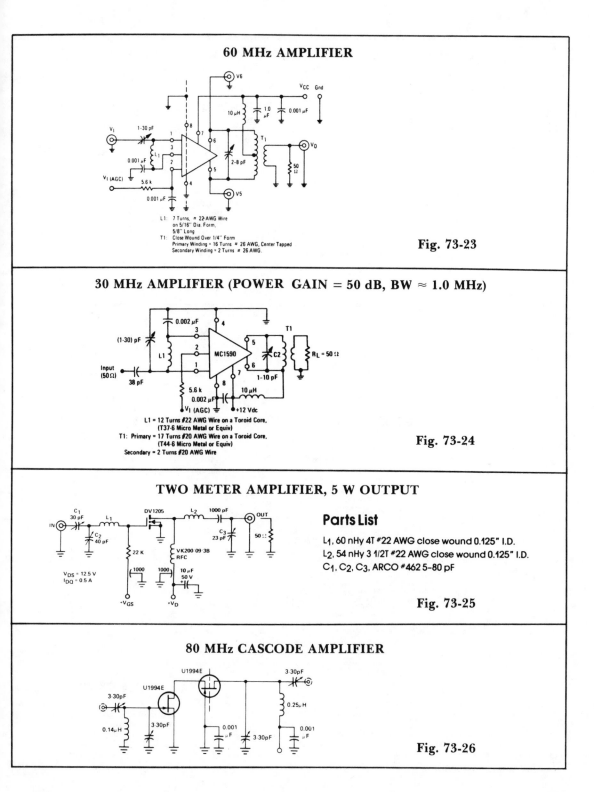

60 MHz AMPLIFIER

L1: 7 Turns, #22 AWG Wire
on 5/16" Dia. Form,
5/8" Long
T1: Close Wound Over 1/4" Form
Primary Winding = 16 Turns #26 AWG, Center Tapped
Secondary Winding = 2 Turns #26 AWG.

Fig. 73-23

30 MHz AMPLIFIER (POWER GAIN = 50 dB, BW ≈ 1.0 MHz)

L1 = 12 Turns #22 AWG Wire on a Toroid Core,
(T37-6 Micro Metal or Equiv)
T1: Primary 17 Turns #20 AWG Wire on a Toroid Core,
(T44-6 Micro Metal or Equiv)
Secondary = 2 Turns #20 AWG Wire

Fig. 73-24

TWO METER AMPLIFIER, 5 W OUTPUT

V_{DS} = 12.5 V
I_{DQ} = 0.5 A

Parts List

L_1, 60 nHy 4T #22 AWG close wound 0.125" I.D.
L_2, 54 nHy 3 1/2T #22 AWG close wound 0.125" I.D.
C_1, C_2, C_3, ARCO #462 5–80 pF

Fig. 73-25

80 MHz CASCODE AMPLIFIER

Fig. 73-26

567

200 MHz NEUTRALIZED COMMON SOURCE AMPLIFIER

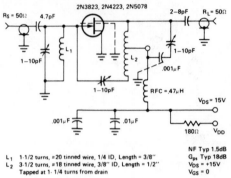

L_1 1-1/2 turns, ≈20 tinned wire, 1/4 ID, Length = 3/8"
L_2 3-1/2 turns, ≈18 tinned wire, 3/8" ID, Length = 1/2"
       Tapped at 1-1/4 turns from drain

NF Typ 1.5dB
G_{ps} Typ 18dB
V_{DS} = +15V
V_{GS} = 0

Fig. 73-27

450 MHz COMMON-SOURCE AMPLIFIER

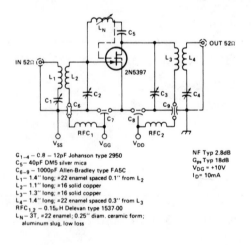

C_{1-4} – 0.8 – 12pF Johanson type 2950
C_5 – 40pF DM5 silver mica
C_{6-9} – 1000pF Allen-Bradley type FA5C
L_1 – 1.4" long; ≈22 enamel spaced 0.1" from L_2
L_2 – 1.1" long; ≈16 solid copper
L_3 – 1.3" long; ≈16 solid copper
L_4 – 1.4" long; ≈22 enamel spaced 0.3" from L_3
$RFC_{1,2}$ – 0.15μH Delevan type 1537-00
L_N – 3T, ≈22 enamel; 0.25" diam. ceramic form;
     aluminum slug, low loss

NF Typ 2.8dB
G_{ps} Typ 18dB
V_{DG} = +10V
I_D = 10mA

Fig. 73-28

74

RF Oscillators

The sources of the following circuits are contained in the Sources section beginning on page 730. The figure number contained in the box of each circuit correlates to the source entry in the Sources section.

500 MHz Oscillator
Low Distortion Oscillator
400 MHz Oscillator
2 MHz Oscillator

1.0 MHz Oscillator
Hartley Oscillator
Colpitts Oscillator
RF Oscillator

500 MHz OSCILLATOR

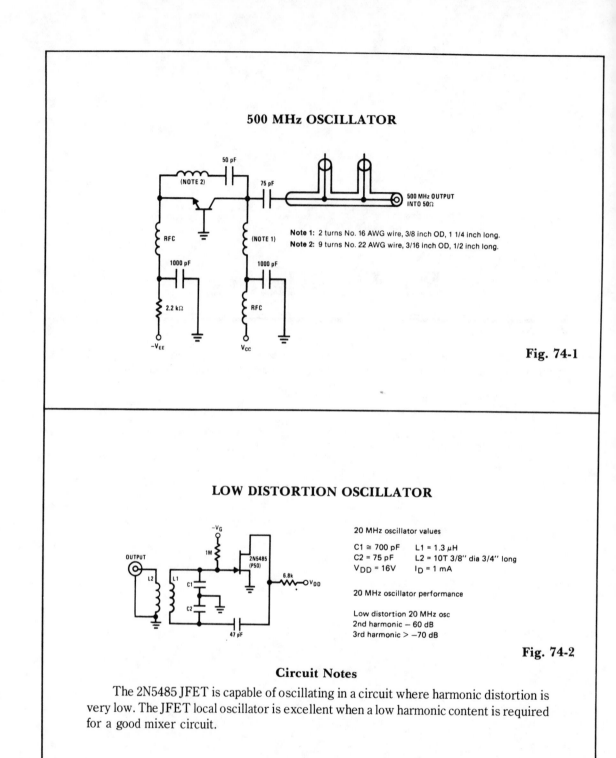

Note 1: 2 turns No. 16 AWG wire, 3/8 inch OD, 1 1/4 inch long.
Note 2: 9 turns No. 22 AWG wire, 3/16 inch OD, 1/2 inch long.

Fig. 74-1

LOW DISTORTION OSCILLATOR

20 MHz oscillator values

C1 ≅ 700 pF   L1 = 1.3 μH
C2 = 75 pF   L2 = 10T 3/8'' dia 3/4'' long
V_{DD} = 16V I_D = 1 mA

20 MHz oscillator performance

Low distortion 20 MHz osc
2nd harmonic − 60 dB
3rd harmonic > −70 dB

Fig. 74-2

Circuit Notes

The 2N5485 JFET is capable of oscillating in a circuit where harmonic distortion is very low. The JFET local oscillator is excellent when a low harmonic content is required for a good mixer circuit.

400 MHz OSCILLATOR

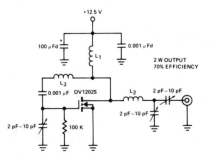

Parts List

L₁ — 8 turns #22 closewound on 1/4" diameter
L₂ — 1/2 inch #16 wire
L₃ — 1 inch #16 wire

Fig. 74-3

1.0 MHz OSCILLATOR

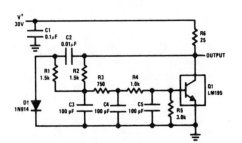

Fig. 74-5

2 MHz OSCILLATOR

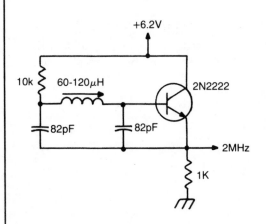

Fig. 74-4

Circuit Notes

Miller 9055 miniature slugtuned coil; all resistors 1/4W 5%; all caps min. 25 V ceramic.

HARTLEY OSCILLATOR

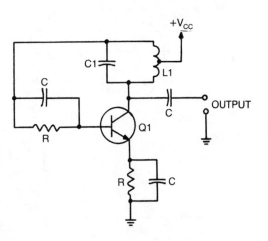

Fig. 74-6

Circuit Notes

Resonant frequency is $\frac{1}{2} \pi \sqrt{L1C1}$.

COLPITTS OSCILLATOR

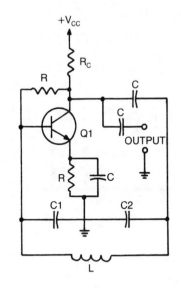

Circuit Notes

When calculating its resonant frequency, use C1C2/C1+C2 for the total capacitance of the L-C circuit.

Fig. 74-7

RF OSCILLATOR

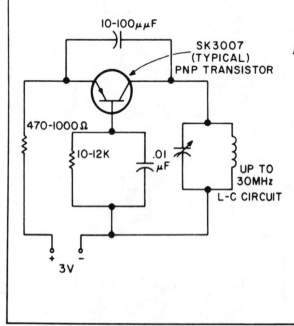

Circuit Notes

This rf oscillator is useful up to 30 MHz. An SK 3007 PNP transistor is recommended.

Fig. 74-8

75

Remote Control Circuits

The sources of the following circuits are contained in the Sources section beginning on page 730. The figure number contained in the box of each circuit correlates to the source entry in the Sources section.

Radio Control Receiver/Decoder
Carrier Operated Relay
Remote Control Servo System

Tone-Actuated Relay
Radio Control Motor Speed Controller
Remote On-Off Switch

Automatic Turn Off for TV Set

RADIO CONTROL RECEIVER/DECODER

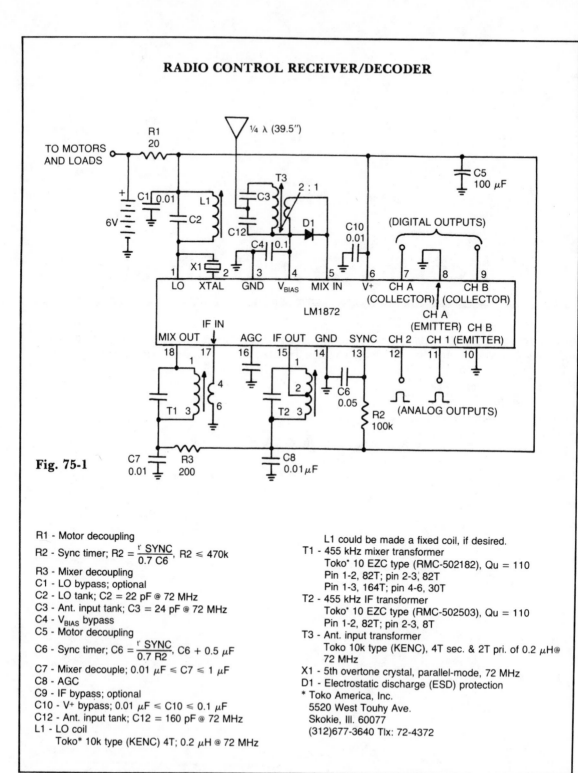

Fig. 75-1

R1 - Motor decoupling
R2 - Sync timer; $R2 = \dfrac{t' \, SYNC}{0.7 \, C6}$, R2 ≤ 470k

R3 - Mixer decoupling
C1 - LO bypass; optional
C2 - LO tank; C2 = 22 pF @ 72 MHz
C3 - Ant. input tank; C3 = 24 pF @ 72 MHz
C4 - V_{BIAS} bypass
C5 - Motor decoupling
C6 - Sync timer; $C6 = \dfrac{t' \, SYNC}{0.7 \, R2}$, C6 + 0.5 μF

C7 - Mixer decouple; 0.01 μF ≤ C7 ≤ 1 μF
C8 - AGC
C9 - IF bypass; optional
C10 - V+ bypass; 0.01 μF ≤ C10 ≤ 0.1 μF
C12 - Ant. input tank; C12 = 160 pF @ 72 MHz
L1 - LO coil
    Toko* 10k type (KENC) 4T; 0.2 μH @ 72 MHz

L1 could be made a fixed coil, if desired.
T1 - 455 kHz mixer transformer
    Toko* 10 EZC type (RMC-502182), Qu = 110
    Pin 1-2, 82T; pin 2-3, 82T
    Pin 1-3, 164T; pin 4-6, 30T
T2 - 455 kHz IF transformer
    Toko* 10 EZC type (RMC-502503), Qu = 110
    Pin 1-2, 82T; pin 2-3, 8T
T3 - Ant. input transformer
    Toko 10k type (KENC), 4T sec. & 2T pri. of 0.2 μH @
    72 MHz
X1 - 5th overtone crystal, parallel-mode, 72 MHz
D1 - Electrostatic discharge (ESD) protection
* Toko America, Inc.
    5520 West Touhy Ave.
    Skokie, Ill. 60077
    (312)677-3640 Tlx: 72-4372

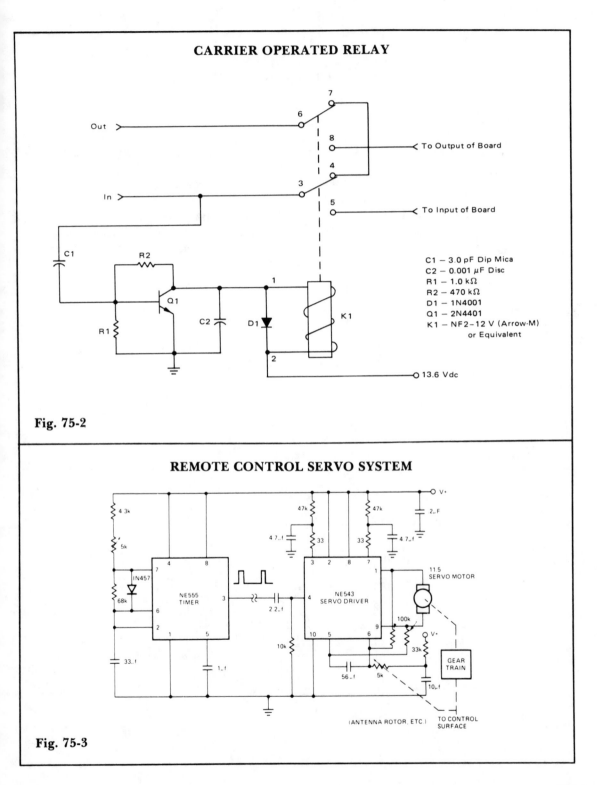

CARRIER OPERATED RELAY

Out >

6
7

8
To Output of Board

4
5
To Input of Board

In >

3

C1

R2

Q1

R1

C2

D1

K1

1

2

13.6 Vdc

C1 — 3.0 pF Dip Mica
C2 — 0.001 μF Disc
R1 — 1.0 kΩ
R2 — 470 kΩ
D1 — 1N4001
Q1 — 2N4401
K1 — NF2–12 V (Arrow-M)
       or Equivalent

Fig. 75-2

REMOTE CONTROL SERVO SYSTEM

4 3k

5k

1N457

68k

NE555
TIMER

7   4   8

3

6

2

1   5

33..f

1..f

2 2..f

47k

4 7..f

33

33

47k

4 7..f

2..F

V•

NE543
SERVO DRIVER

3   2   8   7

4

1

10   5   6

9

10k

100k

56..f

5k

33k

10µf

11.5
SERVO MOTOR

V•

GEAR
TRAIN

(ANTENNA ROTOR, ETC.)

TO CONTROL
SURFACE

Fig. 75-3

TONE-ACTUATED RELAY

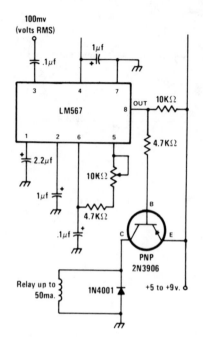

Circuit Notes

The circuit is built around the LM567 tone decoder IC that requires about 100 millivolts at its operating frequency. The frequency is set by a 10 K variable resistor and can be between 700 and 1500 Hz. When a tone at the set frequency is present, the 567's output goes low to energize a relay through a 2N3906 PNP transistor.

Fig. 75-4

RADIO CONTROL MOTOR SPEED CONTROLLER

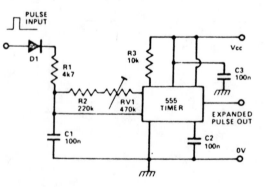

Fig. 75-5

REMOTE ON-OFF SWITCH

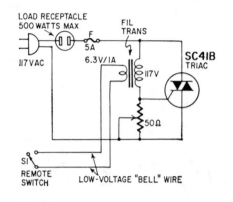

LOAD RECEPTACLE
500 WATTS MAX

FIL
TRANS

F
5A

117 VAC

6.3V/1A

117V

SC41B
TRIAC

50Ω

S1
REMOTE
SWITCH

LOW-VOLTAGE "BELL" WIRE

Fig. 75-6

Circuit Notes

This circuit provides power control without running line-voltage switch leads. The primary of a 6-volt filament transformer is connected between the gate and one of the main terminals of a triac. The secondary is connected to the remote switch through ordinary low-voltage line. With switch open, transformer blocks gate current, prevents the triac from firing and applying power to the equipment. Closing the switch short-circuits the secondary, causing the transformer to saturate and trigger the triac.

AUTOMATIC TURN OFF FOR TV SET

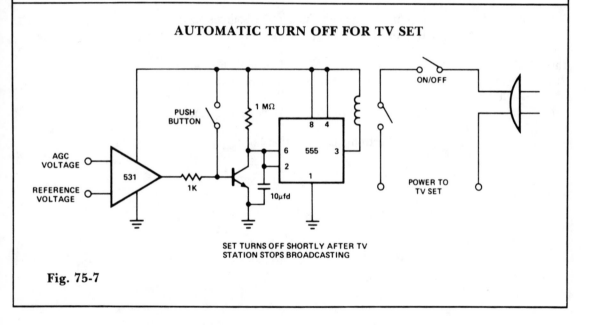

ON/OFF

PUSH
BUTTON

1 MΩ

8   4

AGC
VOLTAGE

531

6   555   3

REFERENCE
VOLTAGE

1K

2

1

10μfd

POWER TO
TV SET

SET TURNS OFF SHORTLY AFTER TV
STATION STOPS BROADCASTING

Fig. 75-7

76

Safety and Security Circuits

$\mathbf{T}$he sources of the following circuits are contained in the Sources section beginning on page 730. The figure number contained in the box of each circuit correlates to the source entry in the Sources section.

Tarry Light
Ground Tester
Ground-Fault Interrupter
Single Source Emergency Lighting System

Power Failure Alarm
Ac Hot Wire Probe
Power Failure Detector
Power-Failure Alarm

Electronic Combination Lock

TARRY LIGHT

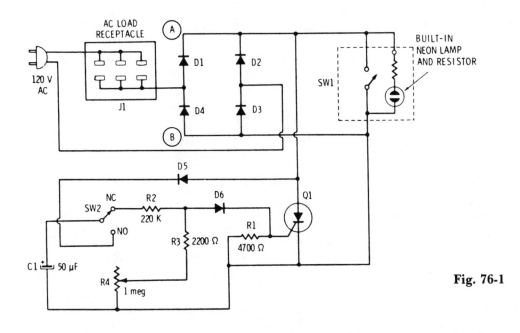

Fig. 76-1

Circuit Notes

The push button and potentiometer initiate a time delay that turns a light on then automatically turns it off again after a predetermined time. The potentiometer can be set for a delay of a few seconds to just under three minutes. When the push-button switch SW2 is pressed, capacitor C1 gets charged through D5 to the full dc voltage developed by the diode bridge. When the button is released, the charged capacitor is connected across the series combination of R2, R3, and potentiometer R4 whose setting determines the total resistance and thereby sets the time it takes for the capacitor to discharge. A steering diode, D6, connected to the junction of R2 and R3, and potentiometer R4 whose setting determines The total resistance and thereby sets the time it takes for the capacitor to discharge. Diode, D6, picks off a portion of this decaying dc voltage and applies it to the gates terminal of Q1, the SCR, triggering it into a conductive state. This SCR will remain on as long as there is sufficient voltage on its gate. As soon as this voltage decays below the minimum holding voltage of the SCR, it will turn off on the next line alternation.

GROUND TESTER

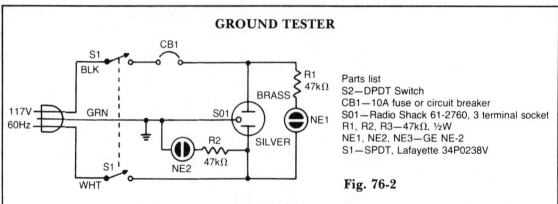

Parts list
S2—DPDT Switch
CB1—10A fuse or circuit breaker
S01—Radio Shack 61-2760, 3 terminal socket
R1, R2, R3—47kΩ, ½W
NE1, NE2, NE3—GE NE-2
S1—SPDT, Lafayette 34P0238V

Fig. 76-2

Circuit Notes

This circuit checks the reliability of appliances so that the equipment may be used safely. The test circuit must be plugged into a properly wired three terminal wall outlet. When a two-lead or three-lead appliance is plugged into circuit outlet S01, neon lamps NE1 and NE2 will light if the appliance is safe. If neon NE2 is lit the appliance is dangerous, because the neutral lead is 110 Vac above ground.

GROUND-FAULT INTERRUPTER
(120 Hz NEUTRAL TRANSFORMER APPROACH)

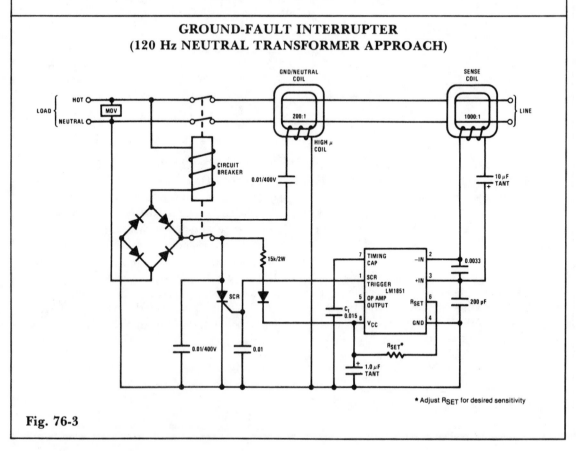

Fig. 76-3

SINGLE SOURCE EMERGENCY LIGHTING SYSTEM

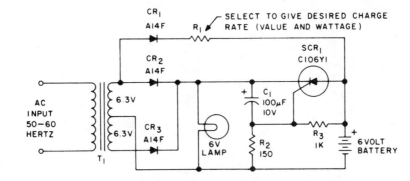

Fig. 76-4

Circuit Notes

This emergency lighting system maintains a 6 volt battery at full charge and switches automatically from the ac supply to the battery.

POWER FAILURE ALARM

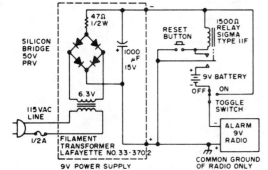

Fig. 76-5

Circuit Notes

If the power fails, the radio alarm goes on. No loud siren, bell, or whistle. Even if the power is restored, the alarm stays on until RESET button is pushed.

AC HOT WIRE PROBE

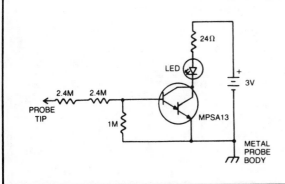

Circuit Notes

Insert the probe tip into either terminal of an ac outlet and hold the probe body against anything that the circuit ground is connected to. The LED will glow when the hot terminal is touched. Two 2.4 M resistors are used in the probe tip for safety (redundancy) reasons.

Fig. 76-6

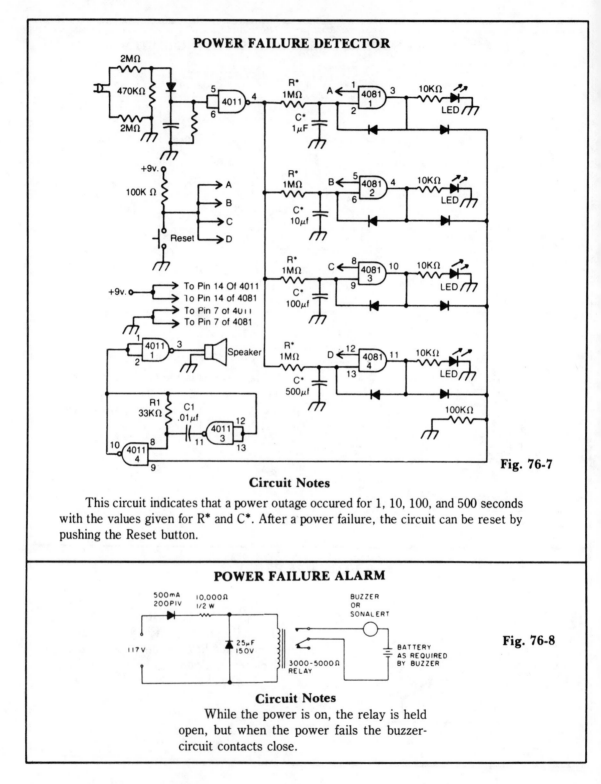

POWER FAILURE DETECTOR

Fig. 76-7

Circuit Notes

This circuit indicates that a power outage occured for 1, 10, 100, and 500 seconds with the values given for R* and C*. After a power failure, the circuit can be reset by pushing the Reset button.

POWER FAILURE ALARM

Fig. 76-8

Circuit Notes

While the power is on, the relay is held open, but when the power fails the buzzer-circuit contacts close.

ELECTRONIC COMBINATION LOCK

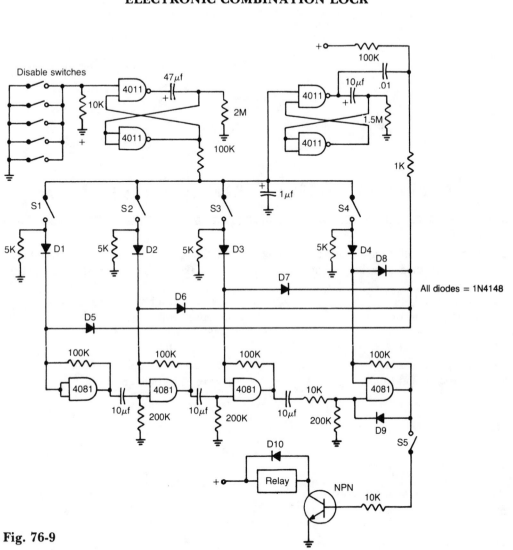

Fig. 76-9

Circuit Notes

Switches S1 through S5 must be operated in rapid sequence to operate the lock. They can be any numbers on a 10-button switch pad. If an incorrect button is pushed, alarm sounds and the circuit is disabled for two minutes.

77

Sample and Hold Circuits

The sources of the following circuits are contained in the Sources section beginning on page 730. The figure number contained in the box of each circuit correlates to the source entry in the Sources section.

Peak Detect and Hold
Low Drift Sample and Hold
JFET Sample and Hold
High Speed Sample and Hold Amplifier
High Speed Sample and Hold
High Speed Sample and Hold

Sample and Hold with Offset Adjustment
Differential Hold
× 1000 Sample and Hold
Sample and Hold
High Accuracy Sample and Hold
High Speed Sample and Hold

PEAK DETECT AND HOLD

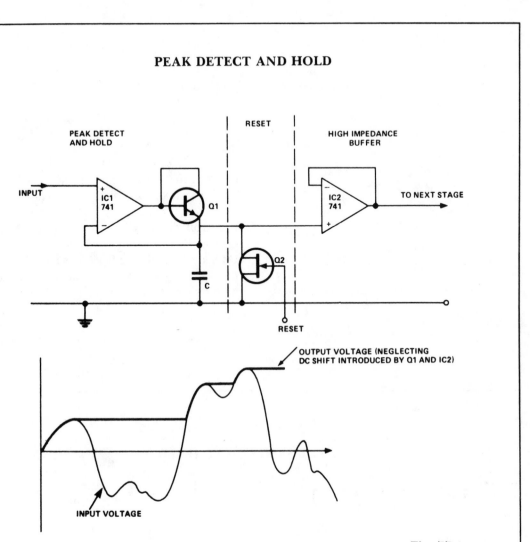

Fig. 77-1

Circuit Notes

If the voltage at the input exceeds the voltage on the capacitor, then the output of the 741 goes positive, the diode conducts, and the capacitor is charged up to the input voltage-forward voltage drop of diode. When the voltage at the input is less than that on the capacitor, the output of the 741 goes negative, and the diode cuts off. To prevent the capacitor from discharging through the input resistance of the next stage, a high input impedance buffer stage (IC2) is used. The circuit can be reset by means of a FET or similar high impedance device connected across the capacitor.

LOW DRIFT SAMPLE AND HOLD

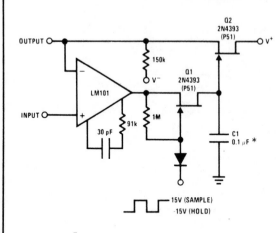

Circuit Notes

The JFETs, Q1 and Q2, provide complete buffering to C1, the sample and hold capacitor. During sample, Q1 is turned on and provides a path, $r_{ds(on)}$, for charging C1. During hold, Q1 is turned off, thus leaving Q1 $I_{D(off)}$ (< 100 pA) and Q2 I_{GSS} (< 100 pA) as the only discharge paths. Q2 serves a buffering function so feedback to the LM101 and output current are supplied from its source.

*Polycarbonate dielectric capacitor

Fig. 77-2

JFET SAMPLE AND HOLD

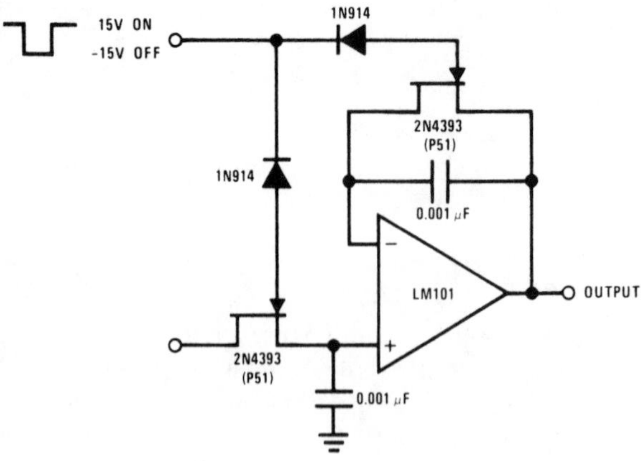

Fig. 77-3

Circuit Notes

The logic voltage is applied simultaneously to the sample and hold JFETs. By matching input impedance and feedback resistance and capacitance, errors due to $r_{ds(on)}$ of the JFETs are minimized.

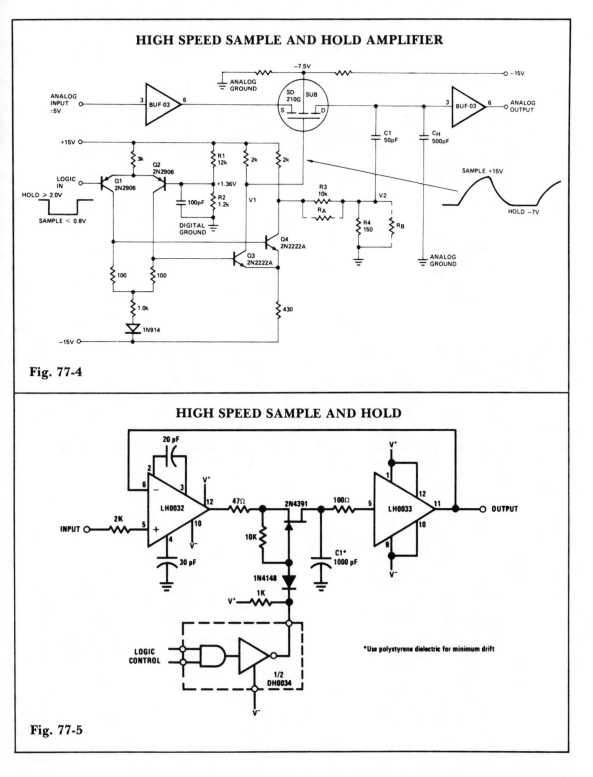

HIGH SPEED SAMPLE AND HOLD AMPLIFIER

Fig. 77-4

HIGH SPEED SAMPLE AND HOLD

*Use polystyrene dielectric for minimum drift

Fig. 77-5

587

HIGH SPEED SAMPLE AND HOLD

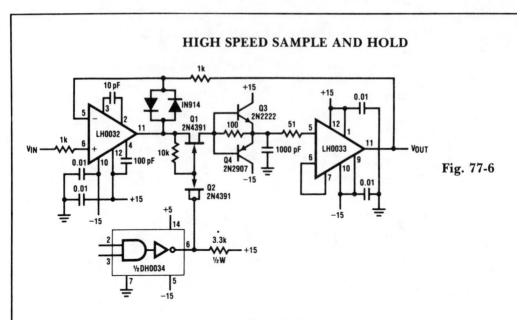

Fig. 77-6

Circuit Notes

This circuit exhibits a 10 V acquisition time of 900 ns to 0.1% accuracy and a droop rate of only 100 μV/ms at 25° C ambient condition. An even faster acquisition time can be obtained using a smaller value hold-capacitor. By decreasing the value from 1000 pF to 220 pF, the acquisition time improves to 500 ns for a 10 V step. However, the droop rate increases to 500 μV/ms.

SAMPLE AND HOLD WITH OFFSET ADJUSTMENT

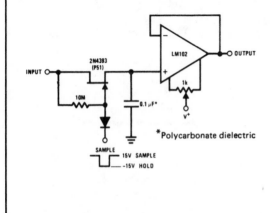

Fig. 77-7

Circuit Notes

The 2N4393 JFET was selected because of its low I_{GSS} (< 100 pA), very low $I_{D(off)}$ (< 100 pA) and low pinchoff voltage. Leakages of this level put the burden of circuit performance on clean, solder-resin free, low leakage circuit layout.

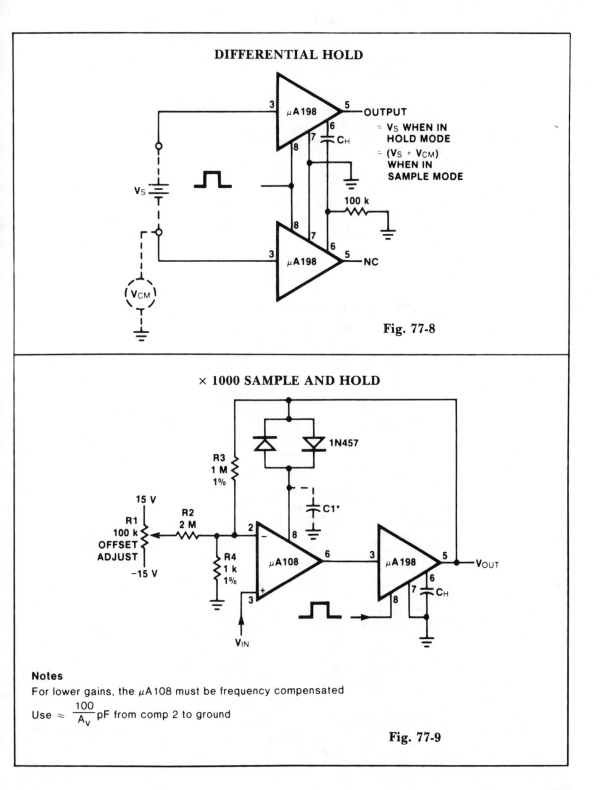

DIFFERENTIAL HOLD

μA198

3   5 OUTPUT
6
7 C_H
8

= V_S WHEN IN
   HOLD MODE
= (V_S + V_{CM})
   WHEN IN
   SAMPLE MODE

V_S

100 k

μA198

8
7
3   6
5 NC

V_{CM}

Fig. 77-8

× 1000 SAMPLE AND HOLD

R3
1 M
1%

1N457

C1*

15 V

R1
100 k
OFFSET
ADJUST

−15 V

R2
2 M

2
−
μA108
8
6

R4
1 k
1%

3
+

V_{IN}

3   5 V_{OUT}
μA198
6
7 C_H
8

Notes
For lower gains, the μA108 must be frequency compensated

Use $\approx \dfrac{100}{A_v}$ pF from comp 2 to ground

Fig. 77-9

SAMPLE AND HOLD

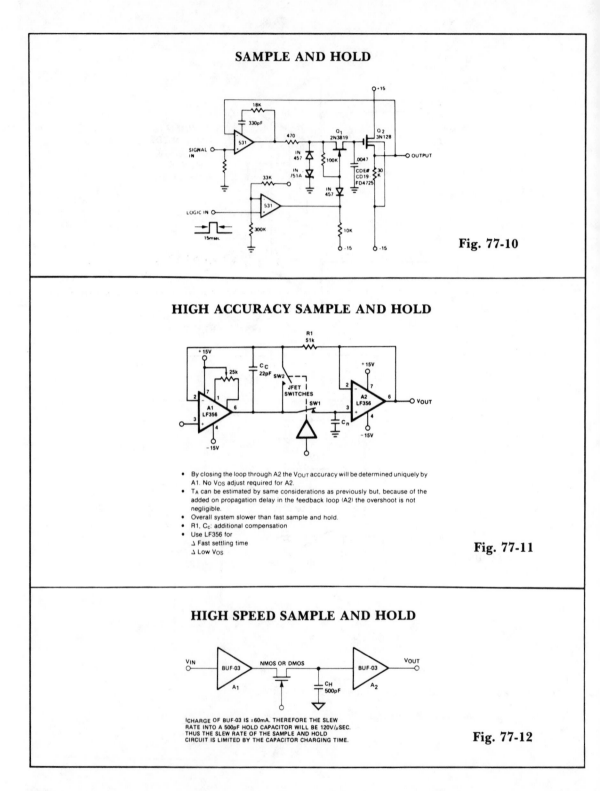

Fig. 77-10

HIGH ACCURACY SAMPLE AND HOLD

- By closing the loop through A2 the V_{OUT} accuracy will be determined uniquely by A1. No V_{OS} adjust required for A2.
- T_A can be estimated by same considerations as previously but, because of the added on propagation delay in the feedback loop (A2) the overshoot is not negligible.
- Overall system slower than fast sample and hold.
- R1, C_C: additional compensation
- Use LF356 for
  Δ Fast settling time
  Δ Low V_{OS}

Fig. 77-11

HIGH SPEED SAMPLE AND HOLD

I_{CHARGE} OF BUF-03 IS ±60mA. THEREFORE THE SLEW RATE INTO A 500pF HOLD CAPACITOR WILL BE 120V/μSEC. THUS THE SLEW RATE OF THE SAMPLE AND HOLD CIRCUIT IS LIMITED BY THE CAPACITOR CHARGING TIME.

Fig. 77-12

78

Schmitt Triggers

The sources of the following circuits are contained in the Sources section beginning on page 730. The figure number contained in the box of each circuit correlates to the source entry in the Sources section.

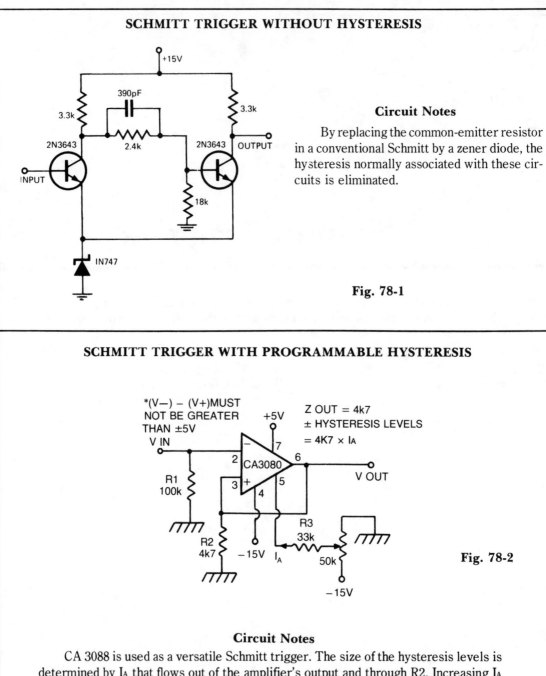

SCHMITT TRIGGER WITHOUT HYSTERESIS

+15V

390pF

3.3k

3.3k

2.4k

2N3643

2N3643   OUTPUT

INPUT

18k

1N747

Circuit Notes

By replacing the common-emitter resistor in a conventional Schmitt by a zener diode, the hysteresis normally associated with these circuits is eliminated.

Fig. 78-1

SCHMITT TRIGGER WITH PROGRAMMABLE HYSTERESIS

*(V—) — (V+)MUST NOT BE GREATER THAN ±5V

V IN

+5V

Z OUT = 4k7
± HYSTERESIS LEVELS
= 4K7 × I_A

2

CA3080

7

6

V OUT

R1
100k

3 +

5

4

R3
33k

R2
4k7

−15V I_A

50k

−15V

Fig. 78-2

Circuit Notes

CA 3088 is used as a versatile Schmitt trigger. The size of the hysteresis levels is determined by I_A that flows out of the amplifier's output and through R2. Increasing I_A increases hysteresis and vice versa. The positive and negative hysteresis levels are symmetrical about 0 V.

SCHMITT TRIGGER (ZERO CROSSING DETECTOR WITH HYSTERESIS)

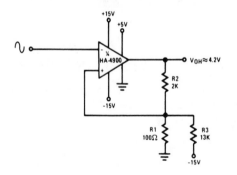

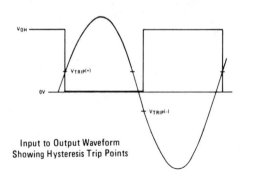

Input to Output Waveform
Showing Hysteresis Trip Points

Fig. 78-3

Circuit Notes

This circuit has a 100 mV hysteresis which can be used in applications where very fast transition times are required at the output even though the signal is very slow. The hysteresis loop also reduces false triggering due to noise on the input. The waveforms show the trip points developed by the hysteresis loop.

SCHMITT TRIGGER

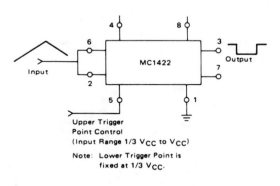

Fig. 78-4

Circuit Notes

The lower trigger point is fixed at ⅓ Vcc, but the upper trigger point is adjustable by means of Pin 5 from ⅓ Vcc to slightly less than Vcc. The Schmitt trigger will operate with input frequencies up to 50 kHz.

79

Smoke and Flame Detectors

The sources of the following circuits are contained in the Sources section beginning on page 730. The figure number contained in the box of each circuit correlates to the source entry in the Sources section.

PHOTOELECTRIC SMOKE DETECTOR (NON-LATCHING)

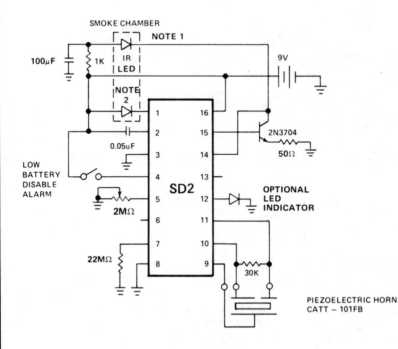

Fig. 79-1

Notes: 1. IR Diode   RCA Type SG 1010A or Spectronics Type SE 5455-4
Clairex Type CLED-1
2. IR Photo detectors   Vactec VTS4085

Circuit Notes

The LED predriver output pulses an external transistor which in turn, switches on the infrared light emitting diode at a very low duty cycle. The desired IR LED pulse period is determined by the value of the external timing resistor. The Smoke Sensitivity is adjustable through a trimmer resistor which varies the IR LED pulse width. The light sensing element is a silicon photovoltaic cell which is held at near zero bias to minimize leakage currents. The circuit can detect signals as low as 1 mV and generate an alarm. The IR LED pulse repetition rate increases when smoke is detected.

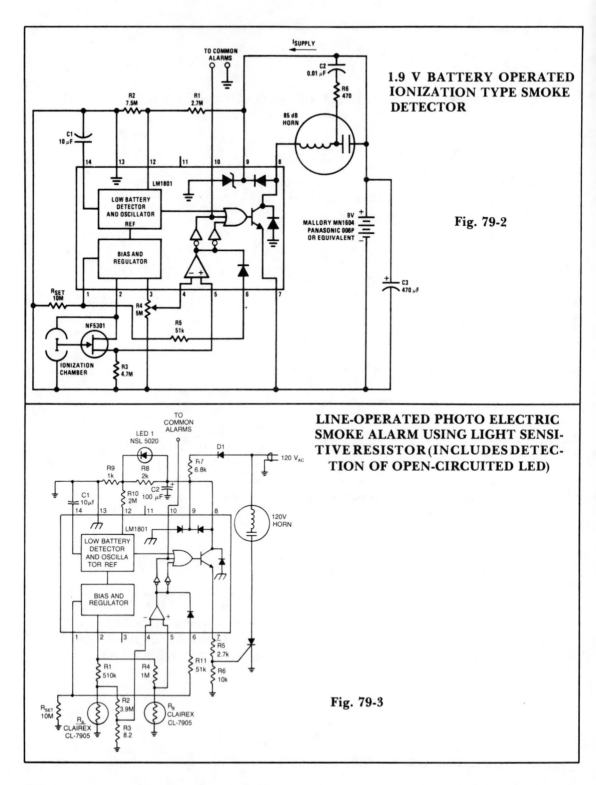

1.9 V BATTERY OPERATED IONIZATION TYPE SMOKE DETECTOR

Fig. 79-2

LINE-OPERATED PHOTO ELECTRIC SMOKE ALARM USING LIGHT SENSITIVE RESISTOR (INCLUDES DETECTION OF OPEN-CIRCUITED LED)

Fig. 79-3

80

Sound Effect Circuits

The sources of the following circuits are contained in the Sources section beginning on page 730. The figure number contained in the box of each circuit correlates to the source entry in the Sources section.

VOLTAGE-CONTROLLED AMPLIFIER OR TREMOLO CIRCUIT

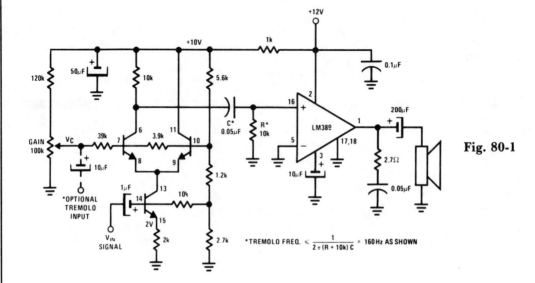

Fig. 80-1

$$*\text{TREMOLO FREQ.} \leq \frac{1}{2\pi(R+10k)C} = 160\,Hz\ \text{AS SHOWN}$$

Circuit Notes

The transistors form a differential pair with an active current-source tail. This configuration, known technically as a variable-transconductance multiplier, has an output proportional to the product of the two input signals. Multiplication occurs due to the dependence of the transistor transconductance on the emitter current bias. Tremolo (amplitude modulation of an audio frequency by a sub-audio oscillator—normally 5-15 Hz) applications require feeding the low frequency oscillator signal into the optional input shown. The gain control pot maybe set for optimum depth.

MUSIC SYNTHESIZER

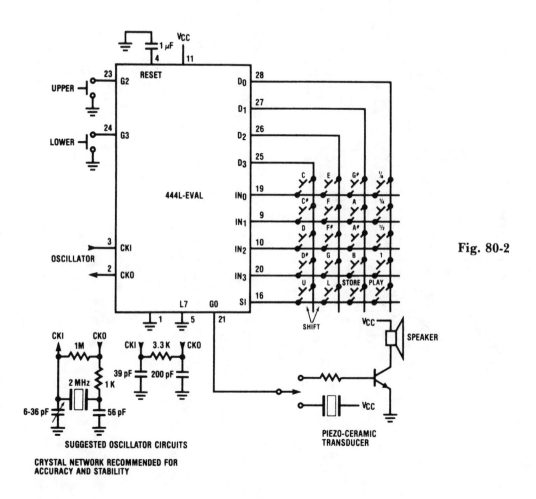

Fig. 80-2

Circuit Notes

Three modes of operation are available in the music synthesizer mode: play a note, play one of four stored tunes, or record a tune for subsequent replay.

PREPROGRAMMED SINGLE-CHIP
MICROCONTROLLER FOR MUSICAL ORGAN

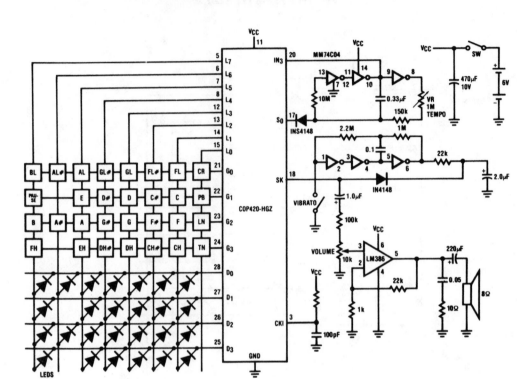

Fig. 80-3

Circuit Notes

Twenty-five musical keys and 25 LEDs are provided to denote F to F″ with half notes in between. Memory can store a played tune. There are ten preprogrammed tunes (each has an average of 55 notes) masked in the chip. Any tune can be recalled by depressing the Tune Button followed by the corresponding Sharp Key. In learn mode, the player can learn the ten preprogrammed tunes.

MUSICAL ENVELOPE GENERATOR AND MODULATOR

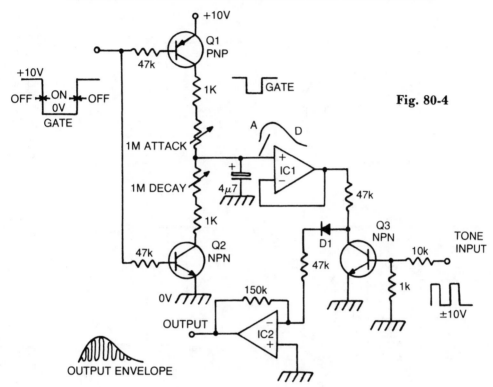

Fig. 80-4

Circuit Notes

When a gate voltage is applied, Q1 is turned on and capacitor C is charged via the attack pot in series with the 1 K resistor varying this pot, attact time constant. A fast attack gives a percussive sound, a slow attack the affect of "backward" sounds. When the gate voltage returns to its off state, Q2 is turned on and capacitor is discharged via decay pot to ground. The envelope is buffered by IC1 and applied to Q3, which is used as a transistor chopper. A musical tone in the form of a squarewave is connected to the base of Q3. This turns the transistor on or off and thus the envelope is chopped up at regular intervals, the intervals being determined by the pitch of the squarewave. The resultant waveform has the amplitude of the envelope and the harmonic structure of the squarewave. IC2 buffers the signal and D1 ensures that the envelope dies away at the end of a note.

STEREO REVERB SYSTEM

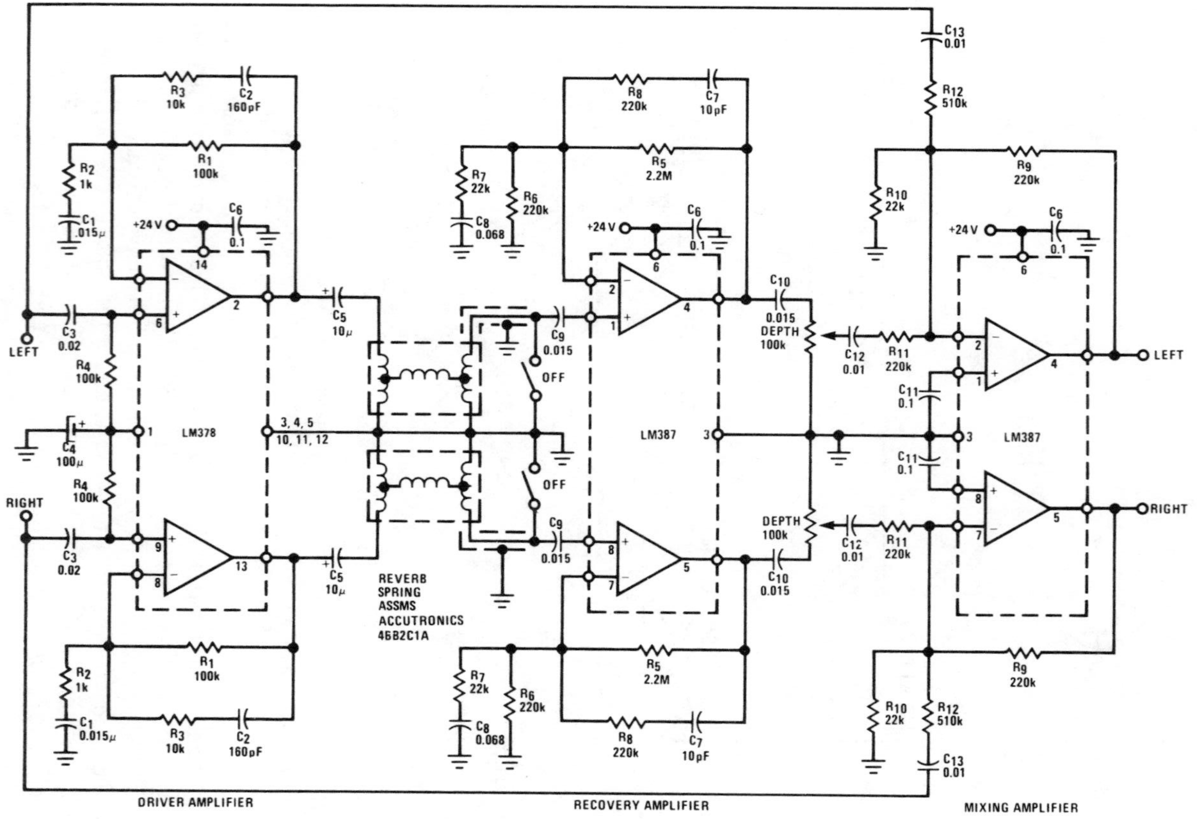

Fig. 80-5

Circuit Notes

The LM378 dual power amplifier is used as the spring driver. The recovery amplifier is a low noise dual preamplifier. Mixing of the delayed signal with the original is done with another LM387 used in an inverting summing configuration.

FOUR CHANNEL SYNTHESIZER

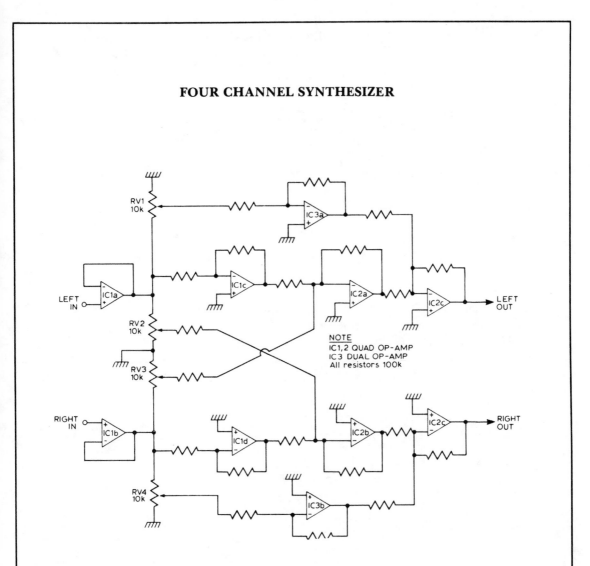

Fig. 80-6

Circuit Notes

This circuit will synthesize two rear channels for quadraphonic sound when fed with a stereo signal. The rear output for the left channel, is a combination of the left channel input 180 out of phase, added to a proportion of the right hand channel (also out of phase). The right hand rear output is obtained in a similar way.

TONE BURST GENERATOR

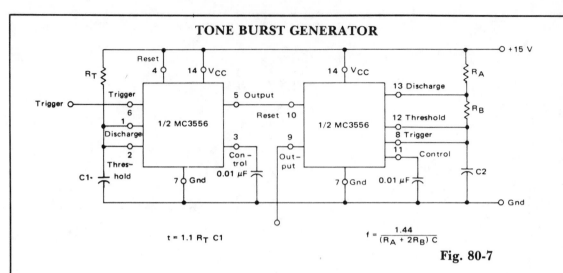

$t = 1.1\ R_T\ C1$

$f = \dfrac{1.44}{(R_A + 2R_B)\ C}$

Fig. 80-7

Circuit Notes

The first timer is used as a monostable and determines the tone duration when triggered by a positive pulse at pin 6. The second timer is enabled by the high output of the monostable. It is connected as an astable and determines the frequency of the tone.

MUSICAL CHIME GENERATOR

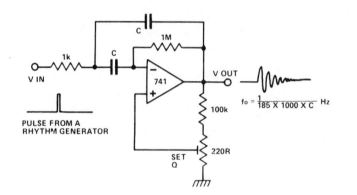

$f_0 = \dfrac{1}{185 \times 1000 \times C}$ Hz

Fig. 80-8

Circuit Notes

The circuit is that of a multiple feedback bandpass filter. A short click (pulse), makes it ring with a frequency which is its natural resonance frequency. Oscillations die away exponentially and closely resemble many naturally occuring percussive or plucked sounds. The higher the Q the longer the decay time constant. High frequency resonances resemble chimes, lower frequencies sound like claves or bongos. Several circuits, all with different tuning, driven by pulses from a rhythm generator can produce an interesting pattern of sounds.

SOUND EFFECT GENERATOR

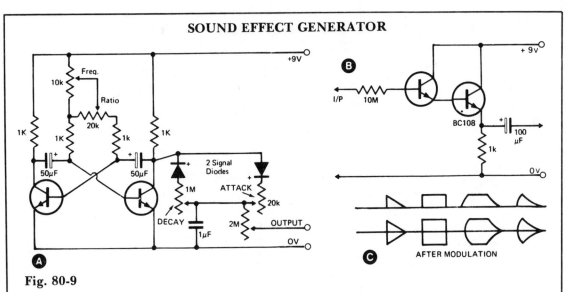

Fig. 80-9

Circuit Notes

This waveshape generator is basically a slow running oscillator with variable attack and decay. A variable amplitude (high impedance) output is available via the 2 M potentiometer. B shows an add-on circuit which should be used if a low impedance output is required. Some of the output waveforms that can be produced are shown in C.

PROGRAMMABLE BIRD SOUNDS

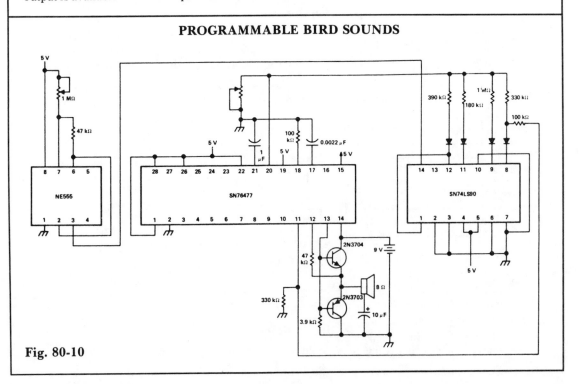

Fig. 80-10

605

STEREO REVERB ENHANCEMENT SYSTEM

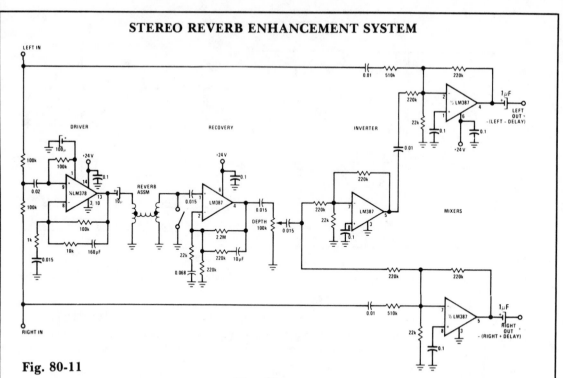

Fig. 80-11

Circuit Notes

The system can be used to synthesize a stereo effect from a monaural source such as AM radio or FM-mono broadcast, or it can be added to an existing stereo (or quad) system where it produces an exciting "opening up" special effect that is truly impressive.

SIREN/SPACE WAR/PHASOR GUN

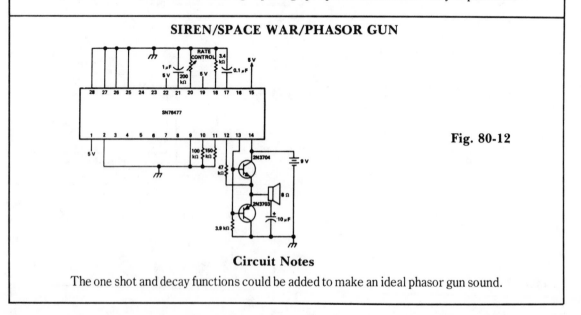

Fig. 80-12

Circuit Notes

The one shot and decay functions could be added to make an ideal phasor gun sound.

81

Sound (Audio)
Operated Circuits

The sources of the following circuits are contained in the Sources section beginning on page 730. The figure number contained in the box of each circuit correlates to the source entry in the Sources section.

Voice Activated Switch and Amplifier
Audio Operated Relay
Sound-Modulated Light Source

Audio-Controlled Lamp
Sound Activated Relay
Sound Operated Two-Way Switch

VOICE ACTIVATED SWITCH AND AMPLIFIER

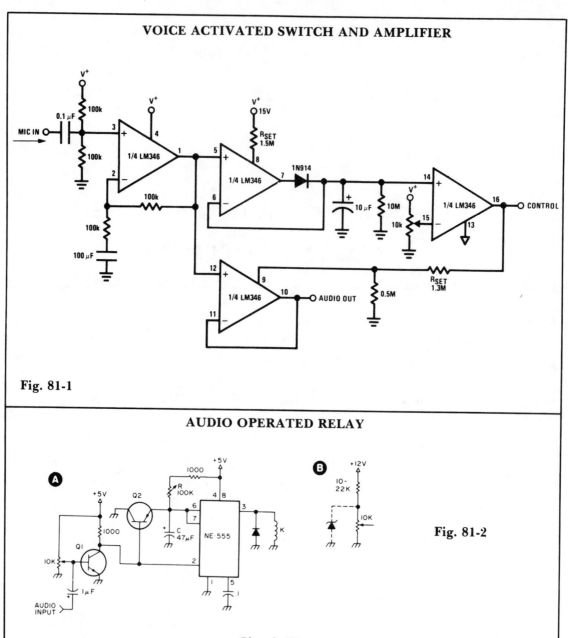

Fig. 81-1

AUDIO OPERATED RELAY

Fig. 81-2

Circuit Notes

Q1 and Q2 are general purpose transistors. The 10 K input pot is adjusted to a point just short of where Q1 turns on as indicated by K pulling in. K is any 5 V reed relay. With the values shown for R (100 K) and C (47 μF), timing values from .05 to slightly over 5 seconds can be achieved. B shows the addition of a 22 K series resistor to the 10 K input pot if a 12 V supply is used. A suitable 12 V reed relay must be used at K.

SOUND-MODULATED LIGHT SOURCE

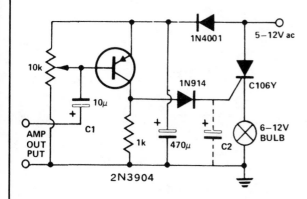

1N4001

5–12V ac

10k

1N914

C106Y

10μ

C1

AMP OUT PUT

1k

470μ

C2

6–12V BULB

2N3904

Circuit Notes

This circuit modulates a light beam with voice or music from the output of an amplifier. If the 10 K pot is adjusted to slightly less than the V_{be} of the transistor, the circuit forms a peak detector. This drives the gate of the SCR, lighting the bulb whose brightness will vary as the sound level varies. C2 may be removed for a faster response.

Fig. 81-3

AUDIO-CONTROLLED LAMP

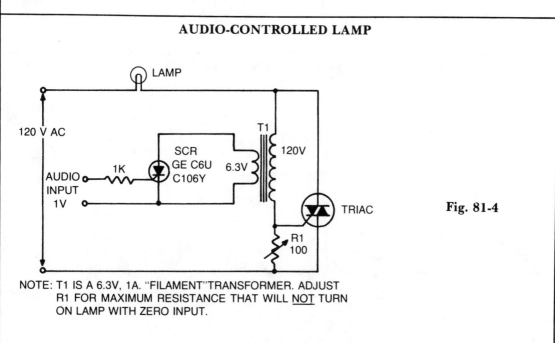

LAMP

120 V AC

SCR GE C6U C106Y

1K

AUDIO INPUT 1V

T1

120V

6.3V

TRIAC

R1 100

Fig. 81-4

NOTE: T1 IS A 6.3V, 1A. "FILAMENT" TRANSFORMER. ADJUST R1 FOR MAXIMUM RESISTANCE THAT WILL <u>NOT</u> TURN ON LAMP WITH ZERO INPUT.

Circuit Notes

This is an on-off control with isolated, low voltage input. Since the switching action is very rapid, compared with the response time of the lamp and the response of the eye, the effect produced with audio input is similar to a proportional control circuit. If the input signal to the SCR consists of phase-controlled pulses, full wave control of the lamp load is obtained.

SOUND ACTIVATED RELAY

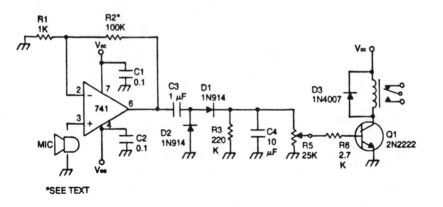

Fig. 81-5

*SEE TEXT

Circuit Notes

The device remains dormat (in an off condition) until some sound causes it to turn on. The input stage is a 741 operational amplifier connected as a noninverting follower audio amplifier. Gain is approximately 100. To increase gain raise the value of R2. The amplified signal is rectified and filtered to a dc level by R4. Then R5 is set to the audio level desired to activate the relay.

SOUND OPERATED TWO-WAY SWITCH

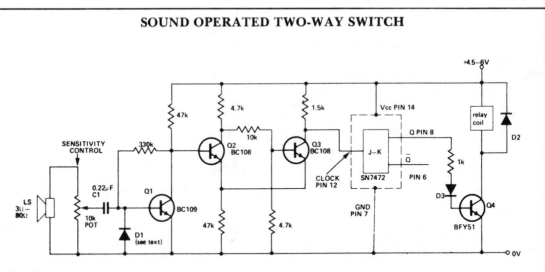

Fig. 81-6

Circuit Notes

This circuit operates a relay each time a sound of sufficient intensity is made, thus one clap of the hands will switch it one way, a second clap will revert the circuit to the original condition. Q2 and Q3 form a Schmitt trigger. The JK flip-flop is used as a bistable whose output changes state every time a pulse is applied to the clock input (pin 12). Q4 allows the output to drive a relay.

82

Square Wave Oscillators

The sources of the following circuits are contained in the Sources section beginning on page 730. The figure number contained in the box of each circuit correlates to the source entry in the Sources section.

R/C Oscillator
1 kHz Square Wave Oscillator
TTL Oscillator
Square Wave Oscillator
Adjustable TTL Clock
Square Wave Oscillator
Oscillator/Clock Generator

CMOS Oscillator
Free-Running Square-Wave Oscillator
Precision Squares
Square Wave Oscillator
0.5 Hz Square-Wave Oscillator
Simple Triangle/Square Wave Oscillator
Squarewave Oscillator

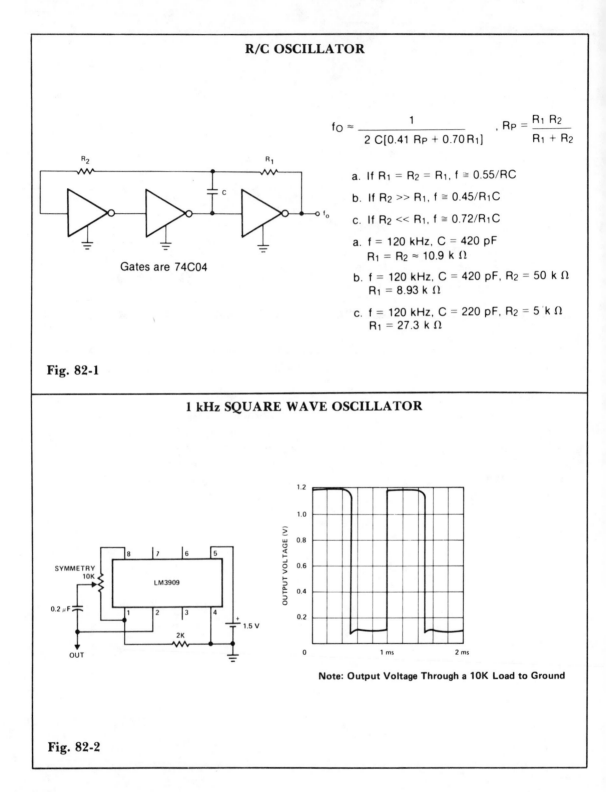

R/C OSCILLATOR

$$f_O \approx \frac{1}{2\,C[0.41\,R_P + 0.70\,R_1]} \quad , R_P = \frac{R_1\,R_2}{R_1 + R_2}$$

a. If $R_1 = R_2 = R_1$, $f \cong 0.55/RC$

b. If $R_2 \gg R_1$, $f \cong 0.45/R_1C$

c. If $R_2 \ll R_1$, $f \cong 0.72/R_1C$

a. $f = 120$ kHz, $C = 420$ pF
   $R_1 = R_2 \approx 10.9$ k Ω

b. $f = 120$ kHz, $C = 420$ pF, $R_2 = 50$ k Ω
   $R_1 = 8.93$ k Ω

c. $f = 120$ kHz, $C = 220$ pF, $R_2 = 5$ k Ω
   $R_1 = 27.3$ k Ω

Gates are 74C04

Fig. 82-1

1 kHz SQUARE WAVE OSCILLATOR

Note: Output Voltage Through a 10K Load to Ground

Fig. 82-2

TTL OSCILLATOR

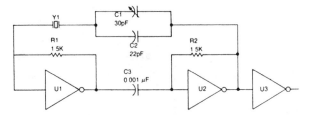

Fig. 82-3

Circuit Notes

TTL inverter stages, U1 and U2, are cross-connected with a crystal Y1. A resistor in each stage biases the normally digital gates into a region where they operate as amplifiers. Inverter stage U3 is used as a buffer.

SQUARE WAVE OSCILLATOR

Oscillator Frequency for Various Capacitor Values

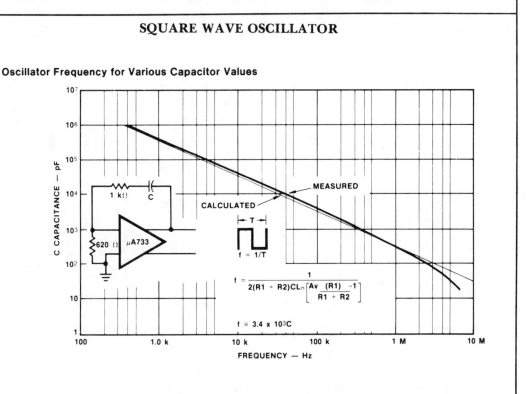

Fig. 82-4

ADJUSTABLE TTL CLOCK (MAINTAINS 50% DUTY CYCLE)

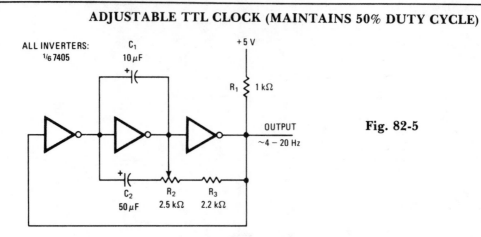

Fig. 82-5

Circuit Notes

Symmetry of the square-wave output is maintained by connecting the right side of R2 through resistor R3 to the output of the third amplifier stage. This changes the charging current to the capacitors in proportion to the setting of frequency-adjusting potentiometer R2. Thus, a duty cycle of 50% is constant over the entire range of oscillation. The lower frequency limit is set by capacitor C2. With the components shown, the frequency of oscillation can be varied by R2 from about 4 to 20 hertz. Other frequency ranges can be obtained by changing the values of C1 and R3, which control the upper limit of oscillation, or C2, which limits the low-frequency end.

SQUARE WAVE OSCILLATOR

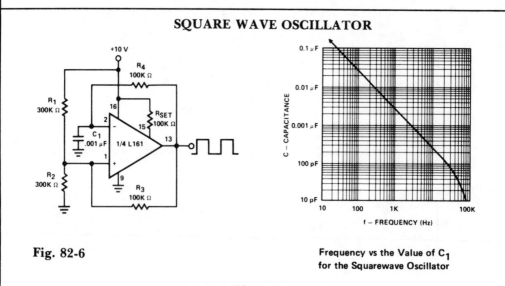

Fig. 82-6

Frequency vs the Value of C_1
for the Squarewave Oscillator

Circuit Notes

This generator is operable to over 100 kHz. The low frequency limit is determined by C1. Frequency is constant for supply voltages down to +5 V.

OSCILLATOR/CLOCK GENERATOR

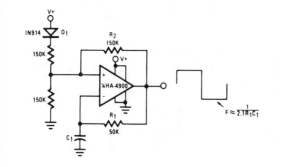

Fig. 82-7

Circuit Notes

This self-starting fixed frequency oscillator circuit gives excellent frequency stability. R1 and C1 comprise the frequency determining network while R2 provides the regenerative feedback. Diode D1 enhances the stability by compensating for the difference between V_{OH} and V_{Supply}. In applications where a precision clock generator up to 100 kHz is required, such as in automatic test equipment, C1 may be replaced by a crystal.

$$F \approx \frac{1}{2.1 R_1 C_1}$$

CMOS OSCILLATOR

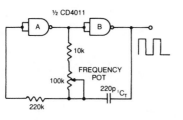

Circuit Notes

Varying the 100 K pot changes the discharge rate of C_T and hence the frequency. A square wave output is generated. The maximum frequency using CMOS is limited to 2 MHz.

Fig. 82-8

FREE-RUNNING SQUARE-WAVE OSCILLATOR

Fig. 82-9

PRECISION SQUARER

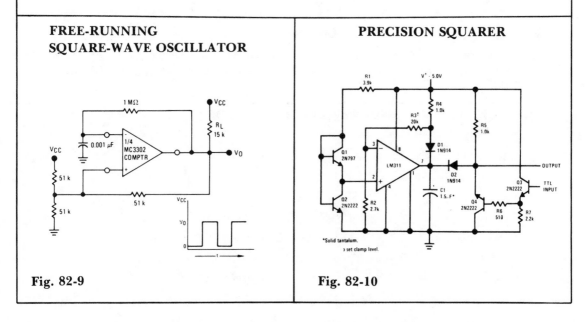

Fig. 82-10

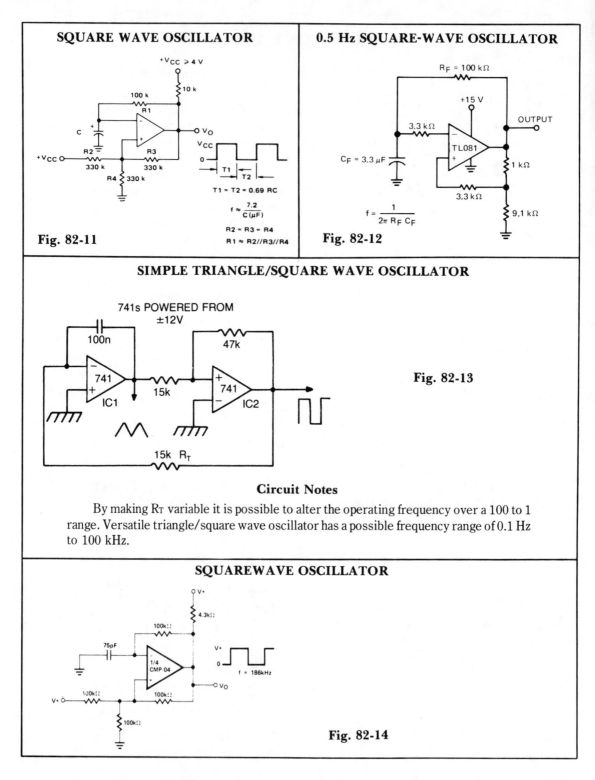

SQUARE WAVE OSCILLATOR

$+V_{CC} \geqslant 4 V$

10 k

100 k
R1

C

$+$

$-$

$+$

V_O

R2
330 k

R3
330 k

$+V_{CC}$

R4 330 k

V_{CC}

0

T1 T2

T1 = T2 = 0.69 RC

$f \approx \dfrac{7.2}{C(\mu F)}$

R2 = R3 = R4

R1 ≈ R2//R3//R4

Fig. 82-11

0.5 Hz SQUARE-WAVE OSCILLATOR

$R_F = 100 \, k\Omega$

+15 V

OUTPUT

$3.3 \, k\Omega$

$C_F = 3.3 \, \mu F$

TL081

$-$

$+$

1 kΩ

$3.3 \, k\Omega$

9.1 kΩ

$f = \dfrac{1}{2\pi \, R_F \, C_F}$

Fig. 82-12

SIMPLE TRIANGLE/SQUARE WAVE OSCILLATOR

741s POWERED FROM
±12V

100n

$-$

741
$+$
IC1

15k

47k

$+$
741
$-$
IC2

15k R_T

Fig. 82-13

Circuit Notes

By making R_T variable it is possible to alter the operating frequency over a 100 to 1 range. Versatile triangle/square wave oscillator has a possible frequency range of 0.1 Hz to 100 kHz.

SQUAREWAVE OSCILLATOR

V+

4.3kΩ

100kΩ

75pF

1/4
CMP-04

V+

0

f = 186kHz

V+

100kΩ

100kΩ

V_O

100kΩ

Fig. 82-14

616

83

Stereo Balance Circuits

The sources of the following circuits are contained in the Sources section beginning on page 730. The figure number contained in the box of each circuit correlates to the source entry in the Sources section.

Stereo Balance Meter   Stereo Balancer

Stereo Balance Meter

STEREO BALANCE METER

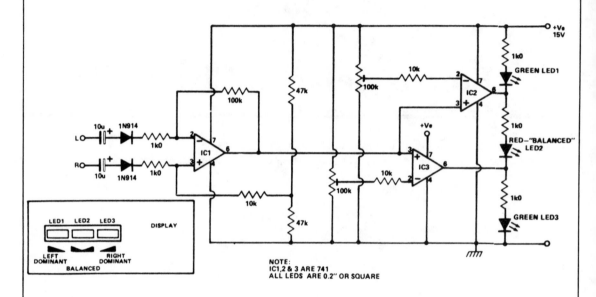

Fig. 83-1

Circuit Notes

Outputs from each channel are fed to the two inputs of IC1 connected as a differential amplifier. IC2 and 3 are driven by the output of IC1. Output of IC1 is connected to the noninverting inputs of IC2 and 3. If the output of IC1 approaches the supply rail, the outputs of ICs 2 and 3 will also go high, illuminating LED3. This would happen if the right channel were dominating. If the left channel was dominant, the outputs of ICs 2 and 3 would be low, illuminating LED1. If the two channels are equal in amplitude, the outputs of ICs 2 and 3 would be high and low respectively, lighting up LED2.

STEREO BALANCER

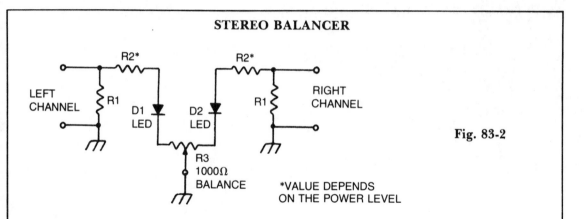

Fig. 83-2

*VALUE DEPENDS ON THE POWER LEVEL

Circuit Notes

This circuit will allow you to set the gain of two stereo channels to the same level. The signal across the two channel-load resistors is sampled by resistors R2. (Values of these resistors will depend upon the power level.) For most 20 milliampere LED, use approximately 2.5 K per watt. (For a 10-watt system use a 25,000 ohm resistor.) To set up, short the two inputs and connect them to one channel of a power amplifier. Apply a signal and adjust R3 until both LEDs glow at the same brightness level. The balancer is ready for use. Connect the inputs of the stereo balancer across the output of the power amplifier, and then turn up either the independent volume controls, or the balance control until both LEDs glow at the same level. To use this circuit in-line with loudspeakers, disconnect both R1s, and use the speakers as the load.

STEREO BALANCE METER

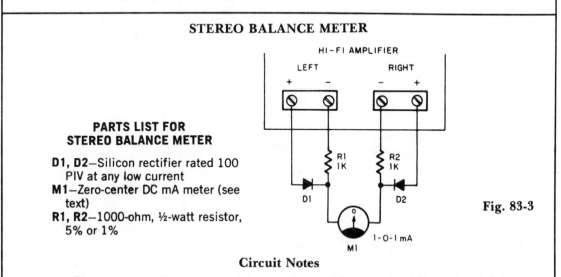

Fig. 83-3

PARTS LIST FOR STEREO BALANCE METER

D1, D2—Silicon rectifier rated 100 PIV at any low current
M1—Zero-center DC mA meter (see text)
R1, R2—1000-ohm, ½-watt resistor, 5% or 1%

Circuit Notes

Play any stereo disc or tape and then set the amplifier to mono. Adjust left and right channel balance until meter M1 indicates zero; then the left and right output level are identical.

84

Switches

The sources of the following circuits are contained in the Sources section beginning on page 730. The figure number contained in the box of each circuit correlates to the source entry in the Sources section.

DTL-TTL Controlled Buffered Analog
    Switch
High Toggle Rate High Frequency Analog
    Switch

Differential Analog Switch
High Frequency Switch
Two-Channel Switch
10 A, 25 VDC Solid State Relays

DTL-TTL CONTROLLED BUFFERED ANALOG SWITCH

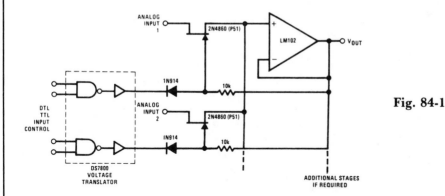

Fig. 84-1

Circuit Notes

This analog switch uses the 2N4860 JFET for its 25 ohm r_{on} and low leakage. The LM102 serves as a voltage buffer. This circuit can be adapted to a dual trace oscilloscope chopper. The DS7800 monolithic IC provides adequate switch drive controlled by DTL/TTL logic levels.

HIGH TOGGLE RATE HIGH FREQUENCY ANALOG SWITCH

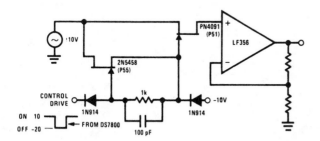

Fig. 84-2

Circuit Notes

Commutator circuit provides low impedance gate drive to the PN4091 analog switch for both on and off drive conditions. This circuit also approaches the ideal gate drive conditions for high frequency signal handling by providing a low ac impedance for off drive and high ac impedance for on drive to the PN4091

DIFFERENTIAL ANALOG SWITCH

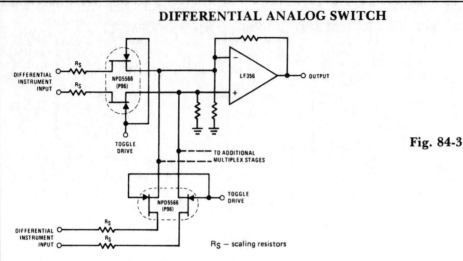

Fig. 84-3

Circuit Notes

The NPD5566 monolithic dual is used in a differential multiplex application where $R_{ds(ON)}$ should be closely matched. Since $R_{ds(ON)}$ for the monolithic dual tracks at better than ±1% over wide temperature ranges ($-25°$ C to $+125°$ C), this makes it an unusual but ideal choice for an accurate multiplexer. This close tracking greatly reduces errors due to common-mode signals.

HIGH FREQUENCY SWITCH

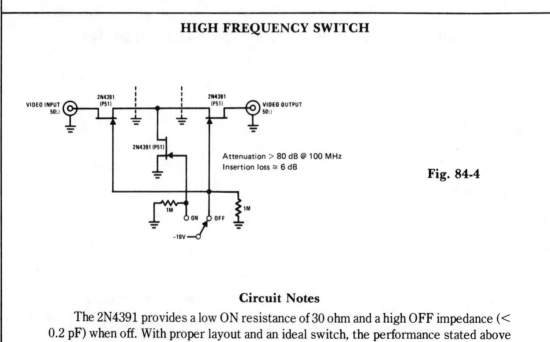

Attenuation > 80 dB @ 100 MHz
Insertion loss ≅ 6 dB

Fig. 84-4

Circuit Notes

The 2N4391 provides a low ON resistance of 30 ohm and a high OFF impedance (< 0.2 pF) when off. With proper layout and an ideal switch, the performance stated above can be readily achieved.

TRIAC ZERO VOLTAGE SWITCHING

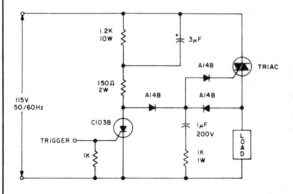

Circuit Notes

The triac will be gated on at the start of the positive half cycle by current flow through the 3 μF capacitor as long as the C103 SCR is off. The load voltage then charges up the 1 μF capacitor so that the triac will again be energized during the subsequent negative half cycle of line voltage. A selected gate triac is required because of the III+ triggering mode.

Fig. 84-5

TWO-CHANNEL SWITCH

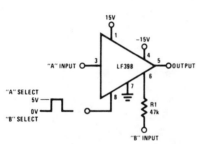

	A	B
Gain	1 ±0.02%	1 ±0.2%
Z_{IN}	$10^{10}\,\Omega$	47 kΩ
BW	≃ 1 MHz	≈ 400 kHz
Crosstalk @ 1 kHz	−90 dB	−90 dB
Offset	≤ 6 mV	≤ 75 mV

Fig. 84-6

10 A, 25 Vdc SOLID STATE RELAYS

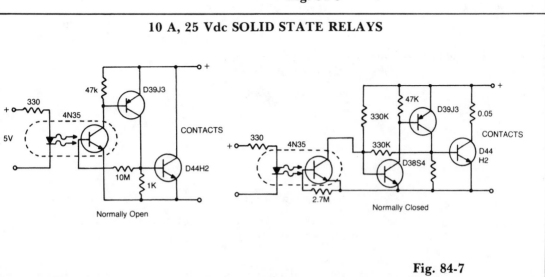

Fig. 84-7

85

Telephone Related Circuits

The sources of the following circuits are contained in the Sources section beginning on page 730. The figure number contained in the box of each circuit correlates to the source entry in the Sources section.

Portable Tone Generator
Telephone Status Monitor Using an Optoisolator
Telephone Tone Ringer
F.C.C. Approved Telephone Tone Ringer
Telephone or Extension Tone Ringer
Telephone Line Monitor
Tone Dial Generator
Tone Dial Encoder
Tone Dial Sequence Decoder
Remote Ring Extender Switch

Tone Dial Decoder
Telephone Relay
Telephone-Controlled Tape Starter (TCTS)
Telephone Line Powered Repertory Dialer
Telephone Off-Hook Indicator
Telephone Handset Tone Dial Encoder
Low Line Loading Ring Detector
Phone Auto Answer and Ring Indicator
Autopatch Telephone Phone Line Interface
Telephone Ringer Uses Piezoelectric Device

Electronic Phone Bell

PORTABLE TONE GENERATOR

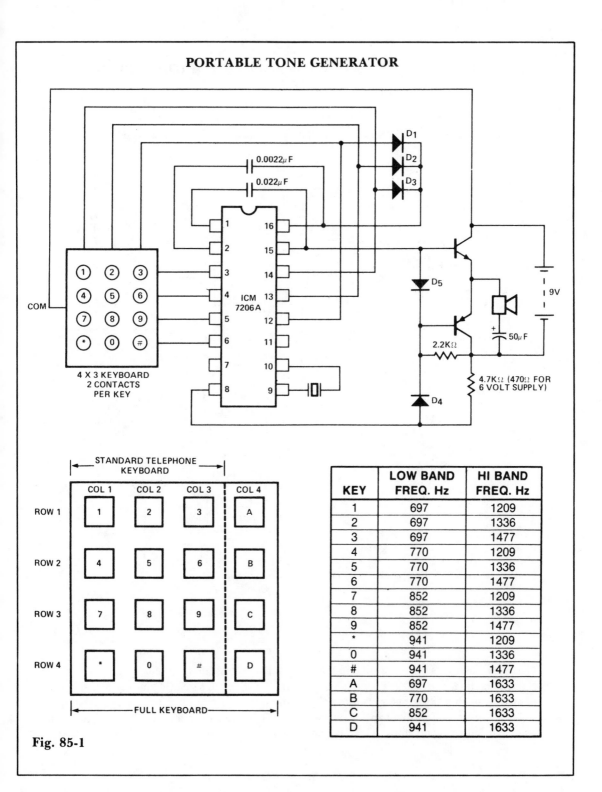

Fig. 85-1

KEY	LOW BAND FREQ. Hz	HI BAND FREQ. Hz
1	697	1209
2	697	1336
3	697	1477
4	770	1209
5	770	1336
6	770	1477
7	852	1209
8	852	1336
9	852	1477
*	941	1209
0	941	1336
#	941	1477
A	697	1633
B	770	1633
C	852	1633
D	941	1633

TELEPHONE STATUS MONITOR USING AN OPTOISOLATOR

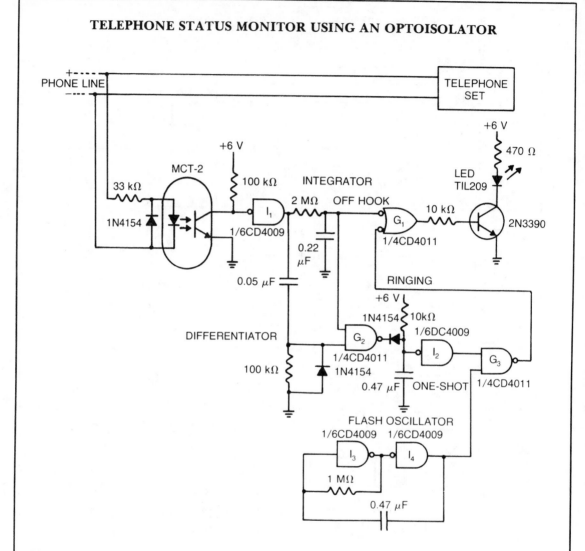

Fig. 85-2

Circuit Notes

The LED indicates the status of a remote telephone. The light is off if the phone is hung up. It shines steadily if the phone is off hook, and it flashes on and off while phone rings and for 5 seconds after ringing stops. The flashing oscillator operates continuously but can drive the LED only when a ringing signal discharges the one shot capacitor to enable NAND gate G3. Thus, one oscillator handles several phone lines.

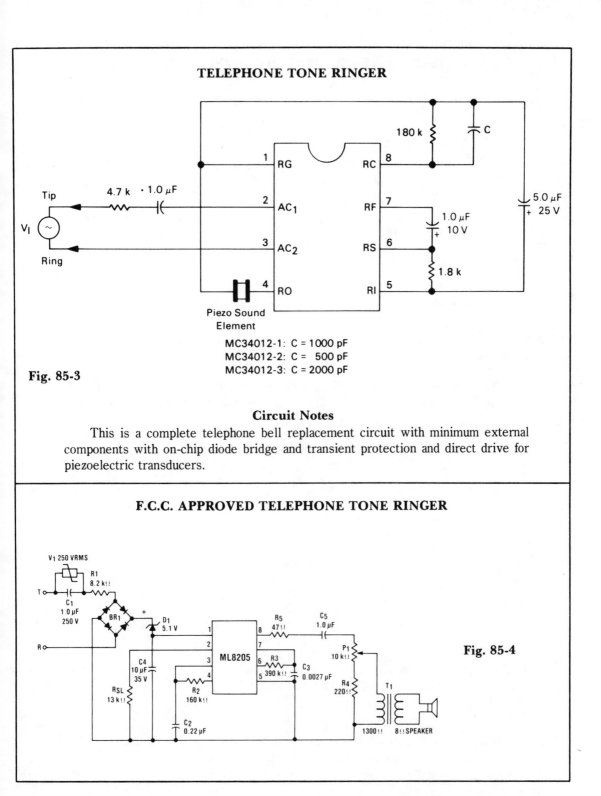

TELEPHONE TONE RINGER

180 k

C

1 RG
2 AC$_1$
3 AC$_2$
4 RO

8 RC
7 RF
6 RS
5 RI

Tip

4.7 k · 1.0 μF

V$_I$ ~

Ring

5.0 μF 25 V

1.0 μF 10 V

1.8 k

Piezo Sound
Element

MC34012-1: C = 1000 pF
MC34012-2: C = 500 pF
MC34012-3: C = 2000 pF

Fig. 85-3

Circuit Notes

This is a complete telephone bell replacement circuit with minimum external components with on-chip diode bridge and transient protection and direct drive for piezoelectric transducers.

F.C.C. APPROVED TELEPHONE TONE RINGER

V$_1$ 250 VRMS

R1 8.2 kΩ

T

C1 1.0 μF 250 V

BR$_1$

D$_1$ 5.1 V

ML8205

R5 47Ω

C5 1.0 μF

P$_1$ 10 kΩ

R

C4 10 μF 35 V

R3 390 kΩ

C3 0.0027 μF

R4 220Ω

T$_1$

RSL 13 kΩ

R2 160 kΩ

C2 0.22 μF

1300Ω

8Ω SPEAKER

Fig. 85-4

TELEPHONE OR EXTENSION TONE RINGER

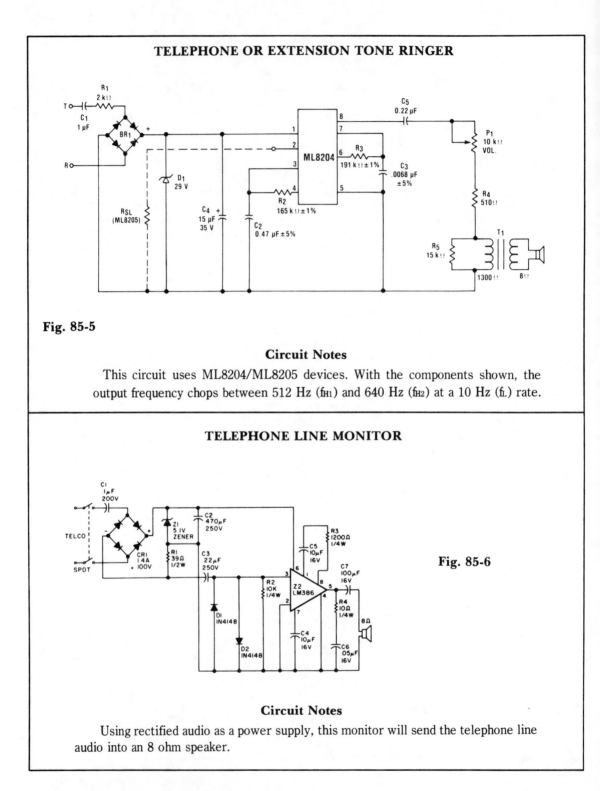

Fig. 85-5

Circuit Notes

This circuit uses ML8204/ML8205 devices. With the components shown, the output frequency chops between 512 Hz (f_{H1}) and 640 Hz (f_{H2}) at a 10 Hz (f_L) rate.

TELEPHONE LINE MONITOR

Fig. 85-6

Circuit Notes

Using rectified audio as a power supply, this monitor will send the telephone line audio into an 8 ohm speaker.

TONE DIAL GENERATOR

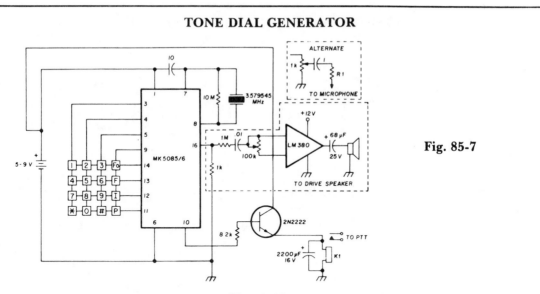

Fig. 85-7

Circuit Notes

The circuit requires a minimum of parts and uses a low cost standard 3.579545-MHz television color-burst crystal. The speaker can be eliminated and the output fed directly into the microphone input of a transmitter.

TONE DIAL ENCODER

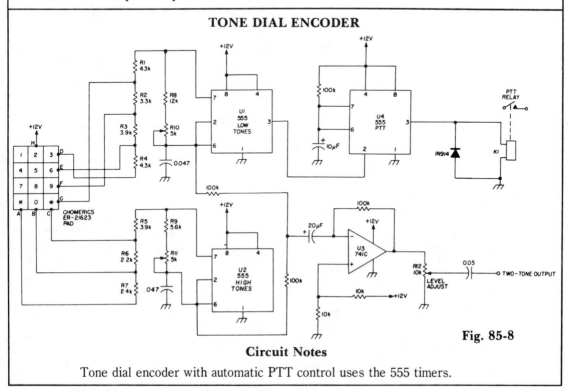

Fig. 85-8

Circuit Notes

Tone dial encoder with automatic PTT control uses the 555 timers.

TONE DIAL SEQUENCE DECODER

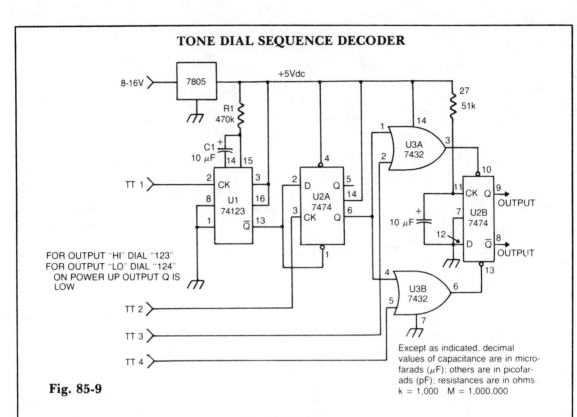

FOR OUTPUT "HI" DIAL "123"
FOR OUTPUT "LO" DIAL "124"
ON POWER UP OUTPUT Q IS
LOW

Except as indicated, decimal
values of capacitance are in microfarads (μF); others are in picofarads (pF); resistances are in ohms.
k = 1,000   M = 1,000,000

Fig. 85-9

Circuit Notes

The circuit takes active low inputs from a Touch Tone decoder and reacts to a proper sequence of digits. The proper sequence is determined by which Touch Tone digits the user connects to the sequence decoder inputs TT1, TT2, TT3, and TT4.

REMOTE RING EXTENDER SWITCH

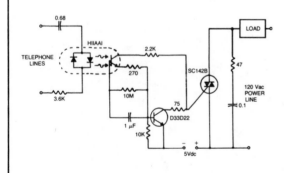

Circuit Notes

The circuit can operate lamps and buzzers from the 120 V, 60 Hz power line while maintaining positive isolation between the telephone line and the power line. Use of the isolated tab triac simplifies heat sinking by removing the constraint of isolating the triac heat sink from the chassis.

Fig. 85-10

TONE DIAL DECODER

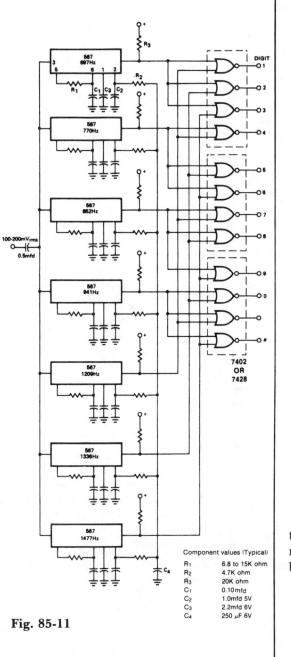

Component values (Typical)	
R₁	6.8 to 15K ohm
R₂	4.7K ohm
R₃	20K ohm
C₁	0.10 mfd
C₂	1.0 mfd 5V
C₃	2.2 mfd 6V
C₄	250 µF 6V

Fig. 85-11

TELEPHONE RELAY

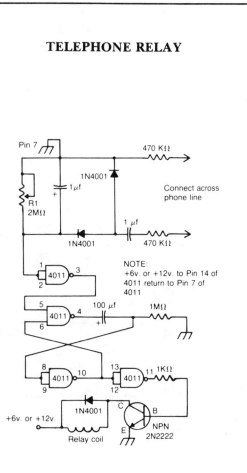

NOTE:
+6v. or +12v. to Pin 14 of 4011 return to Pin 7 of 4011

Circuit Notes

Connected across the bell circuit of phone, this circuit closes a relay when the phone is ringing. Use the delay contacts to actuate any bell, siren, buzzer or lamp.

Fig. 85-12

TELEPHONE-CONTROLLED TAPE STARTER (TCTS)

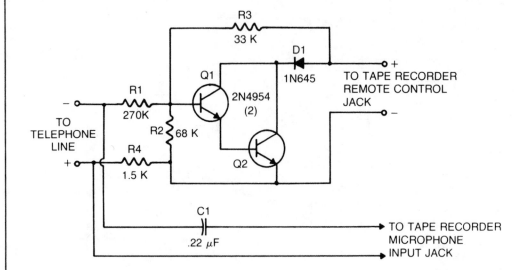

Fig. 85-13

Circuit Notes

This circuit converts a tape recorder into a completely automatic telephone conversation recording instrument that needs no external power source. Voltage at the switch terminals of tape recorder applied to a pair of Darlington-connected transistors, Q1 and Q2, will turn on and start the tape recorder. To turn the transistors off, and thereby stop the machine, apply a negative voltage to the base of Q1 from the phone line. When the telephone receiver is on the hook, there is typically about 50 volts dc across the phone divided across R1, R2, and R4 in such a way that the base of Q1 is sufficiently negative to keep the tape recorder off. When the phone's receiver is picked up, the voltage on the telephone line drops to about 5 volts, which leaves insufficient negative voltage on the base of Q1 to keep it cut off, so the tape recorder starts and begins to record.

TELEPHONE-LINE POWERED REPERTORY DIALER

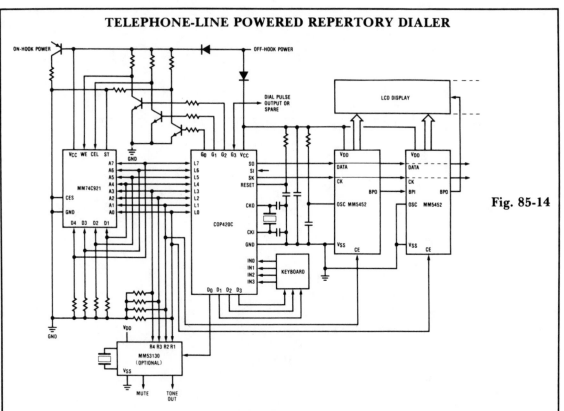

Fig. 85-14

Circuit Notes

Repertory dialer phone has a library of fifteen frequently used numbers, (plus the last number dialed) stored in a standard CMOS RAM. A pushbutton keyboard enables tele-phone numbers to be keyed in and dialed out directly or a telephone number to be stored in the RAM and dialed automatically.

TELEPHONE OFF-HOOK INDICATOR

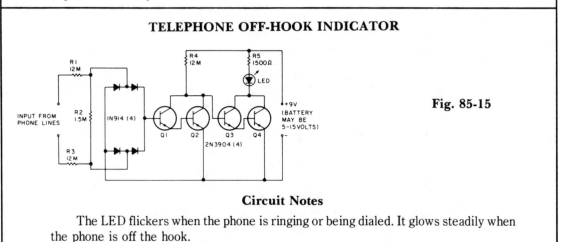

Fig. 85-15

Circuit Notes

The LED flickers when the phone is ringing or being dialed. It glows steadily when the phone is off the hook.

TELEPHONE HANDSET TONE DIAL ENCODER

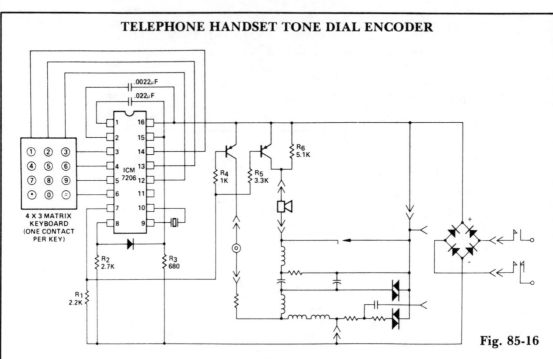

Fig. 85-16

Circuit Notes

This encoder uses a single contact per key keyboard and provides all other switching function electronically. The diode between terminals 8 and 15 prevents the output going more than 1 volt negative with respect to the negative supply V−. The circuit operates over the supply voltage range from 3.5 volts to 15 volts.

LOW LINE LOADING RING DETECTOR

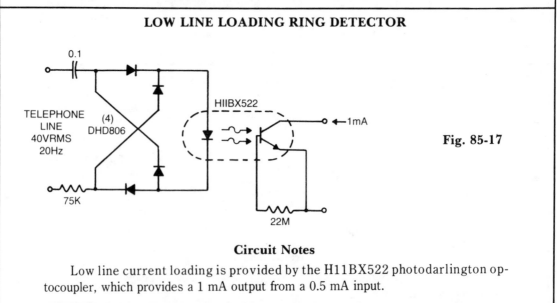

Fig. 85-17

Circuit Notes

Low line current loading is provided by the H11BX522 photodarlington optocoupler, which provides a 1 mA output from a 0.5 mA input.

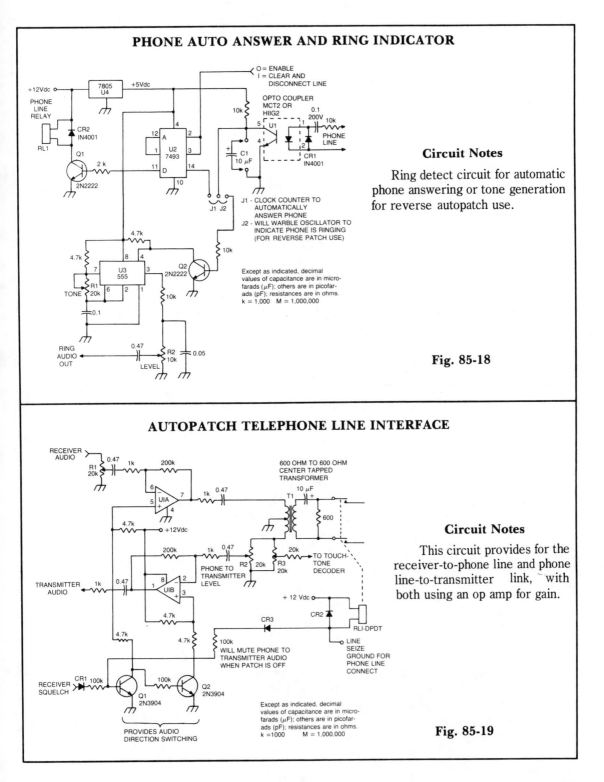

PHONE AUTO ANSWER AND RING INDICATOR

O = ENABLE
I = CLEAR AND
DISCONNECT LINE

OPTO COUPLER
MCT2 OR
HIIG2

Circuit Notes

Ring detect circuit for automatic phone answering or tone generation for reverse autopatch use.

J1 - CLOCK COUNTER TO AUTOMATICALLY ANSWER PHONE
J2 - WILL WARBLE OSCILLATOR TO INDICATE PHONE IS RINGING (FOR REVERSE PATCH USE)

Except as indicated, decimal values of capacitance are in microfarads (μF); others are in picofarads (pF); resistances are in ohms.
k = 1,000   M = 1,000,000

Fig. 85-18

AUTOPATCH TELEPHONE LINE INTERFACE

600 OHM TO 600 OHM CENTER TAPPED TRANSFORMER

Circuit Notes

This circuit provides for the receiver-to-phone line and phone line-to-transmitter link, with both using an op amp for gain.

PHONE TO TRANSMITTER LEVEL

TO TOUCH-TONE DECODER

WILL MUTE PHONE TO TRANSMITTER AUDIO WHEN PATCH IS OFF

LINE SEIZE GROUND FOR PHONE LINE CONNECT

PROVIDES AUDIO DIRECTION SWITCHING

Except as indicated, decimal values of capacitance are in microfarads (μF); others are in picofarads (pF); resistances are in ohms.
k =1000   M = 1.000.000

Fig. 85-19

TELEPHONE RINGER USES PIEZOELECTRIC DEVICE

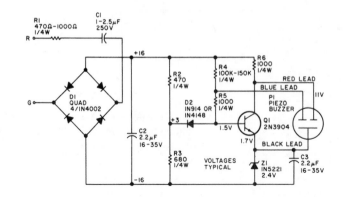

Fig. 85-20

Circait Notes

The electronic bell needs no power supply. Most of the resistors are not critical, although C2, R2, and R3 work best at the values given. Leaving out R1 will make the unit ring louder. The piezo buzzer may vary from store to store. If it has two leads, connect the red lead to the collector and the black lead to the emitter of Q1. If a third (blue) lead is present, connect it to the base of Q1.

ELECTRONIC PHONE BELL

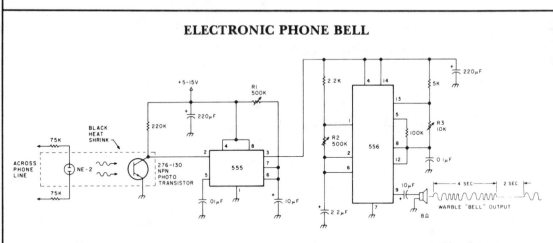

Fig. 85-21

Circuit Notes

The speaker emits a distinctive warble tone when ring pulses are applied to the phone line. Use this circuit as a remote bell or disconnect the phone's ringer for direct use. R1 adjusts the duration of the output; R2 and R3 control the tone's duty cycle and frequency. The transistor is a general-purpose NPN photodevice. The neon bulb and transistor are coupled with the heat-shrink tubing to form an optoisolator.

86

Temperature Controls

The sources of the following circuits are contained in the Sources section beginning on page 730. The figure number contained in the box of each circuit correlates to the source entry in the Sources section.

BOILER CONTROL

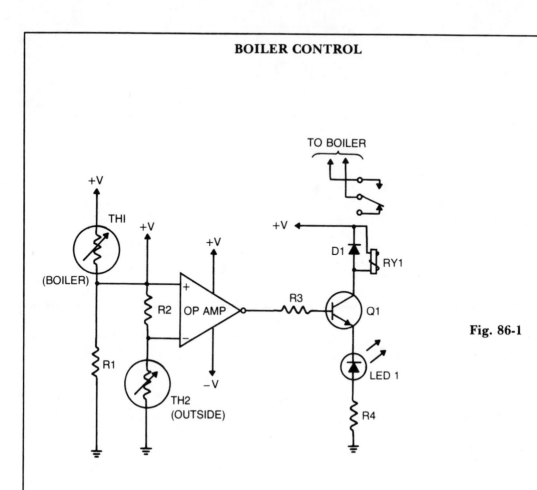

TO BOILER

Fig. 86-1

Circuit Notes

The purpose of this circuit is to control the water temperature in a hot-water heating system. What it does is to lower the boiler temperature as the outside air temperature increases. The op amp is used as a comparator. Thermistor TH2 and R2 form a voltage divider that supplies a reference voltage to the op-amp's inverting input. Thermistor TH2 is placed outdoors, and the values of TH2 and R2 should be chosen so that when the outside temperature is 25 °F, the resistance of the thermistor and resistor are equal. Resistor R1 and thermistor TH1 make up a voltage divider that supplies a voltage to the op amp's noninverting input. Thermistor TH1 is placed inside the boiler and the values of TH1 and R1 should be chosen so that when the boiler's temperature is 160 °F, their resistances are equal. The output of the op amp controls Q1, which is configured as a transistor switch. When the logic output of the op amp is high, Q1 is turned on, energizing relay RY1. The relay's contacts should be wired so that the boiler's heat supply is turned off (relay energized).

HEATER CONTROL

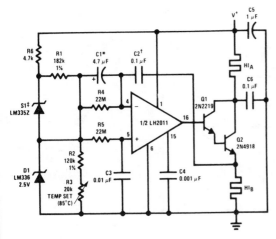

Circuit Notes

This proportional control crystal oven heater uses lead/lag compensation for fast setting. The time constant is changed with R4 and compensating resistor R5. If Q2 is inside the oven, a regulated supply is recommended for 0.1 °C. control.

* solid tantalum
† mylar
‡ close thermal coupling between sensor and oven shell is recommended.

Fig. 86-2

TWO-WIRE REMOTE AC ELECTRONIC THERMOSTAT (GAS OR OIL FURNACE CONTROL)

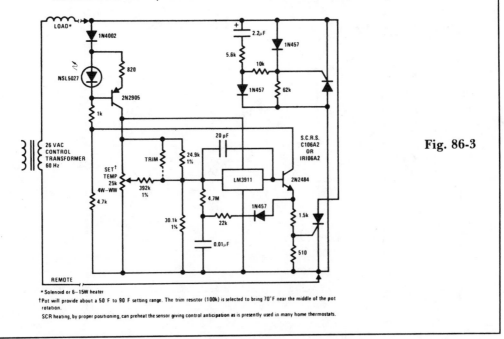

Fig. 86-3

* Solenoid or 6–15W heater
† Pot will provide about a 50 F to 90 F setting range. The trim resistor (100k) is selected to bring 70° F near the middle of the pot rotation.
SCR heating, by proper positioning, can preheat the sensor giving control anticipation as is presently used in many home thermostats.

THREE-WIRE ELECTRONIC THERMOSTAT

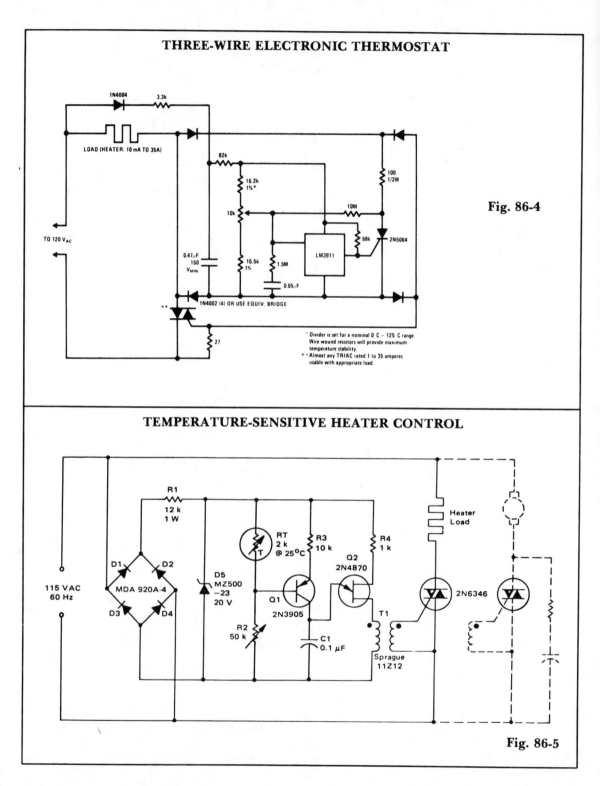

Fig. 86-4

* Divider is set for a nominal 0 C – 125 C range.
  Wire wound resistors will provide maximum
  temperature stability.
** Almost any TRIAC rated 1 to 35 amperes
  usable with appropriate load.

TEMPERATURE-SENSITIVE HEATER CONTROL

Fig. 86-5

TEMPERATURE CONTROLLER

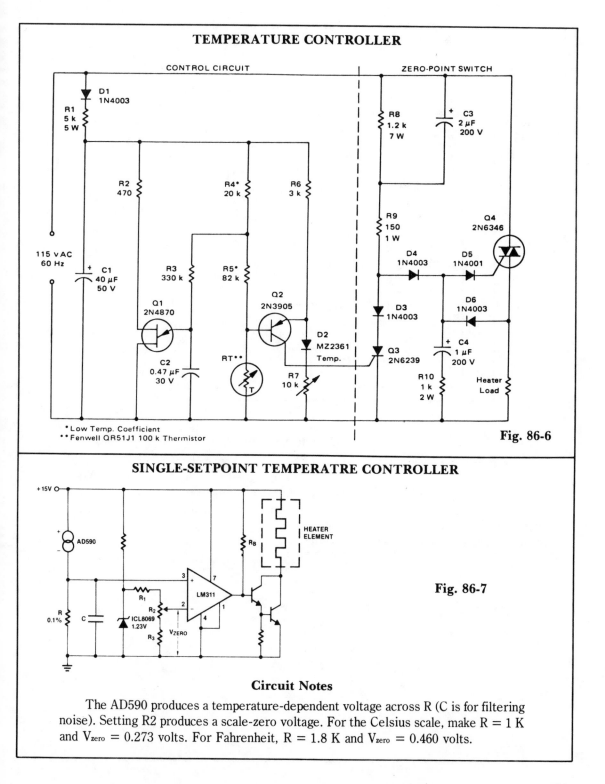

CONTROL CIRCUIT | ZERO-POINT SWITCH

Fig. 86-6

* Low Temp. Coefficient
** Fenwell QR51J1 100 k Thermistor

SINGLE-SETPOINT TEMPERATRE CONTROLLER

Fig. 86-7

Circuit Notes

The AD590 produces a temperature-dependent voltage across R (C is for filtering noise). Setting R2 produces a scale-zero voltage. For the Celsius scale, make R = 1 K and V_{zero} = 0.273 volts. For Fahrenheit, R = 1.8 K and V_{zero} = 0.460 volts.

TEMPERATURE CONTROLLER

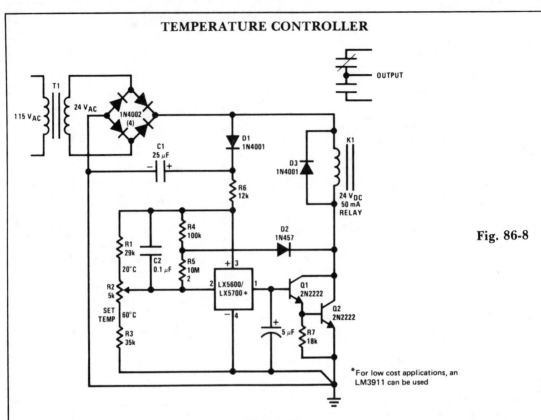

Fig. 86-8

*For low cost applications, an LM3911 can be used

Circuit Notes

The sensor is a standard TO-5 or TO-46 package. For surface or air temperature sensing. Small clip-on heat sinks can be used. A simple probe can be made using heat-shrink tubing and RTV silicon rubber. Three-leads-plus-shield cable is a good choice for wire with the shield connected to pin 4. The controller can be used for baths, ovens, oven-temperature protection, or even home thermostats. Long-term stability and repeatability is better than 0.5 °C.

TEMPERATURE CONTROL

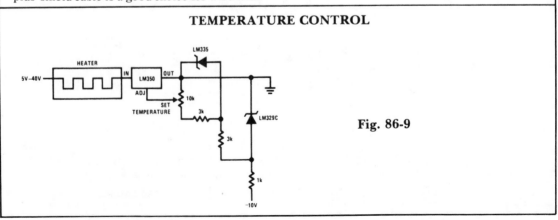

Fig. 86-9

TEMPEATURE CONTROLLER

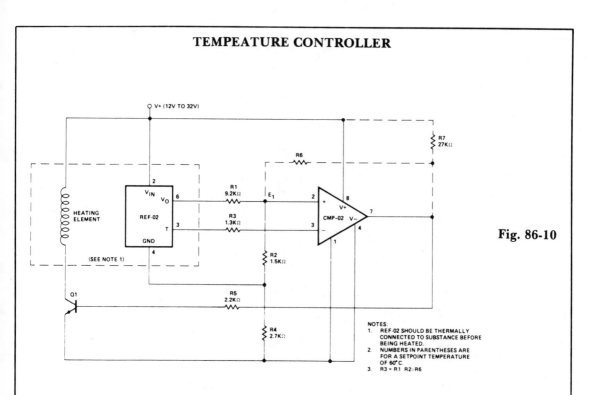

Fig. 86-10

NOTES:
1. REF-02 SHOULD BE THERMALLY CONNECTED TO SUBSTANCE BEFORE BEING HEATED.
2. NUMBERS IN PARENTHESES ARE FOR A SETPOINT TEMPERATURE OF 60°C.
3. R3 = R1·R2/R6

Circuit Notes

Temperature control is achieved using the REF-02 +5 V Reference/Thermometer and a CMP-02 Precision Low Input Current Comparator. The CMP-02 turns on a heating element driver (Q1) whenever the present temperature drops below a setpoint temperature determined by the ratio of R1 to R2. The circuit also provides adjustable hysteresis and single supply operation.

TEMPERATURE CONTROLLER

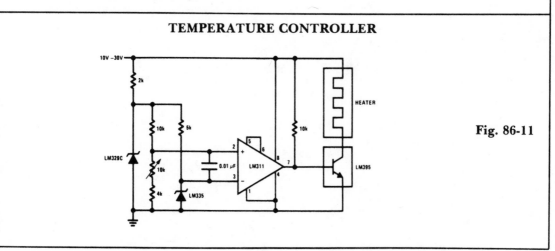

Fig. 86-11

PORTABLE CALIBRATOR

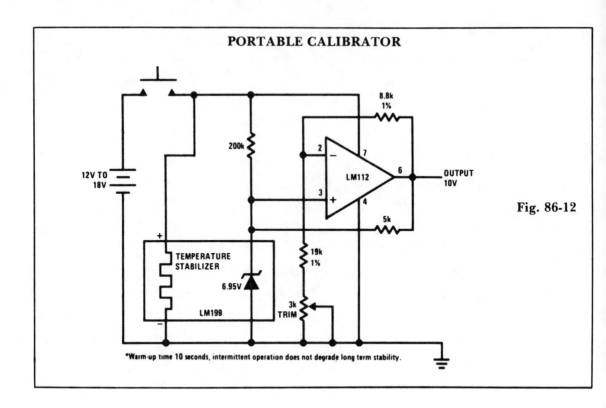

Fig. 86-12

*Warm-up time 10 seconds, intermittent operation does not degrade long term stability.

644

87

Temperature Sensors

The sources of the following circuits are contained in the Sources section beginning on page 730. The figure number contained in the box of each circuit correlates to the source entry in the Sources section.

LINEAR TEMPERATURE-TO-FREQUENCY TRANSCONDUCER

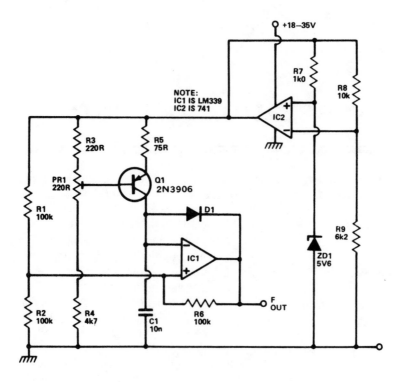

Fig. 87-1

Circuit Notes

This circuit provides a linear increase of frequency of 10 Hz/°C over 0-100 °C and can thus be used with logic systems, including microprocessors. Temperature probes Q1 V_{be} changes 2.2 mV/°C. This transistor is incorporated in a constant current source circuit. Thus, a current proportional to temperature will be available to charge C1. The circuit is powered via the temperature stable reference voltage supplied by the 741. Comparator IC1 is used as a Schmitt trigger whose output is used to discharge C1 via D1. To calibrate the circuit Q1 is immersed in boiling distilled water and PR1 adjusted to give 1 kHz output. The prototype was found to be accurate to within 0.2 °C.

TEMPERATURE METER

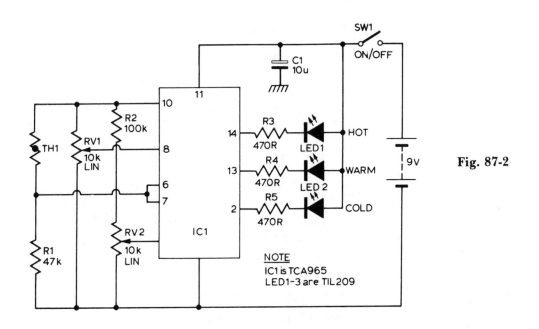

Fig. 87-2

NOTE
IC1 is TCA965
LED1-3 are TIL209

Circuit Notes

TCA965 window discriminator IC allows the potentiometers RV1 and RV2 to set up a window height and window width respectively. R1 and thermistor TH1 for a potential divider connected across the supply lines. R1 is chosen such that at ambient temperature the voltage at the junction of these two components will be approximately half supply. As the temperature of the sensor changes, the voltage will change.

RV1 will set the point which corresponds to the center voltage of a window the width of which is set by RV2. The switching points of the IC feature a Schmitt characteristic with low hysteresis. The outputs of IC1 indicate whether the input voltage is within the window or outside by virtue of being either too high or too low. The outputs of IC1 drive the LEDs via a current limiting resistor.

FOUR-CHANNEL TEMPERATURE SENSOR (0-50 °C)

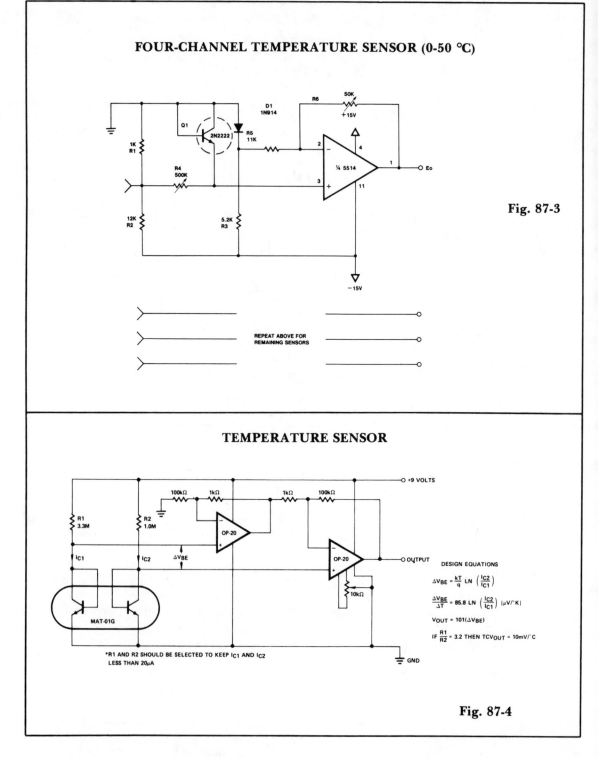

Fig. 87-3

TEMPERATURE SENSOR

DESIGN EQUATIONS

$$\Delta V_{BE} = \frac{kT}{q} \ LN \left(\frac{I_{C2}}{I_{C1}} \right)$$

$$\frac{\Delta V_{BE}}{\Delta T} = 85.8 \ LN \left(\frac{I_{C2}}{I_{C1}} \right) \ |\mu V/°K|$$

$$V_{OUT} = 101(\Delta V_{BE})$$

IF $\frac{R1}{R2}$ = 3.2 THEN TCV$_{OUT}$ = 10mV/°C

*R1 AND R2 SHOULD BE SELECTED TO KEEP I_{C1} AND I_{C2}
LESS THAN 20µA

Fig. 87-4

INTEGRATED CIRCUIT TEMPERATURE SENSOR

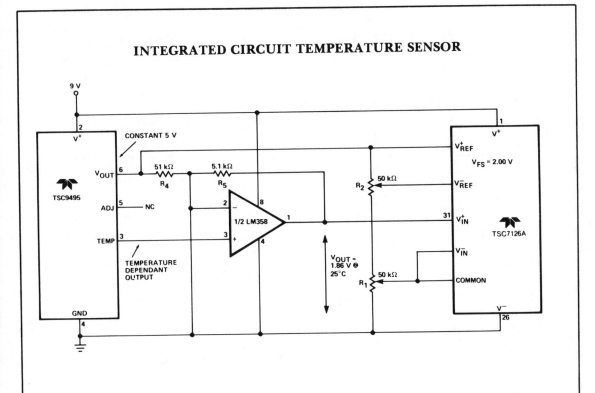

Fig. 87-5

PRECISION TEMPERATURE TRANSDUCER WITH REMOTE SENSOR

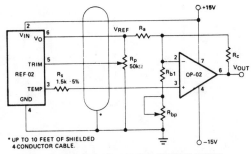

* UP TO 10 FEET OF SHIELDED
4-CONDUCTOR CABLE.

FOR THEORY OF OPERATION AND CALIBRATION PROCEDURE CONSULT
APPLICATION NOTE 18, "THERMOMETER APPLICATIONS OF THE REF-02".

RESISTOR VALUES

	10mV/°C	100mV/°C	10mV/°F
TCV$_{OUT}$ SLOPE (S)	10mV/°C	100mV/°C	10mV/°F
TEMPERATURE RANGE	−55°C to +125°C	−55°C to +125°C	−67°F to +257°F
OUTPUT VOLTAGE RANGE	−0.55V to +1.25V	−5.5V to +12.5V	−0.67V to +2.57V
ZERO SCALE	0V@0°C	0V@0°C	0V@0°F
R$_a$ (±1% resistor)	9.09kΩ	15kΩ	7.5kΩ
R$_{b1}$ (±% resistor)	1.5kΩ	1.82kΩ	1.21kΩ
R$_{bp}$ (Potentiometer)	200Ω	500Ω	200Ω
R$_c$ (±1% resistor)	5.11kΩ	84.5kΩ	8.25kΩ

* For 125°C operation, the op amp output must be able to swing to +12.5V,
increase V$_{IN}$ to +18V from +15V if this is a problem.

Fig. 87-6

CENTIGRADE CALIBRATED THERMOCOUPLE THERMOMETER

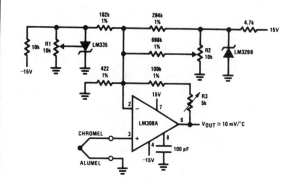

Terminate thermocouple reference junction in close proximity to LM335.

Adjustments:

1. Apply signal in place of thermocouple and adjust R3 for a gain of 245.7.
2. Short non-inverting input of LM308A and output of LM329B to ground.
3. Adjust R1 so that V_{OUT} = 2.982V @ 25°C.
4. Remove short across LM329B and adjust R2 so that V_{OUT} = 246 mV @ 25°C.
5. Remove short across thermocouple.

Fig. 87-7

μP CONTROLLED DIGITAL THERMOMETER

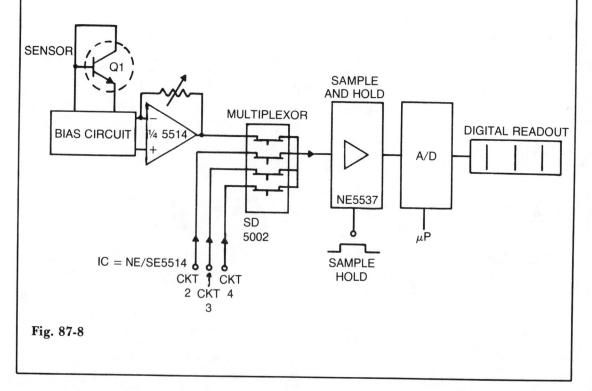

Fig. 87-8

ISOLATED TEMPERATURE SENSOR

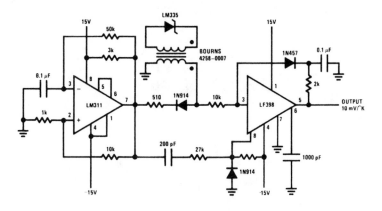

Fig. 87-9

DIGITAL THERMOMETER

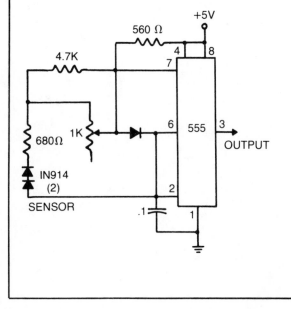

Fig. 87-10

Circuit Notes

The sensor consists of two series-connected 1N914s, part of the circuit of a 555 multivibrator. Wired as shown, the output pulse rate is proportional to the temperature of the diodes. This output is fed to a simple frequency-counting circuit.

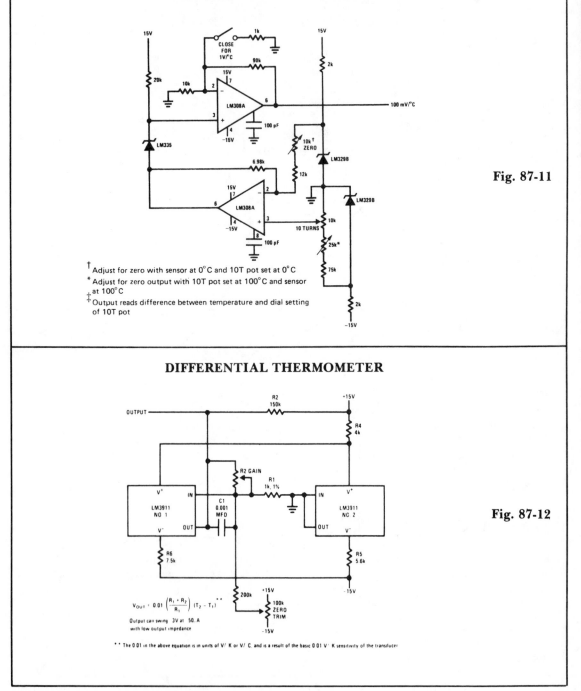

VARIABLE OFFSET THERMOMETER

Fig. 87-11

† Adjust for zero with sensor at 0°C and 10T pot set at 0°C

* Adjust for zero output with 10T pot set at 100°C and sensor at 100°C

‡ Output reads difference between temperature and dial setting of 10T pot

DIFFERENTIAL THERMOMETER

Fig. 87-12

$$V_{OUT} = 0.01 \left(\frac{R_1 + R_2}{R_1} \right) (T_2 - T_1)^{**}$$

Output can swing 3V at 50. A
with low output impedance

** The 0.01 in the above equation is in units of V/°K or V/°C, and is a result of the basic 0.01 V/°K sensitivity of the transducer

652

BASIC DIGITAL THERMOMETER, KELVIN SCALE

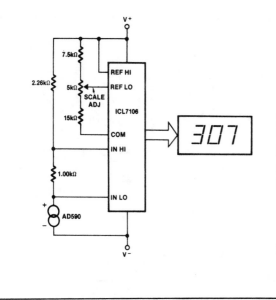

Circuit Notes
The Kelvin scale version reads from 0 to 1999 °K theoretically, and from 223 °K to 473 °K actually. The 2.26 K resistor brings the input within the ICL7106 V$_{CM}$ range: two general-purpose silicon diodes or an LED may be subsituted.

Fig. 87-13

BASIC DIGITAL THERMOMETER, KELVIN SCALE WITH ZERO ADJUST

Fig. 87-14

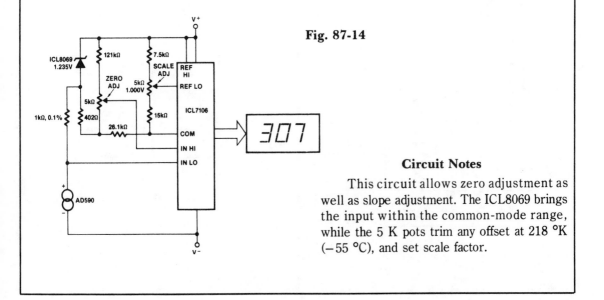

Circuit Notes
This circuit allows zero adjustment as well as slope adjustment. The ICL8069 brings the input within the common-mode range, while the 5 K pots trim any offset at 218 °K (−55 °C), and set scale factor.

653

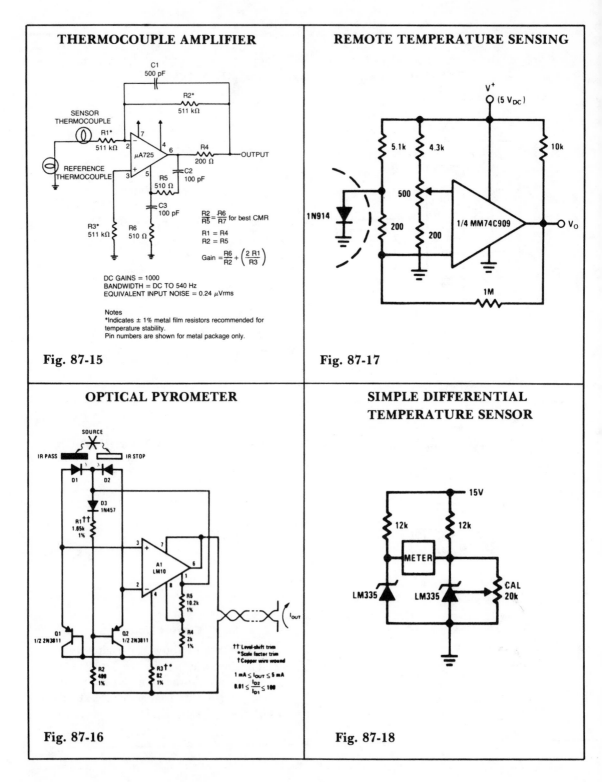

THERMOCOUPLE AMPLIFIER

C1
500 pF

R2*
511 kΩ

SENSOR
THERMOCOUPLE

R1*
511 kΩ

μA725

R4
200 Ω

OUTPUT

REFERENCE
THERMOCOUPLE

R5
510 Ω

C2
100 pF

C3
100 pF

R3*
511 kΩ

R6
510 Ω

$\frac{R2}{R5} = \frac{R6}{R7}$ for best CMR

R1 = R4
R2 = R5

$Gain = \frac{R6}{R2} + \left(\frac{2\ R1}{R3}\right)$

DC GAINS = 1000
BANDWIDTH = DC TO 540 Hz
EQUIVALENT INPUT NOISE = 0.24 μVrms

Notes
*Indicates ± 1% metal film resistors recommended for temperature stability.
Pin numbers are shown for metal package only.

Fig. 87-15

REMOTE TEMPERATURE SENSING

V⁺
(5 V_{DC})

5.1k   4.3k   10k

1N914

500

200   200

1/4 MM74C909

V_O

1M

Fig. 87-17

OPTICAL PYROMETER

SOURCE

IR PASS   IR STOP

D1   D2

D3
1N457

R1††
1.65k
1%

A1
LM10

R5
10.2k
1%

I_OUT

Q1
1/2 2N3811

Q2
1/2 2N3811

R4
2k
1%

R2
400
1%

R3†*
82
1%

†† Level-shift trim
* Scale factor trim
† Copper wire wound

1 mA ≤ I_OUT ≤ 5 mA
$0.01 \le \frac{I_{D2}}{I_{D1}} \le 100$

Fig. 87-16

SIMPLE DIFFERENTIAL
TEMPERATURE SENSOR

15V

12k   12k

METER

LM335   LM335   CAL
20k

Fig. 87-18

654

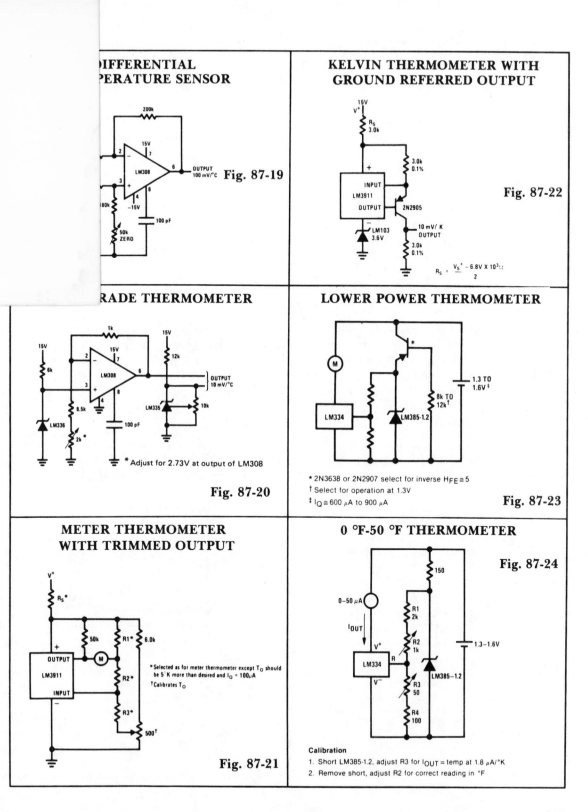

DIFFERENTIAL TEMPERATURE SENSOR

200k

2
−
15V 7
LM308 6 OUTPUT
100 mV/°C **Fig. 87-19**
3
+
8
4
−15V
80k
100 pF
50k
ZERO

KELVIN THERMOMETER WITH GROUND REFERRED OUTPUT

15V
V⁺
Rₛ
3.0k

3.0k
0.1%

+
INPUT
LM3911
OUTPUT 2N2905

10 mV/ K
OUTPUT
LM103
3.6V 3.0k
0.1%

Fig. 87-22

$$R_s = \frac{V_s^+ - 6.8V \times 10^3}{2}$$

CENTIGRADE THERMOMETER

1k

15V
2
−
15V 7
LM308 6
OUTPUT
10 mV/°C
3
+
15V
6k
8
4
LM335 10k
8.5k
100 pF
LM336
2k *

* Adjust for 2.73V at output of LM308

Fig. 87-20

LOWER POWER THERMOMETER

*

M

1.3 TO
1.6V ‡

8k TO
12k †
LM334 LM385-1.2

* 2N3638 or 2N2907 select for inverse Hꜰₑ ≅ 5
† Select for operation at 1.3V
‡ I_Q ≅ 600 μA to 900 μA

Fig. 87-23

METER THERMOMETER WITH TRIMMED OUTPUT

V⁺
Rₛ*

50k R1* 6.0k

+
OUTPUT
LM3911 M

R2*

INPUT
−
R3*

500†

* Selected as for meter thermometer except T_O should
be 5°K more than desired and I_Q = 100μA
† Calibrates T_O

Fig. 87-21

0 °F-50 °F THERMOMETER

150 **Fig. 87-24**

0–50 μA

R1
2k
I_OUT

R2
1k 1.3–1.6V
V⁺
LM334 R
R3
50 LM385-1.2
V⁻

R4
100

Calibration

1. Short LM385-1.2, adjust R3 for I_OUT = temp at 1.8 μA/°K
2. Remove short, adjust R2 for correct reading in °F

655

TEMPERATURE-TO-FREQUENCY CONVERTER

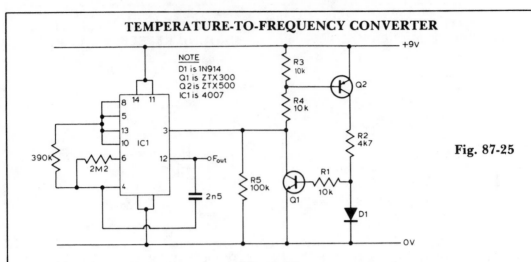

NOTE
D1 is 1N914
Q1 is ZTX300
Q2 is ZTX500
IC1 is 4007

Fig. 87-25

Circuit Notes

The circuit exploits the fact that when fed from a constant current source, the forward voltage of a silicon diode varies with temperature in a reasonably linear way. Diode D1 and resistor R2 form a potential divider fed from the constant current source. As the temperature rises, the forward voltage of D1 falls tending to turn Q1 off. The output voltage from Q1 will thus rise, and this is used as the control voltage for the CMOS VCO. With the values shown, the device gave an increase of just under 3 Hz/°C (between 0 °C and 60 °C) giving a frequency of 470 Hz at 0 °C.

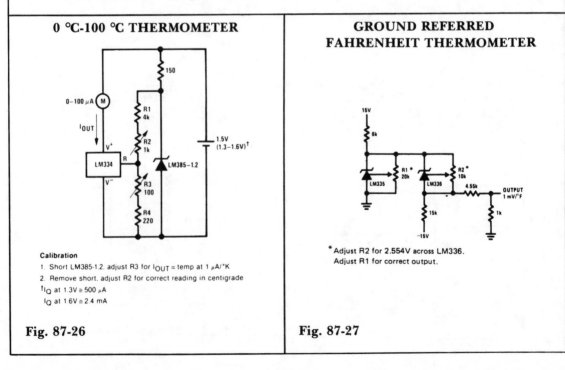

0 °C-100 °C THERMOMETER

Calibration

1. Short LM385-1.2, adjust R3 for I_{OUT} = temp at 1 μA/°K
2. Remove short, adjust R2 for correct reading in centigrade

†I_Q at 1.3V ≅ 500 μA
I_Q at 1.6V ≅ 2.4 mA

Fig. 87-26

GROUND REFERRED FAHRENHEIT THERMOMETER

* Adjust R2 for 2.554V across LM336.
Adjust R1 for correct output.

Fig. 87-27

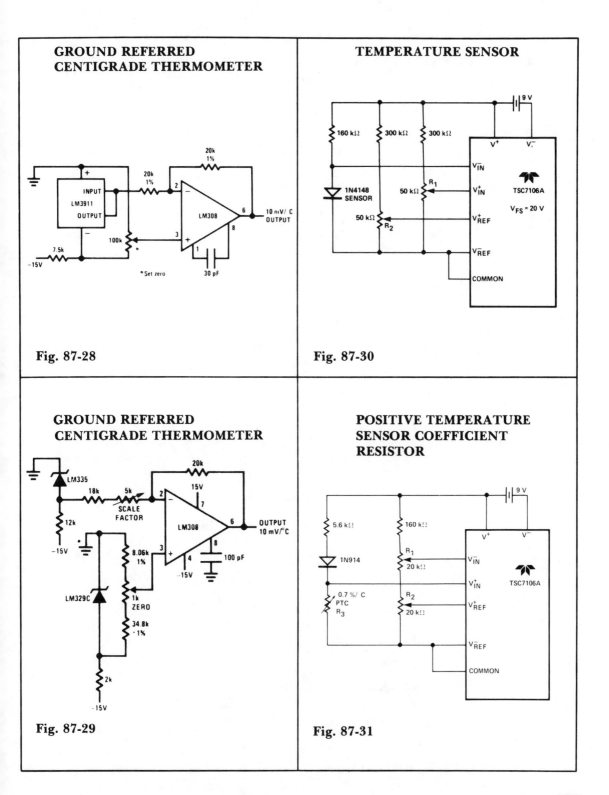

GROUND REFERRED CENTIGRADE THERMOMETER

Fig. 87-28

TEMPERATURE SENSOR

Fig. 87-30

GROUND REFERRED CENTIGRADE THERMOMETER

Fig. 87-29

POSITIVE TEMPERATURE SENSOR COEFFICIENT RESISTOR

Fig. 87-31

BASIC DIGITAL THERMOMETER (CELSIUS AND FAHRENHEIT SCALES)

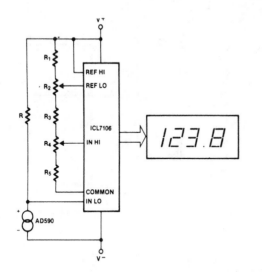

	R	R$_1$	R$_2$	R$_3$	R$_4$	R$_5$
°F	9.00	4.02	2.0	12.4	10.0	0
°C	5.00	4.02	2.0	5.11	5.0	11.8

Fig. 87-32

Circuit Notes

Maximum reading on the Celsius range is 199.9 °C, limited by the (short-term) maximum allowable sensor temperature. Maximum reading on the Fahrenheit range is 199.9 °F (93.3 °C), limited by the number of display digits. V$_{REF}$ for both scales is 500 mV.

FAHRENHEIT THERMOMETER

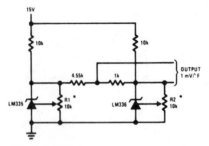

*To calibrate adjust R2 for 2.554V across LM336.
Adjust R1 for correct output.

Fig. 87-33

88

Timers

The sources of the following circuits are contained in the Sources section beginning on page 730. The figure number contained in the box of each circuit correlates to the source entry in the Sources section.

THUMBWHEEL PROGRAMMABLE INTERVAL TIMER

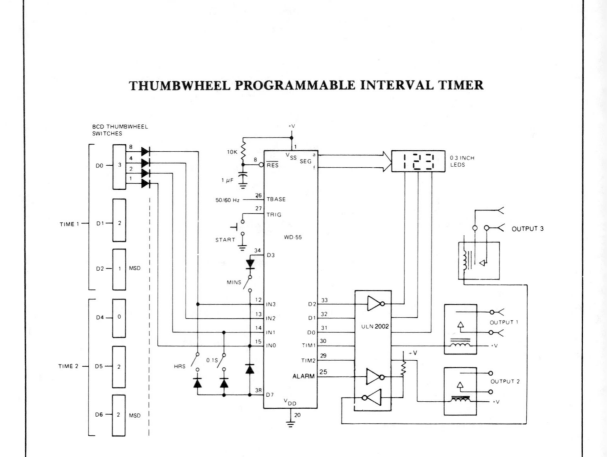

Fig. 88-1

Circuit Notes

Switch programmable on/off or interval timer, has three relay-switched outputs. Output one is active for the duration of time 1, output two is active for the duration of time 2, and output three is active for the duration of both one and two. Timing data is input through 6 BCD-encoded thumbwheel switches. Three SPST switches inform the WD-55 to interpret this data as NNN seconds. NNN seconds, NNN minutes, or NNN hours. The LED display will show the time remaining and the countdown when operating. Since the data is input through switches, the display may be deleted. Also, since the timing information is read from switches, the data is nonvolatile and no battery backup is required.

SEQUENTIAL TIMER

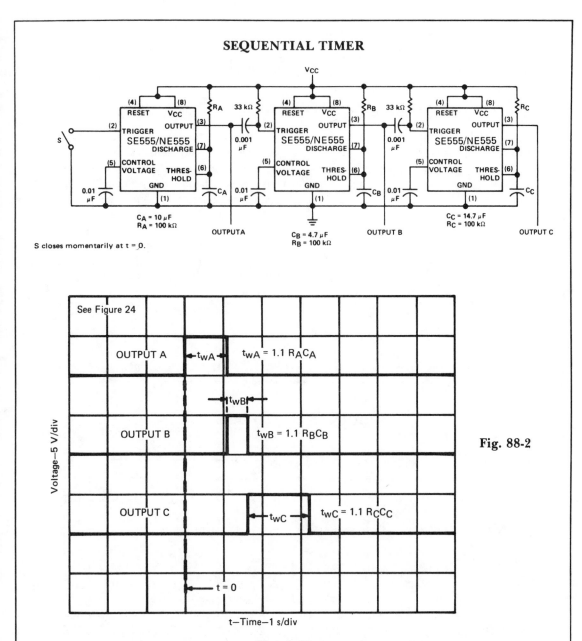

S closes momentarily at t = 0.

Fig. 88-2

Circuit Notes

Many applications, such as computers, require signals for initializing conditions during start-up. Other applications such as test equipment require activation of test signals in sequence. SE555/NE555 circuits may be connected to provide such sequential control. The timers may be used in various combinations of astable or monostable circuit connections, with or without modulation, for extremely flexible waveform control.

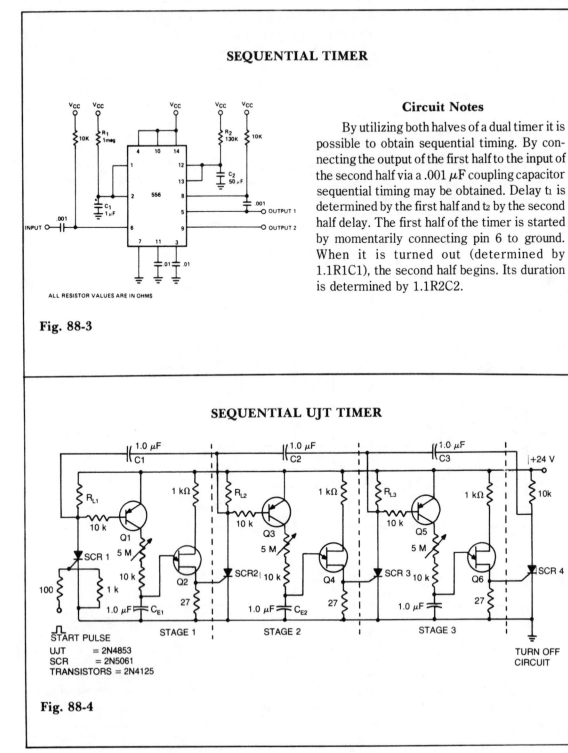

SEQUENTIAL TIMER

Circuit Notes

By utilizing both halves of a dual timer it is possible to obtain sequential timing. By connecting the output of the first half to the input of the second half via a .001 μF coupling capacitor sequential timing may be obtained. Delay t_1 is determined by the first half and t_2 by the second half delay. The first half of the timer is started by momentarily connecting pin 6 to ground. When it is turned out (determined by 1.1R1C1), the second half begins. Its duration is determined by 1.1R2C2.

ALL RESISTOR VALUES ARE IN OHMS

Fig. 88-3

SEQUENTIAL UJT TIMER

START PULSE
UJT   = 2N4853
SCR   = 2N5061
TRANSISTORS = 2N4125

STAGE 1   STAGE 2   STAGE 3

TURN OFF CIRCUIT

Fig. 88-4

TIME-DELAYED RELAY (FOR PATIO-LIGHT, GARAGE LIGHT, ENLARGER PHOTOTIMER, ETC.)

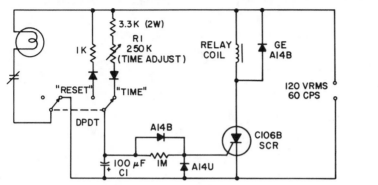

Fig. 88-5

NOTE: ALL RESISTORS 1/2 WATT

Circuit Notes

This simple timing circuit can delay an output switching function from .01 seconds to about 1 minute. The SCR is triggered by only a few microamps from the timing network R1-C1 to energize the output relay.

0.1 TO 90 SECOND TIMER

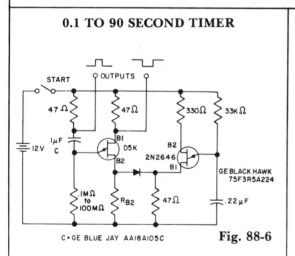

Fig. 88-6

Circuit Notes

The timer interval starts when power is applied to circuit and terminates when voltage is applied to load. 2N2646 is used in oscillator which pulses base 2 of D5K. This reduces the effective I_r of D5K and allows a much larger timing resistor and smaller timing capacitor to be used than would otherwise be possible.

SEQUENTIAL TIMING

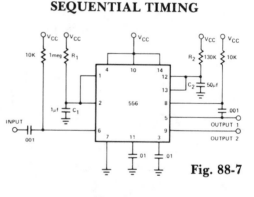

Fig. 88-7

Circuit Notes

By utilizing both halves of the dual timer it is possible to obtain sequential timing. By connecting the output of the first half to the input of the second half via a .001 μF coupling capacitor, sequential timing may be obtained. Delay t_1 is determined by the first half and t_2 by the second half delay. The first half of the timer is started by momentarily connecting pin 6 to ground. When it is timed out (determined by 1.1R1C1) the second half begins. Its time duration is determined by 1.1R2C2.

SOLID-STATE TIMER FOR INDUSTRIAL APPLICATIONS

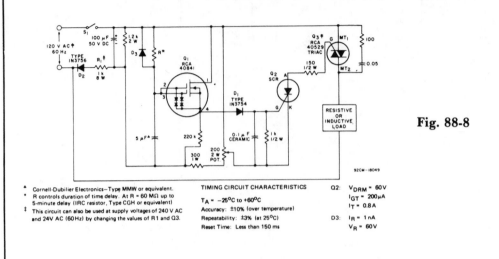

Fig. 88-8

▲ Cornell-Dubilier Electronics—Type MMW or equivalent.
• R controls duration of time delay. At R = 60 MΩ up to
   5-minute delay (IRC resistor, Type CGH or equivalent)
‡ This circuit can also be used at supply voltages of 240 V AC
   and 24V AC (60 Hz) by changing the values of R1 and Q3.

TIMING CIRCUIT CHARACTERISTICS

T_A = −25°C to +60°C

Accuracy: ±10% (over temperature)

Repeatability: ±3% (at 25°C)

Reset Time: Less than 150 ms

Q2: V_{DRM} = 60 V

I_{GT} = 200 μA

I_T = 0.8A

D3: I_R = 1 nA

V_R = 60 V

PRECISION SOLID STATE TIME DELAY CIRCUIT

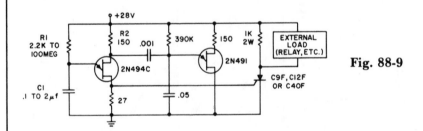

Fig. 88-9

Circuit Notes

Time delays from 0.3 milliseconds to over three minutes are possible with this circuit without using a tantalum or electrolytic capacitor. The timing interval is initiated by applying power to the circuit. At the end of the timing interval, which is determined by the value of R1C1, the 2N494C fires the controlled rectifier. This places the supply voltage minus about one volt across the load. Load currents are limited only by the rating of the controlled rectifier which is from 1 ampere up to 25 amperes for the types specified in the circuit. A calibrated potentiometer could be used in place of R1 to permit setting a predetermined time delay after one initial calibration.

ELECTRONIC EGG TIMER

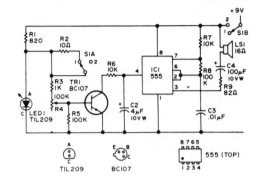

Circut Notes

The IC functions as an af multivibrator which is controlled by the external transistor. S1A/B is the on-off toggle switch.

Fig. 88-10

ON/OFF CONTROLLER

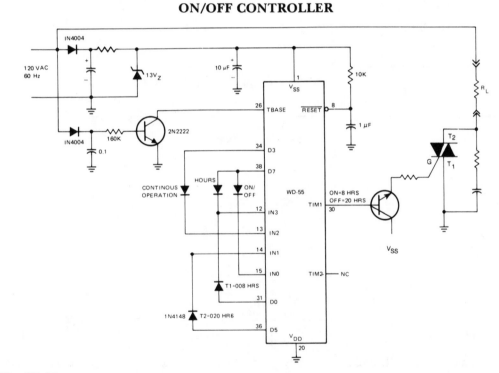

Fig. 88-11

Circuit Notes

The ac line-operated on/off controller is a simple, reliable solid-state alternative to a motive driven cam switch. Time 1 and time 2 are programmed by diodes to be 8 hours and 20 hours respectively. The TIM1 output is buffered by a transistor to supply gate current to a triac which switches the output load. When power is applied to the circuit, the output load is switched on for 8 hours then off for 20 hours repeatedly.

TIMING CIRCUIT

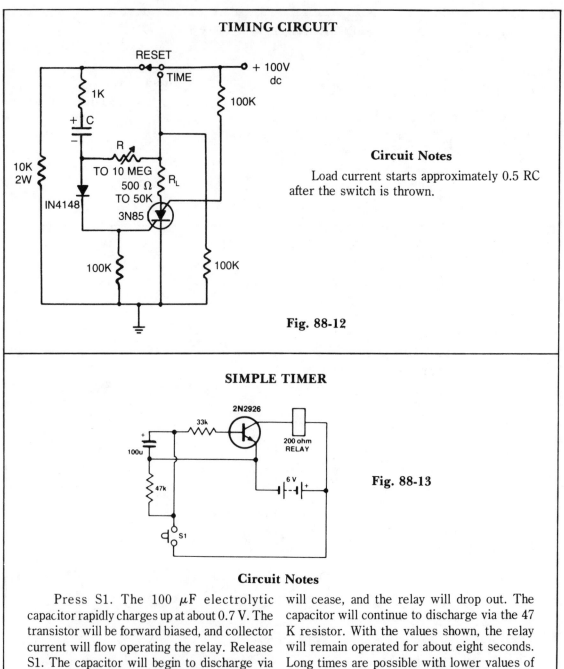

Circuit Notes

Load current starts approximately 0.5 RC after the switch is thrown.

Fig. 88-12

SIMPLE TIMER

Fig. 88-13

Circuit Notes

Press S1. The 100 μF electrolytic capacitor rapidly charges up at about 0.7 V. The transistor will be forward biased, and collector current will flow operating the relay. Release S1. The capacitor will begin to discharge via the 33 K resistor at the base of the transistor. When the voltage across the capacitor gets down to half a volt or so, the transistor base will no longer be forward biased, collector current will cease, and the relay will drop out. The capacitor will continue to discharge via the 47 K resistor. With the values shown, the relay will remain operated for about eight seconds. Long times are possible with lower values of capacitance by substituting a Darlington pair for the 2N2926. In this case, increase the two resistor values into the megohm range.

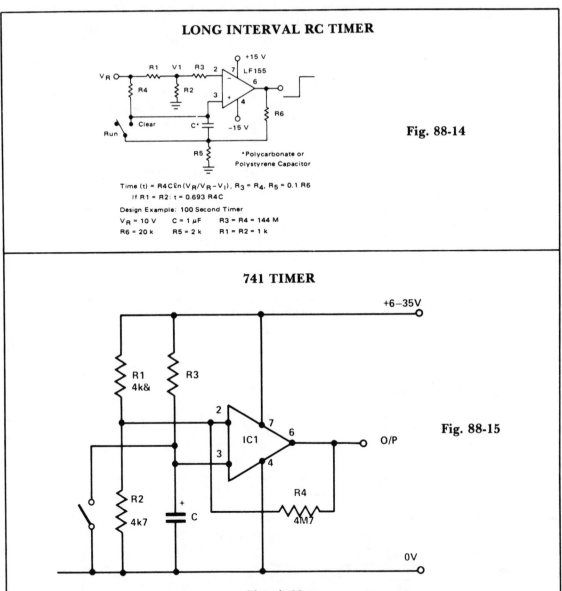

LONG INTERVAL RC TIMER

Fig. 88-14

Time (t) = R4C ℓn($V_R/V_R - V_1$), $R_3 = R_4$, $R_5 = 0.1 R6$
If R1 = R2: t = 0.693 R4C

Design Example: 100 Second Timer
V_R = 10 V   C = 1 μF   R3 = R4 = 144 M
R6 = 20 k   R5 = 2 k   R1 = R2 = 1 k

741 TIMER

Fig. 88-15

Circuit Notes

R1 and R2 hold the inverting input at half supply voltage. R4 applies feedback to increase the input impedance at pin 3. Pin 3, the noninverting input, is connected to the junction of R3 and C. After the switch is opened, C charges via R3. When the capacitor has charged sufficiently for the potential at pin 3 to exceed that at pin 2 the output abruptly changes from 0 V to positive line potential. If reverse polarity operation is required, simply transpose R3 and C. R3 and C can be any values. Time delays from a fraction of a second to several hours can be obtained by judicious selection. The time delay—independent of supply voltage—is 0.7CR seconds where C is in farads.

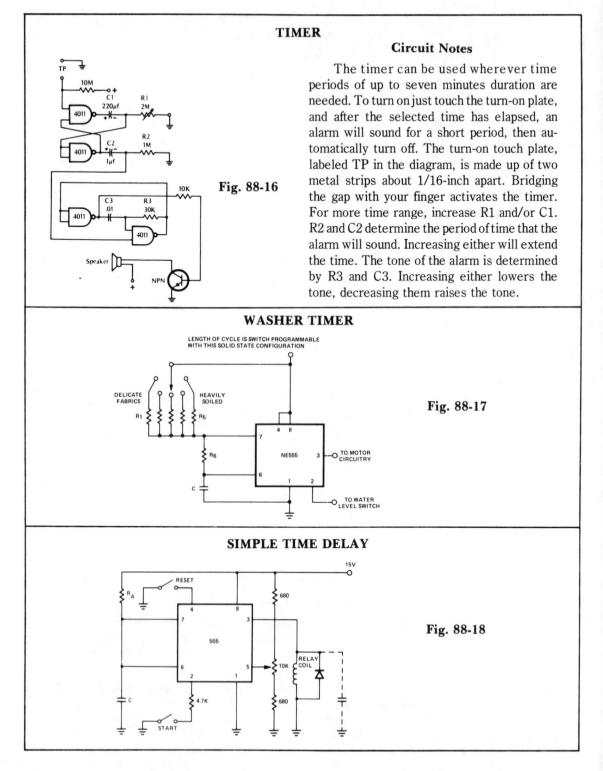

TIMER

Circuit Notes

The timer can be used wherever time periods of up to seven minutes duration are needed. To turn on just touch the turn-on plate, and after the selected time has elapsed, an alarm will sound for a short period, then automatically turn off. The turn-on touch plate, labeled TP in the diagram, is made up of two metal strips about 1/16-inch apart. Bridging the gap with your finger activates the timer. For more time range, increase R1 and/or C1. R2 and C2 determine the period of time that the alarm will sound. Increasing either will extend the time. The tone of the alarm is determined by R3 and C3. Increasing either lowers the tone, decreasing them raises the tone.

Fig. 88-16

WASHER TIMER

LENGTH OF CYCLE IS SWITCH PROGRAMMABLE
WITH THIS SOLID STATE CONFIGURATION

DELICATE FABRICS

HEAVILY SOILED

R_1 R_5

R_6

C

NE555

TO MOTOR CIRCUITRY

TO WATER LEVEL SWITCH

Fig. 88-17

SIMPLE TIME DELAY

15V

RESET

R_A

680

555

10K

RELAY COIL

C

4.7K

680

START

Fig. 88-18

668

89

Tone Controls

The sources of the following circuits are contained in the Sources section beginning on page 730. The figure number contained in the box of each circuit correlates to the source entry in the Sources section.

STEREO PHONOGRAPH AMPLIFIER WITH BASS TONE CONTROL

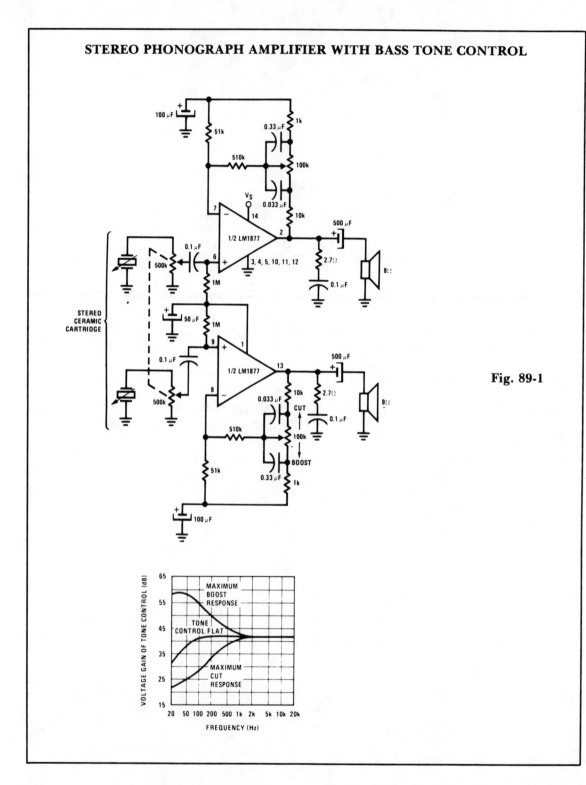

Fig. 89-1

EQUALIZER

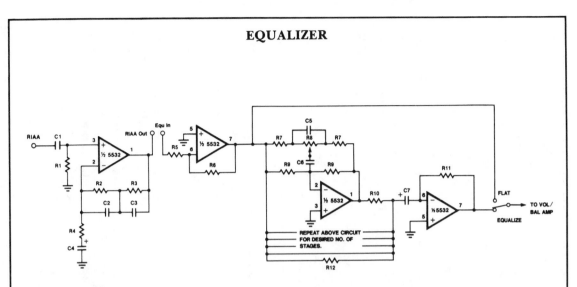

COMPONENT VALUE TABLES

R8 = 25k R7 = 2.4k R9 = 240k			R8 = 50k R7 = 5.1k R9 = 510k			R8 = 100k R7 = 10k R9 = 1meg		
fo	C5	C6	fo	C5	C6	fo	C5	C6
23 Hz	1µF	.1µF	25 Hz	.47µF	.047µF	12 Hz	.47µF	.047µF
50 Hz	.47µF	.047µF	36 Hz	.33µF	.033µF	18 Hz	.33µF	.033µF
72 Hz	.33µF	.033µF	54 Hz	.22µF	.022µF	27 Hz	.22µF	.022µF
108 Hz	.22µF	.022µF	79 Hz	.15µF	.015µF	39 Hz	.15µF	.015µF
158 Hz	.15µF	.015µF	119 Hz	.1µF	.01µF	59 Hz	.1µF	.01µF
238 Hz	.1µF	.01µF	145 Hz	.082µF	.0082µF	72 Hz	.082µF	.0082µF
290 Hz	.082µF	.0082µF	175 Hz	.068µF	.0068µF	87 Hz	.068µF	.0068µF
350 Hz	.068µF	.0068µF	212 Hz	.056µF	.0056µF	106 Hz	.056µF	.0056µF
425 Hz	.056µF	.0056µF	253 Hz	.047µF	.0047µF	126 Hz	.047µF	.0047µF
506 Hz	.047µF	.0047µF	360 Hz	.033µF	.0033µF	180 Hz	.033µF	.0033µF
721 Hz	.033µF	.0033µF	541 Hz	.022µF	.0022µF	270 Hz	.022µF	.0022µF
1082 Hz	.022µF	.0022µF	794 Hz	.015µF	.0015µF	397 Hz	.015µF	.0015µF
1588 Hz	.015µF	.0015µF	1191 Hz	.01µF	.001µF	595 Hz	.01µF	.001µF
2382 Hz	.01µF	.001µF	1452 Hz	.0082µF	820pF	726 Hz	.0082µF	820pF
2904 Hz	.0082µF	820pF	1751 Hz	.0068µF	680pF	875 Hz	.0068µF	680pF
3502 Hz	.0068µF	680pF	2126 Hz	.0056µF	560pF	1063 Hz	.0056µF	560pF
4253 Hz	.0056µF	560pF	2534 Hz	.0047µF	470pF	1267 Hz	.0047µF	470pF
5068 Hz	.0047µF	470pF	3609 Hz	.0033µF	330pF	1804 Hz	.0033µF	330pF
7218 Hz	.0033µF	330pF	5413 Hz	.0022µF	220pF	2706 Hz	.0022µF	220pF
10827 Hz	.0022µF	220pF	7940 Hz	.0015µF	150pF	3970 Hz	.0015µF	150pF
15880 Hz	.0015µF	150pF	11910 Hz	.001µF	100pF	5955 Hz	.001µF	100pF
23820 Hz	.001µF	100pF	14524 Hz	820pF	82pF	7262 Hz	820pF	82pF
			17514 Hz	680pF	68pF	8757 Hz	680pF	68pF
			21267 Hz	560pF	56pF	10633 Hz	560pF	56pF
						12670 Hz	470pF	47pF
						18045 Hz	330pF	33pF

COMPONENT VALUES

R1	1meg	C1	.22µF
R2	100k	C2	750pF
R3	1meg	C3	.0033µF
R4	1.1k	C4	33µF
R5	100k	C5	SEE TABLE
R6	100k	C6	SEE TABLE
R7	SEE TABLE	C7	2.2µF
R8	(pot) SEE TABLE		
R9	SEE TABLE		
R10	100k		
R11	100k		
R12	20k (5 STAGES)		

Fig. 89-2

THREE-CHANNEL TONE CONTROL

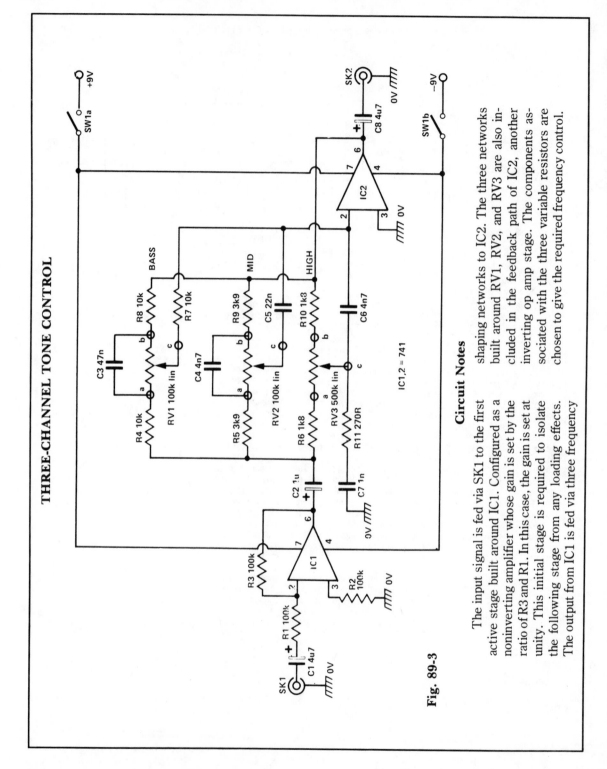

Fig. 89-3

Circuit Notes

The input signal is fed via SK1 to the first active stage built around IC1. Configured as a noninverting amplifier whose gain is set by the ratio of R3 and R1. In this case, the gain is set at unity. This initial stage is required to isolate the following stage from any loading effects. The output from IC1 is fed via three frequency shaping networks to IC2. The three networks built around RV1, RV2, and RV3 are also included in the feedback path of IC2, another inverting op amp stage. The components associated with the three variable resistors are chosen to give the required frequency control.

IC PREAMPLIFIER WITH TONE CONTROL

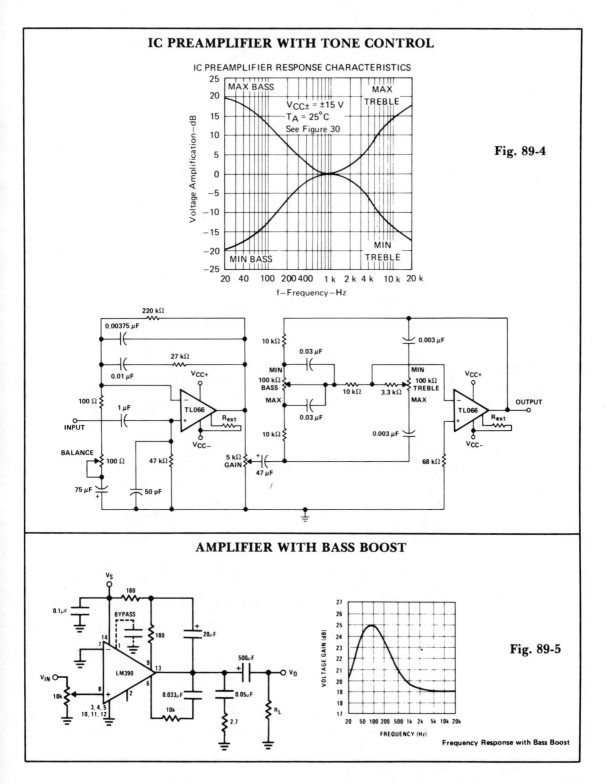

IC PREAMPLIFIER RESPONSE CHARACTERISTICS

Fig. 89-4

AMPLIFIER WITH BASS BOOST

Fig. 89-5

Frequency Response with Bass Boost

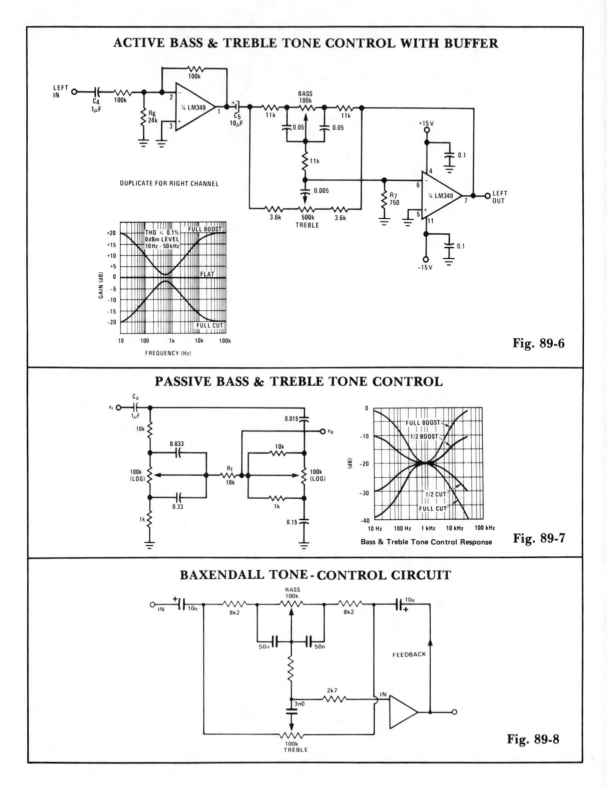

ACTIVE BASS & TREBLE TONE CONTROL WITH BUFFER

LEFT IN

C4 1µF

100k

100k

R6 24k

¼ LM349

DUPLICATE FOR RIGHT CHANNEL

C5 10µF

11k

BASS 100k

0.05   0.05

11k

11k

0.005

3.6k   500k   3.6k
TREBLE

R7 750

+15 V

0.1

¼ LM349

5

11

0.1

−15 V

LEFT OUT

THD ≤ 0.1%
0dBm LEVEL
10 Hz - 50 kHz

+20
+15
+10
+5
0
−5
−10
−15
−20

FULL BOOST

FLAT

FULL CUT

GAIN (dB)

10   100   1k   10k   100k
FREQUENCY (Hz)

Fig. 89-6

PASSIVE BASS & TREBLE TONE CONTROL

e_i

C_0
1µF

10k

0.033

100k
(LOG)

0.33

1k

R_1
10k

10k

1k

0.015

e_0

100k
(LOG)

0.15

0
−10
−20
−30
−40

(dB)

FULL BOOST
1/2 BOOST

1/2 CUT
FULL CUT

10 Hz   100 Hz   1 kHz   10 kHz   100 kHz

Bass & Treble Tone Control Response   **Fig. 89-7**

BAXENDALL TONE-CONTROL CIRCUIT

IN   10u   8k2

BASS 100k

8k2   10u

FEEDBACK

50n   50n

3n0   2k7   IN

100k
TREBLE

Fig. 89-8

674

HIGH QUALITY TONE CONTROL

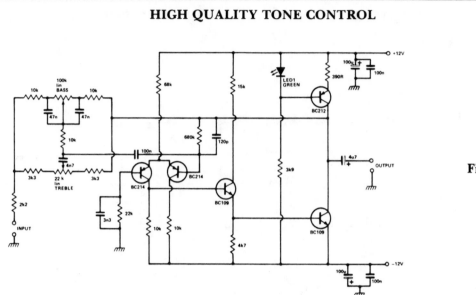

Fig. 89-9

Circuit Notes

The circuit is based on an inverting op amp using discrete transistors to overcome poor slew rate, fairly high distortion, and high noise problems. The output stage is driven by a constant current source, biased by a green LED to provide temperature compensation. With the controls flat, the unit provides unity gain so the stage can be switched in or out. The design is suitable for inputs between 100 mV and 1 V and provides a good overload margin at low distortion for the accurate reproduction of transients. The usual screening precautions against hum should be carried out.

MICROPHONE PREAMPLIFIER WITH TONE CONTROL

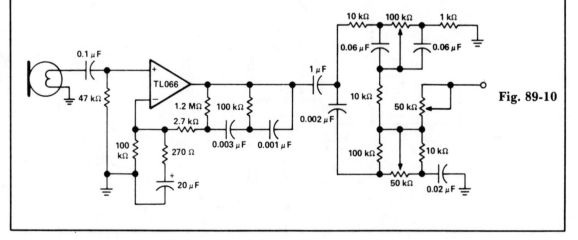

Fig. 89-10

675

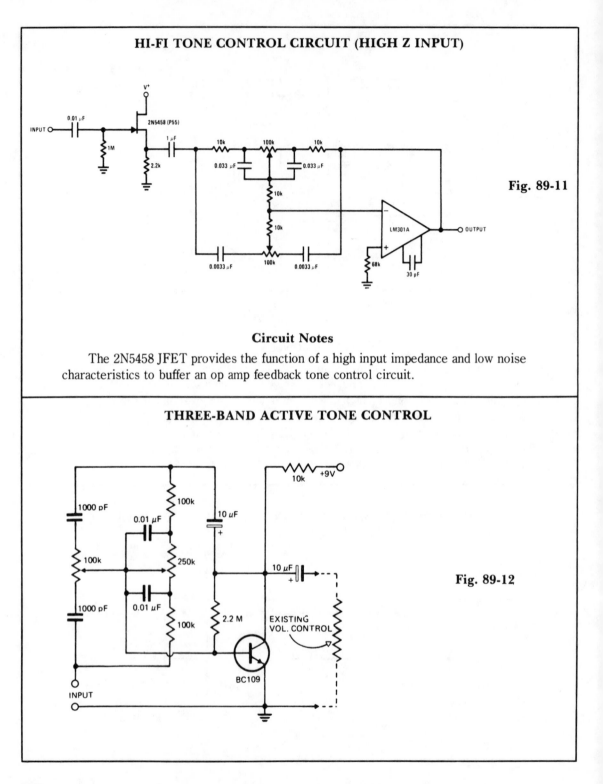

HI-FI TONE CONTROL CIRCUIT (HIGH Z INPUT)

Fig. 89-11

Circuit Notes

The 2N5458 JFET provides the function of a high input impedance and low noise characteristics to buffer an op amp feedback tone control circuit.

THREE-BAND ACTIVE TONE CONTROL

Fig. 89-12

TONE CONTROL CIRCUIT

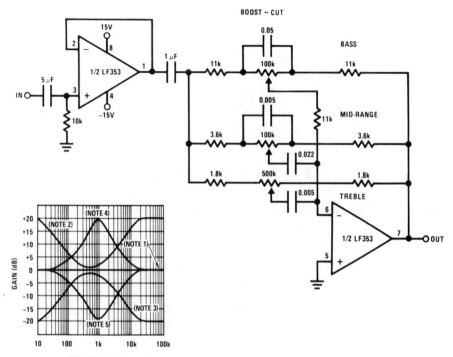

Fig. 89-13

Note 1: All controls flat.
Note 2: Bass and treble boost, mid flat.
Note 3: Bass and treble cut, mid flat.
Note 4: Mid boost, bass and treble flat.
Note 5: Mid cut, bass and treble flat.

- All potentiometers are linear taper
- Use the LF347 Quad for stereo applications

Circuit Notes

A simple single-transistor circuit will give approximately 15 dB boost or cut at 100 Hz and 15 kHz respectively. A low noise audio type transistor is used, and the output can be fed directly into any existing amplifier volume control to which the tone control is to be fitted. The gain of the circuit is near unity when controls are set in the flat position.

90

Transmitters

The sources of the following circuits are contained in the Sources section beginning on page 730. The figure number contained in the box of each circuit correlates to the source entry in the Sources section.

Wireless AM Microphone
27 MHz and 49 MHz RF Oscillator/
    Transmitter

1-2 MHz Broadcaster Transmitter
One Tube, 10 Watt C.W. Transmitter
Simple FM Transmitter

WIRELESS AM MICROPHONE

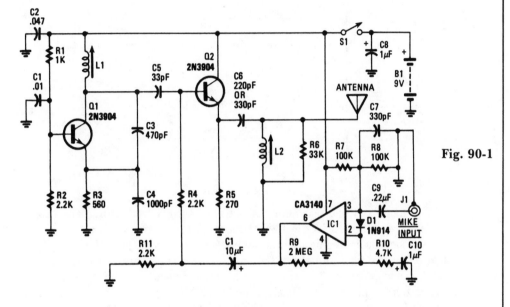

Fig. 90-1

Circuit Notes

Transistor Q1 and its associated components comprise a tuneable rf oscillator. The rf signal is fed to transistor Q2, the modulator. Operational amplifier IC1 increases the audio signal and applies it through resistor R4 to the base of Q2. Tune an AM radio to an unused frequency between 800 to 1600 kHz. Tune L1 for a change in the audio level coming from the radio. Peak the output by adjusting L2. If L1 is disturbed, it may be necessary to readjust L2 for peak performance. Depending on the impedance of the microphone audio sensitivity can be increased by decreasing the value of R10 and vice versa.

27 MHz AND 49 MHz RF OSCILLATOR/TRANSMITTER

Component	27 MHz	49 MHz
T_P	2 Turns	6 Turns
T_S	3 Turns	1 Turn
L1	TOKO KXN K4636 BJF	TOKO KEN K4635 BJE
L_L	MILLER #4611	MILLER #9338-10
C_A	5.4 pF	6.2 pF
R_A	1.15Ω	3.78Ω
C1	1000 pF	220 pF
C2	680 pF	47 pF
C3	20 pF	33 pF
R10	24k	47k

Use TOKO form #51-0116-02 and #30 wir or #51-0178 and #32 wire

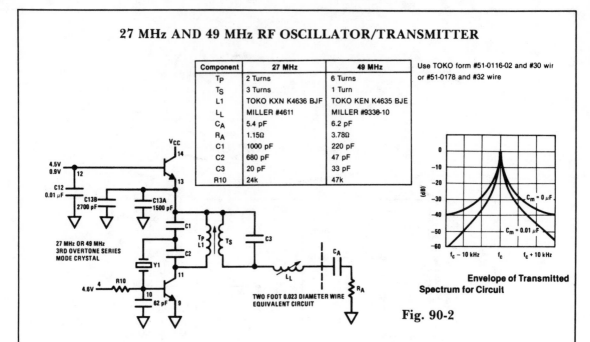

Envelope of Transmitted Spectrum for Circuit

Fig. 90-2

Circuit Notes

The modulator and oscillator consist of two NPN transistors. The base of the modulator transistor is driven by a bidirectional current source with the voltage range for the high condition limited by a saturating PNP collector to the pin 4 V_{REG} voltage and low condition limited by a saturating NPN collector in series with a diode to ground. The crystal oscillator/transmitter transistor is configured to oscillate in a class C mode. Because third overtone crystals are used for 27 MHz or 49 MHz applications a tuned collector load must be used to guarantee operation at the correct frequency.

1-2 MHz BROADCAST TRANSMITTER

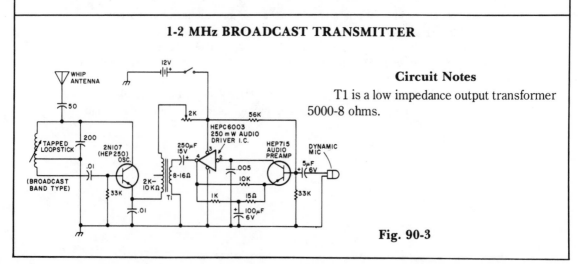

Circuit Notes

T1 is a low impedance output transformer 5000-8 ohms.

Fig. 90-3

680

ONE TUBE, 10 WATT C.W. TRANSMITTER

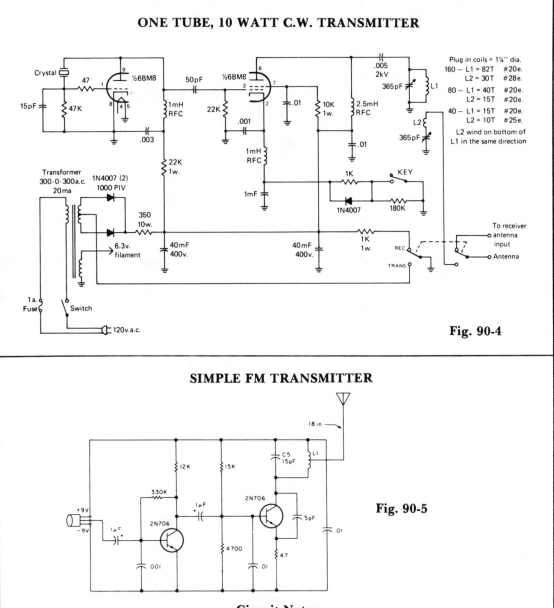

Fig. 90-4

SIMPLE FM TRANSMITTER

Fig. 90-5

Circuit Notes

This transmitter can be tuned to the FM broadcast band, 2 meters, or other VHF bands by changing C5 and L1. The values given for C5 and L1 will place the frequency somewhere in the FM broadcast band. L1 is 4 turns of #20 enameled wire airwound, ¼ inch in diameter, 5mm long and center-tapped. The microphone is an electret type and the antenna is 18 inches of any type of wire. Keep all leads as short as possible to minimize stray capacitance. The range of the transmitter is several hundred yards.

91

Ultrasonics

The sources of the following circuits are contained in the Sources section beginning on page 730. The figure number contained in the box of each circuit correlates to the source entry in the Sources section.

ULTRASONIC SWITCH

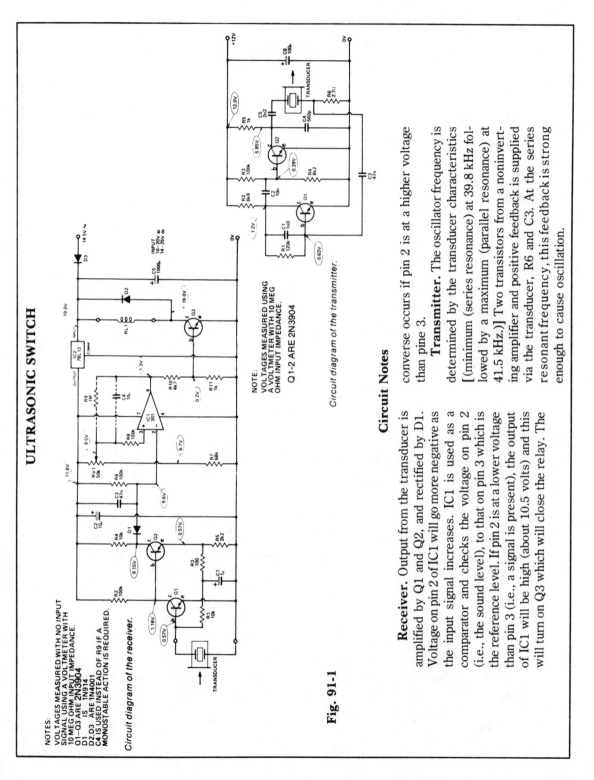

Circuit diagram of the receiver.

NOTE:
VOLTAGES MEASURED USING
A VOLTMETER WITH 10 MEG
OHM INPUT IMPEDANCE.

Q1-2 ARE 2N3904

Circuit diagram of the transmitter.

Fig. 91-1

Circuit Notes

Receiver. Output from the transducer is amplified by Q1 and Q2, and rectified by D1. Voltage on pin 2 of IC1 will go more negative as the input signal increases. IC1 is used as a comparator and checks the voltage on pin 2 (i.e., the sound level), to that on pin 3 which is the reference level. If pin 2 is at a lower voltage than pin 3 (i.e., a signal is present), the output of IC1 will be high (about 10.5 volts) and this will turn on Q3 which will close the relay. The

converse occurs if pin 2 is at a higher voltage than pine 3.

Transmitter. The oscillator frequency is determined by the transducer characteristics [(minimum (series resonance) at 39.8 kHz followed by a maximum (parallel resonance) at 41.5 kHz.)] Two transistors from a noninverting amplifier and positive feedback is supplied via the transducer, R6 and C3. At the series resonant frequency, this feedback is strong enough to cause oscillation.

683

ULTRASONIC BUG-CHASER

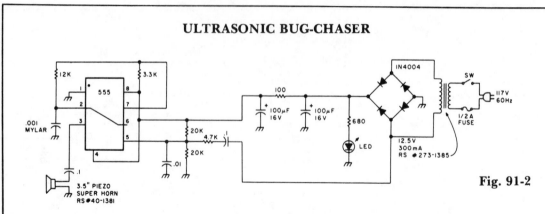

Fig. 91-2

Circuit Notes

Low-intensity ultrasonic sound waves in the 30-45 kHz frequency band repel insects and small rodents. The unit is designed to generate a swept square wave from 30 to 45 kHz. The LM555 IC is wired as an ultrasonic oscillator driving a piezoelectric speaker of the hi-fi super-tweeter type. The output of the oscillator is swept by a 60-Hz signal from the ac input of the bridge rectifier. The LED acts as a pilot.

MOSQUITO-REPELLING CIRCUIT

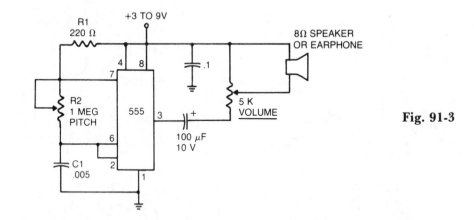

Fig. 91-3

Circuit Notes

In the 555 oscillator circuit, adjusting R2 will provide output frequencies from below 200 Hz to above 62 kHz. Use a good quality miniature speaker so that it will produce frequencies on the order of 20 kHz.

684

ULTRASONIC PEST REPELLER

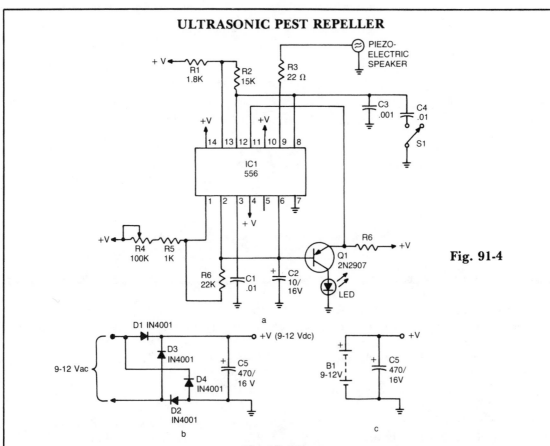

Fig. 91-4

Circuit Notes

The device emits ultrasonic sound waves that sweep between 65,000 and 25,000 hertz. Designed around a 556 dual timer, one half operated as an astable multivibrator with an adjustable frequency of 1 to 3 Hz. The second half is also operated as an astable multivibrator but with a fixed free running frequency around 45,000 Hz. The 25-65 kHz sweep is accomplished by coupling the voltage across C2 (the timing capacitor for the first half of the 556) via Q1 to the control voltage terminal (pin 11) of the second half of the 556. The device that radiates the ultrasonic sound is a piezo tweeter.

40 kHz ULTRASONIC TRANSMITTER

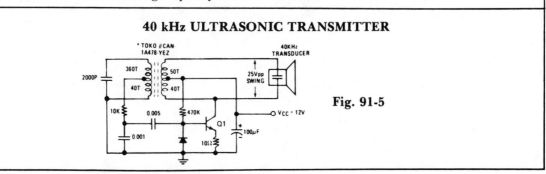

Fig. 91-5

92

Video Amplifiers

The sources of the following circuits are contained in the Sources section beginning on page 730. The figure number contained in the box of each circuit correlates to the source entry in the Sources section.

VIDEO IF AMPLIFIER AND LOW-LEVEL VIDEO DETECTOR CIRCUIT

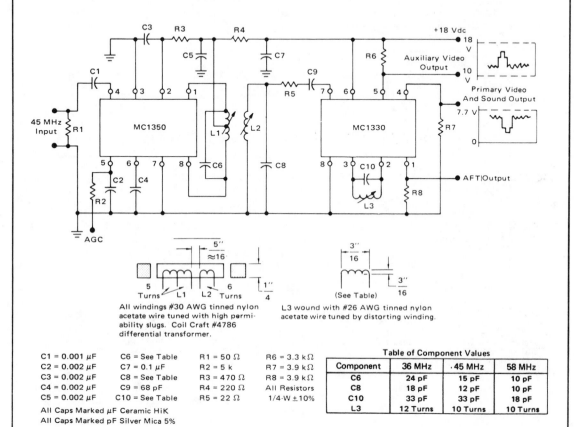

Component	36 MHz	45 MHz	58 MHz
C6	24 pF	15 pF	10 pF
C8	18 pF	12 pF	10 pF
C10	33 pF	33 pF	18 pF
L3	12 Turns	10 Turns	10 Turns

C1 = 0.001 μF   C6 = See Table   R1 = 50 Ω   R6 = 3.3 kΩ
C2 = 0.002 μF   C7 = 0.1 μF   R2 = 5 k   R7 = 3.9 kΩ
C3 = 0.002 μF   C8 = See Table   R3 = 470 Ω   R8 = 3.9 kΩ
C4 = 0.002 μF   C9 = 68 pF   R4 = 220 Ω   All Resistors
C5 = 0.002 μF   C10 = See Table   R5 = 22 Ω   1/4-W ±10%

All Caps Marked μF Ceramic HiK
All Caps Marked pF Silver Mica 5%

Fig. 92-1

Circuit Notes

The circuit has a typical voltage gain of 84 dB and a typical AGC range of 80 dB. It gives very small changes in bandpass shape, usually less than 1 dB tilt for 60 dB compression. There are no shielded sections. The detector uses a single tuned circuit (L3 and C10). Coupling between the two integrated circuits is achieved by a double tuned transformer (L1 and L2).

TELEVISION IF AMPLIFIER
AND DETECTOR USING AN MC1330 AND AN MC1352

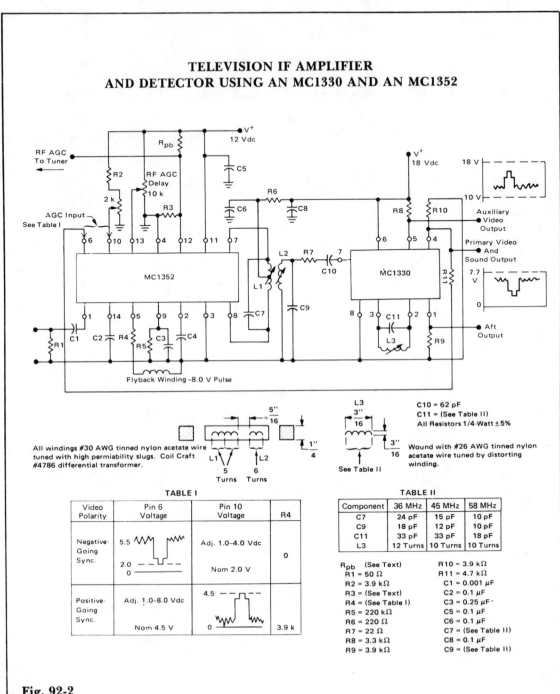

All windings #30 AWG tinned nylon acetate wire tuned with high permiability slugs. Coil Craft #4786 differential transformer.

C10 = 62 pF
C11 = (See Table II)
All Resistors 1/4-Watt ±5%

Wound with #26 AWG tinned nylon acetate wire tuned by distorting winding.

TABLE I

Video Polarity	Pin 6 Voltage	Pin 10 Voltage	R4
Negative-Going Sync.	5.5 ... 2.0 ... 0	Adj. 1.0–4.0 Vdc ... Nom 2.0 V	0
Positive-Going Sync.	Adj. 1.0–8.0 Vdc ... Nom 4.5 V	4.5 ... 0	3.9 k

TABLE II

Component	36 MHz	45 MHz	58 MHz
C7	24 pF	15 pF	10 pF
C9	18 pF	12 pF	10 pF
C11	33 pF	33 pF	18 pF
L3	12 Turns	10 Turns	10 Turns

R_{pb} (See Text)
R1 = 50 Ω
R2 = 3.9 kΩ
R3 = (See Text)
R4 = (See Table I)
R5 = 220 kΩ
R6 = 220 Ω
R7 = 22 Ω
R8 = 3.3 kΩ
R9 = 3.9 kΩ

R10 = 3.9 kΩ
R11 = 4.7 kΩ
C1 = 0.001 µF
C2 = 0.1 µF
C3 = 0.25 µF·
C5 = 0.1 µF
C6 = 0.1 µF
C7 = (See Table II)
C8 = 0.1 µF
C9 = (See Table II)

Fig. 92-2

TWO-STAGE WIDEBAND AMPLIFIER

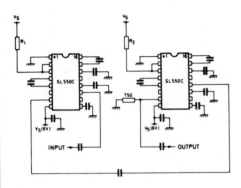

ALL CAPACITORS 1000 pF

Fig 92-3

Circuit Notes

A wideband high gain configuration using two SL550s connected in series. The first stage is connected in common emitter configuration, the second stage is a common base circuit. Stable gains of up to 65 dB can be achieved by the proper choice of R1 and R2. The bandwidth is 5 to 130 MHz, with a noise figure only marginally greater than the 2.0 dB specified for a single stage circuit.

VIDEO IF AMPLIFIER AND LOW-LEVEL VIDEO DETECTOR CIRCUIT

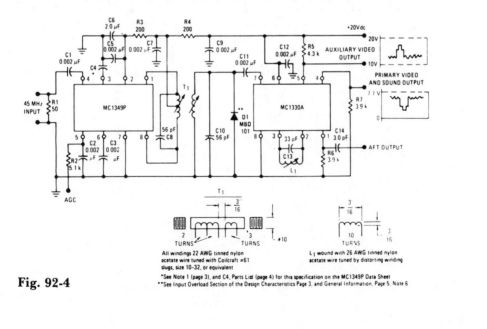

All windings 22 AWG tinned nylon acetate wire tuned with Coilcraft #61 slugs, size 10-32, or equivalent

L_1 wound with 26 AWG tinned nylon acetate wire tuned by distorting winding

*See Note 1 (page 3), and C4, Parts List (page 4) for this specification on the MC1349P Data Sheet
**See Input Overload Section of the Design Characteristics Page 3, and General Information, Page 5, Note 6

Fig. 92-4

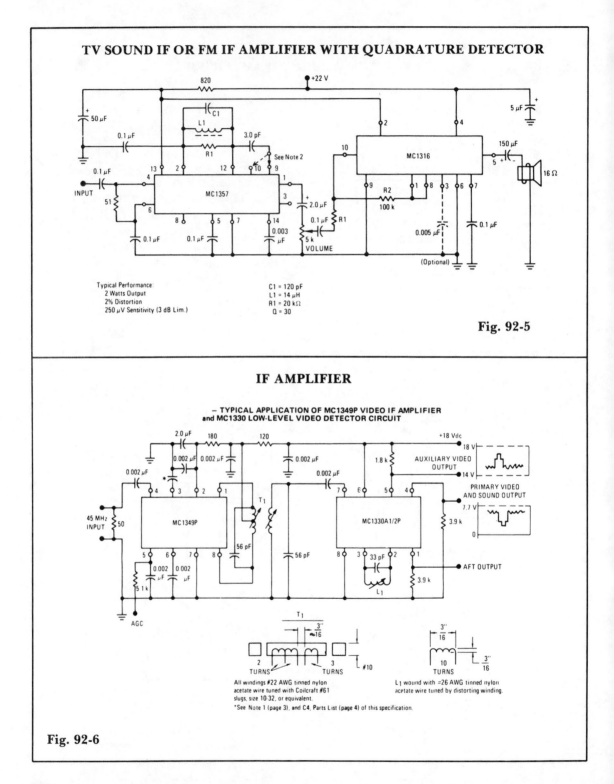

TV SOUND IF OR FM IF AMPLIFIER WITH QUADRATURE DETECTOR

+22 V

820

50 μF

0.1 μF

C1
L1
R1

3.0 pF

See Note 2

5 μF

0.1 μF

INPUT

51

13   2   12   10   9

MC1357

8   6

8   5   7   14

0.1 μF   0.1 μF   0.003 μF

10

MC1316

2   4

9   R2   1 8 3 6 7

100 k

1

3

2.0 μF

0.1 μF   R1

5 k
VOLUME

0.005 μF

(Optional)

0.1 μF

150 μF

5

16 Ω

0.1 μF

Typical Performance:
2 Watts Output
2% Distortion
250 μV Sensitivity (3 dB Lim.)

C1 = 120 pF
L1 = 14 μH
R1 = 20 kΩ
Q = 30

Fig. 92-5

IF AMPLIFIER

— TYPICAL APPLICATION OF MC1349P VIDEO IF AMPLIFIER and MC1330 LOW-LEVEL VIDEO DETECTOR CIRCUIT

2.0 μF   180   120

0.002 μF   0.002 μF   0.002 μF

0.002 μF

0.002 μF

+18 Vdc

18 V

1.8 k

AUXILIARY VIDEO
OUTPUT

14 V

PRIMARY VIDEO
AND SOUND OUTPUT

45 MHz
INPUT

50

4   3   2   1

MC1349P

5   6   7   8

0.002 μF   0.002 μF

5.1 k

AGC

T₁

56 pF

56 pF

7   6   5   4

MC1330A1/2P

8   3   2   1

33 pF

L₁

7.7 V

0

3.9 k

3.9 k

AFT OUTPUT

T₁

3″/16

≈16

2
TURNS   3
TURNS   #10

All windings #22 AWG tinned nylon
acetate wire tuned with Coilcraft #61
slugs, size 10-32, or equivalent.

*See Note 1 (page 3), and C4, Parts List (page 4) of this specification.

3″/16

10
TURNS

3″/16

L₁ wound with ≈26 AWG tinned nylon
acetate wire tuned by distorting winding.

Fig. 92-6

FET CASCODE VIDEO AMPLIFIER

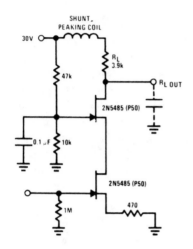

Circuit Notes

The FET cascode video amplifier features very low input loading and reduction of feedback to almost zero. The 2N5485 is used because of its low capacitance and high Y_{fs}. Bandwidth of this amplifier is limited by RL and load capacitance.

Fig. 92-7

HIGH IMPEDANCE LOW CAPACITANCE AMPLIFIER

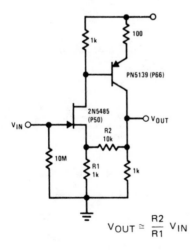

$$V_{OUT} \cong \frac{R2}{R1} \, V_{IN}$$

Fig. 92-8

Circuit Notes

This compound series-feedback circuit provides high input impedance and stable, wide-band gain for general purpose video amplifier applications.

JFET BIPOLAR CASCODE VIDEO AMPLIFIER

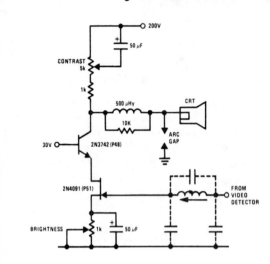

Circuit Notes

The JFET-bipolar cascode circuit will provide full video output for the CRT cathode drive. Gain is about 90. The cascode configuration eliminates Miller capacitance problems with the 2N4091 JFET, thus allowing direct drive from the video detector. An m-derived filter using stray capacitance and a variable inductor prevents 4.5 MHz sound frequency from being amplified by the video amplifier.

Fig. 92-9

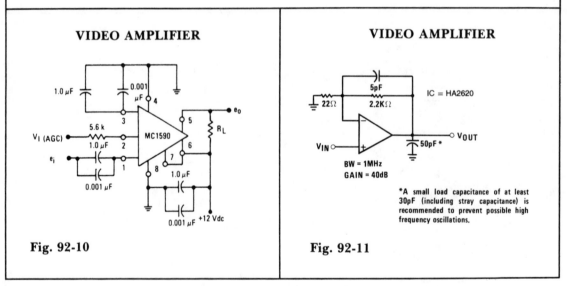

VIDEO AMPLIFIER

Fig. 92-10

VIDEO AMPLIFIER

IC = HA2620

BW = 1MHz
GAIN = 40dB

*A small load capacitance of at least 30pF (including stray capacitance) is recommended to prevent possible high frequency oscillations.

Fig. 92-11

93

Voltage and
Current Sources and References

The sources of the following circuits are contained in the Sources section beginning on page 730. The figure number contained in the box of each circuit correlates to the source entry in the Sources section.

BILATERAL CURRENT SOURCE

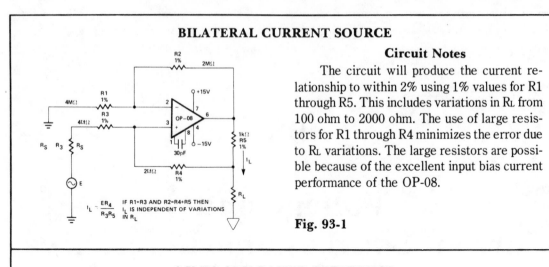

Circuit Notes

The circuit will produce the current relationship to within 2% using 1% values for R1 through R5. This includes variations in R_L from 100 ohm to 2000 ohm. The use of large resistors for R1 through R4 minimizes the error due to R_L variations. The large resistors are possible because of the excellent input bias current performance of the OP-08.

Fig. 93-1

0 V TO 20 V POWER REFERENCE

PROGRAMMABLE VOLTAGE SOURCE

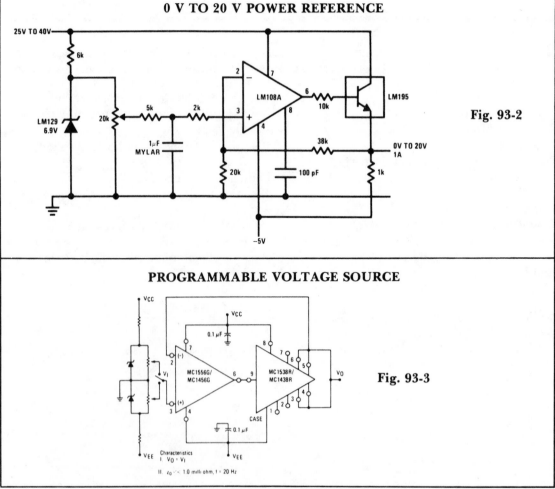

Fig. 93-2

Fig. 93-3

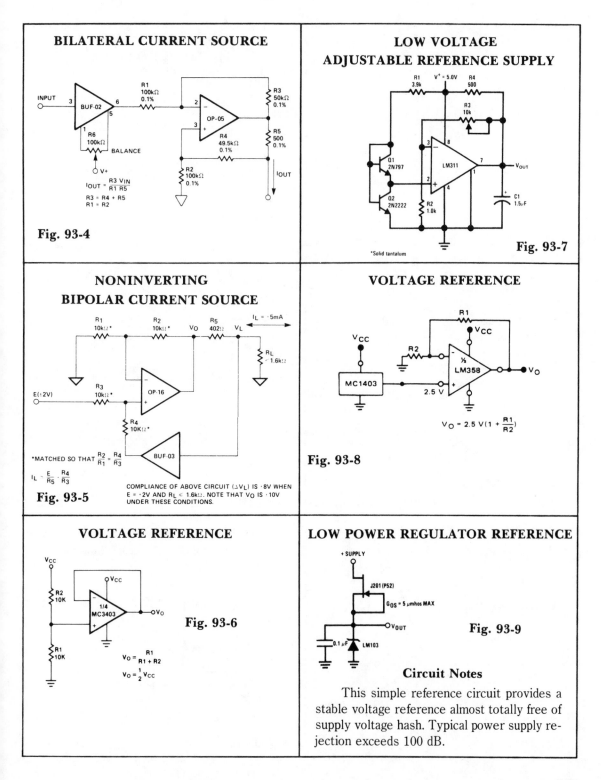

BILATERAL CURRENT SOURCE

INPUT

BUF-02

R1 100kΩ 0.1%

R6 100kΩ

BALANCE

V+

OP-05

R3 50kΩ 0.1%

R4 49.5kΩ 0.1%

R5 500 0.1%

IOUT

R2 100kΩ 0.1%

$I_{OUT} = \frac{R3}{R1} \frac{V_{IN}}{R5}$

R3 = R4 + R5
R1 = R2

Fig. 93-4

LOW VOLTAGE ADJUSTABLE REFERENCE SUPPLY

R1 3.9k

V⁺ = 5.0V

R4 500

R3 10k

Q1 2N797

Q2 2N2222

LM311

R2 1.0k

V_{OUT}

C1 1.5µF

*Solid tantalum

Fig. 93-7

NONINVERTING BIPOLAR CURRENT SOURCE

R1 10kΩ*

R2 10kΩ*

VO

R5 402Ω

VL

$I_L = \cdot 5mA$

RL 1.6kΩ

R3 10kΩ*

OP-16

E(·2V)

R4 10KΩ*

BUF-03

*MATCHED SO THAT $\frac{R2}{R1} = \frac{R4}{R3}$

$I_L \cdot \frac{E}{R5} \cdot \frac{R4}{R3}$

COMPLIANCE OF ABOVE CIRCUIT (·VL) IS ·8V WHEN E = ·2V AND RL < 1.6kΩ. NOTE THAT VO IS ·10V UNDER THESE CONDITIONS.

Fig. 93-5

VOLTAGE REFERENCE

VCC

R2

MC1403

2.5 V

R1

VCC

½ LM358

VO

$V_O = 2.5\ V(1 + \frac{R1}{R2})$

Fig. 93-8

VOLTAGE REFERENCE

VCC

R2 10K

R1 10K

¼ MC3403

VCC

VO

Fig. 93-6

$V_O = \frac{R1}{R1 + R2}$

$V_O = \frac{1}{2} V_{CC}$

LOW POWER REGULATOR REFERENCE

+ SUPPLY

J201 (P52)

$G_{OS} = 5\ \mu mhos\ MAX$

V_{OUT}

0.1 µF

LM103

Fig. 93-9

Circuit Notes

This simple reference circuit provides a stable voltage reference almost totally free of supply voltage hash. Typical power supply rejection exceeds 100 dB.

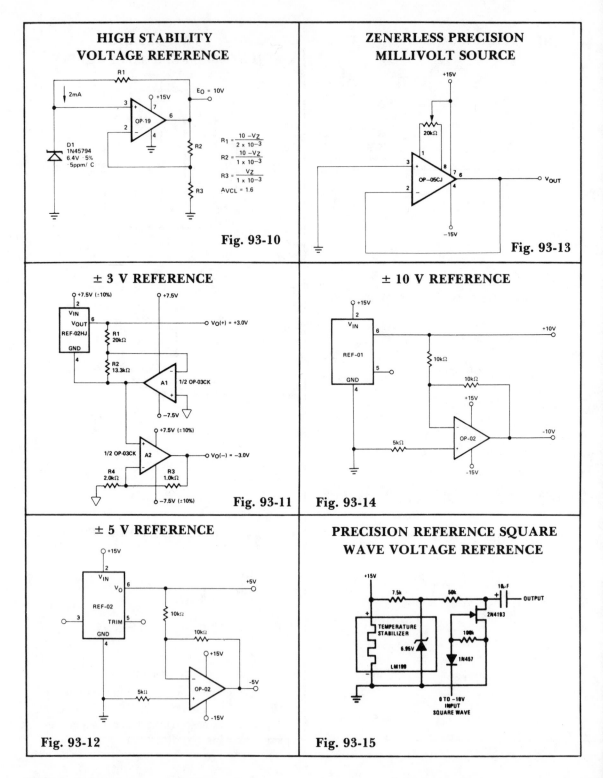

HIGH STABILITY VOLTAGE REFERENCE

$R_1 = \dfrac{10 - V_Z}{2 \times 10^{-3}}$

$R_2 = \dfrac{10 - V_Z}{1 \times 10^{-3}}$

$R_3 = \dfrac{V_Z}{1 \times 10^{-3}}$

$A_{VCL} = 1.6$

Fig. 93-10

ZENERLESS PRECISION MILLIVOLT SOURCE

Fig. 93-13

± 3 V REFERENCE

Fig. 93-11

± 10 V REFERENCE

Fig. 93-14

± 5 V REFERENCE

Fig. 93-12

PRECISION REFERENCE SQUARE WAVE VOLTAGE REFERENCE

Fig. 93-15

696

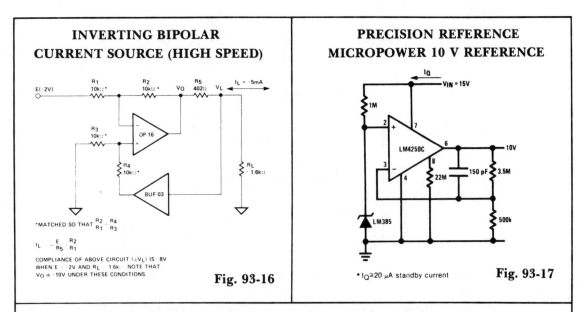

INVERTING BIPOLAR CURRENT SOURCE (HIGH SPEED)

$I_L = \cdot 5mA$

*MATCHED SO THAT $\dfrac{R_2}{R_1} = \dfrac{R_4}{R_3}$

$I_L = \dfrac{E}{R_5} \cdot \dfrac{R_2}{R_1}$

COMPLIANCE OF ABOVE CIRCUIT (ΔV_L) IS ·8V
WHEN E ·2V AND R_L 1.6k. NOTE THAT
V_O IS ·10V UNDER THESE CONDITIONS.

Fig. 93-16

PRECISION REFERENCE MICROPOWER 10 V REFERENCE

* $I_Q \cong 20~\mu A$ standby current

Fig. 93-17

PRECISION REFERENCE LOW NOISE BUFFERED REFERENCE

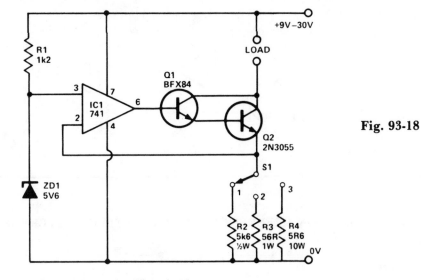

Fig. 93-18

Circuit Notes

The circuit will provide 3 preset currents which will remain constant despite variations of ambient temperature or line voltage. ZD1 produces a temperature stable reference voltage which is applied to the noninverting input of IC1. 100% feedback is applied from the output to the inverting input holding the voltage at Q2s emitter at the same potential as the noninverting input. The current flowing into the load therefore is defined solely by the resistor selected by S1. With the values employed here, a preset current of 10 mA, 100 mA or 1 A can be selected. Q2 should be mounted on a suitable heatsink.

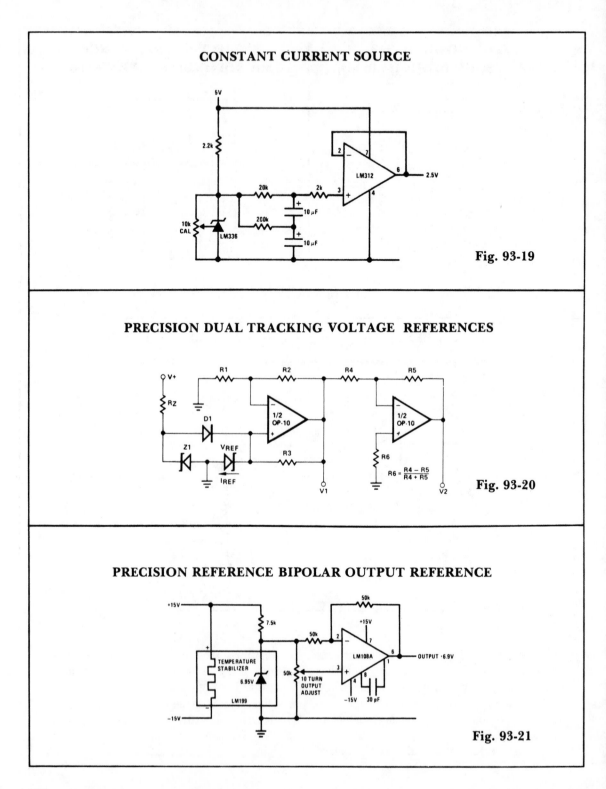

CONSTANT CURRENT SOURCE

Fig. 93-19

PRECISION DUAL TRACKING VOLTAGE REFERENCES

$$R6 = \frac{R4 - R5}{R4 + R5}$$

Fig. 93-20

PRECISION REFERENCE BIPOLAR OUTPUT REFERENCE

Fig. 93-21

PRECISION REFERENCE 0 V TO 20 V POWER REFERENCE

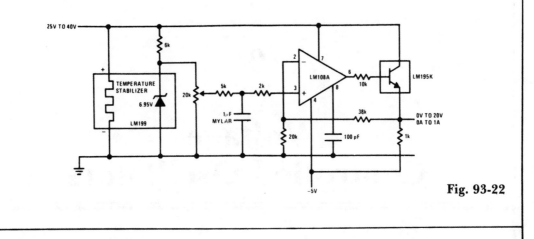

Fig. 93-22

PRECISION REFERENCE STANDARD CELL REPLACEMENT

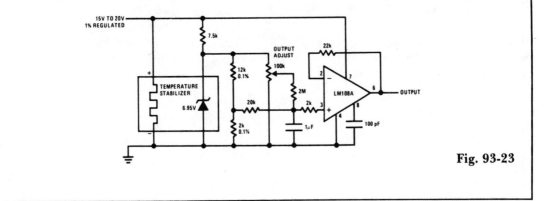

Fig. 93-23

94

Voltage-
Controlled Oscillators

The sources of the following circuits are contained in the Sources section beginning on page 730. The figure number contained in the box of each circuit correlates to the source entry in the Sources section.

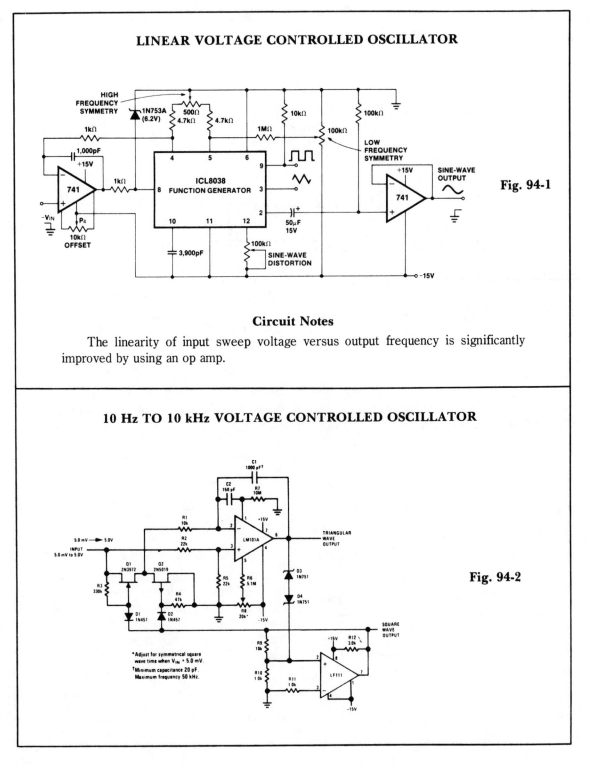

LINEAR VOLTAGE CONTROLLED OSCILLATOR

Fig. 94-1

Circuit Notes

The linearity of input sweep voltage versus output frequency is significantly improved by using an op amp.

10 Hz TO 10 kHz VOLTAGE CONTROLLED OSCILLATOR

Fig. 94-2

PRECISION VOLTAGE CONTROLLED OSCILLATOR

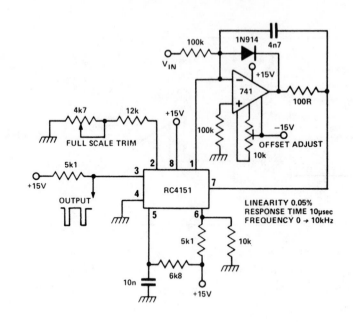

Fig. 94-3

LINEARITY 0.05%
RESPONSE TIME 10μsec
FREQUENCY 0 → 10kHz

Circuit Notes

RC 4151 precision voltage-to-frequency converter generates a pulse train output linearly proportional to the input voltage.

VOLTAGE CONTROLLED OSCILLATOR

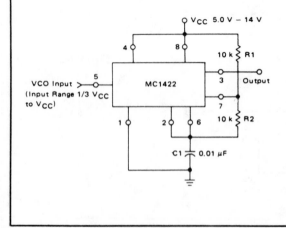

Circuit Notes

The VCO circuit, which has a nonlinear transfer characteristic, will operate satisfactorily up to 200 kHz. The VCO input range is effective from $\frac{1}{3}$ Vcc to Vcc − 2 V, with the highest control voltage producing the lowest output frequency.

Fig. 94-4

SIMPLE VOLTAGE CONTROLLED OSCILLATOR

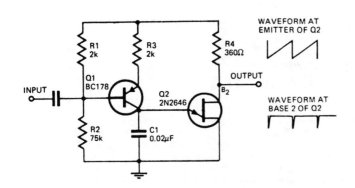

WAVEFORM AT EMITTER OF Q2

WAVEFORM AT BASE 2 OF Q2

Fig. 94-5

Circuit Notes

With the component values shown, the oscillator has a frequency of 8 kHz. When an input signal is applied to the base of Q1 the current flowing through Q1 is varied, thus varying the time required to charge C1. Due to the phase inversion in Q1 the direction of output frequency change is 180 degrees out of phase with the input signal. The output may be used to trigger a bistable flip-flop.

THREE DECADES VCO

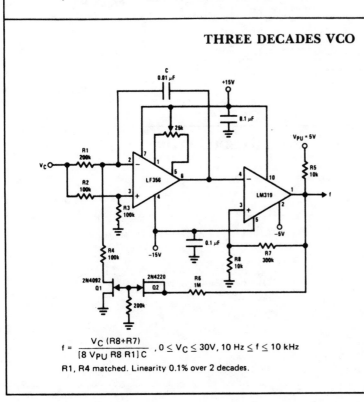

Fig. 94-6

$$f = \frac{V_C\,(R8+R7)}{[8\,V_{PU}\,R8\,R1]\,C}\ ,\ 0 \le V_C \le 30V,\ 10\ Hz \le f \le 10\ kHz$$

R1, R4 matched. Linearity 0.1% over 2 decades.

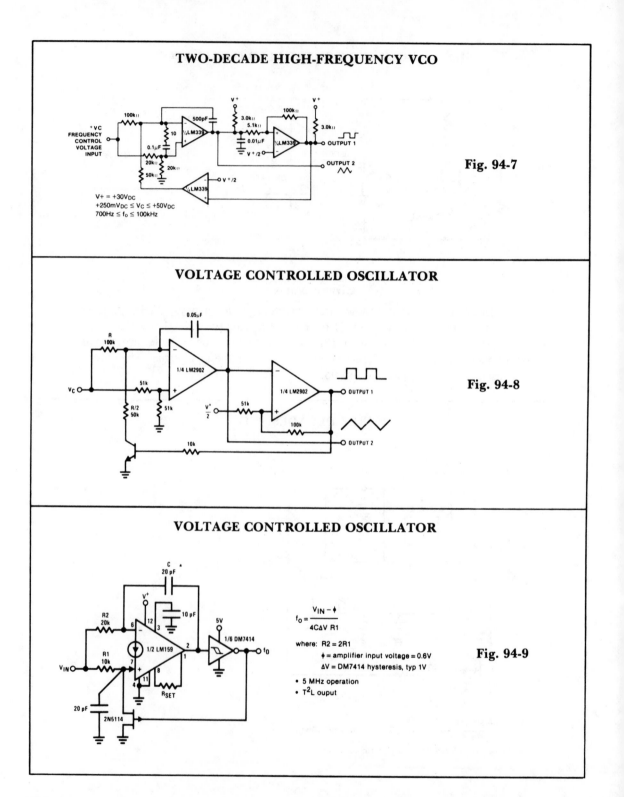

TWO-DECADE HIGH-FREQUENCY VCO

Fig. 94-7

V+ = +30V_{DC}
+250mV_{DC} ≤ V_C ≤ +50V_{DC}
700Hz ≤ f₀ ≤ 100kHz

VOLTAGE CONTROLLED OSCILLATOR

Fig. 94-8

VOLTAGE CONTROLLED OSCILLATOR

$$f_0 = \frac{V_{IN} - \phi}{4C\Delta V \ R1}$$

where: R2 = 2R1
ϕ = amplifier input voltage = 0.6V
ΔV = DM7414 hysteresis, typ 1V

• 5 MHz operation
• T²L ouput

Fig. 94-9

95

Voltage-to-Frequency Converters

The sources of the following circuits are contained in the Sources section beginning on page 730. The figure number contained in the box of each circuit correlates to the source entry in the Sources section.

10 Hz TO 10 kHz VOLTAGE/FREQUENCY CONVERTER

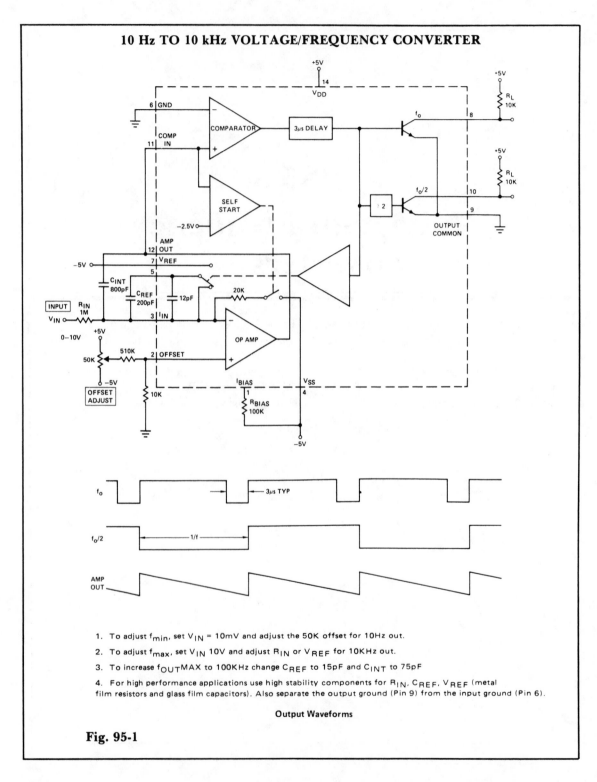

1. To adjust f_{min}, set V_{IN} = 10mV and adjust the 50K offset for 10Hz out.

2. To adjust f_{max}, set V_{IN} 10V and adjust R_{IN} or V_{REF} for 10KHz out.

3. To increase $f_{OUT}MAX$ to 100KHz change C_{REF} to 15pF and C_{INT} to 75pF

4. For high performance applications use high stability components for R_{IN}, C_{REF}, V_{REF} (metal film resistors and glass film capacitors). Also separate the output ground (Pin 9) from the input ground (Pin 6).

Output Waveforms

Fig. 95-1

VOLTAGE-TO-FREQUENCY CONVERTER

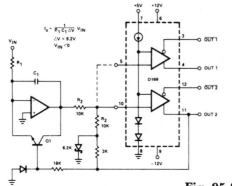

Circuit Notes

The D169 serves as a level detector and provides complementary outputs. The op amp is used to integrate the input signal V_{IN} with a time constant of R1C1. The input (must be negative) causes a positive ramp at the output of the integrator which is summed with a negative zener voltage. When the ramp is positive enough D169 outputs change state and OUT 2 flips from negative to positive. The output pulse repetition rate f_o, is directly proportioned to the negative input voltage V_{IN}.

Fig. 95-2

VOLTAGE-TO-FREQUENCY CONVERTER

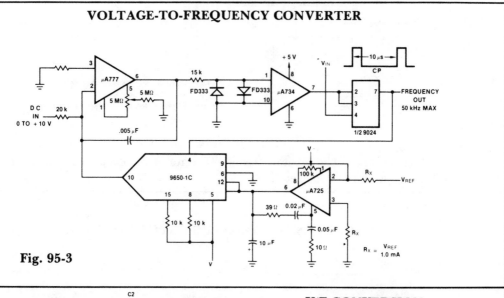

Fig. 95-3

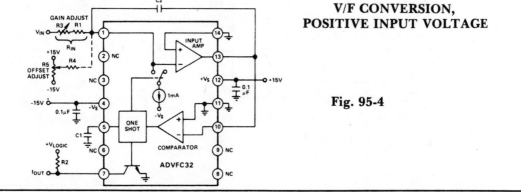

V/F CONVERSION, POSITIVE INPUT VOLTAGE

Fig. 95-4

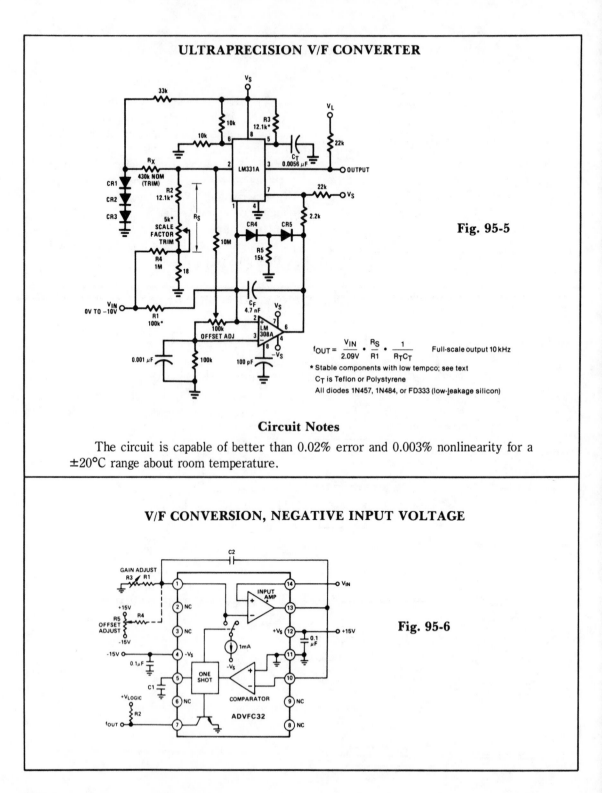

ULTRAPRECISION V/F CONVERTER

Fig. 95-5

$$f_{OUT} = \frac{V_{IN}}{2.09V} \cdot \frac{R_S}{R_1} \cdot \frac{1}{R_T C_T} \qquad \text{Full-scale output 10 kHz}$$

* Stable components with low tempco; see text
  C_T is Teflon or Polystyrene
  All diodes 1N457, 1N484, or FD333 (low-leakage silicon)

Circuit Notes

The circuit is capable of better than 0.02% error and 0.003% nonlinearity for a ±20°C range about room temperature.

V/F CONVERSION, NEGATIVE INPUT VOLTAGE

Fig. 95-6

96
Voltmeters

The sources of the following circuits are contained in the Sources section beginning on page 730. The figure number contained in the box of each circuit correlates to the source entry in the Sources section.

3-¾ Digit DVM, Four Decade, ±0.4 V, ±4 V, ±40 V, and ±400 V Full Scale
Automatic Nulling DVM
3-½ Digit True RMS AC Voltmeter
3-½ Digit DVM Common Anode Display
DVM Auto-Calibrate Circuit
FET Voltmeter

Extended Range VU Meter (Bar Mode)
High Input Impedance Millivoltmeter
Wide Band AC Voltmeter
Suppressed Zero Meter
Ac Millivoltmeter
4½ Digit LCD-DVM
Sensitive Low Cost VTVM

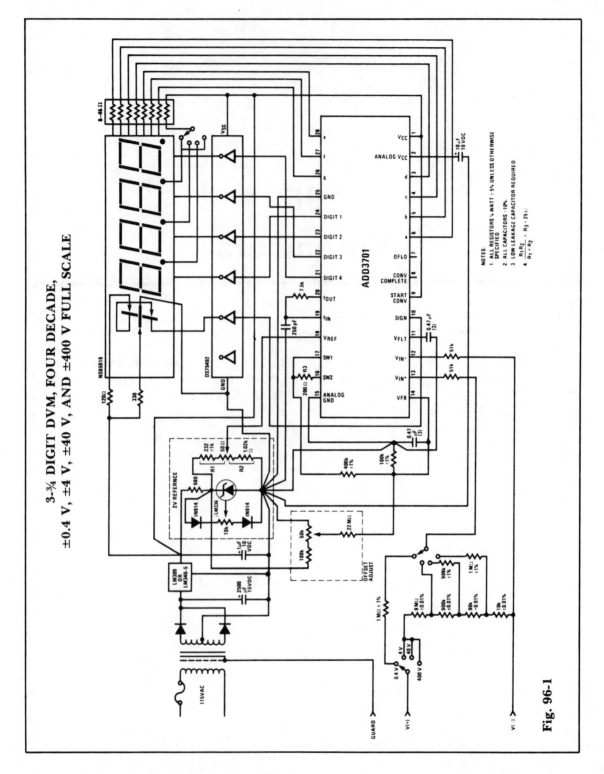

3-3/4 DIGIT DVM, FOUR DECADE, ±0.4 V, ±4 V, ±40 V, AND ±400 V FULL SCALE

Fig. 96-1

AUTOMATIC NULLING DVM

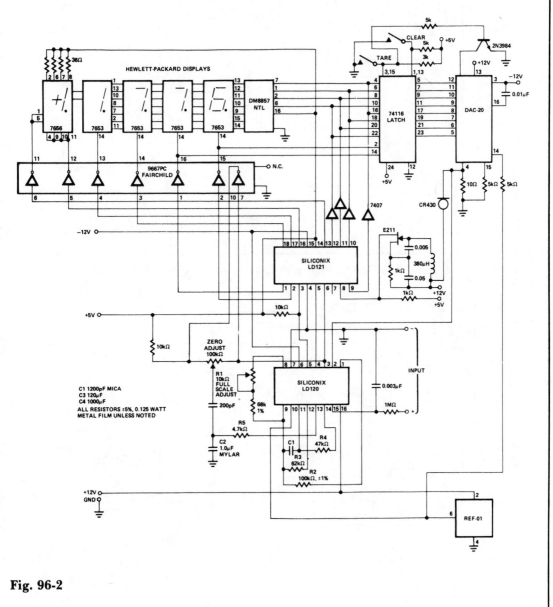

Fig. 96-2

3-½ DIGIT TRUE RMS AC VOLTMETER

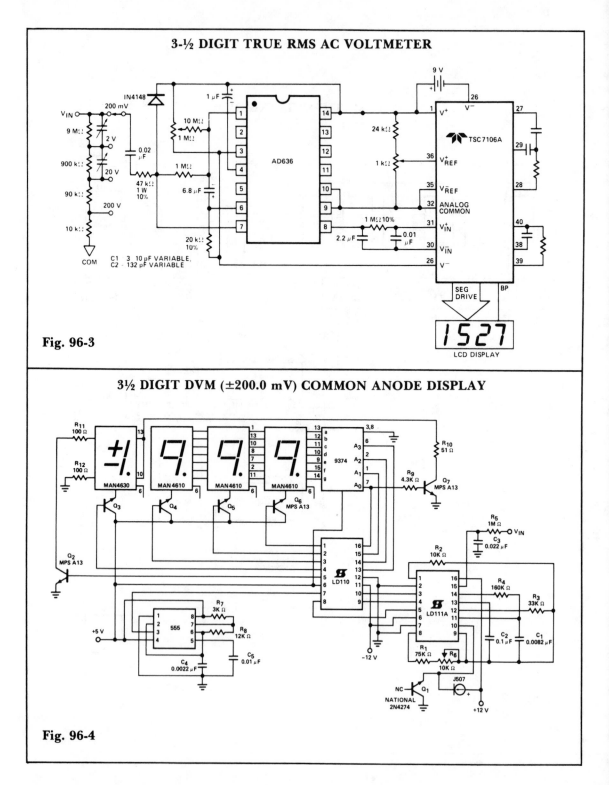

Fig. 96-3

3½ DIGIT DVM (±200.0 mV) COMMON ANODE DISPLAY

Fig. 96-4

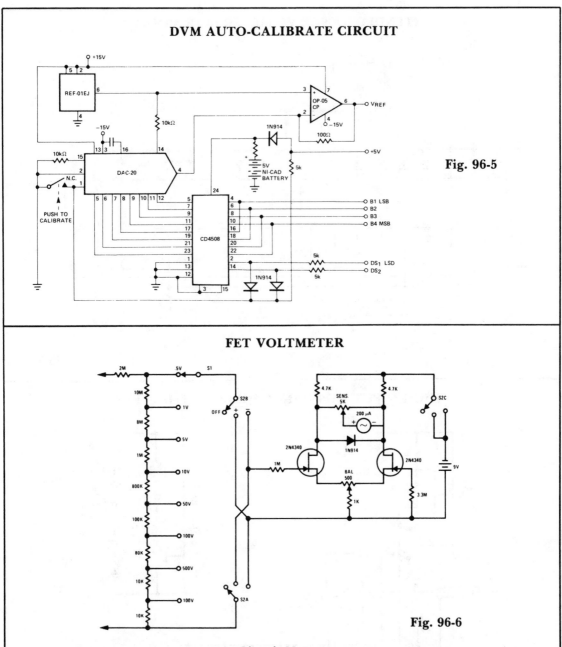

DVM AUTO-CALIBRATE CIRCUIT

Fig. 96-5

FET VOLTMETER

Fig. 96-6

Circuit Notes

This FETVM replaces the function of the VTVM while at the same time ridding the instrument of the usual line cord. In addition, drift rates are far superior to vacuum tube circuits allowing a 0.5 volt full scale range which is impractical with most vacuum tubes. The low-leakage, low-noise 2N4340 is an ideal device for this application.

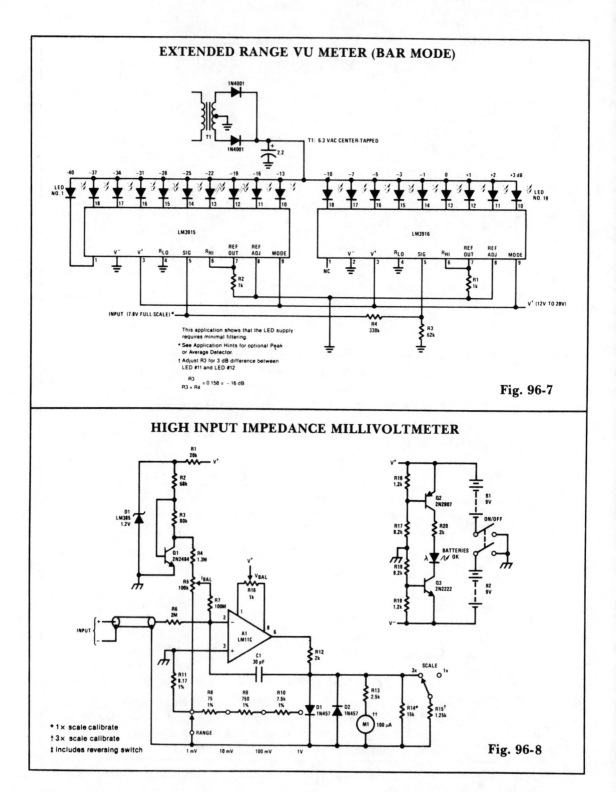

EXTENDED RANGE VU METER (BAR MODE)

Fig. 96-7

HIGH INPUT IMPEDANCE MILLIVOLTMETER

Fig. 96-8

WIDE BAND AC VOLTMETER

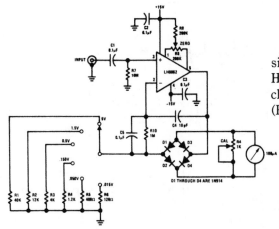

Circuit Notes

This voltmeter is capable of measuring ac signals as low as 15 mV at frequencies from 100 Hz to 500 kHz. Full scale sensitivity may be changed by altering the values R1 through R6 ($R = V_{IN}/100 \ \mu A$).

Fig. 96-9

SUPPRESSED ZERO METER

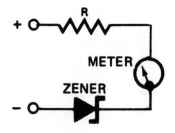

Fig. 96-10

Circuit Notes

A zener diode placed in series with a voltmeter will prevent the meter from reading until the applied voltage exceeds the zener voltage. Thus, a 10 volt zener in series with a 5-volt meter will allow the condition of a 12 V car battery to be monitored with much greater sensitivity than would be possible with a meter reading 0-15 volts.

AC MILLIVOLTMETER

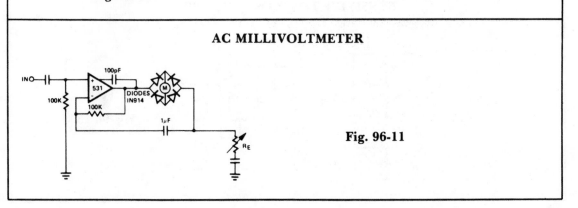

Fig. 96-11

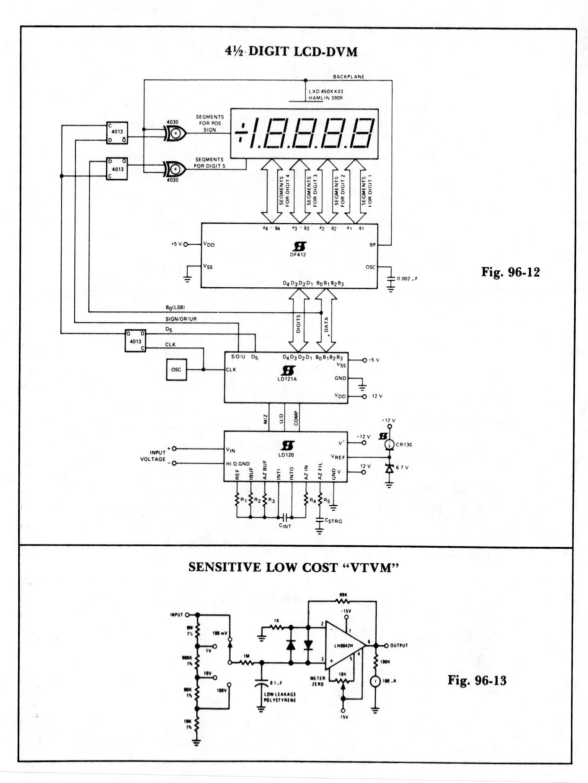

4½ DIGIT LCD-DVM

Fig. 96-12

SENSITIVE LOW COST "VTVM"

Fig. 96-13

97

Waveform and Function Generators

The sources of the following circuits are contained in the Sources section beginning on page 730. The figure number contained in the box of each circuit correlates to the source entry in the Sources section.

LOW COST ADJUSTABLE FUNCTION GENERATOR

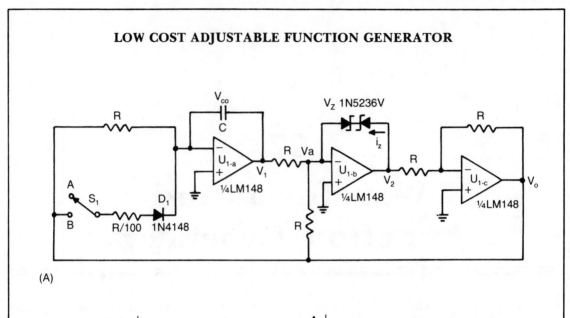

(A)

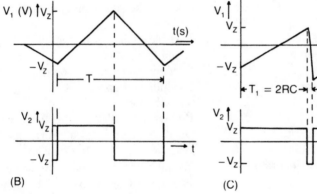

(B)

(C)

Fig. 97-1

Circuit Notes

This low-cost operational-amplifier circuit (A) generates four different functions with adjustable periods. For the components shown here, the period of the output waveforms is given by $T = 4RC$ and $T = 2RC$. With switch S1 in position A, V1 is a triangular waveform, while V2 is a square wave (B). With the switch in position B, a sawtooth waveform is generated at V1 and a pulse at V2(C).

DAC CONTROLLED FUNCTION GENERATOR

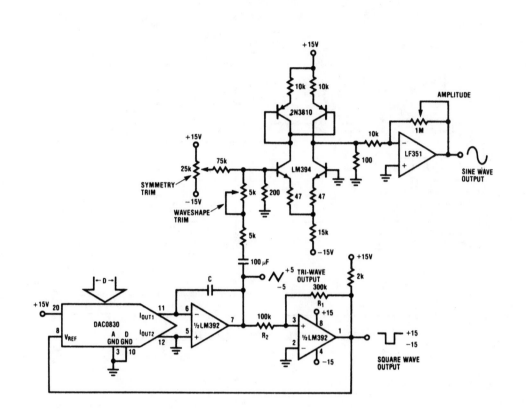

- DAC controls the frequency of sine, square, and triangle outputs.
- $f = \dfrac{D}{256(20k)C}$ for $V_{0MAX} = V_{0MIN}$ of square wave output and $R_1 = 3R_2$.
- 255 to 1 linear frequency range; oscillator stops with D = 0
- Trim symmetry and wave-shape for minimum sine wave distortion.

Fig. 97-2

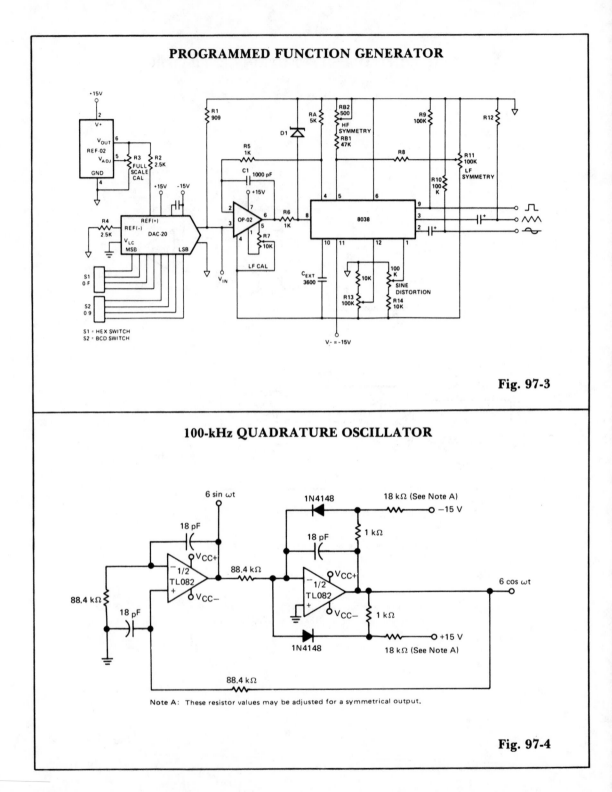

PROGRAMMED FUNCTION GENERATOR

Fig. 97-3

100-kHz QUADRATURE OSCILLATOR

Note A: These resistor values may be adjusted for a symmetrical output.

Fig. 97-4

STROBE-TONE BURST GENERATOR

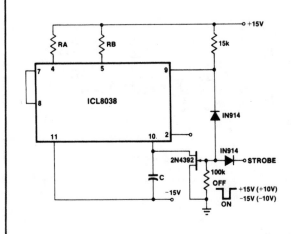

Circuit Notes

With a dual supply voltage, the external capacitor on pin 10 can be shorted to ground to halt the 8038 oscillation. The circuit uses a FET switch and diode ANDed with an input strobe signal to allow the output to always start on the same slope.

Fig. 97-5

LOW COST HIGH FREQUENCY GENERATOR

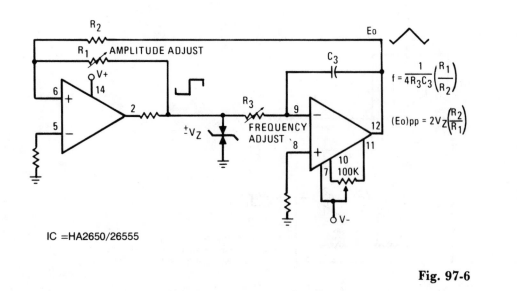

$$f = \frac{1}{4R_3C_3}\left(\frac{R_1}{R_2}\right)$$

$$(Eo)pp = 2V_Z\left(\frac{R_2}{R_1}\right)$$

IC =HA2650/26555

Fig. 97-6

TONE-BURST OSCILLATOR AND DECODER

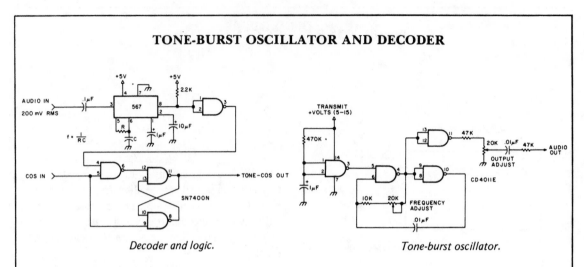

Decoder and logic.

Tone-burst oscillator.

Fig. 97-7

Circuit Notes

A tone burst sent at the beginning of each transmission is decoded (at receiver) by a PLL causing output from pin 3 of logic gate to turn on carrier-operated switch (COS).

TRIANGLE AND SQUARE WAVEFORM GENERATOR

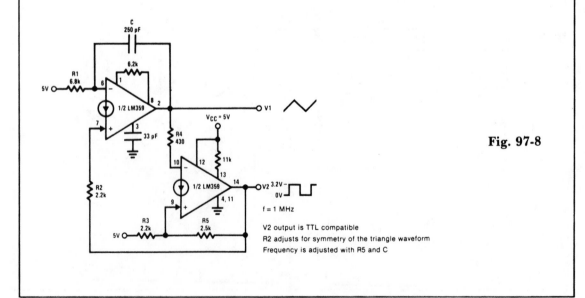

V2 output is TTL compatible
R2 adjusts for symmetry of the triangle waveform
Frequency is adjusted with R5 and C

Fig. 97-8

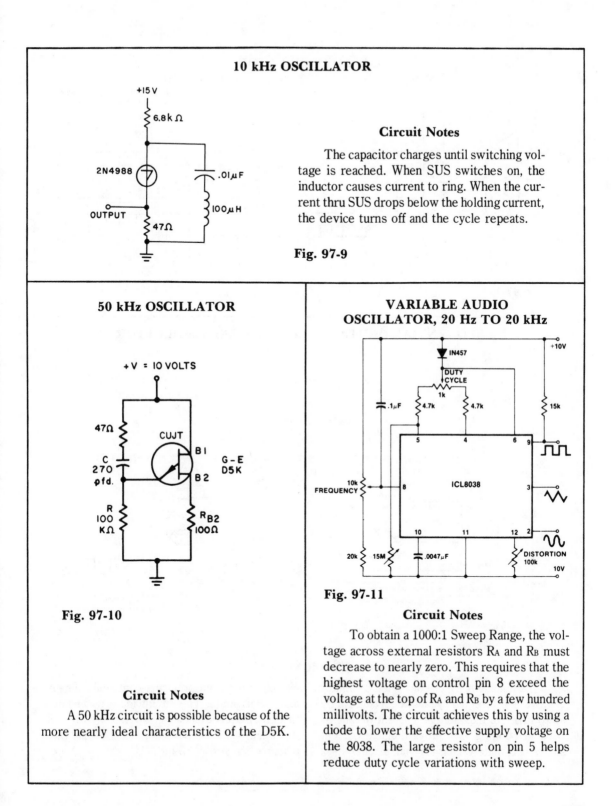

10 kHz OSCILLATOR

+15 V

6.8k Ω

2N4988

.01 μF

OUTPUT

100 μH

47 Ω

Circuit Notes

The capacitor charges until switching voltage is reached. When SUS switches on, the inductor causes current to ring. When the current thru SUS drops below the holding current, the device turns off and the cycle repeats.

Fig. 97-9

50 kHz OSCILLATOR

+V = 10 VOLTS

47 Ω

CUJT

B1

G - E
D5K

C
270
ϕfd.

B2

R
100
KΩ

R_B2
100 Ω

Fig. 97-10

Circuit Notes

A 50 kHz circuit is possible because of the more nearly ideal characteristics of the D5K.

VARIABLE AUDIO OSCILLATOR, 20 Hz TO 20 kHz

IN457

+10V

DUTY CYCLE

1k

.1μF   4.7k   4.7k

15k

5   4   6   9

10k
FREQUENCY

8   ICL8038   3

10   11   12   2

20k   15M   .0047μF

DISTORTION
100k

10V

Fig. 97-11

Circuit Notes

To obtain a 1000:1 Sweep Range, the voltage across external resistors R_A and R_B must decrease to nearly zero. This requires that the highest voltage on control pin 8 exceed the voltage at the top of R_A and R_B by a few hundred millivolts. The circuit achieves this by using a diode to lower the effective supply voltage on the 8038. The large resistor on pin 5 helps reduce duty cycle variations with sweep.

GATED OSCILLATOR

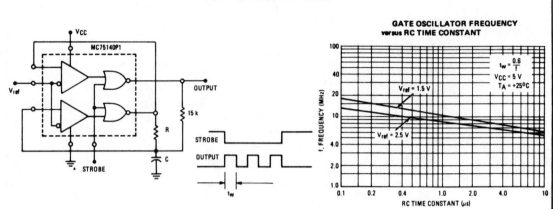

Fig. 97-12

EXPONENTIAL DIGITALLY-CONTROLLED OSCILLATOR

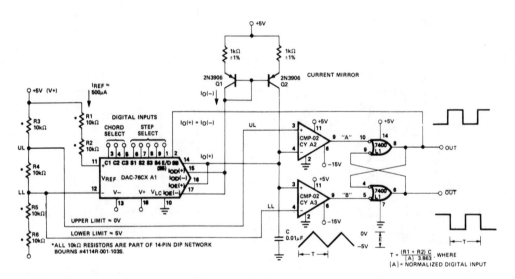

Fig. 97-13

Circuit Notes

The microprocessor-controlled oscillator has a 8159 to 1 frequency range covering 2.5 Hz to 20 kHz. An exponential, current output IC DAC functioning as a programmable current source alternately charges and discharges a capacitor between precisely-controlled upper and lower limits. The circuit features instantaneous frequency change, operates with +5 ±1 V and −15 V ±3 V supplies, and has the dynamic range of a 13-bit DAC.

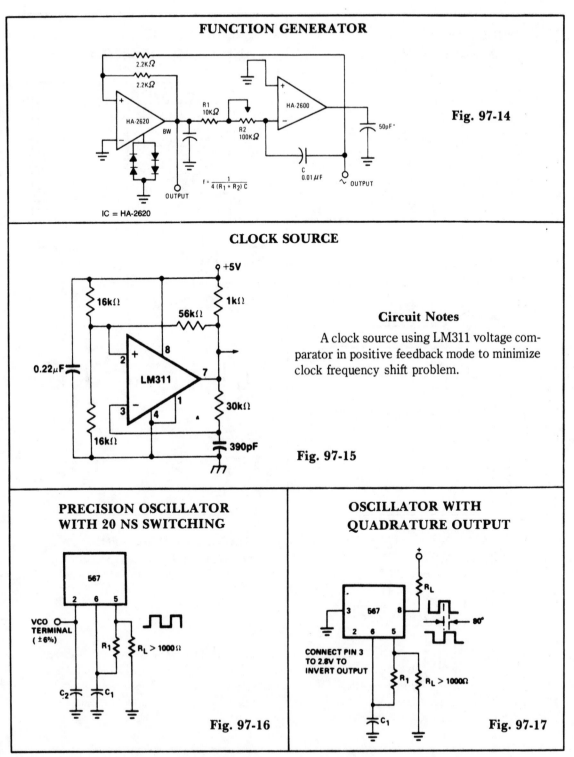

FUNCTION GENERATOR

2.2K Ω

2.2K Ω

HA-2620

BW

R1
10K Ω

R2
100K Ω

HA-2600

50pF

OUTPUT

C
0.01 μF

OUTPUT

$f = \dfrac{1}{4 (R_1 + R_2) C}$

IC = HA-2620

Fig. 97-14

CLOCK SOURCE

+5V

16k Ω

1k Ω

56k Ω

0.22 μF

2

8

LM311

7

3

1

4

16k Ω

30k Ω

390pF

Circuit Notes

A clock source using LM311 voltage comparator in positive feedback mode to minimize clock frequency shift problem.

Fig. 97-15

PRECISION OSCILLATOR WITH 20 NS SWITCHING

567

2   6   5

VCO
TERMINAL
(±6%)

R_1

$R_L > 1000 \Omega$

C_2 C_1

Fig. 97-16

OSCILLATOR WITH QUADRATURE OUTPUT

R_L

3   567   8

90°

2   6   5

CONNECT PIN 3
TO 2.8V TO
INVERT OUTPUT

R_1 $R_L > 1000 \Omega$

C_1

Fig. 97-17

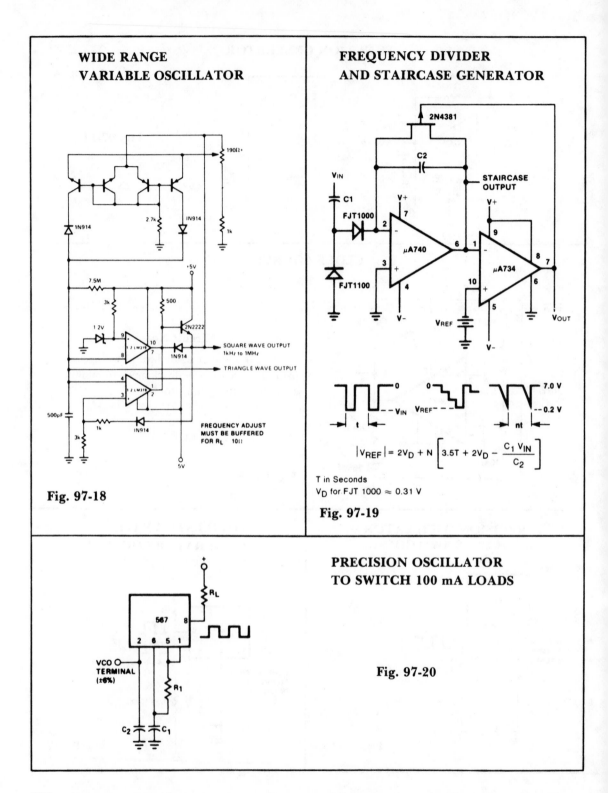

WIDE RANGE VARIABLE OSCILLATOR

190Ω

2.7k

1N914

1N914

1k

1N914

7.5M

+5V

3k

500

1.2V

2N2222

9 1/2 LM319 10
8 7

1N914

SQUARE WAVE OUTPUT
1kHz to 1MHz

TRIANGLE WAVE OUTPUT

4 1/2 LM319 1
3 2

500pF

3k

1k

1N914

FREQUENCY ADJUST
MUST BE BUFFERED
FOR R_L 10Ω

5V

Fig. 97-18

FREQUENCY DIVIDER AND STAIRCASE GENERATOR

2N4381

C2

STAIRCASE
OUTPUT

V_{IN}

C1

FJT1000

V+
7

V+
9

μA740

6 1

μA734

8
7

3

+

4

V-

10

+

5

V_{REF}

6

V_{OUT}

FJT1100

V-

0

V_{IN}

0

V_{REF}

7.0 V

0.2 V

t

nt

$$|V_{REF}| = 2V_D + N\left[3.5T + 2V_D - \frac{C_1 V_{IN}}{C_2}\right]$$

T in Seconds
V_D for FJT 1000 ≈ 0.31 V

Fig. 97-19

PRECISION OSCILLATOR TO SWITCH 100 mA LOADS

R_L

567   8

2 6 5 1

VCO
TERMINAL
(±6%)

R_1

C_2 C_1

Fig. 97-20

98

Zero Crossing Detectors

The sources of the following circuits are contained in the Sources section beginning on page 730. The figure number contained in the box of each circuit correlates to the source entry in the Sources section.

ZERO CROSSING SWITCH

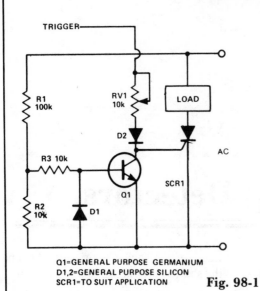

TRIGGER—

R1 100k

RV1 10k

LOAD

D2

R3 10k

AC

SCR1

Q1

R2 10k

D1

Q1=GENERAL PURPOSE GERMANIUM
D1,2=GENERAL PURPOSE SILICON
SCR1=TO SUIT APPLICATION

Fig. 98-1

Circuit Notes

When switching loads with the aid of a thyristor, a large amount of RFI can be generated unless some form of zero crossing switch is used. The circuit shows a simple single transistor zero crossing switch. R1 and R2 act as a potential divider. The potential at their junction is about 10% of the ac voltage. This voltage level is fed, via R3, to the transistor's base. If the voltage at this point is above 0.2, the transistor will conduct, shunting any thyristor gate current to ground. When the line potential is less than about 2 V, it is possible to trigger the thyristor. The diode D1 is to remove any negative potential that might cause reverse breakdown.

ZERO CROSSING DETECTOR

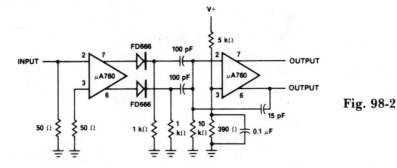

V+

5 kΩ

INPUT

2 7

µA760

3 6

FD666

100 pF

100 pF

FD666

50 Ω   50 Ω

1 kΩ

1 kΩ

10 kΩ

390 Ω

2 7

µA760

3 6

OUTPUT

OUTPUT

15 pF

0.1 µF

Fig. 98-2

Total Delay = 30 ns
Input frequency = 300 Hz to 3 MHz
Minimum input voltage = 20 mVpk-pk

ZERO CROSSING DETECTOR

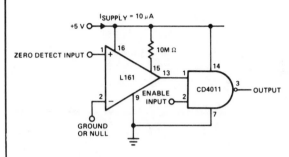

Circuit Notes

This detector is useful in sine wave squaring circuits and A/D converters. The positive input may either be grounded or connected to a nulling voltage which cancels input offsets and enables accuracy to within microvolts of ground. The CMOS output will switch to within a few millivolts of either rail for an input voltage change of less than 200 μV.

Fig. 98-3

ZERO CROSSING DETECTOR WITH TEMPERATURE SENSOR

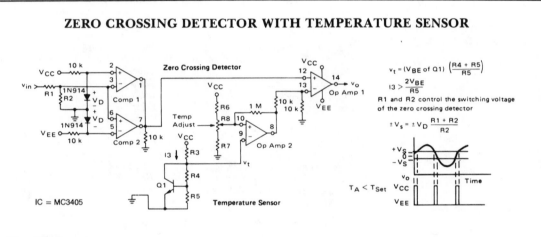

$$v_t = (V_{BE} \text{ of } Q1) \left(\frac{R4 + R5}{R5} \right)$$

$$I3 > \frac{2V_{BE}}{R5}$$

R1 and R2 control the switching voltage of the zero crossing detector

$$\pm V_s = \pm V_D \frac{R1 + R2}{R2}$$

Fig. 98-4

ZERO-CROSSING DETECTOR

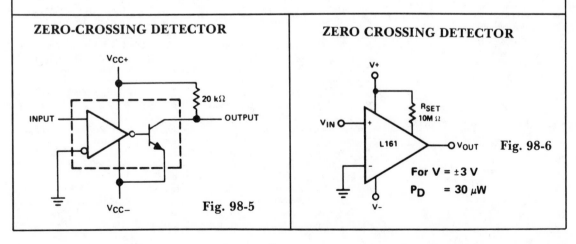

Fig. 98-5

ZERO CROSSING DETECTOR

For V = ±3 V

P_D = 30 μW

Fig. 98-6

Sources

Chapter 1

Fig. 1-1: The Build-It Book Of Electronics Projects, TAB Book No. 1498, p. 73.
Fig. 1-2: QST, 7/81, p. 28.
Fig. 1-3: Radio Electronics, 10/78, p. 41.
Fig. 1-4: '73 Magazine, 10/77, p. 122.
Fig. 1-5: Modern Electronics, 2/78, p. 50.
Fig. 1-6: Electronics Today International, 3/82, p. 69.
Fig. 1-7: Modern Electronics, 7/78, p. 51.
Fig. 1-8: Electronics Today International, 4/83, p. 72.
Fig. 1-9: 101 Electronic Projects, 1977, #64.
Fig. 1-10: Electronics Today International, 10/78, p. 94.
Fig. 1-11: Modern Electronics, 2/78, p. 55.
Fig. 1-12: Modern Electronics, 2/78, p. 48.
Fig. 1-13: Signetics 555 Timers, 1973, p. 26.
Fig. 1-14: Electronics Today International, 3/83, p. 23.
Fig. 1-15: Electronics Today International, 3/83, p. 23.
Fig. 1-16: National Semiconductor, Linear Databook, 1982, p. 3-288.
Fig. 1-17: Electronics Today International, 3/83, p. 23.
Fig. 1-18: Signetics 555 Timers, 1973, p. 22.
Fig. 1-19: 101 Electronic Projects, 1977, #65.
Fig. 1-20: Modern Electronics, 6/78, p. 58.
Fig. 1-21: Modern Electronics, 6/78, p. 55.

Chapter 2

Fig. 2-1: Modern Electronics, 3/78, p. 69.
Fig. 2-2: Electronics Today International, 10/78, p., 30.
Fig. 2-3: CQ, 5/77, p. 50.
Fig. 2-4: Ham Radio, 10/78, p. 34.
Fig. 2-5: Ham Radio, 10/78, p. 89.
Fig. 2-6: 73 Magazine, 7/78, p. 62.
Fig. 2-7: 101 Electronic Projects, 1975, p. 22.
Fig. 2-8: 73 Magazine, 7/82, p. 46.
Fig. 2-9: 73 Magazine, 7/83, p. 103.
Fig. 2-10: 101 Electronic Projects, 1975, p. 13.
Fig. 2-11: Ham Radio, 5/78, p. 87.
Fig. 2-12: 73 Magazine, p. 164.
Fig. 2-13: Modern Electronics, 2/78, p. 16.
Fig. 2-14: 73 Magazine, 10/77, p. 52.
Fig. 2-15: 73 Magazine, 7/77, p. 34.
Fig. 2-16: 104 Weekend Electronics Projects, TAB Book No. 1436, p. 120.
Fig. 2-17: Ham Radio, 10/70, p. 76.
Fig. 2-18: Electronics Today International, 7/77, p. 72.

Chapter 3

Fig. 3-1: Courtesy of Fairchild Camera & Instrument Corporation. Linear Databook, 1982, p. 4-119.
Fig. 3-2: Signetics Analog Data Manual, 1982, p. 3-83.
Fig. 3-3: Teledyne Semiconductor, Data & Design Manual, 1981, p. 11-207.
Fig. 3-4. Signetics Analog Data Manual, 1983, p. 10-99.
Fig. 3-5: Reprinted with the permission of National Semiconductor Corp. Data Conversion/Acquisition Databook, 1980, p. 3-107.
Fig. 3-6: Reprinted with the permission of National Semiconductor Corp. Transistor Databook, 1982, p. 11-29.
Fig. 3-7: Reprinted with the permission of National Semiconductor Corp. Audio/Radio Handbook, 1980, p. 2-67.
Fig. 3-8: Reprinted with the permission of National Semiconductor Corp. Hybrid Products Databook, 1982, p. 7-7.
Fig. 3-9: Electronics Today International, 2/82, p. 58.
Fig. 3-10. Signetics Analog Data Manual, 1983, p. 10-100.
Fig. 3-11: Precision Monolithics Incorporated 1981 Full Line Catalog, p. 12-50.
Fig. 3-12: Courtesy of Fairchild Camera & Instrument Corporation. Linear Databook, 1982, p. 9-17.
Fig. 3-13: Signetics Analog Data Manual, 1977, p. 35.
Fig. 3-14: Courtesy of Fairchild Camera & Instrument Corporation. Linear Databook, 1982, p. 5-39.
Fig. 3-15: Precision Monolithics Incorporated, 1981 Full Line Catalog, p. 6-10.
Fig. 3-16: Courtesy of Motorola Inc. Motorola Semiconductor Library, Volume 6, Series B, p. 8-21.
Fig. 3-17: Signetics Analog Data Manual, 1983, p. 17-17.
Fig. 3-18: Intersil Data Book, 5/83, p. 5-36.
Fig. 3-19: Courtesy of Motorola Inc. Linear Integrated Circuits, 1979, p. 3-17.
Fig. 3-20: Reprinted with the permission of National Semiconductor Corp. Hybrid Products Databook, 1982, p. 1-83.
Fig. 3-21: Precision Monolithics Incorporated, 1981 Full Line Catalog, p. 16-160.
Fig. 3-22: Signetics Analog Data Manual, 1982, p. 3-103.
Fig. 3-23: Precision Monolithics Incorporated, 1981 Full Line Catalog, p. 6-127.
Fig. 3-24: Courtesy of Motorola Inc., Linear Integrated Circuits, 1979, p. 3-83.
Fig. 3-25: Courtesy of Motorola Inc.

Linear Integrated Circuits, 1979, p. 3-131.

Fig. 3-26: Harris Semiconductor, Analog Data Book 1984.

Fig. 3-27: Intersil Data Book, 5/83, p. 5-36.

Fig. 3-28: Precision Monolithics Incorporated, 1981 Full Line Catalog, p. 16-37.

Fig. 3-29: Courtesy of Motorola Inc. Linear Integrated Circuits, 1979, p. 3-31.

Fig. 3-30: Siliconix Analog Switch & IC Product Data Book, 1/82, p. 6-21.

Fig. 3-31: Siliconix Analog Switch & IC Product Data Book, 1/82, p. 6-15.

Fig. 3-32: Precision Monolithics Incorporated, 1981 Full Line Catalog, p. 16-37.

Fig. 3-33: Siliconix Analog Switch & IC Product Data Book, 1/82, p. 7-56.

Fig. 3-34: Reprinted with permission of Analog Devices, Inc. Data Acquisition Databook, 1982, p. 4-119.

Fig. 3-35: Courtesy of Fairchild Camera & Instrument Corporation. Linear Databook, 1982, p. 4-42.

Fig. 3-36: Courtesy of Motorola Inc., Linear Integrated Circuits, p. 3-17.

Fig. 3-37: Courtesy of Motorola Inc. Linear Integrated Circuits, 1979, p. 6-23.

Fig. 3-38: Courtesy of Texas Instruments Incorporated. Linear Control Circuits Data Book, Second Edition, p. 145.

Fig. 3-39: Courtesy of Motorola Inc. Linear Integrated Circuits, 1979, p. 3-83.

Fig. 3-40: Courtesy of Fairchild Camera & Instrument Corporation. Linear Databook, 1982, p. 4-41.

Fig. 3-41: Canadian Projects Number 1, Spring/78, p. 29.

Fig. 3-42: Reprinted with the permission of National Semiconductor Corp. Application Note AN125, p. 2.

Fig. 3-43: Harris Semiconductor, Linear & Data Acquisition Products, p. 2-58.

Fig. 3-44: Reprinted with permission of Analog Devices, Inc. Data Acquisition Databook 1982, p. 4-98.

Fig. 3-45: Reprinted with the permission of National Semiconductor Corp. Application Note AN125, p. 3.

Chapter 4

Fig. 4-1: Courtesy of Fairchild Camera & Instrument Corporation. Linear Databook, 1982, p. 7-8.

Fig. 4-2: Intersil Data Book, 5/83, p. 4-83.

Fig. 4-3: Ferranti, Technical Handbook Vol. 10, Data Converters, 1983, p. 7-10.

Fig. 4-4: Precision Monolithics Incorporated, 1981 Full Line Catalog, p. 16-12.

Fig. 4-5: Reprinted with permission of Analog Devices, Inc. Data Acquisition Databook, 1982, p. 10-241.

Fig. 4-6: Precision Monolithics Incorporated, 1981 Full Line Catalog, p. 8-13.

Fig. 4-7: Reprinted with the permission of National Semiconductor Corp. National Semiconductor CMOS Databook. 1981. p. 3-63.

Fig. 4-8: Reprinted with permission of Analog Devices, Inc. Data Acquisition Databook, 1982, p. 10-240.

Fig. 4-9: Teledyne Semiconductor, Data & Design Manual, 1981, p. 7-39.

Fig. 4-10: Reprinted with permission of Analog Devices, Inc. Data Acquisition Databook, 1982, p. 10-50.

Fig. 4-11: Courtesy of Fairchild Camera & Instrument Corporation. Linear Databook, 1982, p. 5-32.

Fig. 4-12: Precision Monolithics Incorporated 1981 Full Line Catalog, p. 8-13.

Chapter 5

Fig. 5-1: Reprinted with the permission of National Semiconductor Corp. Data Conversion/Acquisition Databook, 1980, p. 3-22.

Fig. 5-2: Reprinted with the permission of National Semiconductor Corp. Transistor Databook, 1982, p. 11-29.

Fig. 5-3: Reprinted with the permission of National Semiconductor Corp. Data Conversion/Acquisition Databook, 1980, p. 8-64.

Fig. 5-4: Precision Monolithics Incorporated, 1981 Full Line Catalog, p. 12-39.

Chapter 6

Fig. 6-1: Electronics Today International, 3/82, p. 66.

Fig. 6-2: 101 Electronic Projects, 1977, IC 23.

Fig. 6-3: Reprinted with the permission of National Semiconductor Corp. Audio/Radio Handbook, 1980, p. 2-66.

Fig. 6-4: Electronics Today International, 10/79, p. 93.

Fig. 6-5: No reference.

Fig. 6-6: No reference.

Fig. 6-7: Electronics Today International, 3/75, p. 66.

Fig. 6-8: Electronics Today International, 3/78, p. 52.

Fig. 6-9: Electronics Today International, 5/78, p. 85.

Fig. 6-10: Modern Electronics, 7/78, p. 58.

Chapter 7

Fig. 7-1: Courtesy of Fairchild Camera & Instrument Corporation. Fairchild Semiconductor Application Note 300.

Fig. 7-2: Ham Radio, 1/78, p. 78.

Fig. 7-3: Courtesy of Motorola Inc. Linear Integrated Circuits, 1979, p. 6-23.

Fig. 7-4: 73 Magazine, 12/76, p. 97.

Fig. 7-5: 73 Magazine, 7/77, p. 34.

Fig. 7-6: Reprinted with the permission of National Semiconductor Corp. Linear Applications Handbook, 1982, p. AN29-9.

Fig. 7-7: Reprinted with the permission of National Semiconductor Corp. Linear Applications Handbook, 1982, P. LB16-1.

Fig. 7-8: Reprinted with the permission of National Semiconductor Corp. Transistor Databook, 1982, p. 11-31.

Fig. 7-9: Reprinted with the permission of National Semiconductor Corp. Linear Databook, 1982, p. 10-25.

Fig. 7-10: How to Design/Build Remote Control Devices TAB Book No. 1277, p. 230.

Fig. 7-11: Radio Electronics, 7/83, p. 7.

Fig. 7-12: Electronics Today International, Summer 1982, p. 45.

Fig. 7-13: 73 Magazine, p. 31.

Fig. 7-14: Reprinted from Electronics, 11/83. Copyright 1983, McGraw Hill Inc. All rights reserved.

Fig. 7-15: Electronics Today International, 7/72, p. 84.

Fig. 7-16: Courtesy of Motorola Inc. Linear Integrated Circuits, 1979, p. 3-42.

Fig. 7-17: Reprinted with the permission of National Semiconductor Corp. Linear Databook, 1982, p. 3-171.

Chapter 8

Fig. 8-1: Courtesy of Fairchild Camera & Instrument Corporation, Fairchild Progress, 11-12/76, p. 26.

Fig. 8-2: Courtesy of Fairchild Camera & Instrument Corporation. Fairchild Progress, 5-6/77, p. 22.

Fig. 8-3: Reprinted with the permission

of National Semiconductor Corp.
Audio/Radio Handbook, 1980, p. 4-44.
Fig. 8-4: Reprinted with the permission
of National Semiconductor Corp.
Audio/Radio Handbook, 1980, p. 4-14.
Fig. 8-5: Reprinted with the permission
of National Semiconductor Corp.
Audio/Radio Handbook, 1980, p. 4-14.
Fig. 8-6: Reprinted with the permission
of National Semiconductor Corp.
Transistor Databook, 1982, p. 7-23.
Fig. 8-7: Reprinted with the permission
of National Semiconductor Corp.
Audio/Radio Handbook, 1980, p. 4-51.
Application Note AN125, p. 7.
Fig. 8-8: Reprinted with the permission
of National Semiconductor Corp.
Audio/Radio Handbook, 1980, p. 4-51.
Application Note AN125, p. 6.
Fig. 8-9: Reprinted with the permission
of National Semiconductor Corp.
Linear Databook, 1982, p. 10-171.
Fig. 8-10: Reprinted with the permis-
sion of National Semiconductor Corp.
Linear Databook, 1982, p. 10-63.
Fig. 8-11: No reference.
Fig. 8-12: Electronics Today Interna-
tional, 3/78, p. 81.
Fig. 8-13: Courtesy of Motorola Inc.
Motorola Semiconductor Library, Vol-
ume 6, Series B, p. 8-21.
Fig. 8-14: Courtesy of Motorola Inc.
Motorola Semiconductor Library, Vol-
ume 6, Series B, p. 8-21.
Fig. 8-15: Courtesy of Motorola Inc.
Motorola Semiconductor Library, Vol-
ume 6, Series B, p. 8-21.
Fig. 8-16: Reprinted with the permis-
sion of National Semiconductor Corp.
National Semiconductor Application
Note AN125, p. 7.
Fig. 8-17: Reprinted with the permis-
sion of National Semiconductor Corp.
Application Note AN69, p. 4.
Fig. 8-18: Reprinted with the permis-
sion of National Semiconductor Corp.
Linear Databook, 1982, p. 10-25.
Fig. 8-19: Courtesy of Motorola Inc.
Linear Integrated Circuits, 1979, p.
5-17.
Fig. 8-20: Reprinted with the permis-
sion of National Semiconductor Corp.
Linear Databook, 1982, p. 10-170.
Fig. 8-21: Reprinted with the permis-
sion of National Semiconductor Corp.
Hybrid Products Databook, 1982, p.
17-170.
Fig. 8-22: Reprinted with permission of
National Semiconductor, Corp. Appli-
cation Note AN69, p. 4.
Fig. 8-23: Courtesy of Fairchild Cam-

era & Instrument Corporation. Linear
Databook, 1982, p. 4-89.
Fig. 8-24: Reprinted with permission of
National Semiconductor Corp. Linear
Databook, 1982, p. 10-203.

Chapter 9

Fig. 9-1: Canadian Projects Number 1,
Spring/78, p. 27.
Fig. 9-2: No reference.
Fig. 9-3: Electronics Today Interna-
tional, 4/79, p. 18.
Fig. 9-4: Reprinted with permission of
National Semiconductor Corp. Linear
Databook, 1982, p. 3-389.
Fig. 9-5: Reprinted with the permission
of National Semiconductor Corp. Tran-
sistor Databook, 1982. p. 11-29.
Fig. 9-6: Reprinted with permission of
National Semiconductor Corp. Data
Conversion/Acquisition Databook,
1980, p. 3-91.
Fig. 9-7: Reprinted with permission of
National Semiconductor Corp. Au-
dio/Radio Handbook, 1980, p. 2-45.
Fig. 9-8: Reprinted with permission of
National Semiconductor Corp. Au-
dio/Radio Handbook, 1980, p. 2-43.
Fig. 9-9: Reprinted with permission of
National Semiconductor Corp. Trans-
istor Databook, 1982, p. 11-28.
Fig. 9-10: Signetics Analog Data Man-
ual, 1982, p. 4-8.
Fig. 9-11: Signetics Analog Data Man-
ual, 1982, p. 15-6.
Fig. 9-12: Signetics Analog Data Man-
ual, 1977, p. 466.
Fig. 9-13: Reprinted with permission of
National Semiconductor Corp. Au-
dio/Radio Handbook, 1980, p. 2-27.
Fig. 9-14: Reprinted with permission of
National Semiconductor Corp.
Audio/Radio Handbook, 1980, p. 2-32.
Fig. 9-15: Signetics Analog Data Man-
ual, 1982, p. 15-6.
Fig. 9-16: Signetics Analog Data Man-
ual, 1977, p. 466.
Fig. 9-17: Reprinted with permission of
National Semiconductor Corp. Data
Conversion/Acquisition Databook,
1980, p. 3-88.
Fig. 9-18: Reprinted with permission of
National Semiconductor Corp. Au-
dio/Radio Handbook, 1980, p. 2-20.
Fig. 9-19: Reprinted with permission of
National Semiconductor Corp. Au-
dio/Radio Handbook, 1980. p. 2-21.
Fig. 9-20: Signetics Analog Data Man-
ual, 1977, p. 466.
Fig. 9-21: Signetics Analog Data Man-
ual, 1983, p. 10-92.

Fig. 9-22: Signetics Analog Data Manual, 1982, p. 15-6.

Chapter 10

Fig. 10-1: Reprinted with the permission of National Semiconductor Corp.
Linear Applications Handbook, 1982,
p. AN162-10.
Fig. 10-2: Electronics Today International, 6/79, p. 75.
Fig. 10-3: Signetics 555 Timers, 1973,
p. 24.
Fig. 10-4: Electronics Today Interna-
tional, 12/75, p. 72.
Fig. 10-5: Electronics Today Interna-
tional, 2/75, p. 51.
Fig. 10-6: Electronics Today Interna-
tional, 7/81, p. 22.
Fig. 10-7: Electronics Today Interna-
tional, 7/77, p. 32.
Fig. 10-8: Reprinted with the permis-
sion of National Semiconductor Corp.
Linear Applications Handbook, 1982,
p. LB33-1.
Fig. 10-9: Reprinted with the permis-
sion of National Semiconductor Corp.
Linear Databook, 1982, p. 9-141.
Fig. 10-10: Courtesy of Motorola Inc.
Linear Integrated Circuits, 1979, p.
3-138.
Fig. 10-11: Reprinted with the permis-
sion of National Semiconductor Corp.
Transistor Databook, 1982, p. 7-31.
Fig. 10-12: 73 Magazine, 7/77, p. 34.
Fig. 10-13: Modern Electronics, 2/78,
p. 56.
Fig. 10-14: Reprinted with the permis-
sion of National Semiconductor Corp.
Linear Databook, 1982, p. 9-140.
Fig. 10-15: The Build-It Book Of
Electronic Projects, TAB Book No.
1498, p. 80.
Fig. 10-16: 73 Magazine, 1/82, p. 41.
Fig. 10-17: Electronics Today Interna-
tional, 10/77, p. 47.
Fig. 10-18: Modern Electronics, 9/78,
p. 37.
Fig. 10-19: Electronics Today Interna-
tional, 10/77, p. 38.
Fig. 10-20: The Build-It Book Of
Electronic Projects, TAB Book No.
1498, p. 111.
Fig. 10-21: Modern Electronics, 5/78,
p. 7.
Fig. 10-22: Reprinted with the permis-
sion of National Semiconductor Corp.
Linear Databook, 1982, p. 9-143.
Fig. 10-23: Reprinted with the permis-
sion from General Electric Semicon-
ductor Department. General Electric
SCR Manual, Sixth Edition, 1979, p.
207.

Fig. 10-24: No reference.

Chapter 11

Fig. 11-1: Reprinted with the permission of National Semiconductor Corp. Voltage Regulator Handbook, p. 7-32.
Fig. 11-2: 101 Electronics Projects, 1977, p. 97.
Fig. 11-3: Courtesy of Motorola Inc. Application Note AN-294, p. 6.
Fig. 11-4: 73 Magazine, 2/79, p. 156.
Fig. 11-5: 73 Magazine, 7/77.
Fig. 11-6: Ham Radio, 12/79, p. 67.
Fig. 11-7: 73 Magazine, 2/83, p. 99.
Fig. 11-8: 44 Electronics Projects For SWLS, CBers & Radio Experimenters, TAB Book No. 1258, p. 153.
Fig. 11-9: Yuasa Battery (America) Inc. Application Manual for NP type battery.
Fig. 11-10: Electronics Today International, 11/80.
Fig. 11-11: 73 Magazine, 7/77.
Fig. 11-12: Reprinted with permission from General Electric Semiconductor Department. General Electric SCR Manual, Sixth Edition, 1979, p. 203.
Fig. 11-14: Reprinted with the permission of National Semiconductor Corp. Linear Databook, 1982, p. 9-31.
Fig. 11-15: Reprinted with the permission of National Semiconductor Corp. Voltage Regulator Handbook, p. 10-141.

Chapter 12

Fig. 12-1: NASA Tech Brief, B73-10249.
Fig. 12-2:Electronics Today International, 1/75, p. 66.
Fig. 12-3: Electronics Australia, 2/76, p. 91.
Fig. 12-4: 73 Magazine, 2/79, p. 78.
Fig. 12-5: Electronics Today International, 6/79, p. 103.
Fig. 12-6: Ham Radio, 9/82, p. 78.
Fig. 12-7: Courtesy of Texas Instruments Incorporated. Optoelectronics Databook, 1983-84, p. 15-5.
Fig. 12-8: 73 Magazine, 2/79, p. 78.
Fig. 12-9: © Siliconix incorporated. Siliconix Analog Switch & IC Product Data Book, 1/82, p. 6-19.
Fig. 12-10: Reprinted with the permission of National Semiconductor Corp. Linear Databook, 1982, p. 3-109.
Fig. 12-11: Reprinted with the permission of National Semiconductor Corp. Linear Databook, 1982, p. 3-109.

Chapter 13

Fig. 13-1: Intersil Data Book, 5/83, p. 5-238.

Fig. 13-2: Reprinted with the permission of National Semiconductor Corp. Hybrid Products Databook, 1982, p. 17-131.
Fig. 13-3: Precision Monolithics Incorporated, 1981 Full Line Catalog, p. 16-160.
Fig. 13-4: Precision Monolithics Incorporated, 1981 Full Line Catalog, p. 7-17.
Fig. 13-5: Reprinted with the permission of National Semiconductor Corp. Transistor Databook, 1982, p. 11-31.
Fig. 13-6: Precision Monolithics Incorporated, 1981 Full Line Catalog. p. 16-159.
Fig. 13-7: Reprinted with the permission of National Semiconductor Corp. Linear Databook, 1982, p. 3-324.
Fig. 13-8: Reprinted with the permission of National Semiconductor Corp. Linear Databook, 1982, p. 3-324.
Fig. 13-9: Precision Monolithics Incorporated, 1981 Full Line Catalog, p. 6-35.
Fig. 13-10: Precision Monolithics Incorporated, 1981 Full Line Catalog, p. 7-11.

Chapter 14

Fig. 14-1: Radio - Electronics, 1/67.
Fig. 14-2: Modern Electronics, 2/78, p. 17.
Fig. 14-3: Electronics Today International, 5/75, p. 68.
Fig. 14-4: Electronics Today International, 4/78, p. 81.
Fig. 14-5: Modern Electronics, 6/78, p. 14.
Fig. 14-6: Reprinted with permission from General Electric Semiconductor Department. General Electric, 2/68.
Fig. 14-7: Electronics Today International, 6/74, p. 67.
Fig. 14-8: Modern Electronics, 2/78, p. 16.
Fig. 14-9:© Siliconix incorporated. T100/T300 Applications.
Fig. 14-10: Reprinted with permission from General Electric Semiconductor Department. General Electric SCR Manual, Sixth Edition, 1979, p. 224.
Fig. 14-11: Reprinted with the permission of National Semiconductor Corp. Linear Databook, 1982, p. 9-143.
Fig. 14-12: Electronics Today International, 6/82, p. 69.
Fig. 14-13: ©Siliconix incorporated. Siliconix Application Note AN154.
Fig. 14-14: Wireless World, 5/78, p. 69.

Fig. 14-15: Reprinted with permission from General Electric Semiconductor Department. General Electric, 2/68.

Chapter 15

Fig. 15-1. Reprinted with the permission of National Semiconductor Corp. Linear Applications Handbook, 1982, p. AN146-1.
Fig. 15-2: Reprinted with the permission of National Semiconductor Corp. Linear Databook, 1982, p. 9-112.
Fig. 15-3: Supertex Data Book, 1983, p. 5-23.
Fig. 15-4: Supertex Data Book, 1983, p. 5-22.
Fig. 15-5: How To Design/Build Remote Control Devices, TAB Book No. 1277, p. 287.
Fig. 15-6: How To Design/Build Remote Control Devices, TAB Book No. 1277, p. 289.
Fig. 15-7: How To Design/Build Remote Control Devices, TAB Book No. 1277, p. 290.
Fig. 15-8: How To Design/Build Remote Control Devices, TAB Book No. 1277, p. 291.
Fig. 15-9: Signetics Analog Data Manual, 1982, p. 16-28.

Chapter 16

Fig. 16-1: Reprinted from Electronics, 6/78, p. 150. Copyright 1978, McGraw Hill Inc. All rights reserved.
Fig. 16-2: Reprinted from Electronics, 5/73, p. 96. Copyright 1973, McGraw Hill Inc. All rights reserved.
Fig. 16-3: 303 Dynamic Electronic Circuits, TAB Book No. 1060, p. 290.
Fig. 16-4: 73 Magazine, 2/79, p. 79.
Fig. 16-5: Wireless World, 12/74, p. 504.
Fig. 16-6: Courtesy of Motorola Inc. Linear Integrated Circuits, 1979, p. 6-123.
Fig. 16-7: Electronics Today International, 3/78, p. 51.
Fig. 16-8: Reprinted with the permission of National Semiconductor Corp. Linear Databook, 1982, p. 10-215.
Fig. 16-9: Courtesy of Motorola Inc. Linear Integrated Circuits, 1979, p. 6-17.
Fig. 16-10: Courtesy of Motorola Inc. Linear Interface Integrated Circuits, 1979, p. 7-8.
Fig. 16-11: Courtesy of Motorola Inc. Linear Interface Circuits, 1979, p. 7-8.
Fig. 16-12: Courtesy of Motorola Inc. Linear Integrated Circuits, 1979, p. 6-123.

Fig. 16-13: Siliconix Application Note AN73-6, p. 5.

Fig. 16-14: Precision Monolithics Incorporated, 1981 Full Line Catalog, p. 8-31.

Fig. 16-15: Precision Monolithics Incorporated 1981 Fall Line Catalog, p. 8-31.

Fig. 16-16: Teledyne Semiconductor, Databook, p. 9.

Fig. 16-17: ©Siliconix incorporated. Siliconix Analog Switch & IC Product Data Book, 1/82, p. 6-4.

Fig. 16-18: Signetics Analog Data Manual, 1982, p. 8-14.

Fig. 16-19: Precision Monolithics Incorporated 1981 Full Line Catalog, p. 8-12.

Fig. 16-20: Signetics Analog Data Manual, 1982, p. 3-38.

Fig. 16-21: Harris Semiconductor, Linear & Data Acquisition Products, p. 2-46.

Fig. 16-22: Harris Semiconductor Application Note 509.

Chapter 17

Fig. 17-1: Reprinted with the permission of National Semiconductor Corp. Linear Applications Handbook, 1982, p. AN240-5.

Fig. 17-2: Electronics Today International, 10/77, p. 45.

Fig. 17-3: ©Siliconix incorporated. Siliconix Analog Switch & IC Product Data Book, 1/82, p. 7-29.

Fig. 17-4: Reprinted with the permission of National Semiconductor Corp. National Semiconductor CMOS Databook, 1981, p. 3-61.

Fig. 17-5: Precision Monolithics Incorporated 1981 Full Line Catalog, p. 16-142.

Fig. 17-6: ™Siliconix incorporated. Siliconix Analog Switch & IC Product Data Book, 1/82, p. 7-29.

Fig. 17-7: Electronics Today International, 10/77, p. 39

Fig. 17-8: Reprinted with the permission of National Semiconductor Corp. Audio/Radio Handbook, 1980, p. 4-28.

Fig. 17-9: ©Siliconix Incorporated. T100/T300 Applications.

Fig. 17-10: Reprinted with the permission of National Semiconductor Corp. Linear Applications Handbook, 1982, p. AN240-2.

Fig. 17-11: © Siliconix incorporated. Siliconix Analog Switch & IC Product Data Book, 1/82, p. 7-30.

Fig. 17-12: Signetics Analog Data Manual, 1982, p. 3-71.

Fig. 17-13: Signetics Analog Data Manual, 1982, p. 6-20.

Fig. 17-14: Signetics Analog Data Manual, 1983, p. 10-99.

Fig. 17-15: Reprinted with permission of Analog Devices, Inc. Data Acquisition Databook, 1982, p. 6-27.

Fig. 17-16: Reprinted with the permission of National Semiconductor Corp. Linear Databook, 1982, p. 8-258.

Fig. 17-17: Reprinted with the permission of National Semiconductor Corp. Linear Databook, 1982, p. 3-50.

Fig. 17-18: Reprinted with the permission of National Semiconductor Corp. Linear Databook, 1982, p. 8-258.

Fig. 17-19: ©Siliconix incorporated. Siliconix Analog Switch & IC Product Data Book, 1/82, p. 7-31.

Fig. 17-20: Signetics Analog Data Manual, 1982, p. 3-15.

Fig. 17-21: RCA Corporation, Solid State Division, Digital Integrated Circuits Application Note ICAN-6346, p. 4.

Fig. 17-22: ©Siliconix incorporated. MOSPOWER Design Catalog, 1/83, p. 6-42.

Fig. 17-23: Signetics Analog Data Manual, 1982, p. 8-14.

Fig. 17-24: Reprinted with permission of Analog Devices, Inc. Data Acquisition Databook, 1982, p. 4-56.

Chapter 18

Fig. 18-1: Reprinted with the permission of National Semiconductor Corp. Audio/Radio Handbook, 1980, p. 5-4.

Fig. 18-2: Reprinted with the permission of National Semiconductor Corp. Audio/Radio Handbook, 1980, p. 5-5.

Fig. 18-3: Reprinted with the permission of National Semiconductor Corp. Audio/Radio Handbook, 1980, p. 5-4.

Chapter 19

Fig. 19-1: Courtesy of Motorola Inc. Application Note AN-417B, p. 5.

Fig. 19-2: Courtesy of Motorola Inc. Application Note AN417B, p. 3.

Fig. 19-3: The Complete Handbook of Amplifiers, Oscillators & Multivibrators, TAB Book No. 1230, p. 326.

Fig. 19-4: Electronics Today International, 1/76, p. 46.

Fig. 19-5: Ham Radio, 2/79, p. 40.

Fig. 19-6. Electronics Today International, 8/83, p. 57.

Fig. 19-7: Electronics Today International, 11/76, p. 44.

Fig. 19-8: Ham Radio, 2/79, p. 40.

Fig. 19-9: Ham Radio, 2/79, p. 42.

Fig. 19-10: Ham Radio, 2/79, p. 41.

Fig. 19-11: Ham Radio, 2/79, p. 43.

Fig. 19-12: Ham Radio, 2/79, p. 43.

Fig. 19-13: Ham Radio, 2/79, p. 43.

Fig. 19-14: Ham Radio, 2/79, p. 43.

Fig. 19-15: Ham Radio, 2/79, p. 38.

Fig. 19-16: Ham Radio, 2/79, p. 39.

Fig. 19-17: Ham Radio, 3/82, p. 66.

Fig. 19-18: Electronics Today International, 8/73, p. 82.

Fig. 19-19: The Complete Handbook of Amplifiers, Oscillators & Multivibrators, TAB Book No. 1230, p. 322.

Fig. 19-20: Ham Radio, 4/78, p. 51.

Fig. 19-21: Modern Electronics, 6/78, p. 57.

Fig. 19-22: The Complete Handbook of Amplifiers, Oscillators & Multivibrators, TAB Book No. 1230, p. 336.

Fig. 19-23: 73 Magazine, 8/78, p. 80.

Fig. 19-24: Third Book Of Electronic Projects, TAB Book No. 1446, p. 22.

Fig. 19-25: CHRYSTAL OSCILLATOR CIRCUITS, Robert J. Matthys, Copyright © 1983, John Wiley & Sons, Inc. Reprinted by permission of John Wiley & Sons, Inc. r.f. Design, 5-6/83, p. 69.

Fig. 19-26: CHRYSTAL OSCILLATOR CIRCUITS, Robert J. Matthys, Copyright © 1983, John Wiley & Sons, Inc. Reprinted by permission of John Wiley & Sons, Inc. r.f. Design, 5-6/83, p. 64.

Fig. 19-27: Ham Radio, 4/78, p. 50.

Fig. 19-28: CHRYSTAL OSCILLATOR CIRCUITS, Robert J. Matthys, Copyright © 1983, John Wiley & Sons, Inc. Reprinted by permission of

Fig. 19-29: CHRYSTAL OSCILLATOR CIRCUITS, Robert J. Matthys, Copyright © 1983, John Wiley & Sons, Inc. Reprinted by permission of John Wiley & Sons, Inc. r.f. Design, 5-6/83, p. 63.

Fig. 19-30: CHRYSTAL OSCILLATOR CIRCUITS, Robert J. Matthys, Copyright © 1983, John Wiley & Sons, Inc. Reprinted by permission of John Wiley & Sons, Inc. r.f. Design, 5-6/83, p. 63.

Fig. 19-31: CHRYSTAL OSCILLATOR CIRCUITS, Robert J. Matthys, Copyright © 1983, John Wiley & Sons, Inc. Reprinted by permission of John Wiley & Sons, Inc. r.f. Design, 5-6/83, p. 63.

Fig. 19-32: CHRYSTAL OSCIL-

LATOR CIRCUITS, Robert J. Matthys, Copyright © 1983, John Wiley & Sons, Inc. Reprinted by permission of John Wiley & Sons, Inc. r.f. Design, 5-6/83, p. 63.

Fig. 19-33: Third Book Of Electronic Projects, TAB Book No. 1446, p. 21.

Fig. 19-34: Intersil.

Fig. 19-35: The Complete Handbook Of Amplifiers, Oscillators & Multivibrators, Tab Book No. 1230, p. 324.

Fig. 19-36: CHRYSTAL OSCILLATOR CIRCUITS, Robert J. Matthys, Copyright © 1983, John Wiley & Sons, Inc. Reprinted by permission of John Wiley & Sons, Inc. r.f. Design, 5-6/83, p. 64.

Fig. 19-37: The Complete Handbook Of Amplifiers, Oscillators & Multivibrators, TAB Book No. 1230, p. 325.

Fig. 19-38: Ham Radio, 2/79, p. 41.

Fig. 19-40. The Complete Handbook Of Amplifiers, Oscillators & Multivibrators, TAB Book No. 1230, p. 330.

Fig. 19-41: The Complete Handbook Of Amplifiers, Oscillators & Multivibrators, TAB Book No. 1230, p. 331.

Fig. 19-42: Ham Radio, 4/78, p. 50.

Fig. 19-43: Ham Radio, 2/79, p. 40.

Fig. 19-44: 73 Magazine.

Fig. 19-45: Reprinted with the permission of National Semiconductor Corp. Linear Databook, 1982, p. 3-241.

Fig. 19-46: Teledyne Semiconductor Databook, p. 9.

Fig. 19-47: Reprinted with the permission of National Semiconductor Corp. Application Note 32, p. 8.

Fig. 19-48: Reprinted with the permission of National Semiconductor Corp. Transistor Databook, 1982, p. 7-26.

Fig. 19-49: Ham Radio, 2/79, p. 40.

Fig. 19-50: CHRYSTAL OSCILLATOR CIRCUITS, Robert J. Matthys, Copyright © 1983, John Wiley & Sons, Inc. Reprinted by permission of John Wiley & Sons, Inc. r.f. Design, 5-6/83, p. 66.

Chapter 20

Fig. 20-1: Reprinted with the permission of National Semiconductor Corp. Linear Databook, 1982, p. 3-123.

Fig. 20-2: Intersil Data Book, 5/83, p. 5-289.

Fig. 20-3: Reprinted with the permission of National Semiconductor Corp. Application Note AN-71, p. 5.

Fig. 20-4: Reprinted with permission from General Electric Semiconductor Department. GE Semiconductor Data Handbook, Third Edition, p. 305.

Fig. 20-5: Reprinted with the permission of National Semiconductor Corp. Transistor Databook, 1982, p. 11-35.

Chapter 21

Fig. 21-1: Reprinted with the permission of National Semiconductor Corp. Linear Databook, 1982, p. 3-123.

Fig. 21-2: Reprinted with the permission of National Semiconductor Corp. Transistor Databook, 1982, p. 11-30.

Fig. 21-3: Reprinted with the permission of National Semiconductor Corp. Voltage Regulator Handbook, p. 10-112.

Fig. 21-4: Reprinted with the permission of National Semiconductor Corp. Transistor Databook, 1982, p. 11-30.

Chapter 22

Fig. 22-1: Electronics Today International, 9/75, p. 65.

Fig. 22-2: Signetics Analog Data Manual, 1982, p. 6-13.

Fig. 22-3: Electronic Today International, 8/79, p. 99.

Fig. 22-4: © Siliconix incorporated. Siliconix Analog Switch & IC Product Data Book, 1/82, p. 6-15.

Fig. 22-5: © Siliconix incorporated. MOSPOWER Design Catalog, 1/83, p. 6-41.

Fig. 22-6: Signetics Analog Data Manual, 1982, p. 6-21.

Fig. 22-7: Signetics Analog Data Manual, 1982, p. 6-21.

Chapter 23

Fig. 23-1: Ham Radio 11/78, p. 64.

Fig. 23-2: Reprinted with the permission of National Semiconductor Corp. Data Conversion/Acquisition Databook, 1980, p. 2-5.

Fig. 23-3: Signetics Analog Data Manual, 1983, p. 11-15.

Fig. 23-4: Signetics Analog Data Manual, 1983, p. 11-10.

Fig. 23-5: Signetics Analog Data Manual, 1982, p. 16-28.

Fig. 23-6: Signetics Analog Manual, 1982, p. 16-28.

Chapter 24

Fig. 24-1: Signetics 555 Timers, 1973, p. 19.

Fig. 24-2: Courtesy of Motorola Inc: Linear Interface Integrated Circuits, 1979, p. 7-30.

Fig. 24-3: Electronics Today International, 1/76, p. 45.

Fig. 24-4: Precision Monolithics Incorporated 1981 Full Line Catalog, p., 8-33.

Fig. 24-5: Reprinted with permission from General Electric Semiconductor Department. General Electric SCR Manual, Sixth Edition, 1979, p. 219.

Fig. 24-6: Reprinted with permission from General Electric Semiconductor Department. General Electric SCR Manual, Sixth Edition, 1979, p. 218.

Fig. 24-7: Courtesy of Motorola Inc. Application Note AN294.

Fig. 24-8: Signetics 555 Timers, 1973, p. 20.

Chapter 25

Fig. 25-1: Radio-Electronics, 2/83, p. 76.

Fig. 25-2: Courtesy of Motorola Inc. Linear Integrated Circuits, 1979, p. 6-98.

Fig. 25-3: Radio-Electronics, 12/78, p. 77.

Fig. 25-4: Precision Monolithics Incorporated, 1981 Full Line Catalog, p. 14-17.

Fig. 25-5: Precision Monolithics Incorporated, 1981 Full Line Catalog, p. 14-17.

Fig. 25-6: Electronics Today International, 3/78, p. 50.

Fig. 25-7: RCA Corp., Solid State Division, Digital Integrated Circuits Application Note ICAN-6346, p. 5.

Fig. 25-8: Reprinted with the permission of National Semiconductor Corp. Linear Databook, 1982, p. 3-97.

Fig. 25-9: Courtesy of Fairchild Camera & Instrument Corporation. Linear Databook, 1982, p. 5-25.

Fig. 25-10: Reprinted with the permission of National Semiconductor Corp. National Semiconductor, Application Note LB-25.

Fig. 25-11: Electronics Today International, 9/72, p. 86.

Fig. 25-12: 104 Weekend Electronics Projects, TAB Book No. 1436, p. 56.

Fig. 25-13: Courtesy of Fairchild Camera & Instrument Corporation. Linear Databook, 1982, p. 4-180.

Fig. 25-14: © Siliconix incorporated. Siliconix Analog Switch & IC Product Data Book, 1/82, p. 6-9.

Fig. 25-15: Signetics Analog Data Manual, 1983, p. 10-100.

Fig. 25-16: © Siliconix incorporated. Siliconix Application Note AN73-6, p. 4.

Fig. 25-17: Signetics Analog Data Manual, 1983, p. 13-6.
Fig. 25-18: Signetics 555 Timers, 1973, p. 17.
Fig. 25-19: Reprinted with permission of Analog Devices, Inc. Data Acquisition Databook, 1982, p. 4-123.
Fig. 25-20: Courtesy of Texas Instruments Incorporated. Linear Control Circuits Data Book, Second Edition, p. 205.
Fig. 25-21: Siliconix incorporated. Siliconix Analog Switch & IC Product Data Book, 1/82, p. 6-14.
Fig. 25-22: Signetics Analog Data Manual, 1983, p. 11-9.
Fig. 25-23: Signetics Analog Data Manual, 1983, p. 11-9.
Fig. 25-24: Signetics Analog Data Manual, 1983, p. 10-100.
Fig. 25-25: Courtesy of Fairchild Camera & Instrument Corporation. Linear Databook, 1982, p. 5-38.
Fig. 25-26: Precision Monolithics Incorporated, 1981 Full Line Catalog, p. 8-12.
Fig. 25-27: Signetics Analog Data Manual, 1977, p. 264.
Fig. 25-28: Reprinted with the permission of National Semiconductor Corp. Linear Databook, 1982, p. 9-31.
Fig. 25-29: Courtesy of Fairchild Camera & Instrument Corporation. Linear Databook, 1982, p. 5-38.

Chapter 26

Fig. 26-1: © Siliconix incorporated. Siliconix Analog Switch & IC Product Data Book, 1/82, p. 8-5.
Fig, 26-2: © Siliconix incorporated. Siliconix Analog Switch & IC Product Data Book, 1/82, p. 8-4.
Fig. 26-3: Precision Monolithics Incorporated, 1981 Full Line Catalog, p. 6-10.
Fig. 26-4. Precision Monolithics Incorporated, 1981 Full Line Catalog, p. 11-55.
Fig. 26-5: Precision Monolithics Incorporated, 1981 Full Line Catalog, p. 16-10.
Fig. 26-6: Ferranti, Technical Handbook Vol. 10, Data Converters, 1983, p. 1-25.
Fig. 26-7: Courtesy of Motorola Inc. Linear Integrated Circuits, 1979, p. 4-50.
Fig. 26-8: ©Siliconix incorporated. Siliconix Analog Switch & IC Product Data Book, 1/82, p. 8-5.
Fig. 26-9. Courtesy of Fairchild Camera & Instrument Corporation. Linear

Databook, 1982, p. 7-7.
Fig. 26-10. Precision Monolithics Incorporated, 1981 Full Line Catalog, p. 11-55.
Fig. 26-11: Reprinted with permission of Analog Devices, Inc. Data Acquisition Databook, 1982, p. 8-20.
Fig. 26-12:Courtesy of Motorola Inc. Linear Integrated Circuits, 1979, p. 3-17.
Fig. 26-13: Reprinted with permission of Analog Devices, Inc. Data Acquisition Databook, 1982, p. 10-50.
Fig. 26-14: Precision Monolithics Incorporated, 1981 Full Line Catalog, p. 11-54.
Fig. 26-15: Precision Monolithics Incorporated, 1981 Full Line Catalog, p. 16-159.

Chapter 27

Fig. 27-1: Ham Radio, 8/81, p. 27.
Fig. 27-2: Ham Radio, 8/81, p. 28.
Fig. 27-3: Ham Radio, 8/81, p. 27.
Fig. 27-4: Ham Radio, 8/81, p. 26.
Fig. 27-5: Ham Radio, 8/81, p. 26.
Fig. 27-6: Ham Radio, 6/77, p. 42.
Fig. 27-7: Ham Radio, 8/81, p. 27.

Chapter 28

Fig. 28-1: Reprinted from Electronics, 12/74. p. 105. Copyright 1974, Mc-Graw Hill Inc. All rights reserved.
Fig. 28-2: Electronics Today International, 10/82, p. 80.
Fig. 28-3: Reprinted with the permission of National Semiconductor Corp. Linear Databook, 1982, p. 9-188.
Fig. 28-4: Reprinted with the permission of National Semiconductor Corp. Linear Databook, 1982, p. 9-172.
Fig. 28-5: Courtesy of Motorola Inc. Linear Interface Integrated Circuits, 1979, p. 5-102.
Fig. 28-6: Intersil Data Book, 5/83, p. 6-52.
Fig. 28-7: Reprinted with the permission of National Semiconductor Corp., Linear Databook, 1982, p. 9-171.
Fig. 28-8: Electronics Today International, 3/78, p. 50.
Fig. 28-9: Intersil Data Book, 5/83, p. 6-34.

Chapter 29

Fig. 29-1: Ham Radio, 1/78, p. 94
Fig. 29-2: Reprinted with permission from General Electric Semiconductor Department GE Semiconductor Data Handbook, Third Edition, p. 577.
Fig. 29-3: Reprinted with permission from General Electric Semiconductor

Department GE Semiconductor Data Handbook, Third Edition, p. 577.
Fig. 29-4: Reprinted with permission from General Electric Semiconductor Department GE Semiconductor Data Handbook, Third Edition, p. 573.
Fig. 29-5: Reprinted with permission from General Electric Semiconductor Department GE Semiconductor Data Handbook, Third Edition, p. 183.

Chapter 30

Fig. 30-1: Reprinted with the permission of National Semiconductor Corp. National Semiconductor CMOS Databook, 1981, p. 8-44.
Fig. 30-2: Electronics Today International, 4/79, p. 22.
Fig. 30-3: SGS-ATES Databook COS/MOS B-Series, 2/82, p. 548.
Fig. 30-4: ©Siliconix incorporated. MOSPOWER Design Catalog, 1/83, p. 6-60.
Fig. 30-5: Reprinted with permission of Analog Devices, Inc. Data Acquisition Databook, 1982, p. 4-81.
Fig. 30-6: Signetics Analog Data Manual, 1982, p. 8-10.
Fig. 30-7. Reprinted with the permission of National Semiconductor Corp. Transistor Databook, 1982, p. 7-19.
Fig. 30-8. Precision Monolithics Incorporated, 1981 Full Line Catalog, p. 16-159.
Fig. 30-9. Precision Monolithics Incorporated, 1981 Full Line Catalog, p. 16-159.
Fig. 30-10: Precision Monolithics Incorporated, 1981 Full Line Catalog, p. 7-11.
Fig. 30-11: Reprinted with the permission of National Semiconductor Corp. Hybrid Products Databook, 1982, p. 1-21.
Fig. 30-12: Reprinted with the permission of National Semiconductor Corp. Hybrid Products Databook, 1982, p. 17-167.
Fig. 30-13: SGS-ATES Databook COS/MOS B-Series, 2/82, p. 548.
Fig. 30-14: Reprinted with permission of Analog Devices, Inc. Data Acquisition Databook, 1982, p. 4-123.
Fig. 30-15: Reprinted with permission of Analog Devices, Inc. Data Acquisition Databook, 1982, p. 4-123.
Fig. 30-16: Courtesy of Fairchild Camera & Instrument Corporation. Linear Databook, 1982, p. 5-39.
Fig. 30-17: SGS-ATES Databook COS/MOS B-Series, 2/82, p. 548.

Chapter 31

Fig. 31-1: Reprinted with permission from General Electric Semiconductor Department. Optoelectronics, Second Edition, p. 113.

Fig. 31-2: Reprinted with the permission of National Semiconductor Corp. Hybrid Products Databook, 1982, p. 13-11.

Fig. 31-3: Reprinted with the permission of National Semiconductor Corp. Hybrid Products Databook, 1982, p. 17-153.

Fig. 31-4: Reprinted with the permission of National Semiconductor Corp. Hybrid Products Databook, 1982, p. 13-14.

Fig. 31-5: Reprinted with the permission of National Semiconductor Corp. Hybrid Products Databook, 1982, p. 13-20.

Fig. 31-6: Reprinted with the permission of National Semiconductor Corp. Hybrid Products Databook, 1982, p. 13-20.

Chapter 32

Fig. 32-1: No reference.
Fig. 32-2: No reference.
Fig. 32-3: Modern Electronics, 2/78, p. 47.
Fig. 32-4: No reference.
Fig. 32-5: The Giant Book Of Electronics Projects, TAB Book No. 1367, p. 480.
Fig. 32-6: The Giant Book Of Electronics Projects, TAB Book No. 1367, p. 114.
Fig. 32-7: The Giant Book Of Electronics Projects, TAB Book No. 1367, p. 114
Fig. 32-8: 73 Magazine.

Chapter 33

Fig. 33-1: Precision Monolithics Incorporated, 1981 Full Line Catalog, p. 6-58.
Fig. 33-2: Intersil Data Book, 5/83, p. 3-135.
Fig. 33-3: Precision Monolithics Incorporated, 1981 Full Line Catalog, p. 16-114.
Fig. 33-4: Reprinted with the permission of National Semiconductor Corp. Linear Databook, 1982, p. 3-50.
Fig. 33-5: Electronics, 9/76, p. 100.
Fig. 33-6: Reprinted with the permission of National Semiconductor Corp. Data Conversion/Acquision Databook, 1980, p. 3-117.
Fig. 33-7: Reprinted from Electronics, 12/78, p. 124. Copyright 1978, Mc-

Graw Hill Inc. All rights reserved.
Fig. 33-8: Reprinted with the permission of National Semiconductor Corp. Hybrid Products Databook, 1982, 17-132.
Fig. 33-9: Reprinted with the permission of National Semiconductor Corp. Application Note LB-5, p. 1.
Fig. 33-10. Electronics Today International, 11/74, p. 67.
Fig. 33-11. Courtesy of Fairchild Camera & Instrument Corporation. Linear Databook, 1982, p. 4-180.
Fig. 33-12. Courtesy of Fairchild Camera & Instrument Corporation. Linear Databook, 1982, p. 4-179.
Fig. 33-13. Courtesy of Fairchild Camera & Instrument Corporation. Linear Databook, 1982, p. 4-41.
Fig. 33-14. Courtesy of Fairchild Camera & Instrument Corporation. Linear Databook, 1982, p. 4-119.
Fig. 33-15: Reprinted with the permission of National Semiconductor Corp. Linear Databook, 1982, p. 3-177.
Fig. 33-16: Courtesy of Fairchild Camera & Instrument Corporation. Linear Databook, 1982, p. 4-178.
Fig. 33-17: 73 Magazine, 4/79, p. 42.
Fig. 33-18: 303 Dynamic Electronic Circuits, TAB Book No. 1060, p. 289.
Fig. 33-19: Reprinted with the permission of National Semiconductor Corp. Data Conversion/Acquisition Databook, 1980, p. 3-15.
Fig. 33-20: Signetics Analog Data Manual, 1982, p. 3-77.
Fig. 33-21: Harris Semiconductor, Linear & Data Acquisition Products, p. 2-85.
Fig. 33-22: © Siliconix incorporated. Siliconix Analog Switch & IC Product Data Book, 1/82, p. 6-9.
Fig. 33-23: Reprinted with permission of Analog Devices, Inc. Data Acquisition Databook, 1982, p. 4-104.
Fig. 33-24: Reprinted with the permission of National Semiconductor Corp. Data Conversion/Acquisition Databook, 1o80, p. 3-23.
Fig. 33-25: Precision Monolithics Incorporated, 1981 Full Line Catalog, p. 16-116.
Fig. 33-26; Signetics Analog Data Manual, 1982, p. 4-8.
Fig. 33-27: Precision Monolithics Incorporated, 1981 Full Line Catalog, p. 16-115.
Fig. 33-28: Precision Monolithics Incorporated, 1981 Full Line Catalog, p. 16-116.

Fig. 33-29: Harris Semiconductor, Linear & Data Acquisition Products, p. 2-84.
Fig. 33-30: Courtesy of Motorola Inc. Motorola Semiconductor Library Vol. 6, Series B, p. 3-126.
Fig. 33-31: Ham Radio, 2/78, p. 72.
Fig. 33-32: Signetics Analog Data Manual, p. 401.
Fig. 33-33: Signetics Analog Data Manual, p. 75.
Fig. 33-34: Reprinted with the permission of National Semiconductor Corp. Audio/Radio Handbook, 1980, p. 2-58.
Fig. 33-35: Reprinted with permission of Analog Devices, Inc. Data Acquisition Databook, 1982, p. 4-97.
Fig. 33-36: Reprinted with the permission of National Semiconductor Corp. Linear Databook, 1982, p. 3-157.
Fig. 33-37: Precision Monolithics Incorporated, 1981 Full Line Catalog, p. 7-11.
Fig. 33-38: Precision Monolithics Incorporated, 1981 Full Line Catalog, p. 16-158.
Fig. 33-39: 73 Magazine, 1/79, p. 127.
Fig. 33-40: Courtesy of Motorola Inc. Linear Integrated Circuits, 1979, p. 3-131.
Fig. 33-41: Reprinted with the permission of National Semiconductor Corp. Audio/Radio Handbook, 1980, p. 2-59.
Fig. 33-42: Reprinted with the permission of National Semiconductor Corp. Audio/Radio Handbook, 1980, p. 2-56.
Fig. 33-43:Reprinted with the permission of National Semiconductor Corp. Audio/Radio Handbook, 1980, p. 2-58.

Chapter 34

Fig. 34-1: Reprinted with permission from General Electric Semiconductor Department. GE Application Note 201.10.
Fig. 34-2: Electronics Today International, 4/75, p. 42.
Fig. 34-3: © Siliconix incorporated, Application Note AN154.
Fig. 34-4: Reprinted with the permission of National Semiconductor Corp. Linear Databook, 1982, p. 3-289.
Fig. 34-5: Reprinted with permission from General Electric Semiconductor Department. GE Semiconductor Data Handbook, Second Edition, p. 905.
Fig. 34-6: Reprinted with permission from General Electric Semiconductor Department. GE Semiconductor Data Handbook, Third Edition, p. 573.
Fig. 34-7: Radio-Electronics, 5/79, p. 84.

Fig. 34-8: 49 Easy To Build Electronic Projects, TAB Book No. 1337, p. 22.
Fig. 34-9: 49 Easy To Build Electronic Projects, TAB Book No. 1337, p. 98.
Fig. 34-10: Electronics Today International, 12/74, p. 66.
Fig. 34-11: No reference.
Fig. 34-12: Electronics Today International, 5-75, p. 67.
Fig. 34-13: Reprinted with permission from General Electric Semiconductor Department. General Electric SCR Manual, Sixth Edition, 1979, p. 205.
Fig. 34-14: Reprinted with permission from General Electric Semiconductor Department. General Electric SCR Manual, Sixth Edition, 1979, p. 207.
Fig. 34-15: Reprinted vith the permission of National Semiconductor Corp. Linear Databook, 1982, p. 12-14.
Fig. 34-16: © Siliconix incorporated, Application Note AN154.
Fig. 34-17: © Siliconix incorporated, Application Note AN154.
Fig. 34-18: ©Siliconix incorporated, Application Note AN154.
Fig. 34-19: 0 Siliconix incorporated, Application Note AN154.
Fig. 34-20: © Siliconix incorporated, Application Note AN154.
Fig. 34-21: © Siliconix incorporated, Application Note AN154.
Fig. 34-22: © Siliconix incorporated, Application Note AN154.
Fig. 34-23: © Siliconix incorporated, Application Note AN154.
Fig. 34-24:© Siliconix incorporated, Application Note AN154.
Fig. 34-25: © Siliconix incorporated, Application Note AN154.
Fig. 34-26: © Siliconix incorporated, Application Note AN154.

Chapter 35

Fig. 35-1: Intersil Data Book, 5/83, p. 6-49.
Fig. 35-2: The Giant Book Of Electronic Projects, TAB Book No. 1367, p. 109.
Fig. 35-3: 73 Magazine, 6/83, p. 106.
Fig. 35-4: 104 Weekend Electronic Projects, TAB Book No. 1436, p. 166.

Chapter 36

Fig. 36-1: Reprinted with the permission of National Semiconductor Corp. Linear Databook, 1982, p. 10-110.
Fig. 36-2: Reprinted with the permission of National Semiconductor Corp. Linear Databook, 1982, p. 5-9.
Fig. 36-3: Courtesy of Motorola Inc.

Linear Integrated Circuits, 1979, p. 6-99.
Fig. 36-4: Courtesy of Motorola Inc. Linear Integrated Circuits, p. 6-99.
Fig. 36-5: Signetics Analog Data Manual, 1982, p. 16-29.

Chapter 37

Fig. 37-1: Teledyne Semiconductor Publication DG-114-87, p. 7.
Fig. 37-2: ©Siliconix incorporated, Analog Switch & IC Product Data Book, 1/82, p. 7-30.
Fig. 37-3: Reprinted with the permission of National Semiconductor Corp. Linear Databook, 1982, p. 9-140.
Fig. 37-4: Reprinted with the permission of National Semiconductor Corp. Linear Databook, 1982, p. 8-257.
Fig. 37-5: Reprinted with permission of Analog Devices, Inc. Data Acquisition Databook, 1982, p. 12-20.
Fig. 37-6: Reprinted with the permission of National Semiconductor Corp. Linear Databook, 1982, p. 9-143.
Fig. 37-7: Reprinted with the permission of National Semiconductor Corp. Linear Databook, 1982, p. 8-257.

Chapter 38

Fig. 38-1: Electronics Today International, 1/77, p. 83.
Fig. 38-2: 101 Electronic Projects, 1975, #32.
Fig. 38-3: Electronics Today International, 10/76, p. 66.
Fig. 38-4: Electronics Today International, 4/75, p. 67.
Fig. 38-5: Canadian Project Number 1, Spring 78, p. 55.
Fig. 38-6: Electronics Today International, 11/76, p. 44.

Chapter 39

Fig. 39-1: Modern Electronics, 2/78, p. 49.
Fig. 39-2: Electronics Today International, 10/78, p. 103.
Fig. 39-3: Radio-Electronics, 3/78, p. 76.
Fig. 39-4: Popular Mechanics, 5/78, p. 45.
Fig. 39-5: 303 Dynamic Electronic Circuits, TAB Book No. 1060, p. 36.
Fig. 39-6: Electronics Today International, 9/82, p. 70.
Fig. 39-7: Electronics Today International, 4/78, p. 77.
Fig. 39-8: 73 Magazine.
Fig. 39-9: No reference
Fig. 39-10: Electronics Today International, 2/77, p. 73.

Chapter 40

Fig. 40-1: Reprinted with permission of Control Engineering, 1301 S. Grove Ave. Barrington, Illinois 12/73, p. 43.
Fig. 40-2: Courtesy of Motorola Inc. Communications Engineering Bulletin EB-33.
Fig. 40-3: Courtesy of Motorola Inc. Communications Engineering Bulletin EB-33.

Chapter 41

Fig. 41-1: Courtesy of Texas Instruments Incorporated. Optoelectronics Databook, 1983-84, p. 15-12.
Fig. 41-2: 73 Magazine, 7/77, p. 35.
Fig. 41-3: Electronics Today International, 6/76, p. 40.
Fig. 41-4: Reprinted with the permission of National Semiconductor Corp. Linear Databook, 1982, p. 9-172.
Fig. 41-5: Courtesy of Texas Instruments Incorporated. Optoelectronics Databook, 1983-84, p. 15-11.
Fig. 41-6: Signetics Analog Data Manual, 1982, p. 8-14.
Fig. 41-7: ©Siliconix incorporated, Analog Switch & IC Product Data Book, 1/82, p. 6-14.
Fig. 41-8: 73 Magazine.
Fig. 41-9: Reprinted from Electronics, 3/73, p. 119. Copyright 1973, McGraw Hill Inc. All rights reserved.

Chapter 42

Fig. 42-1: Reprinted with the permission of National Semiconductor Corp. Linear Databook, 1982, p. 9-127.
Fig. 42-2: Supertex Data Book, 1983, p. 5-20.
Fig. 42-3: Plessey Semiconductors, Linear IC Handbook, 5/82, p. 86.
Fig. 42-4: Plessey Semiconductors, Linear IC Handbook, 5/82, p. 91.
Fig. 42-5: Reprinted with the permission of National Semiconductor Corp. Hybrid Products Databook, 1982, p. 1-74.
Fig. 42-6: Electronics Today International, 6/82, p. 70.

Chapter 43

Fig. 43-1: Harris Semiconductor, Linear & Data Acquisition Products, 1977, p. 2-85.
Fig. 43-2: Precision Monolithics Incorporated, 1981 Full Line Catalog, p. 6-77.
Fig. 43-3: Courtesy of Fairchild Camera & Instrument Corporation. Linear Databook, 1982, p. 4-178.
Fig. 43-4: Courtesy of Fairchild Cam-

era & Instrument Corporation. Linear Databook, 1982, p. 4-43.

Fig. 43-5: Reprinted with the permission of National Semiconductor Corp. Application Note 32, p. 5.

Fig. 43-6: Reprinted with the permission of National Semiconductor Corp. Application Note LB1, p. 2.

Fig. 43-7: Courtesy of Texas Instruments Incorporated. Linear Control Circuits Data Book, Second Edition, p. 120.

Fig. 43-8: ©Siliconix incorporated. T100/T300 Applications.

Fig. 43-9: Reprinted with the permission of National Semiconductor Corp. Data Conversion/Acquisition Databook, 1980, p. 4-27.

Fig. 43-10: Reprinted with the permission of National Semiconductor Corp. Linear Applications Handbook, 1982, p. AN242-15.

Fig. 43-11: Signetics Analog Data Manual, 1982, p. 3-71.

Fig. 43-12: ©Siliconix incorporated. Application Note, AN73-6, p. 3.

Fig. 43-13: Reprinted with the permission of National Semiconductor Corp. Hybrid Products Databook, 1982, p. 3-7.

Fig. 43-14: Reprinted with the permission of National Semiconductor Corp. Data Conversion/Acquisition Databook, 1980, p. 3-102.

Fig. 43-15: Courtesy of Motorola Inc. Linear Integrated Circuits, 1979, p. 3-82.

Fig. 43-16: Precision Monolithics Incorporated, 1981 Full Line Catalog, p. 6-171.

Fig. 43-17: Courtesy of Texas Instruments Incorporated. Linear Control Circuits Data Book, Second Edition, p. 122.

Fig. 43-18: Precision Monolithics Incorporated, 1981 Full Line Catalog, p. 7-11.

Fig. 43-19: Precision Monolithics Incorporated, 1981 Full Line Catalog, p. 7-6.

Fig. 43-20: Precision Monolithics Incorporated, 1981 Full Line Catalog, p. 16-159.

Fig. 43-21: Reprinted with permission of Analog Devices, Inc. Data Acquisition Databook, 1982, p. 4-56.

Fig. 43-22: Reprinted with permission of Analog Devices, Inc. Data Acquisition Databook, 1982, p. 4-92.

Fig. 43-23: Precision Monolithics Incorporated, 1981 Full Line Catalog, p. 6-50.

Fig. 43-24: Precision Monolithics Incorporated, 1981 Full Line Catalog, p. 16-37.

Fig. 43-25: Signetics Analog Data Manual, 1982, p. 3-15.

Chapter 44

Fig. 44-1: Courtesy of Texas Instruments Incorporated. Optoelectronics Databook, 1983, p. 15-13.

Fig. 44-2: CQ, 3/78, p. 72.

Fig. 44-3: Signetics Analog Data Manual, 1982, p. 3-76.

Fig. 44-4: Courtesy of Texas Instruments Incorporated. Linear Control Circuits Data Book, Second Edition, p. 207.

Fig. 44-5: Reprinted with permission from General Electric Semiconductor Department. General Electric Newsletter, Vol. 11. No. 1, p. 5.

Fig. 44-6: Reprinted with permission from General Electric Semiconductor Department. Optoelectronics, Second Edition, p. 112.

Fig. 44-7: Courtesy of Fairchild Camera & Instrument Corporation. Linear Databook, 1982, p. 4-42.

Fig. 44-8: Electronics Today International, 5/77, p. 77.

Fig. 44-9: Reprinted from Computers & Electronics. Copyright Ziff-Davis Publishing Company. 4/83, p. 109.

Fig. 44-10: The Build-It Book Of Electronic Projects, TAB Book No. 1498, p. 42.

Fig. 44-11: Copyright by Computer Design. All rights reserved. Reprinted by permission. 1/83, p. 77.

Fig. 44-12: Reprinted with permission from General Electric Semiconductor Department. General Electric SCR Manual, Sixth Edition, 1979, p. 440.

Fig. 44-13: Copyright by Computer Design. All rights reserved. Reprinted by permission. 1/83, p. 77.

Fig. 44-14: Reprinted with permission from General Electric Semiconductor Department. GE Semiconductor Data Handbook, Third Edition, p. 1371-4.

Fig. 44-15: Precision Monolithics Incorporated, Linear & Conversion IC Products, 7/78, p. 7-12.

Fig. 44-16: Electronic Projects, 1977, p. 82.

Fig. 44-17: Reprinted with the permission of National Semiconductor Corp. Linear Databook, 1982, p. 3-109.

Fig. 44-18: Reprinted with permission from General Electric Semiconductor Department. Optoelectronics, Second Edition, p. 111

Fig. 44-19: Reprinted with the permission of National Semiconductor Corp. Data Conversion/Acquisition Databook, 1980, p. 3-88.

Chapter 45

Fig. 45-1: RCA Corporation, RCA Solid-State Devices Manual, 1975, p. 734.

Fig. 45-2: Reprinted with permission from General Electric Semiconductor Department. GE Project H5, p. 157.

Fig. 45-3: Solid State Products, New Design Idea, No. 5.

Fig. 45-4: Reprinted from Electronics, 12/74, p. 111. Copyright 1974, McGraw Hill Inc. All rights reserved.

Fig. 45-5: Electronics Today International, 12/72, p. 86.

Fig. 45-6: Reprinted with permission from General Electric Semiconductor Department. GE Semiconductor Data Handbook. Second Edition, p. 585.

Fig. 45-7: 101 Electronic Projects, 1975.

Fig. 45-8: Courtesy of Motorola Inc. Motorola Semiconductor Products. Circuit Applications for the Triac (AN-466), p. 12.

Fig. 45-9: Courtesy of Motorola Inc. Motorla Semiconductor Products Circuit Applications for the Triac (AN-466), p. 5.

Fig. 45-10: Electronics Today International, 7/75, p. 41.

Fig. 45-11: Reprinted with permission from General Electric Semiconductor Department. General Electric SCR Manual Sixth Edition, 1979, p. 264.

Fig. 45-12: Courtesy of Motorola Inc. Motorola Semiconductor Products Circuit Applications for the Triac (AN-466), p. 6.

Fig. 45-13: Reprinted with permission from General Electric Semiconductor Department. General Electric SCR Manual, Sixth Edition, 1979, p. 443.

Fig. 45-14: Reprinted with permission from General Electric Semiconductor Department. General Electric SCR Manual Sixth Edition, 1979, p. 114.

Fig. 45-15: Reprinted with permission from General Electric Semiconductor Department. GE Semiconductor Data Handbook, Third Edition. p. 64.

Fig. 45-16: Reprinted with permission from General Electric Semiconductor Department. GE Semiconductor Data Handbook, Second Edition, p. 727.

Fig. 45-17: Solid State Products, New Design Idea, No. 9.

Fig. 45-18: Reprinted with the permis-

sion of National Semiconductor Corp. Transistor Databook, 1982, p. 7-35.
Fig. 45-19: Reprinted with permission from General Electric Semiconductor Department. GE Semiconductor Data Handbook, Second Edition, p. 727.
Fig. 45-20: Reprinted with the permission of National Semiconductor Corp. Linear Databook, 1982, p. 3-111.
Fig. 45-21: SGS-ATES Databook COS/MOS B-Series, 2/82, p. 548.

Chapter 46

Fig. 46-1: Machine Design, 9/80, p. 126.
Fig. 46-2: Machine Design, 9/80, p. 127.
Fig. 46-3: Reprinted with the permission of National Semiconductor Corp. linear Databook, 1982, p. 9-191.
Fig. 46-4: Reprinted with the permission of National Semiconductor Corp. Data Conversion/Acquisition Databook, 1980, p. 3-91.
Fig. 46-5: Reprinted with the permission of National Semiconductor Corp. Hybrid Products Databook, 1982, p. 1-89.
Fig. 46-6: Reprinted with the permission of National Semiconductor Corp. Data Conversion/Acquisition Databook, 1980, p. 13-50.

Chapter 47

Fig. 47-1: NASA Tech Briefs, Spring 1983, p. 249.
Fig. 47-2: Courtesy of Texas Instruments Incorporated. Optoelectronics Databook, 1983-84, p. 15-9.
Fig. 47-3: Reprinted with the permission of National Semiconductor Corp. Linear Databook, 1982, p. 9-93.
Fig. 47-4: Reprinted with permission from General Electric Semiconductor Department. General Electric SCR Manual, Sixth Edition, 1979, p. 226.
Fig. 47-5: Modern Electronics, 7/78, p. 55.
Fig. 47-6: Electronics Today International, 8/74, p. 66.
Fig. 47-7: Reprinted with permission from General Electric Semiconductor Department. GE Semiconductor Application Note, 200.35, p. 14.
Fig. 47-8: Modern Electronics, 3/78, p. 68.
Fig. 47-9: Modern Electronics, 7/78, p. 55.
Fig. 47-10: Reprinted with the permission of National Semiconductor Corp. Linear Databook, 1982, p. 9-93.

Chapter 48

Fig. 48-1: Reprinted with permission from General Electric Semiconductor Department. General Electric SCR Manual, Sixth Edition, 1979, p. 438.
Fig. 48-2: Electronics Today International, 1/78, p. 83.
Fig. 48-3: Reprinted with the permission of National Semiconductor Corp. Transistor Databook, 1982, p. 11-29.
Fig. 48-4: Courtesy of Motorola Inc. Linear Integrated Circuits, 1979, p. 3-138.
Fig. 48-5: Courtesy of Fairchild Camera & Instrument Corporation. Linear Databook, 1982, p. 5-46.
Fig. 48-6: Courtesy of Fairchild Camera & Instrument Corporation. Linear Databook, 1982, p. 5-48.
Fig. 48-7: Precision Monolithics Incorporated, 1981 Full Line Catalog, p. 8-32.
Fig. 48-8: Courtesy of Motorola Inc. Linear Integrated Circuits, 1979, p. 3-139.
Fig. 48-9: Courtesy of Fairchild Camera & Instrument Corporation. Linear Databook, 1982, p. 5-46.

Chapter 49

Fig. 49-1: Reprinted with the permission of National Semiconductor Corp. Transistor Databook, 1982, p. 11-49.
Fig. 49-2: Reprinted with the permission of National Semiconductor Corp. National Semiconductor CMOS Databook, 1981, p. 8-124.
Fig. 49-3: Intersil Data Book, 1978.
Fig. 49-4: Reprinted with the permission of National Semiconductor Corp. Linear Databook, 1982, p. 3-86.
Fig. 49-5: Radio-Electronics, 10/77, p. 72.
Fig. 49-6: Electronics Today International, 8/78, p. 91.
Fig. 49-7: Third Book Of Electronic Projects, TAB Book No. 1446, p. 40.
Fig. 49-8: Electronics Today International, 8/73, p. 82.
Fig. 49-9: 303 Dynamic Electronic Circuits, TAB Book No. 1060, p. 153.
Fig. 49-10: Electronics Today International, 10/78, p. 97.
Fig. 49-11: Radio-Electronics, 1/80, p. 68.
Fig. 49-12: Signetics Analog Data Manual, 1983, p. 9-40.
Fig. 49-13: Signetics Analog Data Manual, 1983, p. 9-38.
Fig. 49-14: Reprinted with the permission of National Semiconductor Corp.

Linear Databook, 1982, p. 9-187.
Fig. 49-15: Electronics Today International, 1/76, p. 47.
Fig. 49-16: Reprinted with the permission of National Semiconductor Corp. Linear Databook, 1982, p. 9-140.
Fig. 49-17: Courtesy of Fairchild Camera & Instrument Corporation. Linear Databook, 1982, p. 5-25.
Fig. 49-18: Precision Monolithics Incorporated, 1981 Full Line Catalog, p. 10-8.
Fig. 49-19: Electronics Today International, 7/75, p. 40.

Chapter 50

Fig. 50-1: Reprinted from Electronics, 12/77, p. 78. Copyright 1978, McGraw Hill Inc. All rights reserved.
Fig. 50-2: 101 Electronic Projects, 1977, p. 48.

Chapter 51

Fig. 51-1: ETI Canada, 7/78, p. 46.
Fig. 51-2: The Build-It Book Of Electronic Projects, TAB Book No. 1498, p. 131.
Fig. 51-3: Modern Electronics, 3/78, p. 7.

Chapter 52

Fig. 52-1: Reprinted with the permission of National Semiconductor Corp. Application Note AN69, p. 6.
Fig. 52-2: Courtesy of Texas Instruments Incorporated. Complex Sound Generator, Bulletin No. DL-S 12612, p. 13.
Fig. 52-3: ©Siliconix incorporated. MOSPOWER Design Catalog, 1/83, p. 6-60.
Fig. 52-4: Signetics Analog Data Manual, 1983, p. 10-99.
Fig. 52-5: Signetics Analog Data Manual, 1983, p. 10-99.
Fig. 52-6: Precision Monolithics Incorporated, 1981 Full Line Catalog, p. 16-157.
Fig. 52-7: ©Siliconix incorporated. MOSPOWER Design Catalog, 1/83, p. 6-42.
Fig. 52-8: Reprinted with permission from General Electric Semiconductor Department. GE Semiconductor Data Handbook. Second Edition, p. 727.-
Fig. 52-9: Reprinted with the permission of National Semiconductor Corp. Audio/Radio Handbook, 1980, p. 4-37.
Fig. 52-10: Courtesy of Motorola Inc. Linear Integrated Circuits, 1979, p. 3-139.
Fig. 52-11: Electronics Today International, 6/82, p. 64.

Fig. 52-12: Courtesy of Motorola Inc. Linear Integrated Circuits, 1979, p. 3-139.
Fig. 52-13: Precision Monolithics Incorporated, 1981 Full Line Catalog, p. 16-163.
Fig. 52-14: ©Siliconix incorporated. Application Note AN154.
Fig. 52-15: Signetics Analog Data Manual, 1982, p. 3-50.
Fig. 52-16: Signetics Analog Data Manual, 1983, p. 10-20.
Fig. 52-17: Precision Monolithics Incorporated, 1981 Full Line Catalog, p. 6-10.
Fig. 52-18: FERRANTI, Technical Handbook, Vol. 10, Data Converters, 1983, p. 7-26.
Fig. 52-19: Reprinted with the permission of National Semiconductor Corp. Voltage Regulator Handbook, p. 10-60.
Fig. 52-20: Reprinted with permission of Analog Devices, Inc. Data Acquisition Databook, 1982, p. 4-56.
Fig. 52-21: Signetics Analog Data Manual, 1982, p. 4-8.
Fig. 52-22: Courtesy of Fairchild Camera & Instrument Corporation. Linear Databook, 1982, p. 5-38.

Chapter 53

Fig. 53-1: ©Siliconix incorporated. Siliconix Analog Switch & IC Product Data Book, 1/82, p. 4-24.
Fig. 53-2: ©Siliconix incorporated. Siliconix Analog Switch & IC Product Data Book, 1/82, p. 4-23.
Fig. 53-3: Courtesy of Motorola Inc. Linear Integrated Circuits, 1979, p. 6-99.
Fig. 53-4: Teledyne Semiconductor, Data & Design Manual, 1981, p. 11-178.
Fig. 53-5: Courtesy of Motorola Inc. Motorola Semiconductor Library, Vol. 6, Series B, p. 8-58.
Fig. 53-6: Reprinted with the permission of National Semiconductor Corp. Data Conversion/Acquisition Databook, 1980, p. 4-26.
Fig. 53-7: Reprinted with the permission of National Semiconductor Corp. Transistor Databook, 1982, p. 11-34.

Chapter 54

Fig. 54-1: Modern Electronics, 3/78, p. 6.
Fig. 54-2: 101 Electronic Projects, 1977, p. 25.
Fig. 54-3: 101 Electronic Projects, 1975, p. 53.

Chapter 55

Fig. 55-1: Courtesy of Motorola Inc. Application Note AN-829.
Fig. 55-2: Radio-Electronics, 8/78, p. 41.
Fig. 55-3: Courtesy of Texas Instruments Incorporated. Linear Control Circuits Data Book, Second Edition, p. 288.
Fig. 55-4: Courtesy of Motorola Inc. Linear Integrated Circuits, 1979, p. 6-137.
Fig. 55-5: Courtesy of Motorola Inc. Linear Integrated Circuits, 1979, p. 6-122.
Fig. 55-6: 44 Electronics Projects for Hams, SWLs, CBers, & Radio Experimenters, TAB Book No. 1258, p. 133.
Fig. 55-7: Signetics 555 Timers, 1973, p. 23.
Fig. 55-8: Courtesy of Motorola Inc. Linear Integrated Circuits, 1979, p. 3-17.
Fig. 55-9: Electronics Australia, 4/78, p. 51.
Fig. 55-10: Signetics Analog Data Manual, 1983, p. 11-9.
Fig. 55-11: Courtesy of Texas Instruments Incorporated. Linear Control Circuits Data Book, Second Edition, p. 288.
Fig. 55-12: Courtesy of Motorola Inc. Linear Integrated Circuits, 1979, p. 6-98.
Fig. 55-13: Electronics Today International, 8/83, p. 57.
Fig. 55-14: Courtesy of Fairchild Camera & Instrument Corporation. Linear Databook, 1982, p. 4-81.
Fig. 55-15: Courtesy of Motorola Inc. Linear Integrated Circuits, 1979, p. 6-16.
Fig. 55-16: The Giant Book Of Electronics Projects, TAB Book No. 1367.

Chapter 56

Fig. 56-1: Electronics Today International, 4/78, p. 63.
Fig. 56-2: Modern Electronics, 5/78, p. 6.
Fig. 56-3: Electronics Today International, 8/78, p. 61.
Fig. 56-4: Electronics Today International, 12/78, p. 93.

Chapter 57

Fig. 57-1: Reprinted with the permission of National Semiconductor Corp. Voltage Regulator Handbook, p. 10-201.

Fig. 57-2: Reprinted with permission from General Electric Semiconductor Department. Project H13, p. 191.
Fig. 57-3. Courtesy of Motorola Inc. Circuit Applications for the Triac, AN-466, p. 7.
Fig. 57-4: Courtesy of Motorola Inc. AN-443.
Fig. 57-5: Courtesy of Motorola Inc. AN-198.
Fig. 57-6: Reprinted with permission from General Electric Semiconductor Department. GE Semiconductor Data Handbook, Third Edition, p. 573.
Fig. 57-7: Intersil Data Book, 5/83, p. 5-261.
Fig. 57-8: 101 Electronic Projects, 1977, p.98.
Fig. 57-9: Reprinted with permission from General Electric Semiconductor Department, GE Application Note 201.7.
Fig. 57-10: Courtesy of Motorola Inc. Linear Interface Integrated Circuits, p. 5-145.
Fig. 57-11: Reprinted with the permission of National Semiconductor Corp. Hybrid Products Databook, 1982, p. 17-167.
Fig. 57-12: 101 Electronic Projects, 1975, p. 55.
Fig. 57-13: Electronics Today International. 6/75.
Fig. 57-14: RCA Solid State Devices Manual, 1975, p. 501.
Fig. 57-15: Modern Electronics, 6/78, p. 56.
Fig. 57-16: Reprinted with permission from General Electric Semiconductor Department. GE Project H16, p. 203.
Fig. 57-17: Electronics Today International, 4/75, p. 65.
Fig. 57-18: Courtesy of Motorola Inc. AN-443.
Fig. 57-19: Reprinted with the permission of National Semiconductor Corp. Application Note AN125, p. 9.
Fig. 57-20: Courtesy of Fairchild Camera & Instrument Corporation. Linear Databook, 1982, p. 4-114.
Fig. 57-21: Reprinted with permission from General Electric Semiconductor Department. GE Semiconductor Data Handbook, Third Edition, p. 964.
Fig. 57-22: 101 Electronic Projects, 1977, p. 93.
Fig. 57-23: Courtesy of Fairchild Camera & Instrument Corporation. Linear Databook, 1982, p. 4-114.

Chapter 58

Fig. 58-1: Courtesy of Texas Instruments Incorporated. *Linear Control Circuits Data Book, Second Edition*, p. 285.

Fig. 58-2: Courtesy of Texas Instruments Incorporated. Linear Control Circuits Data Book, Second Edition, p. 286.

Fig. 58-3: RCA Corporation, Solid State Division, Digital Integrated Circuits Application Note, ICAN-6346, p. 5.

Fig. 58-4: Precision Monolithics Incorporated, 1981 Full Line Catalog, p. 16-154.

Fig. 58-5: Courtesy of Motorola Inc. Linear Integrated Circuits, p. 6-136.

Fig. 58-6: Courtesy of Motorola Inc. Application Note, AN294.

Fig. 58-7: Courtesy of Fairchild Camera & Instrument Corporation. Linear Databook, 1982, p. 5-47.

Fig. 58-8: Signetics 555 Timers, 1973, p. 22.

Fig. 58-9: Signetics Analog Data Manual, 1983, p. 15-6.

Fig. 58-10: Precision Monolithics Incorporated, 1981 Full Line Catalog, p. 8-32.

Fig. 58-11: Courtesy of Fairchild Camera & Instrument Corporation. Linear Databook, 1982, p. 5-46.

Fig. 58-12: Courtesy of Fairchild Camera & Instrument Corporation. Linear Databook, 1982, p. 5-46.

Fig. 58-13: Reprinted with the permission of National Semiconductor Corp. Linear Databook, 1982, p. 5-7.

Chapter 59

Fig. 59-1: Electronics Today International, 4/76, p. 23.

Fig. 59-2: Popular Electronics, 4/75, p. 87.

Fig. 59-3: Electronics Today International, 4/78, p. 30.

Fig. 59-4: Popular Electronics, 12/76, p. 28.

Fig. 59-5: The Radio Hobbyist's Handbook, TAB Book No. 1346, p. 256.

Chapter 60

Fig. 60-1: Reprinted from Electronics, 7/72, p. 77. Copyright 1972, McGraw Hill Inc. All rights reserved.

Fig. 60-2: Reprinted from Electronics, 10/73, p. 125. Copyright 1973, McGraw Hill Inc. All rights reserved.

Fig. 60-3: 73 Magazine, 12/76, p. 170.

Fig. 60-4: Electronics Today International, 1978.

Fig. 60-6: CQ, 11/83, p. 72.

Fig. 60-7: Electronics Today International, 7/77, p. 77.

Chapter 61

Fig. 61-1: Machine Design, 7/75, p. 39.

Fig. 61-2: Electronics Today International, 4/73, p. 89.

Fig. 61-3: Signetics Analog Data Manual, 1982, p. 16-28.

Fig. 61-4: Teledyne Semiconductor Data & Design Manual, 1981, p. 11-207.

Fig. 61-5: ©Siliconix incorporated, Analog Switch & IC Product Data Book, 1/82, p. 6-4.

Fig. 61-6: Reprinted with the permission of National Semiconductor Corp. Application Note 32, p. 8.

Chapter 62

Fig. 62-1: Electronics Today International, 4/82, p. 39.

Fig. 62-2: Western Digital, Components Handbook, 1983, p. 577.

Fig. 62-3: Modern Electronics, 2/78, p. 72.

Fig. 62-4: Canadian Projects Number 1, Spring 1978, p. 78.

Fig. 62-5: 101 Electronic Projects, 1977, p. 49.

Fig. 62-6: Electronics Today International, 10/74, p. 67.

Fig. 62-8: 44 Electronics Projects For The Darkroom, TAB Book No. 1248, p. 282.

Fig. 62-9: 44 Electronics Projects For The Darkroom, TAB Book No. 1248, p. 284.

Fig. 62-10: Signetics 555 Timers, 1973, p. 23.

Chapter 63

Fig. 63-1: Reprinted with the permission of National Semiconductor Corp. Linear Databook, 1982, p. 9-205.

Fig. 63-2: Reprinted with the permission of National Semiconductor Corp. Linear Databook, 1982, p. 9-191.

Fig. 63-3: Courtesy of Texas Instruments Incorporated. Linear Control Circuits Data Book, Second Edition, p. 374.

Fig. 63-4: Reprinted with the permission of National Semiconductor Corp. Application Note 222.

Fig. 63-5: Courtesy of Motorola Inc. Motorola Semiconductor Library, Vol. 6, Series B, p. 8-58.

Chapter 64

Fig. 64-1: ©Siliconix incorporated, MOSPOWER Design Catalog, 1/83, p. 6-71.

Fig. 64-2: Ferranti Semiconductors,

Technical Handbook, Volume 10, Data Converters, 1983. p. 3-12.

Fig. 64-3: Courtesy of Motorola Inc. Linear Integrated Circuits, 1979, p. 5-144.

Fig. 64-4: Intersil Data Book, 5/83, p. 5-201.

Fig. 64-5: Signetics 555 Timers, 1973, p. 27.

Fig. 64-6: Signetics Analog Data Manual, 1982, p. 6-21.

Fig. 64-7: Signetics Analog Data Manual, 1983, p. 12-36.

Fig. 64-8: Signetics Analog Data Manual, 1983, p. 12-26.

Fig. 64-9: Signetics Analog Data Manual, 1983, p. 12-22.

Fig. 64-10: Electronics Today International, 7/75, p. 39.

Fig. 64-11: Courtesy of Motorola Inc. Circuit Applications for the Triac, AN-466, p. 12.

Fig. 64-13: Electronics Today International, 3/75, p. 67.

Fig. 64-14: Courtesy of Motorola Inc. Linear Integrated Circuits, 1979, p. 4-50.

Fig. 64-15: 73 Magazine, 3/77, p. 152.

Fig. 64-16: Intersil Data Book, 5/83, p. 5-77.

Fig. 64-17: Intersil Data Book, 5/83, p. 5-77.

Fig. 64-18: Intersil Data Book, 5/83, p. 5-77.

Fig. 64-19: Intersil Data Book, 5/83, p. 5-77.

Fig. 64-20: Intersil Data Book, 5/83, p. 5-76.

Fig. 64-21: Courtesy of Motorola Inc. Linear Integrated Circuits, 1979, p. 4-105.

Fig. 64-22: Reprinted with the permission of National Semiconductor Corp. Voltage Regulator Handbook, p. 10-15.

Fig. 64-23: Reprinted with the permission of National Semiconductor Corp. Voltage Regulator Handbook, p. 10-77.

Fig. 64-24: Courtesy of Motorola Inc. Linear Integrated Circuits, 1979, p. 4-105.

Fig. 64-25: Courtesy of Motorola Inc. Linear Integrated Circuits, 1979, p. 4-105.

Fig. 64-26: Electronics Today International, 6/77, p. 77.

Fig. 64-27: Courtesy of Motorola Inc. Linear Integrated Circuits, 1979, p. 4-15.

Fig. 64-28: Courtesy of Motorola Inc. Linear Integrated Circuits, 1979, p. 4-15.

Fig. 64-29: *Signetics Analog Data Manual, 1982, p. 6-14.*

Fig. 64-30: *Courtesy of Motorola Inc. Linear Integrated Circuits, 1979, p. 3-147.*

Fig. 64-31: *Electronics Today International, 3/75, p. 67.*

Fig. 64-32: *Reprinted with the permission of National Semiconductor Corp. Voltage Regulator Handbook, p. 10-179.*

Fig. 64-33: *Signetics Analog Data Manual, 1983, p. 12-28.*

Chapter 65

Fig. 65-1: *Reprinted with the permission of National Semiconductor Corp. Linear Databook, 1982, p. 2-8.*

Fig. 65-2: *Courtesy of Motorola Inc. Linear Integrated Circuits, 1979, p. 4-23.*

Fig. 65-3: *Courtesy of Motorola Inc. Linear Integrated Circuits, 1979, p. 4-152.*

Fig. 65-4: *101 Electronic Projects, 1975, p. 49.*

Fig. 65-5: *Electronics Today International, 9/75, p. 64.*

Fig. 65-6: *Electronics Today International, 3/75, p. 68.*

Fig. 65-7: *Electronics Today International, 1/75, p. 67.*

Fig. 65-8: *Reprinted with the permission of National Semiconductor Corp. Voltge Regulator Handbook, p. 10-15.*

Fig. 65-9: *Electronics Today International, 4/82, p. 29.*

Fig. 65-10: *Reprinted with the permission of National Semiconductor Corp. Voltage Regulator Handbook, p. 10-142.*

Fig. 65-11: *Signetics Analog Data Manual, 1982, p. 6-25.*

Fig. 65-12: *Reprinted with the permission of National Semiconductor Corp. Voltage Regulator Handbook, p. 10-77.*

Fig. 65-13: *Reprinted with the permission of National Semiconductor Corp. Voltage Regulator Handbook, p. 10-15.*

Fig. 65-14: *Reprinted with the permission of National Semiconductor Corp. Linear Databook, 1982, p. 1-68.*

Fig. 65-15: *Reprinted with the permission of National Semiconductor Corp.*

Fig. 65-16: *Signetics Analog Data Manual, 1982, p. 6-25.*

Fig. 65-17: *Signetics Analog Data Manual, 1982, p. 6-25.*

Fig. 65-18: *Electronics Today International, 8/78, p. 91.*

Fig. 65-19: *Courtesy of Motorola Inc. Linear Integrated Circuits, 1979, p. 4-15.*

Fig. 65-20: *Courtesy of Motorola Inc. Linear Integrated Circuits, 1979, p. 5-147.*

Fig. 65-21: *Reprinted with the permission of National Semiconductor Corp. CMOS Databook, 1981, p. 6-38.*

Chapter 66

Fig. 66-1: *No reference.*

Fig. 66-2: *73 Magazine.*

Fig. 66-3: *Electronics Today International, 3/77, p. 71.*

Fig. 66-4: *Courtesy of Motorola Inc. Circuit Applications for the Triac, AN-466, p. 14.*

Fig. 66-5: *Electronics Today International, 1/79, p. 95.*

Fig. 66-6: *Electronics Today International, 8/76, p. 66.*

Fig. 66-7: *Reprinted with the permission of National Semiconductor Corp. Linear Databook, 1982, p. 2-39.*

Chapter 67

Fig. 67-1: *Ham Radio, 8/80, p. 18.*

Fig. 67-2: *Canadian Projects Number 1. p. 86.*

Fig. 67-3: *Electronics Today International, 5/77, p. 37.*

Fig. 67-4: *Electronics Today International, 3/81, p.19.*

Fig. 67-5: *101 Electronic Projects, 1975, p. 47.*

Fig. 67-6: *Electronics Today International, 1/76, p. 52.*

Fig. 67-7: *Electronics Today International, 1/76, p. 51.*

Fig. 67-8: *Electronics Today International, 11/75, p. 74.*

Fig. 67-9: *Ham Radio, 2/73, p. 56.*

Fig. 67-10: *73 Magazine, 10/83, p. 66.*

Fig. 67-11: *Electronics Today International, 6/79, p. 103.*

Fig. 67-12: *Electronics Today International, 1/76, p. 44.*

Fig. 67-13: *Reprinted from Electronics, 7/76, p. 121. Copyright 1976, McGraw Hill Inc. All rights reserved.*

Chapter 68

Fig. 68-1: *©Siliconix incorporated, Analog Switch & IC Product Data Book, 1/82, p. 6-20.*

Fig. 68-2: *Electronics Today International, 6/79, p. 17.*

Fig. 68-3: *Courtesy of Motorola Inc. Motorola Semiconductor Library, Volume 6, Series B, p. 5-52.*

Fig. 68-4: *Reprinted with permission from General Electric Semiconductor Department. General Electric SCR Manual, Sixth Edition, 1979, p. 445.*

Fig. 68-5: *Reprinted with the permission of National Semiconductor Corp. Linear Databook, 1982, p. 3-241.*

Fig. 68-6: *Courtesy of Fairchild Camera & Instrument Corporation. Linear Databook, 1982, p. 5-48.*

Fig. 68-7: *Courtesy of Fairchild Camera & Instrument Corporation. Linear Databook, 1982, p. 5-24.*

Fig. 68-8: *Signetics Analog Data Manual, 1982, p. 16-29.*

Fig. 68-9: *Signetics Analog Data Manual, 1982, p. 16-29.*

Fig. 68-10: *Teledyne Semiconductor, Databook, p. 8.*

Fig. 68-11: *© Siliconix incorporated. Analog Switch & IC Product Data Book, 1/82, p. 6-20.*

Chapter 69

Fig. 69-1: *Reprinted from Electronics, 3/75, p. 117. Copyright 1975, McGraw Hill Inc. All rights reserved.*

Fig. 69-2: *Reprinted from Electronics, 8/78, p. 106. Copyright 1978, McGraw Hill Inc. All rights reserved.*

Fig. 69-3: *Reprinted with the permission of National Semiconductor Corp. Hybrid Products Databook, 1982, p. 2-15.*

Fig. 69-4: *49 Easy To Build Projects, TAB Book No. 1337, p. 77.*

Fig. 69-5: *Electronics Today International, 1/79, p. 97.*

Fig. 69-6: *Reprinted with the permission of National Semiconductor Corp. Hybrid Products Databook, 1982, p. 2-16.*

Chapter 70

Fig. 70-1: *Reprinted with the permission of National Semiconductor Corp. Linear Databook, 1982, p. 7-12.*

Fig. 70-2: *Courtesy of Motorola Inc. Linear Integrated Circuits, p. 6-49.*

Fig. 70-3: *Ferranti, Technical Handbook Vol. 10, Data Converters, 1983, p. 7-13.*

Fig. 70-4: *Reprinted with the permission of National Semiconductor Corp. Hybrid Products Databook, 1982, p. 4-23.*

Chapter 71

Fig. 71-1: *Intersil Data Book, 5/83, p. 7-83.*

Fig. 71-2: *Reprinted with the permission of National Semiconductor Corp. Transistor Databook, 1982, p. 7-67.*

Fig. 71-3: *Reprinted with the permission of National Semiconductor Corp. Audio/Radio Handbook, 1980, p. 4-37.*

Fig. 71-4: Reprinted with the permission of National Semiconductor Corp. Audio/Radio Handbook, 1980, p. 3-16.
Fig. 71-5: Reprinted with the permission of National Semiconductor Corp. Hybrid Products Databook, 1982, p. 13-17.
Fig. 71-6: Courtesy of Motorola Inc. Linear Integrated Circuits, 1979, p. 5-77.
Fig. 71-7: 73 Magazine.
Fig. 71-8: ©Siliconix incorporated, Analog Switch & IC Product Data Book, 1/82, p. 6-18.
Fig. 71-9: Courtesy of Motorola Inc. Linear Integrated Circuits, 1979, p. 6-123.
Fig. 71-10: Ham Radio, 7/76, p. 69.

Chapter 72

Fig. 72-1: 73 Magazine.
Fig. 72-2: CQ, 6/78, p. 32.
Fig. 72-3: Teledyne Semiconductor, Databook, p. 11.
Fig. 72-4: Reprinted from Electronics 4/76, p. 104. Copyright   , McGraw Hill Inc. All rights reserved.
Fig. 72-5: Reprinted by permission from the Aug. 1981 issue of Insulation/ Circuits magazine. Copyright 1981, Lake Publishing Corporation, Libertyville, Illinois, 60048-9989, USA.
Fig. 72-6: ©Siliconix incorporated, Application Note AN154.
Fig. 72-7: Electronics Today International, 11/78, p. 68.
Fig. 72-8: CQ, 6/78, p.33.

Chapter 73

Fig. 73-1: Courtesy of Motorola Inc. Communications Engineering Bulletin EB-67.
Fig. 73-2: Courtesy of Motorola Inc. Communications Engineering Bulletin EB-63.
Fig. 73-3: Courtesy of Motorola Inc. Application Note AN593, p. 3.
Fig. 73-4: Courtesy of Motorola Inc. Application Note AN-593, p. 6.
Fig. 73-5: Courtesy of Motorola Inc. Communications Engineering Bulletin EB-46.
Fig. 73-6: Microwaves & RF, 1/83, p. 89.
Fig. 73-7: ©Siliconix incorporated, Small Signal FET Design Catalog, 7/83, p. 5-52.
Fig. 73-8: Harris Semiconductor, Linear & Data Acquisition Products, 1977, p. 7-54.
Fig. 73-9: Wireless World, 11/79, p. 76.

Fig. 73-10: 101 Electronic Projects, 1975, p. 3.
Fig. 73-11: Ham Radio, 10/78, p. 38.
Fig. 73-12: 73 Magazine, 4/83, p. 106.
Fig. 73-13: Ham Radio, 1/74, p. 67.
Fig. 73-14: Courtesy of Motorola Inc. Motorola Semiconductor Library, Vol. 6, Series B, p. 8-59.
Fig. 73-15: © Siliconix incorporated. MOSPOWER Design Catalog, 1/83, p. 5-36.
Fig. 73-16: Reprinted with the permission of National Semiconductor Corp. Transistor Databook, 1982, p. 11-33.
Fig. 73-17: © Siliconix incorporated. MOSPOWER Design Catalog, 1/83, p. 5-10.
Fig. 73-18: Reprinted with the permission of National Semiconductor Corp. Application Note 32, p. 9.
Fig. 73-19: Teledyne Semiconductor, Data & Design Manual, 1981, p. 11-178.
Fig. 73-20: Signetics Analog Data Manual, 1983, p. 17-13.
Fig. 73-21: Signetics Analog Data Manual, 1983, p. 17-15.
Fig. 73-22: 73 Magazine.
Fig. 73-23: Courtesy of Motorola Inc. Motorola Semiconductor Library, Vol. 6, Series B, p. 8-58.
Fig. 73-24: Courtesy of Motorola Inc. Motorola Semiconductor Library, Vol. 6, Series B, p. 8-58.
Fig. 73-25: © Siliconix incorporated. MOSPOWER Design Catalog, 1/83, p. 5-10.
Fig. 73-26: Teledyne Semiconductor, Data & Design Manual, 1981, p. 11-178.
Fig. 73-27: Teledyne Semiconductor, Data & Design Manual, 1981, p. 11-178.
Fig. 73-28: Teledyne Semiconductor, Data & Design Manual, 1981, p. 11-178.

Chapter 74

Fig. 74-1: Reprinted with the permission of National Semiconductor Corp. Transistor Databook, 1982, p. 8-63.
Fig. 74-2: Reprinted with the permission of National Semiconductor Corp. Transistor Databook, 1982, p. 11-32.
Fig. 74-3: © Siliconix incorporated. MOSPOWER Design Catalog, 1/83, p. 5-6.
Fig. 74-4: The Giant Book Of Electronics Projects, TAB Book No. 1367.
Fig. 74-5: Reprinted with the permission of National Semiconductor Corp.

Linear Databook, 1982, p. 12-14.
Fig. 74-6: Radio-Electronics, 7/83, p. 7.
Fig. 74-7: Radio-Electronics, 7/83, p. 7.
Fig. 74-8: 73 Magazine, 7/77, p. 35.

Chapter 75

Fig. 75-1: Reprinted with the permission of National Semiconductor Corp. Linear Databook, 1982, p. 9-126.
Fig. 75-2: Courtesy of Motorola Inc. Communications Engineering Bulletin, EB-46.
Fig. 75-3: Signetics Analog Data Manual, p. 556.
Fig. 75-4: Modern Electronics, 7/78, p. 55.
Fig. 75-5: Electronics Today International, 6/79, p. 43.
Fig. 75-6: Radio-Electronics, 8/69, p. 74.
Fig. 75-7: Signetics 555 Timers, 1973, p. 25.

Chapter 76

Fig. 76-1: The Build-It Book Of Electronic Projects, TAB Book No. 1498, p. 20.
Fig. 76-2: 303 Dynamic Electronic Circuits, TAB Book No. 1060, p. 153.
Fig. 76-3: Reprinted with the permission of National Semiconductor Corp. Linear Databook, 1982, p. 9-100.
Fig. 76-4: Reprinted with permission from General Electric Semiconductor Department. General Electric SCR Manual, Sixth Edition, 1979, p. 225.
Fig. 76-5: '73 Magazine, 9/75, p. 105.
Fig. 76-6: Howard S. Leopold.
Fig. 76-7: Modern Electronics, 3/78, p. 50.
Fig. 76-8: 73 Magazine, 6/83, p. 106.
Fig. 76-9: Modern Electronics, 2/78, p. 50.

Chapter 77

Fig. 77-1: Electronics Today International.
Fig. 77-2: Reprinted with the permission of National Semiconductor Corp. Transistor Databook, 1982, p. 11-30.
Fig. 77-3: Reprinted with the permission of National Semiconductor Corp. Transistor Databook, 1982, p. 11-31.
Fig. 77-4: Precision Monolithics Incorporated, 1981 Full Line Catalog, p. 7-18.
Fig. 77-5: Reprinted with the permission of National Semiconductor Corp. Linear Databook, 1982, p. 3-325.
Fig. 77-6: Reprinted with the permis-

sion of National Semiconductor Corp. Hybrid Products Databook, 1982, p. 17-152.
Fig. 77-7: Reprinted with the permission of National Semiconductor Corp. Transistor Databook, 1982, p. 11-25.
Fig. 77-8: Courtesy of Fairchild Camera & Instrument Corporation. Linear Databook, 1982, p. 7-25.
Fig. 77-9: Courtesy of Fairchild Camera & Instrument Corporation. Linear Databook, 1982, p. 7-25.
Fig. 77-10: Signetics Analog Data Manual, 1982, p. 3-50.
Fig. 77-11: Signetics Analog Data Manual, 1982, p. 3-15.
Fig. 77-12: Precision Monolithics Incorporated, 1981 Full Line Catalog, p. 16-159.

Chapter 78

Fig. 78-1: Electronics Today International, 9/72, p. 86.
Fig. 78-2: Electronics Today International, 1978.
Fig. 78-3: Reprinted with the permission of National Semiconductor Corp. Linear Applications Handbook, 1982, p. 9-76.
Fig. 78-3: Harris Semiconductor, Linear & Data Acquisition Products, 1977, p. 2-96.
Fig. 78-4: Courtesy of Motorola Inc. Linear Integrated Circuits, 1979, p. 6-17.

Chapter 79

Fig. 79-1: Supertex Data Book, 1983, p. 5-26.
Fig. 79-2: Reprinted with the permission of National Semiconductor Corp. Linear Applications Handbook, 1982, p. 9-75.
Fig. 79-3: Reprinted with the permission of National Semiconductor Corp. Linear Applications Handbook, 1982, p. 9-76.

Chapter 80

Fig. 80-1: Reprinted with the permission of National Semiconductor Corp. Audio/Radio Handbook, 1980, p. 4-40.
Fig. 80-2: Reprinted with the permission of National Semiconductor Corp. COPS Microcontrollers Databook, 1982, p. 9-123.
Fig. 80-3: Reprinted with the permission of National Semiconductor Corp. COPS Microcontrollers Databook, 1982, p. 10-3.
Fig. 80-4: Electronics Today International, 4/78, p. 31.
Fig. 80-5: Reprinted with the permis-

sion of National Semiconductor Corp. Audio/Radio Handbook, 1980, p. 5-8.
Fig. 80-6: Electronics Today International, 1/79, p. 68.
Fig. 80-7: Courtesy of Motorola Inc. Linear Integrated Circuits, 1979, p. 6-136.
Fig. 80-8: Electronics Today International, 4/78, p. 29.
Fig. 80-9: Electronics Today International, 1/76, p. 49.
Fig. 80-10: Courtesy of Texas Instruments Incorporated. Bulletin No. DL-S 12612, p. 14.
Fig. 80-11: Reprinted with the permission of National Semiconductor Corp. Audio/Radio Handbook, 1980, p. 5-9.
Fig. 80-12: Courtesy of Texas Instruments Incorporated. Bulletin No. DL-S 12612, p. 12.

Chapter 81

Fig. 81-1: Reprinted with the permission of National Semiconductor Corp. Linear Databook, 1982, p. 3-204.
Fig. 81-2: 73 Magazine, 10/77, p. 115.
Fig. 81-3: Electronics Today International, 7/81, p. 75.
Fig. 81-4: Reprinted with permission from General Electric Semiconductor Department. GE Application Note 200.35, 3/66, p. 14.
Fig. 81-5: 104 Weekend Electronics Projects, TAB Book No. 1436, p. 64.
Fig. 81-6: Electronics Today International, 1975, p. 72.

Chapter 82

Fig. 82-1: Teledyne Semiconductor, Databook, p. 8.
Fig. 82-2: ©Siliconix incorporated. Application Note AN154.
Fig. 82-3: The Complete Handbook of Amplifiers, Oscillators & Multivibrators, TAB Book No. 1230, p. 335.
Fig. 82-4: Courtesy of Fairchild Camera & Instrument Corporation. Linear Databook, 1982, p. 9-28.
Fig. 82-5: Reprinted from Electronics, 2/77, p. 107. Copyright 19 , McGraw Hill Inc. All rights reserved.
Fig. 82-6: © Siliconix incorporated. Analog Switch & IC Product Data Book, 1/82, p. 6-19.
Fig. 82-7: Harris Semiconductor, Linear & Data Acquisition Products, 1977, p. 2-96.
Fig. 82-8: Electronics Today International, 7/78, p. 16.
Fig. 82-9: Courtesy of Motorola Inc. Linear Interface Integrated Circuits, p. 7-30.

Fig. 82-10: Reprinted with the permission of National Semiconductor Corp. Data Conversion/Acquisition Databook, 1980, p. 13-50.
Fig. 82-11: Courtesy of Motorola Inc. Linear Interface Integrated Circuits, 1979, p. 7-9.
Fig. 82-12: Courtesy of Texas Instruments Incorporated. Linear Control Circuits Data Book, Second Edition, p. 145.
Fig. 82-13: Electronics Today International, 7/78, p. 16.
Fig. 82-14: Precision Monolithics Incorporated, 1981 Full Line Catalog, p. 8-31.

Chapter 83

Fig. 83-1: Electronics Today International, 7/81, p. 72.
Fig. 83-2: 104 Weekend Electronics Projects, TAB Book No. 1436, p. 233.
Fig. 83-3: 101 Electronic Projects, 1977, p. 40.

Chapter 84

Fig. 84-1: Reprinted with the permission of National Semiconductor Corp. Transistor Databook, 1982, p. 11-32.
Fig. 84-2: Reprinted with the permission of National Semiconductor Corp. Transistor Databook, 1982, p. 11-33.
Fig. 84-3: Reprinted with the permission of National Semiconductor Corp. Transistor Databook, 1982, p. 11-28.
Fig. 84-4: Reprinted with the permission of National Semiconductor Corp. Transistor Databook, 1982, p. 11-29.
Fig. 84-5: Reprinted with permission from General Electric Semiconductor Department. General Electric SCR Manual, Sixth Edition, 1979, p. 313.
Fig. 84-6: Reprinted with the permission of National Semiconductor Corp. Data Conversion/Acquisition Databook, p. 11-10.
Fig. 84-7: Reprinted with permission from General Electric Semiconductor Department. Optoelectronics, Second Edition, p. 141.

Chapter 85

Fig. 85-1: Intersil Data Book, 5/83, p. 7-48.
Fig. 85-2: Reprinted from Electronics, 11/75, p. 120. Copyright 1975, McGraw Hill Inc. All rights reserved.
Fig. 85-3: Courtesy of Motorola Inc.
Fig. 85-4: Mitel Databook, p. 2-17.
Fig. 85-5: Mitel Databook, p. 2-13.
Fig. 85-6: 73 Magazine, 12/83, p. 115.
Fig. 85-7: Ham Radio, 2/77, p. 70.
Fig. 85-8: Ham Radio, 8/77, p. 41.

Fig. 85-9: Ham Radio, 1/84, p. 94.

Fig. 85-10: Reprinted with permission from General Electric Semiconductor Department. Optoelectronics, Second Edition, p. 119.

Fig. 85-11: Signetics Analog Data Manual, 1982, p. 16-27.

Fig. 85-12: Modern Electronics, 7/78, p. 56.

Fig. 85-13: The Build-It Book Of Electronic Projects, TAB Book No. 1498, p. 3.

Fig. 85-14: Reprinted with the permission of National Semiconductor Corp. COPS Microcontrollers Databook, 1982, p. 9-118.

Fig. 85-15: 73 Magazine, 1/84, p. 115.

Fig. 85-16: Intersil Data Book, 5/83, p. 7-47.

Fig. 85-17: Reprinted with permission from General Electric Semiconductor Department Optoelectronics, Second Edition, p. 119.

Fig. 85-18: Ham Radio, 1/84, p. 93.

Fig. 85-19: Ham Radio, 1/84, p. 91.

Fig. 85-20: 73 Magazine, 4/83.

Fig. 85-21: 73 Magazine, 9/82, p. 92.

Chapter 86

Fig. 86-1: Radio-Electronics, 7/81, p. 73.

Fig. 86-2: Reprinted with the permission of National Semiconductor Corp. Hybrid Products Databook, 1982, p. 1-87.

Fig. 86-3: Reprinted with the permission of National Semiconductor Corp. Data Conversion/Acquisition Databook, 1980, p. 12-17.

Fig. 86-4: Reprinted with the permission of National Semiconductor Corp. Linear Databook, 1982, p. 9-162.

Fig. 86-5: Courtesy of Motorola Inc. Circuit Applications for the Triac (AN-466), p. 9.

Fig. 86-6: Courtesy of Motorola Inc. Circuit Applications for the Triac, AN-466, p. 13.

Fig. 86-7: Intersil Data Book, 5/83, p. 5-68.

Fig. 86-8: Reprinted with the permission of National Semiconductor Corp. Linear Applications Handbook, 1982, p. LB36-2.

Fig. 86-9: Reprinted with the permission of National Semiconductor Corp. Linear Databook, 1982, p. 9-29.

Fig. 86-10: Precision Monolithics Incorporated, 1981 Full Line Catalog, p. 16-6.

Fig. 86-11: Reprinted with the permission of National Semiconductor Corp.

Linear Databook, 1982, p. 9-29.

Fig. 86-12: Reprinted with the permission of National Semiconductor Corp. Hybrid Products Databook, 1982, p. 7-33.

Chapter 87

Fig. 87-1: Electronics Today International, 4/81, p. 86.

Fig. 87-2: Electronics Today International, 12/78, p. 32.

Fig. 87-3: Signetics Analog Data Manual, 1983, p. 10-65.

Fig. 87-4: Precision Monolithics Incorporated, 1981 Full Line Catalog, p. 6-147.

Fig. 87-5: Teledyne Semiconductor, Databook, p. 12.

Fig. 87-6: Precision Monolithics Incorporated, 1981 Full Line Catalog, p. 10-16.

Fig. 87-7: Reprinted with the permission of National Semiconductor Corp. Data Conversion/Acquisition Databook, 1980, p. 12-9.

Fig. 87-8: Signetics Analog Data Manual, 1982, p. 3-78.

Fig. 87-9: Reprinted with the permission of National Semiconductor Corp. Data Conversion/Acquisition Databook, 1980, p. 12-7.

Fig. 87-10: Radio-Electronics, 3/80, p. 60.

Fig. 87-11: Reprinted with the permission of National Semiconductor Corp. Data Conversion/Acquisition Databook, 1980, p. 12-10.

Fig. 87-12: Reprinted with the permission of National Semiconductor Corp. Linear Databook, 1982, p. 9-162.

Fig. 87-13: Intersil Data Book, 5/83, p. 5-71.

Fig. 87-14: Intersil Data Book, 5/83, p. 5-71.

Fig. 87-15: Courtesy of Fairchild Camera & Instrument Corporation. Linear Databook, 1982, p. 4-42.

Fig. 87-16: Reprinted with the permission of National Semiconductor Corp. Linear Databook, 1982, p. 3-108.

Fig. 87-17: Reprinted with the permission of National Semiconductor Corp. CMOS Databook, 1981, p. 6-7.

Fig. 87-18: Reprinted with the permission of National Semiconductor Corp. Linear Databook, 1982, p. 9-31.

Fig. 87-19: Reprinted with the permission of National Semiconductor Corp. Linear Databook, 1982, p. 9-31.

Fig. 87-20: Reprinted with the permission of National Semiconductor Corp. Linear Databook, 1982, p. 9-29.

Fig. 87-21: Reprinted with the permission of National Semiconductor Corp. Linear Databook, 1982, p. 9-160.

Fig. 87-22: Reprinted with the permission of National Semiconductor Corp. Linear Databook, 1982, p. 9-162.

Fig. 87-23: Reprinted with the permission of National Semiconductor Corp. Voltage Regulator Handbook, p. 10-107.

Fig. 87-24: Reprinted with the permission of National Semiconductor Corp. Linear Databook, 1982, p. 2-46.

Fig. 87-25: Electronics Today International, 10/78, p. 101.

Fig. 87-26: Reprinted with the permission of National Semiconductor Corp. Linear Databook, 1982, p. 2-46.

Fig. 87-27: Reprinted with the permission of National Semiconductor Corp. Linear Databook, 1982, p. 9-29.

Fig. 87-28: Reprinted with the permission of National Semiconductor Corp. Linear Databook, 1982, p. 9-160.

Fig. 87-29: Reprinted with the permission of National Semiconductor Corp. Linear Databook, 1982, p. 9-31.

Fig. 87-30: Teledyne Semiconductor, Databook, p. 11.

Fig. 87-31: Teledyne Semiconductor, Databook, p. 11.

Fig. 87-32: Intersil Data Book, 5/83, p. 5-70.

Fig. 87-33: Reprinted with the permission of National Semiconductor Corp. Linear Databook, 1982, p. 9-29.

Chapter 88

Fig. 88-1: Western Digital, Components Handbook, 1983, p. 579.

Fig. 88-2: Courtesy of Texas Instruments Incorporated. Linear Control Circuits Data Book, Second Edition, p. 289.

Fig. 88-3: Signetics Analog Data Manual, 1983, p. 15-11.

Fig. 88-4: Courtesy of Motorola Inc. Application Note AN-294, p. 6.

Fig. 88-5: Reprinted with permission from General Electric Semiconductor Department. Application Note 201.11.

Fig. 88-6: Reprinted with permission from General Electric Semiconductor Department. GE Semiconductor Data Handbook, Third Edition, p. 1183.

Fig. 88-7: Signetics 555 Timers, 1973, p. 19.

Fig. 88-8: RCA Corporation, Linear Integrated Circuits And MOS/FETS, p. 437.

Fig. 88-9: Reprinted with permission from General Electric Semiconductor

Department. GE Semiconductor Data Handbook, Second Edition, p. 412.
Fig. 88-10: 73 Magazine, 8/75, p. 140.
Fig. 88-11: Western Digital, Components Handbook, 1983, p. 581.
Fig. 88-12: Reprinted with permission from General Electric Semiconductor Department. GE Semiconductor Data Handbook, Second Edition, p. 727.
Fig. 88-13: Electronics Today International, 3/82, p. 67.
Fig. 88-14: Courtesy of Motorola Inc. Linear Integrated Circuits, 1979, p. 3-17.
Fig. 88-15: Electronics Today International, 1/76, p. 52.
Fig. 88-16: Modern Electronics, 2/78, p. 49.
Fig. 88-17: Signetics 555 Timers, 1973, p. 26.
Fig. 88-18: Signetics 555 Timers, 1973, p. 20.

Chapter 89

Fig. 89-1: Reprinted with the permission of National Semiconductor Corp. Linear Databook, 1982, p. 10-170.
Fig. 89-2: Signetics Analog Data Manual, 1982, p. 3-89.
Fig. 89-3: Electronics Today International, 10/77, p. 34.
Fig. 89-4: Courtesy of Texas Instruments Incorporated. Linear Control Circuits Data Book, Second Edition, p. 130.
Fig. 89-5: Reprinted with the permission of National Semiconductor Corp. Linear Databook, 1982, p. 10-63.
Fig. 89-6: Reprinted with the permission of National Semiconductor Corp. Audio/Radio Handbook, 1980, p. 2-53.
Fig. 89-7: Reprinted with the permission of National Semiconductor Corp. Audio/Radio Handbook, 1980, p. 2-49.
Fig. 89-8: Electronics Today International, 6/79, p. 105.
Fig. 89-9: Electronics Today International, 6/82, p. 66.
Fig. 89-10: Courtesy of Texas Instruments Incorporated. Linear Control Circuits Data Book, Second Edition, p. 130.
Fig. 89-11: Reprinted with the permission of National Semiconductor Corp. Transistor Databook, 1982, p. 7-27.
Fig. 89-12: Reprinted with the permission of National Semiconductor Corp. Linear Databook, 1982, p. 3-48.
Fig. 89-13: Electronics Today International.

Chapter 90

Fig. 90-1: Radio-Electronics, 12/81, p. 52.
Fig. 90-2: Reprinted with the permission of National Semiconductor Corp. Linear Databook, 1982, p. 9-108.
Fig. 90-3: 73 Magazine, 6/77, p. 49.
Fig. 90-4: CQ, 6/83, p. 46.
Fig. 90-5: 73 Magazine, 8/83, p. 100.

Chapter 91

Fig. 91-1: Electronics Today International, 6/78, p. 29.
Fig. 91-2: 73 Magazine, 2/83, p. 90.
Fig. 91-3: Radio-Electronics, 3/80, p, 60.
Fig. 91-4: Radio-Electronics, 8/83, p. 96.
Fig. 91-5: Reprinted with the permission of National Semiconductor Corp. Transistor Databook, 1982, p. 7-11.

Chapter 92

Fig. 92-1: Courtesy of Motorola Inc. Application Note AN-545A, p. 7.
Fig. 92-2: Courtesy of Motorola Inc. Application Note AN-545A, p. 12.
Fig. 92-3: Plessey Semiconductors, Linear IC Handbook, 5/82, p. 129.
Fig. 92-4: Courtesy of Motorola Inc. Linear Integrated Circuits, 1979, p. 5-50.
Fig. 92-5: Courtesy of Motorola Inc. Linear Integrated Circuits, 1979, p. 5-73.
Fig. 92-6: Courtesy of Motorola Inc. Linear Integrated Circuits, 1979, p. 5-51.
Fig. 92-7: Reprinted with the permission of National Semiconductor Corp. Transistor Databook, 1982, p. 7-26.
Fig. 92-8: Reprinted with the permission of National Semiconductor Corp. Transistor Databook, 1982, p. 11-31.
Fig. 92-9: Reprinted with the permission of National Semiconductor Corp. Transistor Databook, 1982, p. 11-30.
Fig. 92-10: Courtesy of Motorola Inc. Motorola Semiconductor Library, Volume 6, Series B.
Fig. 92-11: Harris Semiconductor, Linear & Data Acquisition Products, 1977, p. 2-46.

Chapter 93

Fig. 93-1: Precision Monolithics Incorporated, 1981 Full Line Catalog, p. 6-59.
Fig. 93-2: Reprinted with the permission of National Semiconductor Corp. Voltage Regulator Handbook, p. 10-47.
Fig. 93-3: Courtesy of Motorola Inc. Linear Integrated Circuits, 1979, p. 6-23.
Fig. 93-4: Precision Monolithics Incor-
porated, 1981 Full Line Catalog, p. 7-11.
Fig. 93-5: Precision Monolithics Incorporated, 1981 Full Line Catalog, p. 16-158.
Fig. 93-6: Signetics Analog Data Manual, 1982, p. 3-38.
Fig. 93-7: Reprinted with the permission of National Semiconductor Corp. Data Conversion/Acquisition Databook, 1980, p. 13-50.
Fig. 93-8: Courtesy of Motorola Inc., Linear Integrated Circuits, 1979, p. 3-42.
Fig. 93-9: Reprinted with the permission of National Semiconductor Corp. Transistor Databook, 1982, p. 11-25.
Fig. 93-10: Precision Monolithics Incorporated, 1981 Full Line Catalog, p. 6-142.
Fig. 93-11: Precision Monolithics Incorporated, 1981 Full Line Catalog, p. 10-18.
Fig. 93-12: Precision Monolithics Incorporated, 1981 Full Line Catalog, p. 10-15.
Fig. 93-13: Precision Monolithics Incorporated, 1981 Full Line Catalog, p. 16-16.
Fig. 93-14: Precision Monolithics Incorporated, 1981 Full Line Catalog, p. 10-8.
Fig. 93-15: Reprinted with the permission of National Semiconductor Corp. Data Conversion/Acquisition Databook, 1980, p. 14-52.
Fig. 93-16: Precision Monolithics Incorporated, 1981 Full Line Catalog, p. 16-158.
Fig. 93-17: Reprinted with the permission of National Semiconductor Corp. Data Conversion/Acquisition Databook, 1980, p. 14-44.
Fig. 93-18: Electronics Today International, 8/78, p. 91.
Fig. 93-19: Reprinted with the permission of National Semiconductor Corp. Data Conversion/Acquisition Databook, 1980, p. 14-41.
Fig. 93-20: Precision Monolithics Incorporated, 1981 Full Line Catalog, p. 6-78.
Fig. 93-21: Reprinted with the permission of National Semiconductor Corp. Data Conversion/Acquisition Databook, 1980, p. 14-53.
Fig. 93-22: Reprinted with the permission of National Semiconductor Corp. Data Conversion/Acquisition Databook, 1980, p. 14-53.
Fig. 93-23: Reprinted with the permission of National Semiconductor Corp. Data Conversion/Acquisition Data-

book, 1980, p. 14-51.

Chapter 94

Fig. 94-1: Intersil Data Book, 5/83, p. 5-238.
Fig. 94-2: Reprinted with the permission of National Semiconductor Corp. Data Databook, 1982, p. 5-9.
Fig. 94-3: Electronics Today International, 12/78, p. 20.
Fig. 94-4: Courtesy of Motorola Inc. Linear Integrated Circuits, 1979, p. 6-17.
Fig. 94-5: Electronics Today International, 7/72, p. 84.
Fig. 94-6: Reprinted with the permission of National Semiconductor Corp. Data Conversion/Acquisition Databook, 1980, p. 3-13.
Fig. 94-7: Signetics Analog Data Manual, 1982, p. 8-14.
Fig. 94-8: Reprinted with the permission of National Semiconductor Corp. Linear Databook, 1982, p. 3-179.
Fig. 94-9: Reprinted with the permission of National Semiconductor Corp. Linear Databook, 1982, p. 3-238.

Chapter 95

Fig. 95-1: Teledyne Semiconductor, Publication DG-114-87, p. 3.
Fig. 95-2: ©Siliconix incorporated. Analog Switch & IC Product Data Book, 1/82, p. 1-25.
Fig. 95-3: Courtesy of Fairchild Camera & Instrument Ctrporation. Linear Databook, 1982, p. 7-7.
Fig. 95-4: Reprinted with the permission of Analog Devices, Inc. Data Acquisition Databook, 1982, p. 12-19.
Fig. 95-5: Reprinted with the permission of National Semiconductor Corp. Linear Applications Handbook, 1982, p. D-7.
Fig. 95-6: Reprinted with permission of Analog Devices, Inc. Data Acquisition Databook, 1982, p. 12-20.

Chapter 96

Fig. 96-1: Reprinted with the permission of National Semiconductor Corp. National Semiconductor CMOS Databook, 1981, p. 3-50.
Fig. 96-2: Precision Monolithics Incorporated, 1981, Full Line Catalog, p. 16-138.

Fig. 96-3: Teledyne Semiconductor, Databook, p. 11.
Fig. 96-4: ©Siliconix incorporated, Analog Switch & IC Product Data Book, 1/82, p. 7-21.
Fig. 96-5: Precision Monolithics Incorporated, 1981 Full Line Catalog, p. 16-141.
Fig. 96-6: Reprinted with the permission of National Semiconductor Corp. Application Note 32, p. 2.
Fig. 96-7: Reprinted with the permission of National Semiconductor Corp. Linear Databook, 1982, p. 9-204.
Fig. 96-8: Reprinted with the permission of National Semiconductor Corp. Data Conversion/Acquisition Databook, 1980, p. 3-103.
Fig. 96-9: Reprinted with the permission of National Semiconductor Corp. Hybrid Products Databook, 1982, p. 17-54.
Fig. 96-10: Electronics Today International, 7/72, p. 83.
Fig. 96-11: Signetics Analog Data Manual, 1982, p. 3-50.
Fig. 96-12: Siliconix Analog Switch & IC Product Data Book, 1/82, p. 1-7.
Fig. 96-13: Reprinted with the permission of National Semiconductor Corp. Hybrid Products Databook, 1982, p. 1-27.

Chapter 97

Fig. 97-1: Reprinted from Electronics, 7/83, p. 135. Copyright 1983, McGraw Hill Inc. All rights reserved.
Fig. 97-2: Reprinted with the permission of National Semiconductor Corp. Data Conversion/Acquisition Databook, 1980, p. 8-33.
Fig. 97-3: Precision Monolithics Incorporated, 1981 Full Line Catalog, p. 16-173.
Fig. 97-4: Courtesy of Texas Instruments Incorporated. Linear Control Circuits Data Book, Second Edition, p. 145.
Fig. 97-5: Intersil Data Book, 5/83, p. 5-238.
Fig. 97-6: Harris Semiconductor, Linear & Data Acquisition Products, p. 2-58.

Fig. 97-7: 73 Magazine, 8/78, p.132.
Fig. 97-8: Reprinted with the permission of National Semiconductor Corp. Linear Databook, 1982, p. 3-241.
Fig. 97-9: Reprinted with permission from General Electric Semiconductor Department. GE Semiconductor Data Handbook, Third Edition, p. 577.
Fig. 97-10: Reprinted with permission from General Electric Semiconductor Department. GE Semiconductor Data Handbook, Third Edition, p. 1183.
Fig. 97-11: Intersil Data Book, 5/83, p. 5-238.
Fig. 97-12: Courtesy of Motorola Inc. Linear Interface Integrated Circuits, 1979, p. 5-119.
Fig. 97-13: Precision Monolithics Incorporated, 1981 Full Line Catalog, p. 16-81.
Fig. 97-14: Harris Semiconductor Linear – Data Acquisition Products, p. 2-46.
Fig. 97-15: Intersil Data Book 5/83, p. 4-93.
Fig. 97-16: Signetics Analog Data Manual, 1982, p. 16-29.
Fig. 97-17: Signetics Analog Data Manual, 1982, p. 16-29.
Fig. 97-18: Signetics Analog Data Manual, 1977, p. 264.
Fig. 97-19: Courtesy of Fairchild Camera & Instrument Corporation. Linear Databook, 1982, p. 5-25.
Fig. 97-20: Signetics Analog Data Manual, 1982, p. 16-29.

Chapter 98

Fig. 98-1: Electronics Today International, 8/78, p. 69.
Fig. 98-2: Courtesy of Fairchild Camera & Instrument Corporation. Linear Databook, 1982, p. 5-32.
Fig. 98-3: ©Siliconix incorporated. Analog Switch & IC Product Data Book, 1/82, p. 6-18.
Fig. 98-4: Courtesy of Motorola Inc. Linear Integrated Circuits, 1979, p. 6-123.
Fig. 98-5: Courtesy of Texas Instruments Incorporated. Linear Control Circuis Data Book, Second Edition, p. 205.
Fig. 98-6: ©Siliconix incorporated. Analog Switch & IC Product Data Book, 1/82, p. 6-14.

Index

757